SECOND EDITION
Introduction to
SPINTRONICS

SECOND EDITION
Introduction to
SPINTRONICS

Supriyo Bandyopadhyay
Marc Cahay

CRC Press
Taylor & Francis Group
Boca Raton London New York

CRC Press is an imprint of the
Taylor & Francis Group, an **informa** business

The cover artwork shows two figures. On the left is a scanning electron micrograph of two closely spaced nanomagnets—one more shape-anisotropic than the other – which together act as a nanomagnetic inverter or NOT gate because of their mutual dipole interaction. These nanomagnets are made of FeGa alloy and have been delineated on a piezoelectric substrate with electron beam lithography at Virginia Commonwealth University. The magnetizations of both nanomagnets are bistable and can orient along either of the two mutually anti-parallel directions along the ellipses' major axes. These two orientations encode the binary bits 0 and 1. The magnetization of the more shape anisotropic nanomagnet encodes the input bit of the inverter and that of the other encodes the output bit. Because of dipole coupling between them, the magnetizations of the two nanomagnets will be anti-parallel in the ground state, which makes the output bit the logic complement of the input bit, thus realizing a nanomagnetic inverter. Micrograph is provided by Hasnain Ahmad.

On the right is a schematic of an all-electric spin polarizer based on a dual quantum point contact formed in a two-dimensional electron gas. A set of four in-plane side gates is used to control the amount of spin polarization in the narrow portion of the device. The blue and red regions represent accumulations of spin-down and spin-up electrons. The spin polarization configuration can be altered by changing the bias configurations on the four side gates. The onset of spin polarization is accompanied by the presence of anomalies in the conductance of the dual quantum point contact. This figure is the result of simulations and theoretical calculations carried out at University of Cincinnati.

CRC Press
Taylor & Francis Group
6000 Broken Sound Parkway NW, Suite 300
Boca Raton, FL 33487-2742

First issued in paperback 2020

© 2016 by Taylor & Francis Group, LLC
CRC Press is an imprint of Taylor & Francis Group, an Informa business

No claim to original U.S. Government works

ISBN-13: 978-1-4822-5556-0 (hbk)
ISBN-13: 978-0-367-65644-7 (pbk)

Visit the Taylor & Francis Web site at
http://www.taylorandfrancis.com

and the CRC Press Web site at
http://www.crcpress.com

S. B. dedicates this book to Bimalendu, Bela, Anuradha and Saumil Bandyopadhyay.

M. C. dedicates this book to his wife Janie, thanking her for her patience and encouragement, and to the memory of his sister Michèle.

Contents

Preface

This is a textbook intended to introduce a student of engineering, materials science and/or applied physics to the field of *spintronics*. While the term "spintronics" may have different connotations for different people, in this textbook it deals primarily with the science and technology of using the spin degree of freedom of a charge carrier to store, encode, access, process and/or transmit information in some way. That role had been traditionally delegated to the "charge" of an electron, not its "spin." Over the last two decades or so, there has been burgeoning interest in augmenting the role of charge with spin, or even replacing charge with spin in information processing devices.

Interest in spintronics was motivated by a longstanding tacit belief that replacing charge with spin may yield some advantages in terms of increased processing speed, lower power consumption, and/or increased device density on a chip. While this may not always be true, there are some scenarios where it *may* become true in the near future. In this textbook, we place particular emphasis on identifying situations where "spin" may have an advantage over "charge" and where it may not (see, in particular, Chapters 13–15).

The advent of quantum computing has added a new dimension to all this. The spin polarization of a single electron can exist in a coherent superposition of two orthogonal spin polarizations (i.e., mutually anti-parallel spin orientations) for a relatively long time without losing phase coherence. The charge degree of freedom, on the other hand, loses phase coherence much faster. Therefore, spin has become the preferred vehicle to host a quantum bit (or "qubit"), which is a coherent superposition of two orthogonal states of a quantum mechanical entity representing classical logic bits 0 and 1. The potential application of spin to scalable quantum logic processors has a short history, but has provided a tremendous boost to spintronics.

This textbook is expected to equip the reader with sufficient knowledge and understanding to conduct research in the field of spintronic devices, particularly semiconductor-based spintronic devices. We assume that readers have first-year graduate-level knowledge of device engineering, solid state physics, and quantum mechanics.

The first edition of this book was organized into fifteen chapters and the second edition contains eighteen chapters. The first chapter provides a historical perspective to those who have had little or no exposure to this field. It traces the early history of spin, the anomalous Zeeman effect, and ends with an account of the accidental discovery of "spin" by Stern and Gerlach in 1922.

Chapter 2 introduces the quantum mechanics machinery needed to understand spin physics, as well as analyze spin transport and general spin dynamics in solid state structures. It also introduces the concept of Pauli spin matrices, the Pauli equation, and finally its relativistic refinement—the Dirac equation. Since, in this textbook, we will never encounter any situation where relativistic corrections become important, we will not have any occasion to use the Dirac equation. The Pauli equation will be sufficient for all scenarios. Nonetheless, it is important to gain an appreciation for the Dirac equation, since the quantum mechanical nuances associated with spin cannot be absorbed without an understanding of Dirac's seminal work.

Chapter 3 introduces the Bloch sphere concept, since it is a very useful tool to visualize the dynamics of a spin-1/2 particle (e.g., an electron), or qubit encoded in the spin of an electron, under the action of external magnetic fields. Applications of the Bloch sphere concept are elucidated with a number of examples. A spinor, representing an electron's spin, is viewed as a radial vector in the Bloch sphere, and this serves as a nice visualization tool for students interested in quantum computing and other applications of spintronics. All coherent motions of the spinor (where spin does not relax) are essentially excursions on the surface of the Bloch sphere.

Chapter 4 deals with an important application of the Bloch sphere concept, namely, the derivation of Rabi oscillation and the Rabi formula for coherent spin rotation or spin flip. These have important applications in many spin-related technologies such as electron spin resonance spectroscopy, nuclear magnetic resonance, and ultimately solid state versions of quantum computing. This chapter is somewhat mathematical and "seasons" the student to deal with the algebra (and recipes) necessary for calculating quantities that are important in spintronics. This chapter can be skipped at first and revisited later.

Chapter 5 introduces the concept of the "density matrix," pure and mixed states, Bloch equations (that describe the temporal relaxation of spin), the Bloch ball concept, and the notion of the longitudinal (T_1) and transverse (T_2) relaxation times. Several numerical examples are also presented to strengthen key concepts. Since here we allow the dynamics of spin to be incoherent, the motion of the spinor is no longer constrained to the Bloch sphere. The "Bloch sphere" actually refers only to the surface of the sphere and excludes the interior. If spin relaxes so that the norm of the sphere's radius is no longer conserved, then we have to allow excursions into the interior of the Bloch sphere. Therefore, we extend the Bloch sphere concept to the "Bloch ball" concept. This chapter contains advanced concepts and may also be skipped at first reading.

Chapter 6 introduces the rather important topic of spin–orbit interaction which is at the heart of many spintronic devices, since it offers a "handle" to manipulate spins. We introduce the general notion of spin–orbit interaction and then focus on the two special types of spin–orbit interactions that are predominant in the conduction band of most direct-gap semiconductors: the Rashba interaction arising from structural inversion asymmetry, and the

Dresselhaus interaction arising from bulk (crystallographic) inversion asymmetry. These two interactions form the basis of spintronic field effect transistors where the current flowing between two of the transistor's terminals is modulated by influencing the spin–orbit interaction in the device via a potential applied to the third terminal. Therefore, it is particularly important for applied physicists, materials scientists, and engineers to understand these interactions.

In Chapter 7, we derive the electron dispersion relations (energy versus wavevector) of electrons in quasi two- and one-dimensional structures (quantum wells and wires) in the presence of Rashba and Dresselhaus spin–orbit interactions, as well as an external magnetic field. We also derive the spin eigenstates, which allows us to deduce the spin polarization of carriers in any band. All this is accomplished by solving the Pauli equation. This is an example of how the Pauli equation is applied to solve a real life problem. We place special emphasis on how the dispersion relations are modified by an external magnetic field. This is important since it ultimately helps the student to appreciate how an external magnetic field can affect the performance of spin-based devices.

Chapter 8 discusses spin relaxation of conduction electrons in metals and semiconductors. We focus on four primary spin relaxation mechanisms: the D'yakonov–Perel, the Elliott–Yafet, the Bir–Aronov–Pikus and hyperfine interactions with nuclear spins, since these are dominant in the conduction band of semiconductors and therefore are most important in device contexts. Because spin relaxation limits the performance of most, if not all, spin-based devices, it is a vital issue. Ultimately, the aim of all device engineers and physicists is to reduce the rate of spin relaxation in spin devices, in order to make them more robust and useful. Spin relaxation also has peculiarities that are completely unexpected and without parallel in solid state physics. We present one example where spin can relax in *time* but not in *space*.

Chapter 9 is a new addition to the second edition and was not included in the first. Since the publication of the first edition in 2008, there has been an explosion in the study of spin-related physical phenomena, particularly those that may have applications in spintronic devices. In this chapter, we also discuss seven important physical phenomena that all have device applications: the extrinsic and intrinsic spin Hall effect along with the inverse spin Hall effect and the giant spin Hall effect, the spin Hanle effect, the spin capacitor effect, the spin-torque effect, the spin Galvanic effect, the spin Seebeck effect, and the inverse spin Seebeck effect or spin Peltier effect.

Chapter 10 introduces the more advanced concepts of exchange and spin–spin interaction. These form the basis of ferromagnetism and also the basis of single-spin computing schemes that are dealt with in Chapters 15 and 16.

Chapter 11 is an introduction to spin transport in solid state structures in the presence of spin relaxation. We focus on two basic models: the drift–diffusion model of spin transport and the semi-classical model that goes beyond the drift–diffusion model. The "spin" drift–diffusion model is very similar to the "charge" drift–diffusion model applied to bipolar transport; the "up-spin" and "down-spin" carriers assume roles analogous to electrons and holes. However, it has limitations. One limitation that we emphasize with specific examples is that it fails to describe essential features of spin transport, even qualitatively, if electrons are traveling "upstream" against the force exerted on them by an electric field. In this chapter, we present many examples of how spin relaxes in time and space in quasi one-dimensional structures in the presence of the D'yakonov–Perel' spin relaxation mechanism, since it is usually the dominant mechanism for spin relaxation in semiconductor structures. These examples are based on the semi-classical model and therefore applicable to both low field transport and high field (hot electron) transport. The semiclassical model is based on combining the Liouville equation for the time evolution of the spin density matrix with the Boltzmann transport equation for time evolution of the carrier momentum in the presence of scattering and external electric fields.

In Chapter 12, we discuss *passive* spintronic devices such as spin valves and devices based on the giant magnetoresistance effect. Most commercial spintronic products that are currently available (magnetic read heads for reading data in computer hard disks or entertainment systems such Apple iPods, and magnetic random access memory) utilize these passive devices. Therefore, an adequate understanding of these devices is vital for engineers. We also discuss the important notions of spin injection efficiency, spin extraction, and the recently discussed spin blockade. This is a long chapter with many topics and it is intended to introduce the reader to important concepts encountered in the modern spintronics literature. We also discuss three very specific devices that are spintronic "sensors"; one is a magnetic field sensor, another is a light sensor (photodetector), and the third is a mechanical strain sensor. These are discussed to show the reader that spintronics has myriad applications in magnetics, mechanics and optics.

Chapter 13 introduces *active* spintronic devices, such as spin field effect transistors and spin bipolar transistors. We explain the physical basis of how these devices operate and what their shortcomings are. We make a simple estimate of their performance figures in order to project a realistic picture of whether they are or are not competitive with traditional electronic devices that are currently extant. Regardless of their actual device potential, these devices are standard-bearers that aroused early interest in the field among engineers and applied physicists. This chapter discusses only the early variants of spin transistors because of their pedagogical importance. New twists to spin transistors appear in the literature frequently and it was not possible to do justice to them within the limited space available. The only way the reader can keep pace with this field is to follow the literature closely.

Chapter 14 discusses the recently discovered field of "spintronics without magnetism," which allows one to manipulate spin currents by purely electrical means. The reader is introduced to lateral spin–orbit interaction, and its many nuances, and the possibility of implementing spin polarizers and analyzers using quantum point contacts. This too is a new addition to the second edition.

Chapter 15 introduces more exotic concepts dealing with single-spin processors. Here, a single electron spin acts as the primitive bistable "switch" with two stable (mutually anti-parallel spin orientation) states that encode classical logic bits 0 and 1. Switching between the bits is accomplished by flipping the electron's spin without moving the electron in space and causing current flow. This chapter addresses fundamental notions like the ultimate limits of dissipation in performing Boolean logic operations and has relevance to the celebrated *Moore's law* scaling. Another distinguishing feature is that this chapter addresses spin-based architectures and not discrete devices like transistors. For example, it describes combinational logic circuits implemented with single-spin-switches that communicate with each other via exchange interaction and not physical wires. This is an area that has remained neglected, but is really no more challenging than spin-based quantum computing, since phase coherence of spin is not required. Being classical, it does not have the promise of quantum speedup of computation, or the ability to solve intractable problems, but it may provide valuable insights into the limits of classical computation.

Chapter 16 is an introduction to the field of spin-based reversible logic gates (that can, in principle, compute without dissipating energy) and spintronic embodiments of quantum computers. This is a rapidly advancing field, extremely popular among many spintronic researchers, and discoveries are made at a fast pace. This chapter is written mostly for engineers and applied physicists (not computer scientists or theoretical physicists), and should provide them with the preliminary knowledge required to delve further into this field. We have also focused on electrical manipulation of spin qubits rather than optical manipulation since this book is almost entirely devoted to electro-spintronics rather than opto-spintronics. Needless to say, because of the rapid advances in this area, it is impossible to address this field comprehensively. The reader is provided with a few examples to whet her/his appetite and is urged to follow the literature closely to keep abreast of the most recent developments.

Chapter 17 introduces the concept of "single-domain-nanomagnet" based computing and is a more practical rendition of the single-spin logic architecture ideas of Chapter 15. This is a new addition to the second edition. In a single domain ferromagnet, all the spins rotate in unison under an external influence because of strong exchange interaction between spins, making the entire ferromagnet behave like a giant classical spin. This chapter is focused primarily on logic architectures and discusses two main variants: dipole coupled nanomagnetic logic (also known as magnetic quantum cellular automata) and magneto-tunneling junction logic. Particular emphasis is placed on various magnet switching methodologies (magnetic field, spin-torque, spin-Hall effect, toplogical insulators, and magneto-elastic switching) since they determine the energy efficiency of nanomagnetic architectures. Much of the material presented in this chapter, dealing with magneto-elastic devices, was the result of collaborative research with Prof. Jayasimha Atulasimha at Virginia Commonwealth University.

Chapter 18 is a stand-alone chapter that can be treated as an appendix. At first sight, it will appear unrelated to spintronics, which it is, but it has been included for a reason. There are many instances in this book when a student will have to recollect or refamiliarize herself/himself with some key results of quantum mechanics. Rather than making a trip to the nearest library, it would be more convenient to have a "quantum mechanics primer" handy where these key results have been re-derived. This chapter is included for completeness and comprehensiveness. The reader can refer to it if and when necessary.

At the time of writing the second edition, this book is still the only known "textbook" in spintronics written in English. By its very nature, it must be incomplete and omit many topics that are both important and interesting. We have focused mostly on electron spin, and, with the sole exception of discussing hyperfine nuclear interactions, we have ignored nuclear spin altogether. Hence, we do not discuss such well-known phenomena as the Overhauser effect, which is more relevant to nuclear spin. Another area that we have intentionally not covered in any detail is *organic spintronics*. We omitted any discussion of this field (it is still in its infancy) and do not discuss it primarily because we feel that this is very much in evolution. Organic semiconductors (mostly hydrocarbons) have weak spin–orbit interactions, so that spin relaxes slowly in these materials compared to inorganic semiconductors. Hence, they have a major advantage over inorganics when it comes to applications where spin relaxation must be suppressed, such as in spin-based classical or quantum computing. Some reviews have appeared in the literature covering organic spintronics and an edited book is available from this publisher.

This textbook also heavily emphasizes transport phenomena as opposed to optical phenomena dealing with the interaction of polarized photons with spin-polarized electrons and holes. Hence, we do not discuss such devices as spin-light-emitting diodes. Delving into "opto-spintronics" would have easily added a couple hundred pages to the 600-odd pages in this textbook. Our own expertise is more in transport phenomena, which has led us to focus more on transport. However, there are many excellent books (although not necessarily "textbooks") available that deal with opto-spintronics, and the interested reader can easily find an assortment of literature in that area.

Table of Universal Constants

Free electron mass (m_0)	9.1×10^{-31} kilograms
Dielectric constant of free space (ϵ_0)	8.854×10^{-12} Farads/meter
Electronic charge (e)	1.61×10^{-19} Coulombs
Reduced Planck constant ($\hbar$)	1.05×10^{-34} Joules-sec
Bohr radius of ground state in H atom (a_0)	0.529 Å$= 5.29 \times 10^{-11}$ meters
Bohr magneton (μ_B)	9.27×10^{-24} Joules/Tesla

Acknowledgments

Some acknowledgments are due. Many of our associates have contributed indirectly to this book. They are our students, laboratory interns and post-doctoral research associates, past and present. They include Prof. Sandipan Pramanik of the University of Alberta, Canada, who was a graduate student and then a post-doctoral researcher at Virginia Commonwealth University (VCU) at the time this book was composed; Dr. Bhargava Kanchibotla, an erstwhile graduate student at VCU who provided the T_2 data in cadmium sulfide nanostructures from his experiments (Figure 8.5); Dr. Sivakumar Ramanathan and Dr. Sridhar Patibandla, two graduate students at VCU working with S. B. who took the first spintronics graduate course offered at VCU by S. B. and provided valuable feedback; Harsh Agarwal, a summer undergraduate intern visiting VCU from the Banaras Hindu University Institute of Technology, Varanasi, India, who computed and generated some of the plots in Chapter 15; and Dr. Amit Trivedi, another summer undergraduate intern from the Indian Institute of Technology, Kanpur, who performed some of the calculations; and Saumil Bandyopadhyay, who performed some of the coherent room temperature spin transport experiments in single subband quantum wires.

At the University of Cincinnati, a former graduate student, Dr. Junjun Wan, and intern Lindsay Ficke, contributed immensely to the generation of data and plots. We also thank graduate student Nishant Vepachedu for proofreading the manuscript. We remain grateful to all of them.

In spite of all our best efforts, quite a few typographical errors made their way into the first edition. We corrected as many of them as we could catch, but some may have still eluded us. Matthew David Mower, a student from the University of Missouri, brought some typographical errors to our notice. We thank him. As always, we will remain grateful to any reader who points out such errors to us. Our e-mails are sbandy@vcu.edu and marc.cahay@uc.edu.

Welcome to the world of spintronics!

1

Early History of Spin

1.1   Spin

Most students of science and engineering know that every elementary particle, such as electrons, neutrons, photons, neutrinos, etc., has a quantum mechanical property called "spin" which can be measured (perhaps not easily, but at least in principle) and has a quantized value, including zero. The vast majority of these students mentally visualize spin as the angular momentum associated with the elementary particle spinning or rotating about its own axis (like a top or a planetary object). This mental picture, although convenient and comforting, is actually somewhat crude and certainly incomplete. Landau and Lifshitz, in their classic textbook on quantum mechanics [1], wrote "[the spin] property of elementary particles is peculiar to quantum theory. [It] has no classical interpretation... It would be wholly meaningless to imagine the 'intrinsic' angular momentum of an elementary particle as being the result of its rotation about its own axis."

The simplistic notion of self-rotation about an axis, shown in Fig. 1.1, cannot explain many features of spin, such as why its magnitude cannot assume continuous values and why it is quantized to certain specific values. It also causes serious problems if taken too literally. As we will see later (Problem 1.2), if we think of an electron as a solid sphere of radius equal to the Lorentz radius $e^2/(4\pi\epsilon_0 m_0 c^2)$ (where e is the electron's charge, m_0 is the mass, c is the speed of light in vacuum, and ϵ_0 is the dielectric constant of vacuum), then the velocity on the surface of a rotating electron would have to be many times the velocity of light in vacuum if such a rotation were to generate an angular momentum equal to the electron's spin. Obviously that would not be permitted by the theory of relativity. Indeed, a deep understanding of quantum mechanics is required to understand how the spin property comes about. Its origin is in relativistic quantum mechanics and really was first appreciated by Paul Andrew Maurice Dirac when he derived the Dirac equation which is the cornerstone of relativistic quantum mechanics. Richard Feynman, noted for his distaste for mysticism in physics, wrote about the notion of "spin": "It appears to be one of the few places in physics where there is a rule which can be stated very simply, but for which no one has found a simple and easy explanation. The explanation is down deep in relativistic quantum mechanics.

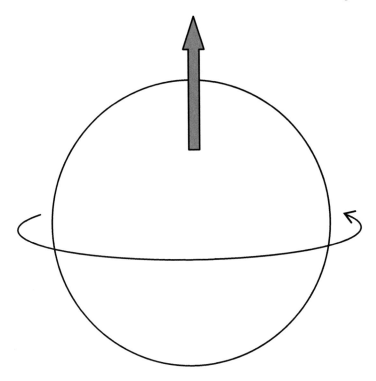

FIGURE 1.1

A mental picture to visualize spin as the angular momentum associated with self-rotation of a particle about an axis. This picture, although certainly incomplete and heuristic, is adequate for many situations that the reader will encounter in this book.

This probably means that we do not have a complete understanding of the fundamental principle involved" [2]. We do not intend to mystify "spin" any more than it already is, but rather we wish to underscore the fact that spin is *not* a classical property easily explained in terms of everyday experience. It is a property associated with relativistic quantum mechanics, for which there is no everyday experience. This is the only place in this textbook where we will mention this, since in the rest of this textbook we will have no occasion to reflect on this truism. We will be primarily involved with practical and applied aspects of spin, without relishing the fact that it is indeed an exotic property far outside what we normally deal with in the applied sciences and engineering.

1.2 Bohr planetary model and space quantization

The history of how the concept of spin was established is somewhat tortuous. In 1913, when Niels Bohr published his theory of the hydrogen atom, he had thought of a planetary model where the electron orbits around the nucleus as shown in Fig. 1.2. The radii of the allowed orbits are quantized. Each orbit has a radius $n^2 a_0$ where a_0 is the Bohr radius of the ground state in the hydrogen atom $(=4\pi\epsilon_0\hbar^2/m_0 e^2)$, which is 0.529 Å. Here $\hbar$ is the reduced Planck constant, m_0 is the electron's mass, ϵ_0 is the permittivity (dielectric constant) of free space, and e is the magnitude of its charge. The quantity n is an integer called the "principal quantum number," and it takes positive non-zero values 1, 2, 3, etc. Later Arnold Sommerfeld (in 1916), and independently Peter Debye, introduced two more quantum numbers l and m, which were called the "orbital" and "magnetic" quantum numbers. While the principal quantum number determines the radius (or size) of the orbit, the angular quantum number l determines its shape. It also determines the angular momentum associated with the orbital motion in units of $\hbar$. The integers n and l obey the relation $n \geq l$. If the atom is placed in a magnetic field, the component of its angular momentum along the field takes on quantized values of $m\hbar$. The number m is an integer and satisfies the inequality $-l \leq m \leq l$. This last inequality limits the number of m values to $2l + 1$, and accordingly, the number of allowed directions of the angular momentum vector in a magnetic field is $2l + 1$. This is known as *space quantization of angular momentum.*

The energy of an electron in an atom was thought to be determined by the three quantum numbers n, l and m. When an electron makes a transition from one energy state to another, the transition involves a change in one or more of these quantum numbers. In the process of transition, the electron absorbs or emits light of a particular frequency ν which is determined by the energy difference between the initial and final states in accordance with conservation of energy:

$$E_{final} - E_{initial} = 2\pi\hbar\nu = h\nu, \tag{1.1}$$

where $E_{initial}$ and E_{final} are the electron's energy in the initial and final states, respectively.

However, when an atom is placed in a magnetic field and the spectra of emitted and absorbed light are measured, it is found that the multiplicity of the spectra (meaning all the observed frequencies) cannot be explained by the space quantization rules (that means allowed values of n, l, and m) alone. In 1920, Sommerfeld tried to explain the multiplicity by invoking yet another quantum number j that he called the *inner* quantum number. However, this was not able to completely explain multiplicity. Additional frequencies (where each line split into two) were observed in a strong magnetic field; this was

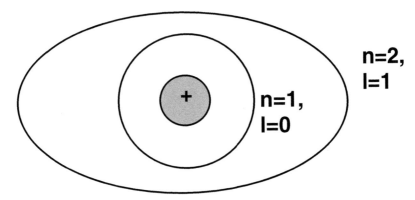

FIGURE 1.2
The Bohr planetary model of electron orbits around the nucleus of an atom (not to scale).

referred to as the anomalous Zeeman effect. This effect could not be explained by space quantization.

1.3   Birth of "spin"

In 1925, the anomalous Zeeman effect led a young American scientist, Ralph de Laer Kronig, to postulate that in addition to the orbital angular momentum, an electron has an additional angular momentum caused by spinning about its own axis. Kronig postulated that the angular momentum associated with this self-rotation has a fixed magnitude of $(1/2)\hbar$, although he obviously had no explanation as to why the angular momentum should have a fixed value, let alone why that value should be $(1/2)\hbar$ (such is the nature of a "postulate"). The correct explanation for these two features had to wait for Dirac's formulation, which is discussed later. However, such an angular momentum will cause multiplicity in atomic spectra due to a relativistic effect. The rotating electron in an atom's orbit experiences an electric field due to the positively charged nucleus, which will transform into a magnetic field $\vec{H}$ via Lorentz transformation in the rest frame of the electron. The angular momentum will give rise to a magnetic moment $\vec{M}$ related to the angular momentum $\vec{W}$ by the Landé relation $\vec{M} = g_0\vec{W}$, where g_0 is the so-called Landé g-factor (Kronig assumed it to be 2, which is the correct value for vacuum). The magnetic moment will interact with the magnetic field and the energy of that interaction will be $E_{int} = -\vec{M} \cdot \vec{H}$. Since the angular momentum has two possible values $\pm(1/2)\hbar$, this will result in energy splitting of $2\left|\vec{M} \cdot \vec{H}\right|$. With this additional angular momentum, he was able to explain the multiplicity of spectra (within a factor of 2), and presented his idea to Wolfgang Pauli, who was not impressed by it (Pauli had his own ideas to explain the multiplicity of spectra, which later turned out to be wrong). Kronig himself was not very confident because his calculations based on this spinning electron model still did not completely explain every feature of the experimentally observed spectra. More important, the idea of a self-rotating electron presents a conundrum within the framework of classical theory. If one thinks of the electron as a sphere of radius $r_e = e^2/(4\pi\epsilon_0 mc^2)$, as considered by Lorentz, then the rotation rate required to produce an angular momentum of $(1/2)\hbar$ is so high that the electron's surface reaches a speed more than 130 times that of light in vacuum, in stark violation of Einstein's theory of relativity! Because of this apparent contradiction, Kronig never published his ideas. Six months later, Uhlenbeck and Goudsmit published essentially the same spinning electron idea that Kronig had come up with in the journal *Naturwissenschaften*. Actually, when they realized the problem with the surface speed (see Problem 1.2), they tried to hurriedly withdraw their paper, but it was too late. The paper appeared in print. Kronig sent a letter to the British journal *Nature* criticizing the idea of Uhlenbeck and Goudsmit, pointing out the problems with the spinning electron model. Meanwhile, in a second paper that appeared in *Nature*, Uhlenbeck and Goudsmit pointed out that their theory did not quite explain the experimental observations of atomic spectra. In fact, there was a discrepancy by a factor of 2 with the experimental results. Later L. H. Thomas showed that this discrepancy comes about because of an incorrect definition of the electron rest system. An electron in an atom moves in a closed orbit around the nucleus. Therefore, there is always a component of the field perpendicular to the instantaneous velocity which causes an additional acceleration in the direction perpendicular to the velocity. This is what causes the electron trajectory to be curved. As a result, the electron is moving in a rotating frame of reference and when this is ocrrectly taken into account, it leads to the "factor of 2 correction". Thomas published his findings in a letter to *Nature* in February 1926, following which, all discrepancies between theory and experiment could be resolved. This made a convert out of Pauli, who ultimately endorsed the spinning electron idea of Kronig, Uhlenbeck and Goudsmit (KUG).

The spinning electron picture remains in vogue, although it obviously raises unanswered questions, such as why the spin angular momentum is quantized, or why the idea cannot be reconciled with the postulates of the special theory

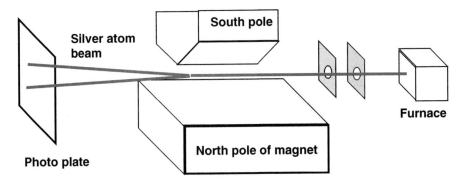

FIGURE 1.3
Experimental arrangement of the Stern-Gerlach apparatus (not to scale).

of relativity. These unanswered questions merely reveal the inadequacy of the model. The reader should be aware of this inadequacy, but that should not affect her/his understanding of any of the topics discussed in this book.

1.4   The Stern-Gerlach experiment

At least three years before the KUG ideas were published, the spin of an electron was already *unwittingly* measured in the famous Stern-Gerlach experiment, which remains a watershed event in the history of spin. Otto Stern and Walther Gerlach in Frankfurt, Germany, took the space quantization business literally and designed an experiment to verify it. To them, the space quantization business involved only the quantum numbers m, l, and n, but no spin, since the spin idea had not yet been thought of by KUG. Space quantization will dictate that the angular momentum of an electron in the Bohr atom will be quantized to $l\hbar$, and, because an orbiting electron will give rise to a magnetic moment proportional to the angular momentum, the magnetic moment will be quantized as well. Therefore, by measuring the magnetic moment of atoms and showing that it can assume only discrete values, it should be possible to demonstrate space quantization.

Consider a beam of atoms coming out of an effusion furnace, going through collimators and a magnetic field, and finally impinging on a detection plate, as shown in Fig. 1.3. Assuming that an atom possesses a non-zero magnetic moment $\vec{\mu}$ due to the oribital angular momenta of its electrons, the magnetic field will produce two effects. First, the magnetic field will exert a torque on the magnetic dipole of the atom and make the magnetic moment vector precess about the magnetic field [3]. Second, the potential energy of the atom will be $U = -\vec{\mu} \cdot \vec{B} = -\mu_z B$, if we assume that the magnetic field is directed along the z-direction. Because of the spatial *non-uniformity* of the magnetic field in the z-direction, the atoms will experience a z-directed force given by [3]

$$F_z = -\frac{\partial U}{\partial z} = \mu_z \frac{\partial B(z)}{\partial z}. \qquad (1.2)$$

Classically, one would expect that, owing to random thermal effects, the atoms in a beam will have their magnetic moment vectors pointing in different directions, resulting in a continuous spread in the value of μ_z from $-|\mu_z|$ to $+|\mu_z|$. This will cause a continuous spread in F_z and therefore a continuous spread in the deflection of the beam in the z-direction. If the beam then impinges on a detection plate, one should see a continuous line on the plate in the z-direction.

However, if space quantization holds sway, then the magnetic moment μ_z is quantized and cannot assume any arbitrary value between $-|\mu_z|$ and $+|\mu_z|$. It is not clear why he thought this, but Stern was convinced that the magnetic moment in the direction of the magnetic field (i.e., μ_z) will be quantized to two values of opposite sign and therefore every atom will experience the same magnitude of the deflecting force, except some will experience it in one direction and the rest in the opposite direction. Therefore, an atomic beam of hydrogen would split into two beams in a magnetic field, and, despite the smearing effect of the inter-atomic collisions that exert a random force on the atoms, they should split so far apart in a strong field that the oppositely directed components would be deflected outside the width of the original beam. If that happens, then one should observe *two* distinct beams that will not produce a line on the detecting plate, but just two spots. This would have been a remarkable experiment anyway, since classical theory predicted a line. Thus, any birefringence (and observation of just two distinct spots) would demonstrate that quantum physics supersedes classical physics, and it would be a *literal* demonstration of space quantization, which is a quantum mechanical phenomenon outside the realm of classical physics.

Bohr himself was not so convinced of his own model of the atom and did not think one could take quantum physics so literally. He thought that space quantization was a symbolic expression and convenient for calculating atomic spectra, but could not be taken literally. However, Stern was fortunately not dissuaded by Bohr and found a convert in Gerlach, who, until that time, had apparently not heard of space quantization!

The actual Stern-Gerlach experiment took place a year after it was conceived by Stern. It was not hydrogen atoms, but a beam of silver atoms produced by the effusion of the metal from an oven heated to 1000°C, that was collimated by two narrow slits 0.03 mm wide. It traversed a magnet 3.5

cm long that produced a magnetic flux density of 0.1 Tesla and a field gradient of 10 Tesla/cm. The calculated splitting of the beam was only 0.2 mm so that extreme alignment accuracy was required. When the experiment was completed, Stern and Gerlach could not even see the traces of the silver beam on the collector plate, so that it was impossible to determine if the splitting was there. In an episode that has now become famous, Stern's breath on the plate made history. He used to smoke cigars and the sulfur from his breath turned the silver into silver sulfide, which is jet black and is easily visible. The results of this experiment, however, were still inconclusive and would not have convinced a sceptic. After many efforts, Stern and Gerlach met in Göttingen in 1922 and decided to give up. But a railroad strike detained Gerlach and he attempted the experiment once again with improved accuracy of alignment. Finally a clear signature of beam splitting was observed [4].

We now know that the Stern-Gerlach experiment had nothing to do with space quantization, contrary to what they believed. In fact, Gerlach could never reproduce the experiment with sodium atoms instead of silver atoms, and Einstein questioned their interpretation (correctly). The agreement of the experiment with Stern's and Gerlach's picture of space quantization (or the Bohr model) was nothing but a coincidence. The net angular momentum of silver atoms is actually *zero*, contrary to what was presumed by Stern and Gerlach. The magnetic moment of Silver atoms is therefore due solely to the *spin angular momentum* (and has nothing to do with the orbital angular momentum), which accounts for the observed splitting of the beam into two. Stern and Gerlach had *unknowingly measured the spin angular momentum*. This is the first convincing experimental observation of "spin."

It is surprising that, even though the Stern-Gerlach experiment was reported in 1922 and was widely known among physicists, the postulation of the electron spin in 1925 did not immediately lead to a re-interpretation of the experiment as being a demonstration of the spin. The earliest attribution of the splitting to spin was reported in 1927, when Ronald Fraser noted that the ground state orbital angular momentum and associated magnetic moments of silver, hydrogen and sodium atoms are zero. Therefore, obviously the splitting could not be due to orbital angular momentum; it had to be due to "spin".

In February, 2002, a team of physicists and chemists tried to re-enact the Stern-Gerlach experiment with sulfur tainted breath and all [5]. For a hilarious account of this re-enactment, see the December 2003 issue of the magazine *Physics Today*. Mere sulfurous breath was not sufficient to reveal traces of silver on a detecting plate, but direct exposure to cigar smoke was enough to reveal the traces.

1.5   Advent of spintronics

Although "spin" plays a fundamental role in explaining the multiplicity of atomic spectra, this is not its most important role. It was realized in the middle of the 20th century that spin plays a fundamental role in magnetism. Every theoretical model postulated to explain the physical origin of magnetism invoked "spin" in some way or the other. This included the Bloch model, the Heisenberg model, the Stoner model, and every other model advanced. While magnetism remains the domain of spin, in the late 20th century, it was realized that spin, alone or in conjunction with charge, can be harnessed to process information, particularly digital information encoded with binary bits 0 and 1. This is the central theme in the field of *spintronics* today (see, for example, [6, 7]). Of course, spintronics is much more than just information rendition and includes the more traditional areas of *magnetoelectronics*, which deals with magnetic or magnetoresistive effects for sensing and storing information. Early successes in this area include the developments of read heads for sensing massively dense magnetic storage media (these read heads are now routinely used in laptop computers and entertainment systems such as Apple iPods), non-volatile magnetic random access memory (MRAM) [8], programmable spintronic logic devices based on magnetic tunnel junction elements [9], rotational speed control systems [10], positioning control devices in robotics and related systems (such as automobile braking systems) [11], perimeter defense systems, magnetometers, and high current monitoring devices for power systems [12], etc. Many of these developments were fueled by the investigation of how spin-polarized electric currents can be injected into ferromagnetic/paramagnetic multilayers which, in the 1980s, had led to the important discovery of the phenomenon of giant magnetoresistance (GMR) [13, 14]. Several books and review articles have been written on magneto-electronics [15, 16, 17, 18, 19, 20, 21]. The reader is referred to these books to gain an understanding of magnetoelectronics. In this book, we will focus more on the "information technology" (IT) applications of spin, which are the basis of modern spintronics.

The application of spintronics in information technology, particularly for computing and signal processing, is a relatively new field. Early efforts in this area were concerned with developing spin-based analogs of conventional signal processing devices such as transistors - both the field effect type [22, 23] and the bipolar junction type [24, 25]. In Spin Field Effect Transistors (SPINFETs) or Spin Bipolar Junction Transistors (SBJTs), information is still processed by modulating the charge current flowing between two terminals via the application of either a voltage or a current at the third terminal. However, the process by which the voltage or current at the third terminal exercises control over device behavior is spin mediated. Therefore "spin" plays somewhat of a secondary role in these devices, while "charge" retains the primary role. These devices are interesting and demonstrate how spin can play a role in information handling. They are discussed in Chapter 14.

The more radical branch of spintronics is what we call *single-spin spintronics*, where charge has no direct role and information is encoded entirely in the spin polarization of a single electron, which is made to have only two values, "up-spin" and "down-spin", by placing the electron in a magnetic field. The down-spin orientation could correspond to polarization parallel to the magnetic field and the up-spin orientation is anti-parallel to the magnetic field. These two values could represent binary bits 0 and 1. Boolean logic circuits can be fashioned from interacting single electron spins by properly engineering the interactions. An embodiment of this approach is the so-called "single spin logic" (SSL) idea [28, 29] discussed in Chapter 15. What makes this idea attractive is that bits can be flipped by simply toggling an electron's spin, without physically moving it in space and causing a current flow. This results in much reduced heat dissipation in the circuits.

The ultimate rendition of spintronic computing circuits is spin-based quantum computers, which dissipate no energy at all to complete a logic operation, since they operate on the basis of *reversible* quantum dynamics. Recently, there has been a great deal of interest in encoding a quantum bit (*qubit*) using the spin degree of freedom of a single electron confined in a quantum dot [30, 31, 32, 33] or bound to a donor atom [34] or housed in a nitrogen vacancy center in diamond [35, 36] to implement a quantum logic gate. The quantum mechanical phase coherence of "spin" is much longer lived than that of "charge"; consequently, spin is a natural choice for building solid state scalable quantum logic processors. Spin-based quantum computing is a rapidly expanding field of research endeavor and Chapter 16 in this textbook discusses some of the basic ideas in this field. Unfortunately, Chapter 16 will be almost certainly outdated very quickly, perhaps even by the time this book appears in print, since extremely rapid strides are being made in this field. Suffice it to say then that spintronics is now poised at a critical juncture where technological breakthroughs may be just around the corner. Therefore, understanding the science and technology of spintronics has become imperative for students of electrical engineering, physics, and materials science.

1.6   Problems

- **Problem 1.1**

  Using the Bohr model of the hydrogen atom, show that the magnetic moment associated with an electron moving in the lowest circular orbit is the Bohr magneton given by

$$\mu_B = \frac{e\hbar}{2m_0} \; . \tag{1.3}$$

Show also that the magnetic flux density B_n associated with the n-th orbital is given by

$$\mu_B B_n = \frac{1}{4}(\frac{\alpha^4 Z^4}{n^5})m_o c^2 \; , \tag{1.4}$$

where m_o is the rest mass of the electron, c is the speed of light in vacuum, Z is the atomic number of the atom, n is the principal quantum number and α is the fine structure constant.

Calculate B_n in Tesla for the electron in the first orbit of the hydrogen atom, i.e., $Z=1$, $n=1$. Use $m_o c^2 \approx 0.5\text{MeV}$.

Solution

Consider a nucleus with Z protons. The total (kinetic + potential) energy of an electron moving around this nucleus is conserved and is given by:

$$E = \frac{1}{2}m_o v^2 - \frac{Ze^2}{4\pi\epsilon_o} \; , \tag{1.5}$$

where e is the magnitude of the charge of the electron, v is the electron's orbital velocity, and ϵ_o is the dielectric constant of vacuum. The first term in the right hand side is the kinetic energy of the electron and the second term is the (electrostatic) potential energy. We ignore any gravitational potential energy since it is negligible.

According to Newton's Law, the centripetal force should be equal to the Coulomb force; hence, we have:

$$m_o \frac{v^2}{r} = \frac{Ze^2}{4\pi\epsilon_o r^2} \; . \tag{1.6}$$

Using Bohr's quantization condition for the angular momentum around the nth orbit:

$$L_n = m_o v_n r = n\hbar \; , \tag{1.7}$$

where the subscript n is the principal quantum number.

We get from Equations (1.6) and (1.7)

$$L_n v_n = \frac{Ze^2}{4\pi\epsilon_o} \; . \tag{1.8}$$

This yields

$$v_n = \frac{Ze^2}{4\pi\epsilon_o\hbar}\frac{1}{n} \ .$$ (1.9)

We can compare this velocity to that of light and get

$$\frac{v_n}{c} = \frac{Z}{n}\frac{e^2}{4\pi\epsilon_o\hbar c} \ ,$$ (1.10)

where the constant $\frac{e^2}{4\pi\epsilon_o\hbar c} = \alpha$ and is referred to as the fine structure constant.

Equation (1.10) shows that an electron in the first orbit (ground state) of the hydrogen atom ($Z = n = 1$) has a velocity equal to $\alpha \approx \frac{1}{137}$ times the speed of light.

The radius of the various orbits allowed by Bohr's space quantization condition can then be obtained from Equation (1.7)

$$m_o v_n r_n = n\hbar \ .$$ (1.11)

We get

$$r_n = \left(\frac{4\pi\epsilon_o}{Ze^2}\right)\left(\frac{n^2\hbar^2}{m_o}\right) \ .$$ (1.12)

This is also written as: $r_n = n^2 a_Z$, where a_Z is the effective Bohr radius given by

$$a_Z = \frac{4\pi\epsilon_o\hbar^2}{m_o Ze^2} \ .$$ (1.13)

The latter is equal to 0.529 Å for the hydrogen atom ($Z = 1$).

The total energy of an electron in the n-th orbit is found from Equation (1.5) and (1.9) to be

$$E_n = \frac{1}{2}m_o v_n{}^2 - \frac{Ze^2}{4\pi\epsilon_o r_n} = -\frac{m_o e^4 Z^2}{32\epsilon_o{}^2\hbar^2} \ .$$ (1.14)

This is the *binding energy* of the electron. It is also called the *ionization energy*, since this is the minimum amount of energy that will be required to liberate the electron from the nuclear attraction and make it free, thereby ionizing the parent atom. This ionization energy is 13.6 eV for the ground state of the hydrogen atom.

According to Biot-Savart's law, an electron moving around the nucleus in one of these orbits will feel a magnetic field given by

$$\vec{B} = \frac{\vec{\mathcal{E}} \times \vec{v}}{2c^2} , \tag{1.15}$$

where $\mathcal{E}$ is the electric field experienced by the moving electron and the factor 2 is due to Thomas's correction for a rotating frame of reference.

From Coulomb's law, the electric field in the nth orbital is given by

$$\mathcal{E}_n = \frac{Ze}{4\pi\epsilon_o r_n{}^2} . \tag{1.16}$$

Using Equations (1.9), (1.15), and (1.16), the magnitude of the magnetic field in the n-th orbit is

$$|\vec{B_n}| = \frac{\mathcal{E}_n v_n}{2c^2} = \frac{1}{2c^2}(\frac{Ze}{4\pi\epsilon_o r_n{}^2})v_n \tag{1.17}$$

and it is directed perpendicular to the orbital plane of the electron.

Finally, using Equations (1.12), (1.9) and (1.17) and the expression for the Bohr magneton given, one finds that

$$\mu_B B_n = \frac{1}{4}(\frac{\alpha^4 Z^4}{n^5})m_o c^2 . \tag{1.18}$$

From the above, the magnetic flux density associated with an electron orbiting the hydrogen atom in the first orbit is 6.16 Tesla.

- **Problem 1.2**

Show that in the classical spinning electron model, the electron's surface speed must be more than 60 times the speed of light in order to produce an angular momentum of $(1/2)\hbar$. This is why, it is inappropriate to think in classical terms that the spin of an electron is associated with rotation about its own axis.

Solution

The angular momentum is

$$(1/2)m_0 v_s r_e = (1/2)\hbar , \tag{1.19}$$

where v_s is the speed on the surface of the electron and r_e is the Lorentz radius of the electron. Solving the above equation with universal constants $m_0 = 9.1 \times 10^{-31}$ Kg, $\hbar = 1.05 \times 10^{-34}$ Joules-sec, and $r_e = e^2/(4\pi\epsilon_0 m_0 c^2) = 2.8$ femtometers,

$$v_s \approx 134c. \tag{1.20}$$

Therefore, the speed of rotation on the surface of the electron is more than 130 times the speed of light.

What does this apparent fallacy imply? It tells us that the concept of spin is inherently quantum mechanical and cannot be described within the framework of classical mechanics as done here. Furthermore, the electron cannot be visualized as a nearly point charge with the Lorentz radius.

1.7   References

[1] L. D. Landau and E. M. Lifshitz, *Quantum Mechanics (Non-relativistic Theory)*. 3rd. edition (Pergamon, New York), p. 198.

[2] R. P. Feynman, R. B. Leighton and M. Sands, *The Feynman Lectures on Physics*, Vol. 3, (Addison-Wesley, Reading, MA, 1965), p. 4-3.

[3] J. D. Cresser, *Quantum Physics Notes*, Macquarie University, Australia. Chapter 6.

[4] W. Gerlach and O. Stern, "Der experimentalle nachweis der richtungsquantelung in magnetic field", Z Phys., **9**, pp. 349-352 (1922).

[5] B. Friedrich and D. Hershbach, "Stern and Gerlach: How a bad cigar helped reorient atomic physics", *Physics Today*, **56**, pp. 53-59 (December 2003).

[6] D. D. Awschalom, M. E. Flatte and N. Samarth, "Spintronics", Scientific American, **286**, 66, June 2002.

[7] S.A. Wolf and D.M. Treger, "Special issue on spintronics", Proc. IEEE, **91**, 647 (2003).

[8] P.P. Freitas, F. Silva, N.J. Oliveira, L.V. Melo, L. Costa and N. Almeida, "Spin valve sensors", Sensors and Actuators A, **81**, 2 (2000).

[9] J. Wang, H. Meng and J.P. Wang, "Programmable spintronics logic device based on a magnetic tunnel junction element", J. Appl. Phys., **97**, 100509 (2005).

[10] P.P. Freitas, L. Costa, N. Almeida, L.V. Melo, F. Silva, J. Bernardo and C. Santos, "Giant magnetoresistive sensors for rotational speed control", J. Appl. Phys., **85**, 5459 (1999).

[11] W.J. Ku, P.P. Freitas, P. Compadrinho and J. Barata, "Precision X-Y robotic object handling using a dual GMR bridge sensor", IEEE Trans. Magn., **36**, 2782 (2000).

[12] J. Pelegri, J.B. Egea, D. Ramirez and P.P. Freitas, "Design, fabrication, and analysis of a spin-valve base current sensor", Sensors and Actuators A, **105**, 132 (2003).

[13] M.N. Baibich, J. M. Broto, A. Fert, F. Nguyen Van Dau, F. Petroff, P. Eitenne, G. Creuzet, A. Friederich and J. Chazelas, "Giant magnetoresistance of (001)Fe/(001)Cr magnetic superlattices", Phys. Rev. Lett., **61**, 2472 (1988).

[14] G. Binasch, P. Grünberg, F. Saurenbach and W. Zinn, "Enhanced magnetoresistance in layered magnetic structures with antiferromagnetic interlayer exchange", Phys. Rev. B, **39**, 4828 (1989).

[15] A detailed review of the work on giant magnetoresistance using magnetic multilayers is given by M.A.M. Gijs and G.E.W. Bauer, "Perpendicular giant magnetoresistance of magnetic multilayers", Adv. Phys., **46**, 285 (1997).

[16] S.S.P. Parkin, X. Jiang, C. Kaiser, A. Panchula, K. Roche and M. Samant, "Magnetically engineered spintronic sensors and memory", Proc. IEEE, **91**, 661 (2003).

[17] D.D. Awschalom and J.M. Kikkawa, "Electron spin and optical coherence in semiconductors", Physics Today, **52**, 33 (1999).

[18] *Magnetic Multilayers and Giant Magnetoresistance*, Ed. U. Hartman (Springer, Berlin, 2000).

[19] *Spin Electronics*, Eds. M. Ziese and M. J. Thornton, (Springer, New York, 2001).

[20] *Spin Dependent Transport in Magnetic Nanostructures*, Eds. S. Maekawa and T. Shinjo, (Taylor and Francis, New York, 2002).

[21] *Magnetic Interactions and Spin Transport*, Eds. A. Chtchelkanova, S. Wolf and Y. Idzerda, (Kluwer Academic Press, Dordrecht, 2003).

[22] S. Datta and B. Das, "Electronic analog of the electro-optic modulator", Appl. Phys. Lett., **56**, 665 (1990).

[23] S. Bandyopadhyay and M. Cahay, "Alternate spintronic analog of the electro-optic modulator", Appl. Phys. Lett., **85**, 1814 (2004).

[24] J. Fabian, I Zutic and S. Das Sarma, "Magnetic bipolar transistors", Appl. Phys. Lett., **84**, 85 (2004).

[25] M. E. Flatté and G. Vignale, "Unipolar spin diodes and transistors", Applied Physics Letters, Vol. 78, 1273 (2001); M. E. Flatté, Z. G. Yu, E. Johnston-Halperin and D. D. Awschalom, "Theory of semiconductor magnetic bipolar transistors", Appl. Phys. Lett., **82**, 4740 (2003); M. E. Flatté and G. Vignale, "Heterostructure unipolar spin transistors", J. Appl. Phys., **97**, 104508 (2005).

[26] S. Bandyopadhyay and M. Cahay, "Re-examination of some spintronic field effect device concepts", Appl. Phys. Lett., **85**, 1433 (2004).

[27] S. Bandyopadhyay and M. Cahay, "Are spin junction transistors useful for signal processing?", Appl. Phys. Lett., **86**, 133502 (2005).

[28] S. Bandyopadhyay, B. Das and A. E. Miller, "Supercomputing with spin polarized single electrons in a quantum coupled architecture", Nanotechnology, **5**, 113 (1994).

[29] S. Bandyopadhyay, "Computing with spins: from classical to quantum computing", Superlat. Microstruct., **37**, 77 (2005).

[30] S. Bandyopadhyay and V. P. Roychowdhury, "Switching in a reversible spin logic gate", Superlat. Microstruct., **22**, 411 (1997).

[31] A. M. Bychkov, L. A. Openov and I. A. Semenihin, "Single electron computing without dissipation", JETP Lett., **66**, 298 (1997).

[32] D. Loss and D.P. DiVincenzo, "Quantum computation with quantum dots", Phys. Rev. A, **57**, 120 (1998).

[33] S. Bandyopadhyay, "Self assembled nanoelectronic quantum computer based on the Rashba effect in quantum dots", Phys. Rev. B, **61**, 13813 (2000).

[34] B. E. Kane, "A silicon-based nuclear spin quantum computer", Nature (London), **393**, 133 (1998).

[35] M. V. G. Dutt, et al., "Quantum register basd on individual electronic and nuclear spin qubits in diamond", Science, **316**, 1312 (2007).

[36] J. R. Weber, W. F. Koehl, J. B. Varley, A. Janotti, B. B. Buckley, C. G. Van de Walle and D. D. Awschalom, "Quantum computing with defects", Proc. National Academy of Sciences of the United States of America, **107**, 8513 (2010).

2

Quantum Mechanics of Spin

In the 1920s, the old quantum theory was gradually being superseded by the new quantum theory. The cornerstone of the old theory was Bohr's model of the hydrogen atom, which predicted that an electron cannot orbit the proton in the hydrogen atom in any arbitrary fashion. Orbits are "quantized," meaning that only certain sizes, shapes, and magnetic properties are allowed. The principal quantum number n determined the allowed radii of the orbits, the orbital quantum number l determined the allowed shapes, and the magnetic quantum number m determined the magnetic behavior. Additionally, there is a fourth quantum number s which denotes the fact that the electron has an additional angular momentum, loosely associated with self rotation about its own axis, and that is quantized in units of $(1/2)\hbar$. The old quantum theory was useful to infer the existence of discrete energy levels in atoms, calculate energy spacings between these levels, and therefore allowed one to interpret atomic spectra.

The new quantum theory appeared to be more revolutionary and more powerful. It was triggered by Heisenberg's discovery of *matrix mechanics* and Schrödinger's discovery of *wave mechanics*. These two formalisms would not only predict the quantization of energy and provide a prescription to determine the energy difference between the levels (and thus explain the multiplicity of atomic spectra), but also allow one to calculate easily *probabilities of transitions* between different quantized energy states. At first, matrix mechanics and wave mechanics looked entirely different in their mathematical appearance and physical meaning. However, Schrödinger and Eckart [1] independently showed that the two theories are mathematically equivalent. Toward the end of 1926, Dirac unified the two theories using the concept of state vector and thus established the *transformation theory* of quantum mechanics. This ultimately had a profound implication for the quantum mechanical (mathematical) recipe to treat spin, as we will show in this chapter.

The transformation theory is the mathematical recipe to handle modern quantum mechanics. In Heisenberg's matrix mechanics, a physical quantity is expressed by a matrix, whereas in Schrödinger's wave mechanics, a physical quantity is expressed by a linear operator. In the unified transformation theory, physical quantities are represented by abstract linear operators called Dirac's *q-numbers*, which are linear operators in an infinite-dimensional linear space. Depending upon which types of orthogonal coordinate systems are used in this linear space, either matrix mechanics or wave mechanics emerges.

In other words, by using coordinate transformation in this linear space, we can derive matrix mechanics from wave mechanics and vice versa. Therefore, this unified theory was named transformation theory. The state of a quantum mechanical object is represented by a so-called *state vector*, which is an abstract vector in this linear space (the "wavefunction" in wave mechanics is an example of this) and the linear space is called the *state space*.

Earlier D. Hilbert and J. von Neumann had introduced the notion of a linear space that could absorb the mathematics of matrices and vectors, as well as the mathematics of linear operators and functions. This so-called *Hilbert space* admitted a finite or denumerably (countably) infinite number of co-ordinate axes. Therefore, a state vector in the Hilbert space could have at most a denumerably infinite number of mutually orthogonal components. Dirac extended this concept to a non-denumerably infinite number of coordinate axes in his linear space via the introduction of his famous δ-function [4]. The state vector therefore could have a non-denumerably infinite number of mutually orthogonal components and could be expressed as

$$\psi(q), \quad q \in [q_1, q_2] , \tag{2.1}$$

where the variable q is a continuous variable in the domain $[q_1, q_2]$. On the other hand, if the coordinate axes were countable, then the state vector would be expressed as

$$\psi_n, \quad n = 1, 2, 3, \dots \tag{2.2}$$

where the variable n is an integer.

According to Dirac's transformation theory, the state vector (i) evolves in time according to a unitary transformation, and (ii) satisfies a *first order* differential equation with respect to time. This second property is very important, as we shall see later. Depending on whether the physical quantity* represented by the state vector will yield discrete or continuous values upon measurement, the eigenvalues of the linear operator describing this physical quantity will have discrete or continuous values. Accordingly, the coordinate axes in the linear space will be discrete or continuous, and the state vector will be ψ_n or $\psi(q)$. The magnitude squared of the component of the state vector, i.e., $|\psi_n|^2$ or $|\psi(q)|^2$ gives the probability of the physical quantity taking on the n-th (or q-th) value when the quantity is measured. This is the physical interpretation (or significance) of the state vector. Therefore, each component of the state vector is called a *probability amplitude*. The familiar "wavefunction" in the Schrödinger formalism of wave mechanics is the probability amplitude where the physical quantities corresponding to the coordinate axes in linear space are the position coordinates and time, i.e.,,

$$\psi(\vec{r}) = \psi(x, y, z, t) . \tag{2.3}$$

---

*A physical quantity, by definition, is anything that can be measured, even if by a gedanken experiment only.

In wave mechanics, the Schrödinger equation for a single particle tells us how the wavefunction evolves in time and space:

$$i\hbar\frac{\partial\psi(\vec{r})}{\partial t} = H_0\psi(\vec{r}) \ . \tag{2.4}$$

If we neglect spin, then

$$H_0 = \frac{|\vec{p}|^2}{2m} + V(\vec{r})$$

$$\vec{p} = p_x\hat{\mathbf{x}} + p_y\hat{\mathbf{y}} + p_z\hat{\mathbf{z}} = -i\hbar\frac{\partial}{\partial x}\hat{\mathbf{x}} - i\hbar\frac{\partial}{\partial y}\hat{\mathbf{y}} - i\hbar\frac{\partial}{\partial z}\hat{\mathbf{z}},$$

$$\vec{r} = [x\hat{\mathbf{x}} + y\hat{\mathbf{y}} + z\hat{\mathbf{z}}]\,, t \tag{2.5}$$

where the quantities with "hats" are unit vectors along the coordinate axes.

Solution of Equation (2.4) yields the wavefunction $\psi(\vec{r})$. The quantity H_0 is the so-called Hamiltonian whose first term is the kinetic energy and second term is the potential energy. The only restriction is that the potential energy term should be a real quantity so that the Hamiltonian remains a Hermitian operator, which guarantees that its eigenvalue (which is its expected value and therefore the expected value of the energy) remains a real quantity.

The question now is how to include "spin" in Equation (2.4)?

This was investigated by Wolfgang Pauli. He derived an equation to replace Equation (2.4) which bears his name and is known as the *Pauli Equation*. But before we discuss this equation, we need to understand an important concept, namely, *Pauli spin matrices*, since they appear in the Pauli equation.

2.1 Pauli spin matrices

In quantum mechanics, any physical observable is associated with an operator (which would be a linear operator in the Schrödinger formalism, or a matrix in the Heisenberg formalism). The eigenvalues of the linear operator, or the eigenvalues of the matrix, are the expectation values of the physical quantity, i.e.,, the values we expect to find if we measure the physical quantity in an experiment[†]. Spin is a physical observable since the associated angular momentum can be measured, as was done unwittingly by Stern and Gerlach. Consequently, there must be a quantum mechanical operator associated with spin. Pauli derived the quantum mechanical operators for the spin components along three orthogonal axes – S_x, S_y and S_z. They are 2×2 complex

---

[†]Repeated measurements of a physical observable will produce a distribution of values whose average will be the expectation value.

matrices that came to be known as the Pauli spin matrices. Pauli's approach was based on the premise that: (1) the measurement of the spin angular momentum component along any coordinate axis for an electron should give the results $+\frac{\hbar}{2}$ or $-\frac{\hbar}{2}$, and (2) the operators for spin components along three mutually orthogonal axes should obey commutation rules similar to those obeyed by the operators associated with components of the orbital angular momentum. This would put spin angular momentum and orbital angular momentum on the same footing.

The operators (matrices) for the orbital angular momentum are known to satisfy the commutation relations

$$L_y L_z - L_z L_y = i\hbar L_x,$$
$$L_z L_x - L_x L_z = i\hbar L_y,$$
$$L_x L_y - L_y L_x = i\hbar L_z, \tag{2.6}$$

which merely reflect the fact that the orbital angular momenta along any two mutually orthogonal axes cannot be simultaneously measured with absolute precision unless the orbital angular momentum along a third axis, perpendicular to both the other two axes, vanishes.

Pauli adopted similar commutation relations for the spin angular momentum operators S_x, S_y and S_z:

$$S_y S_z - S_z S_y = i\hbar S_x,$$
$$S_z S_x - S_x S_z = i\hbar S_y,$$
$$S_x S_y - S_y S_x = i\hbar S_z. \tag{2.7}$$

Now, in the Stern-Gerlach experiment, assuming that the z-axis is the axis joining the south to north pole of the magnet, the observation of two traces on the photographic plate was interpreted as being caused by a spin angular momentum $\vec{S}$ whose z-component has two values $\pm\frac{\hbar}{2}$. Therefore, the matrix operator S_z must be (i) a 2×2 matrix (because such a matrix has *two* eigenvalues), and (ii) these eigenvalues must be $\pm\frac{\hbar}{2}$.

A 2×2 matrix that has eigenvalues of $\pm\frac{\hbar}{2}$ is the matrix

$$M_{2\times2} = \frac{\hbar}{2}\begin{pmatrix} 1 & 0 \\ 0 & -1 \end{pmatrix}. \tag{2.8}$$

This is not the only 2×2 matrix with eigenvalues $\pm\frac{\hbar}{2}$ – there could be many others – but this is the matrix that Pauli chose as a start for the operator S_z.

Next, he had to find appropriate matrices to serve as operators S_x and S_y. Pauli realized that since the choice of the z-axis as the axis joining the north and south poles of the magnet is completely arbitrary, the result of the Stern-Gerlach measurement should not be affected if he had chosen this axis to be the x- or y-axis, instead. This means that the expectation values of S_x and

S_y, i.e.,, their eigenvalues, should also be $\pm\frac{\hbar}{2}$. Moreover, all three matrices – S_x, S_y and S_z – must satisfy the commutation relations in Equation (2.7).

Pauli first defined three dimensionless matrices σ_x, σ_y and σ_z such that

$$S_x = \frac{\hbar}{2}\sigma_x,$$
$$S_y = \frac{\hbar}{2}\sigma_y,$$
$$S_z = \frac{\hbar}{2}\sigma_z. \tag{2.9}$$

Since S_x, S_y and S_z must have eigenvalues of $\pm\frac{\hbar}{2}$, it is obvious that the σ-matrices must have eigenvalues of ± 1. Furthermore, Equation (2.7) mandates that

$$\sigma_y\sigma_z - \sigma_z\sigma_y = 2i\sigma_x,$$
$$\sigma_z\sigma_x - \sigma_x\sigma_z = 2i\sigma_y,$$
$$\sigma_x\sigma_y - \sigma_y\sigma_x = 2i\sigma_z. \tag{2.10}$$

According to Equations (2.8) and (2.9),

$$\sigma_z = \begin{pmatrix} 1 & 0 \\ 0 & -1 \end{pmatrix}. \tag{2.11}$$

So now Pauli needed to pick two matrices σ_x and σ_y such that they have eigenvalues of ± 1 and obey Equation (2.10). Since these matrices will be operators for physical observables (spin components), they must be Hermitian as well. It is easy to verify that σ_z is Hermitian.

We can start our search for σ_x and σ_y with Hermitian matrices that have off-diagonal elements only, i.e.,

$$\sigma_x = \begin{pmatrix} 0 & a \\ a^* & 0 \end{pmatrix}, \tag{2.12}$$

and

$$\sigma_y = \begin{pmatrix} 0 & b \\ b^* & 0 \end{pmatrix}. \tag{2.13}$$

Since the eigenvalues of these matrices are ± 1, we must have $|a|^2 = |b|^2 = 1$, which leads to the possible choices for a and $b = \pm 1$ or $\pm i$.

Next, we must satisfy Equation (2.10) and that mandates

$$Im(ab^*) = 1, \tag{2.14}$$

where *Im* stands for imaginary part.

Therefore, if we select $a = +1$, then we must choose $b = -i$, and this yields

$$\sigma_x = \begin{pmatrix} 0 & 1 \\ 1 & 0 \end{pmatrix}, \tag{2.15}$$

and

$$\sigma_y = \begin{pmatrix} 0 & -i \\ i & 0 \end{pmatrix}. \tag{2.16}$$

This is how Pauli came up with expressions for σ_x, σ_y and σ_z. These matrices are called *Pauli spin matrices* and serve as operators for the spin components according to Equation (2.9).

It is obvious that Pauli's choice was by no means unique. There are other legitimate choices (e.g., we could have chosen $a = -i$ and $b = +1$), but Pauli's choice is now history and universally adopted.

From the expressions for the Pauli spin matrices, we notice that the square of each of the Pauli matrices is the 2×2 unit matrix $[I]$. Hence

$$|S|^2 = S_x^2 + S_y^2 + S_z^2 = \frac{3}{4}\hbar^2[I] = \bar{s}(\bar{s}+1)\hbar^2[I], \tag{2.17}$$

with $\bar{s} = 1/2$. This should be compared with the equivalent relation for the orbital angular momentum operator

$$|L|^2 = m(m+1)\hbar^2[I], \quad m = 1, 2, 3... \tag{2.18}$$

2.1.1 Eigenvectors of the Pauli matrices: Spinors

The eigenvalues of the Pauli spin matrices are ±1. We now evaluate the corresponding eigenvectors that we denote as $|\pm>$.

Matrix σ_z: The eigenvectors of σ_z must satisfy

$$\sigma_z|\pm>_z = \pm1|\pm>_z . \tag{2.19}$$

These eigenvectors (with unit norm) will be

$$|+>_z = \begin{pmatrix} 1 \\ 0 \end{pmatrix}, \tag{2.20}$$

and

$$|->_z = \begin{pmatrix} 0 \\ 1 \end{pmatrix}. \tag{2.21}$$

It is easy to verify that these two eigenvectors are orthonormal, as they must be since they are eigenvectors of a Hermitian matrix corresponding to distinct (non-degenerate) eigenvalues.

Matrix σ_x: The eigenvectors of σ_x must satisfy

$$\sigma_x|\pm>_x = \pm1|\pm>_x . \tag{2.22}$$

Starting with Equation (2.15), these eigenvectors are found to be

$$|+>_x = \frac{1}{\sqrt{2}} \begin{pmatrix} 1 \\ 1 \end{pmatrix}, \tag{2.23}$$

and

$$|->_x= \frac{1}{\sqrt{2}} \begin{pmatrix} 1 \\ -1 \end{pmatrix}. \tag{2.24}$$

Once again, the two eigenvectors are orthonormal. As can be easily checked, these eigenvectors can also be expressed as

$$|\pm>_x= \frac{1}{\sqrt{2}}[|+>_z \pm|->_z]. \tag{2.25}$$

Matrix σ_y: The eigenvectors of σ_y must satisfy

$$\sigma_y|\pm>_y= \pm 1|\pm>_y. \tag{2.26}$$

Using Equation (2.16), these eigenvectors are found to be

$$|+>_y= \frac{1}{\sqrt{2}} \begin{pmatrix} 1 \\ i \end{pmatrix}, \tag{2.27}$$

and

$$|->_y= \frac{1}{\sqrt{2}} \begin{pmatrix} 1 \\ -i \end{pmatrix}. \tag{2.28}$$

These eigenvectors are orthonormal and can be expressed as

$$|\pm>_y= \frac{1}{\sqrt{2}}[|+>_z \pm i|->_z]. \tag{2.29}$$

The eigenvectors of the Pauli spin matrices are examples of "spinors" which are 2×1 column vectors that represent the spin state of an electron. If we know the spinor associated with an electron in a given state, we can deduce the electron's spin orientation, i.e., find the quantities $< S_x >$, $< S_y >$ and $< S_z >$, where the angular brackets $< ... >$ denote expectation values. We will see this later.

2.2   The Pauli equation and spinors

We can absorb the space and time dependent part of an electron's wavefunction in the spinor, so that the general form of a spinor will be

$$[\psi(\mathbf{x})] = \begin{bmatrix} \phi_1(\mathbf{x}) \\ \phi_2(\mathbf{x}) \end{bmatrix}, \tag{2.30}$$

where $\mathbf{x} \equiv (x,y,z,t)$, and ϕ_1 and ϕ_2 are the two components of the spinor wavefunction (assumed to be properly normalized).

With a 2-component wavefunction, the Schrödinger equation must be recast as

$$\left\{ [H] + \frac{\hbar}{i} \frac{\partial}{\partial t} [I] \right\} [\psi(\mathbf{x})] = [0] , \qquad (2.31)$$

where the Hamiltonian is a 2×2 matrix (since it may contain the 2×2 Pauli spin matrices), $[I]$ is the 2×2 identity matrix, and $[0]$ is the 2×1 null vector. Equation (2.31) is a set of two simultaneous differential equations for the two components of the spinor wavefunction – ϕ_1 and ϕ_2. Equation (2.31) is referred to as the *Pauli equation* [2].

Solution of the Pauli equation yields the two-component spinor wavefunction $[\psi(\mathbf{x})]$. Its practical use is in calculating the expected value of the spin angular momentum of an electron along any coordinate axis. The expected value along the n-th coordinate axis at location $(\vec{r}_0 = x_0, y_0, z_0)$ at an instant of time t will be $[\psi(\vec{r}_0, t)]^{\dagger}[S_n][\psi(\vec{r}_0, t)]$, where $S_n = (\hbar/2)\sigma_n$ and the superscript † (dagger) represents the Hermitian conjugate. Using Equation (2.9), we get

$$S_x(\vec{r}_0, t) = (\hbar/2)[\psi(\vec{r}_0, t)]^{\dagger} [\sigma_x] [\psi(x_0, y_0, z_0, t)]$$
$$= (\hbar/2) [\phi_1^*(\vec{r}_0, t) \ \phi_2^*(\vec{r}_0, t)] \begin{bmatrix} 0 & 1 \\ 1 & 0 \end{bmatrix} \begin{bmatrix} \phi_1(\vec{r}_0, t) \\ \phi_2(\vec{r}_0, t) \end{bmatrix}$$
$$= \hbar Re \, [\phi_1^*(\vec{r}_0, t)\phi_2(\vec{r}_0, t)] ,$$
$$S_y(\vec{r}_0, t) = (\hbar/2) [\psi(\vec{r}_0, t)]^{\dagger} [\sigma_y] [\psi(\vec{r}_0, t)]$$
$$= (\hbar/2) [\phi_1^*(\vec{r}_0, t) \ \phi_2^*(\vec{r}_0, t)] \begin{bmatrix} 0 & -i \\ i & 0 \end{bmatrix} \begin{bmatrix} \phi_1(\vec{r}_0, t) \\ \phi_2(\vec{r}_0, t) \end{bmatrix}$$
$$= \hbar Im \, [\phi_1^*(\vec{r}_0, t)\phi_2(\vec{r}_0, t)] ,$$
$$S_z(\vec{r}_0, t) = (\hbar/2) [\psi(\vec{r}_0, t)]^{\dagger} [\sigma_z] [\psi(\vec{r}_0, t)]$$
$$= (\hbar/2) [\phi_1^*(\vec{r}_0, t) \ \phi_2^*(\vec{r}_0, t)] \begin{bmatrix} 1 & 0 \\ 0 & -1 \end{bmatrix} \begin{bmatrix} \phi_1(\vec{r}_0, t) \\ \phi_2(\vec{r}_0, t) \end{bmatrix}$$
$$= (\hbar/2)\{|\phi_1(\vec{r}_0, t)|^2 - |\phi_2(\vec{r}_0, t)|^2\} , \qquad (2.32)$$

where *Re* stands for the real part, *Im* stands for the imaginary part and the superscript * (asterisk) represents complex conjugate.

Therefore, if we can find the 2-component wavefunction in Equation (2.30) by solving the Pauli equation (2.31), then we can find the three components of the expected value of the spin angular momentum at any location at any instant of time. This is why the Pauli equation and the spinor concept are useful and important. In Chapter 7, we will also show how the Pauli equation can be used to derive the energy dispersion relations (relation between the energy and the wavevector) of an electron in a solid in the presence of spin-dependent effects.

2.3 More on the Pauli equation

Referring to Equation (2.31), we ask what terms will the 2×2 Hamiltonian $[H]$ contain. Normally, it will consist of three types of terms:

$$[H] = H_0[I] + [H_B] + [H_{SO}], \qquad (2.33)$$

where H_0 is the spin-independent Hamiltonian, and $[H_B]$, $[H_{SO}]$ are 2×2 matrices that depend on spin and will therefore involve the Pauli spin matrices.

To understand where $[H_B]$ comes from, consider the fact that if we view "spin" as being associated with self-rotation of an electron about its axis, then the self-rotation of the charged entity will give rise to a magnetic moment $\vec{\mu}_e$. This magnetic moment will interact with any externally applied magnetic field, if such a field is present. Let us say that the flux density associated with the external field is $\vec{B}$. Then the energy of interaction of $\vec{\mu}_e$ with $\vec{B}$ is

$$E_{int} = -\vec{\mu}_e \cdot \vec{B}. \qquad (2.34)$$

Landé had shown that the ratio of the magnetic moment $\vec{\mu}_e$ (in units of the Bohr magneton μ_B) to the angular momentum of self-rotation $\vec{S}$ (in units of $\hbar$) is the so-called gyromagnetic factor g [3]. Therefore, the operator associated with E_{int} is

$$[H_B] = -(g/2)\mu_B \vec{B} \cdot \vec{\sigma}, \qquad (2.35)$$

since $\vec{S} = (\hbar/2)\vec{\sigma}$, where $\vec{\sigma} = \sigma_x \hat{x} + \sigma_y \hat{y} + \sigma_z \hat{z}$.

Obviously, the two eigenvalues of the matrix $[H_B]$ will not be the same, meaning that the eigenenergies associated with this Hamiltonian will *not* be degenerate. Therefore, this term will cause spin-splitting, or lift the degeneracy between the two spin states. This splitting is the *Zeeman splitting*. The Hamiltonian $[H_B]$ is called the *Zeeman Hamiltonian* or the *Zeeman interaction* term.

The Hamiltonian $[H_{SO}]$ is associated with spin-orbit interaction which also lifts the spin degeneracy. This interaction is discussed in Chapter 6.

Finally, the general form of the Pauli equation is

$$\left\{ H_0[I] + [H_B] + [H_{SO}] + \frac{\hbar}{i} \frac{\partial}{\partial t} [I] \right\} [\psi(\mathbf{x})] = [0]. \qquad (2.36)$$

Its solution yields the 2-component wavefunction $[\psi(\mathbf{x})]$, which then yields the spin components of an electron at any position and at any instant of time from Equation (2.32). The Pauli Equation of course has many other uses, as we will see later in Chapter 7.

2.4   Extending the Pauli equation - the Dirac equation

The Pauli equation is completely non-relativistic and Pauli never found the avenue to reconcile it with relativity. That task was completed by Paul Andrew Maurice Dirac.

Both Schrödinger and two physicists, O. Klein and W. Gordon, had independently derived a relativistic equivalent of the Schrödinger equation. This is known as the *Klein-Gordon equation* [5]. A free particle not subjected to any force has a constant potential energy which can be taken to be zero (since potential is always undefined to the extent of an arbitrary constant). According to Einstein's special theory of relativity, such a particle obeys the relation

$$\overline{E}^2 = p^2 c^2 + m_0^2 c^4 \ , \tag{2.37}$$

where $\overline{E}$ is the total energy and p is the momentum.

According to De Broglie[‡],

$$\overline{E} = h\nu,$$
$$p = h/\lambda \ , \tag{2.38}$$

where ν is the frequency and λ is the wavelength of the De Broglie wave associated with the particle ($\lambda = c\nu$). Therefore, the last equation can be re-written as

$$\nu^2 - \left(\frac{c}{\lambda}\right)^2 = \left(\frac{m_0 c^2}{h}\right)^2 \ . \tag{2.39}$$

The above equation is known as the *Einstein-De Broglie equation*.

In the operator version of quantum mechanics, the energy operator is $i\hbar(\partial/\partial t)$, and the momentum operator (describing momentum along the x_r axis) is $-i\hbar(\partial/\partial x_r)$. Therefore, the quantum mechanical representation of Equation (2.37) is

$$\left[\left(i\hbar\frac{\partial}{c\partial t}\right)^2 - \sum_{r=1}^{3}\left(-i\hbar\frac{\partial}{\partial x_r}\right)^2 - m_0^2 c^2\right]\psi(x,y,z,t) = 0 \ . \tag{2.40}$$

The above equation has the solution of a plane wave

$$\psi(x,y,z,t) = e^{i(\vec{k}\cdot\vec{r}-\omega t)},$$
$$\omega = 2\pi\nu,$$
$$\vec{k} = (2\pi)/\vec{\lambda},$$
$$\vec{r} = x\hat{x} + y\hat{y} + z\hat{z} \ . \tag{2.41}$$

[‡]The De Broglie relation is reviewed in Chapter 18.

Substituting this solution into Equation (2.40) immediately yields Equation (2.39).

Equation (2.40) is valid for a free particle. Klein and Gordon extended it to a particle subjected to a (time-dependent) force field. Let the time dependent vector potential associated with the force field be $\vec{A} = (A_0, A_x, A_y, A_z)$, where we have treated space and time on the same footing as mandated by the theory of relativity. Klein and Gordon modified Equation (2.40) as

$$\left[\left(i\hbar \frac{\partial}{c\partial t} + eA_0 \right)^2 - \sum_{r=1}^{3} \left(-i\hbar \frac{\partial}{\partial x_r} + eA_r \right)^2 - m_0^2 c^2 \right] \psi(x, y, z, t) = 0.$$

(2.42)

The above equation is the *Klein-Gordon Equation*. For a time, it was thought to be the fundamental equation of relativistic quantum mechanics.

Dirac, however, questioned the Klein-Gordon construct. This equation is a second order differential equation with respect to time, and, according to Dirac's transformation theory, all meaningful equations of quantum mechanics must be first order with respect to time (the Schrödinger and Pauli equations are).

Dirac insisted on an equation that will be first order with respect to time. Now, in relativity, space and time are treated as equivalent and therefore the desired equation must also be first order with respect to space. Accordingly, the sought after equation needs to have a form

$$\left[\left(i\hbar \frac{\partial}{c\partial t} + eA_0 \right) - \sum_{r=1}^{3} \alpha_r \left(-i\hbar \frac{\partial}{\partial x_r} + eA_r \right) - \alpha_0 m_0 c \right] \psi(x, y, z, t) = 0.$$

(2.43)

When Dirac postulated the above equation, he simultaneously came up with four new quantities α_0 and α_r ($r = 1,2,3$), and he also told us how to determine them. He realized that a free particle without the vector potentials A_0 and A_r has to satisfy the Einstein-De Broglie Equation (Equation(2.39)) and therefore must satisfy Equation (2.40). In other words, the wavefunction $\psi(x, y, z, t)$ must be a solution of Equation (2.40). The latter equation is second order in space and time, whereas Dirac's Equation (Equation(2.43)) for a free particle,

$$\left[\left(i\hbar \frac{\partial}{c\partial t} \right) - \sum_{r=1}^{3} \alpha_r \left(-i\hbar \frac{\partial}{\partial x_r} \right) - \alpha_0 m_0 c \right] \psi(x, y, z, t) = 0, \qquad (2.44)$$

is first order.

In order to take the above equation for a free particle and make it second order to match Equation (2.40), we apply it to the operator

$$\left[\left(i\hbar \frac{\partial}{c\partial t} \right) + \sum_{r=1}^{3} \alpha_r \left(-i\hbar \frac{\partial}{\partial x_r} \right) + \alpha_0 m_0 c \right], \qquad (2.45)$$

to yield

$$\left[\left(i\hbar\frac{\partial}{c\partial t}\right) + \sum_{r=1}^{3}\alpha_r\left(-i\hbar\frac{\partial}{\partial x_r}\right) + \alpha_0 m_0 c\right]$$

$$\cdot\left[\left(i\hbar\frac{\partial}{c\partial t}\right) - \sum_{r=1}^{3}\alpha_r\left(-i\hbar\frac{\partial}{\partial x_r}\right) - \alpha_0 m_0 c\right]\psi(x,y,z,t) = 0. \quad (2.46)$$

Dirac insisted that the above equation be Equation (2.40). This can only happen if the quantities α_0 and α_r are not ordinary numbers, but *matrices*. In that case, Equation (2.46) will become

$$[\left(i\hbar\frac{\partial}{c\partial t}\right)^2 - \sum_{r=1}^{3}\{\alpha_r\}^2\left(-i\hbar\frac{\partial}{\partial x_r}\right)^2 -$$

$$\sum_{m<n}(\{\alpha_m\}\{\alpha_n\} - \{\alpha_n\}\{\alpha_m\})(i\hbar)^2\frac{\partial^2}{\partial x_m x_n} - \{\alpha_0\}^2 m_0^2 c^2]\psi(x,y,z,t) = 0,$$

$$(2.47)$$

where the matrices $\{\alpha_0\}$ and $\{\alpha_r\}$ ($r = 1, 2, 3$) have the properties

$$\{\alpha_m\}^2 = [I] \qquad (m = 0,1,2,3),$$
$$\{\alpha_m\}\{\alpha_n\} + \{\alpha_n\}\{\alpha_m\} = [0] \ (m \neq n; m, n = 0,1,2,3). \quad (2.48)$$

The simplest matrices that possess the properties listed in Equation (2.48) are

$$\{\alpha_0\} = \begin{bmatrix} 1 & 0 & 0 & 0 \\ 0 & 1 & 0 & 0 \\ 0 & 0 & -1 & 0 \\ 0 & 0 & 0 & -1 \end{bmatrix},$$

$$\{\alpha_1\} = \begin{bmatrix} 0 & 0 & 0 & 1 \\ 0 & 0 & 1 & 0 \\ 0 & 1 & 0 & 0 \\ 1 & 0 & 0 & 0 \end{bmatrix},$$

$$\{\alpha_2\} = \begin{bmatrix} 0 & 0 & 0 & -i \\ 0 & 0 & i & 0 \\ 0 & -i & 0 & 0 \\ i & 0 & 0 & 0 \end{bmatrix},$$

$$\{\alpha_3\} = \begin{bmatrix} 0 & 0 & 1 & 0 \\ 0 & 0 & 0 & -1 \\ 1 & 0 & 0 & 0 \\ 0 & -1 & 0 & 0 \end{bmatrix}. \quad (2.49)$$

When these 4×4 matrices are introduced into Equation (2.47), the wave-function $\psi(x, y, z, t)$ becomes a 4×1 column vector

$$\psi(x, y, z, t) = \begin{bmatrix} \psi_1 \\ \psi_2 \\ \psi_3 \\ \psi_4 \end{bmatrix}, \tag{2.50}$$

so that Equation (2.47) becomes a set of four coupled differential equations with respect to the four variables ψ_1, ψ_2, ψ_3 and ψ_4.

A little bit of inspection will quickly reveal that the Dirac matrices in Equation (2.49) can be written in terms of the Pauli spin matrices as

$$\{\alpha\}_0 = \begin{bmatrix} \mathbf{I} & \mathbf{0} \\ \mathbf{0} & -\mathbf{I} \end{bmatrix},$$

$$\{\alpha\}_1 = \begin{bmatrix} \mathbf{0} & \sigma_1 \\ \sigma_1 & \mathbf{0} \end{bmatrix},$$

$$\{\alpha\}_2 = \begin{bmatrix} \mathbf{0} & \sigma_2 \\ \sigma_2 & \mathbf{0} \end{bmatrix},$$

$$\{\alpha\}_3 = \begin{bmatrix} \mathbf{0} & \sigma_3 \\ \sigma_3 & \mathbf{0} \end{bmatrix}, \tag{2.51}$$

where $\mathbf{I}$ is a 2×2 identity matrix and $\mathbf{0}$ is a 2×2 null matrix.

Dirac applied his equation to the hydrogen atom and showed that the correct atomic level spacings can be obtained. More importantly, he showed that the orbital angular momentum *alone* is not a conserved quantity, but when the quantity represented by the operator

$$\frac{1}{2} \begin{bmatrix} \mathbf{0} & \sigma \\ \sigma & \mathbf{0} \end{bmatrix} \tag{2.52}$$

is added to it, the total quantity is conserved. Viewed from the perspective of conservation of total angular momentum, this shows that an electron has spin angular momentum given by the operator in Equation (2.52). *This is the first convincing theoretical demonstration of the existence of spin* and hence Dirac is credited with establishing the concept of spin rigorously. In the process, he also demonstrated that spin angular momentum must be quantized to two distinct values since the matrix in Equation (2.52) has two distinct and discrete eigenvalues. Therefore, Dirac was able to explain spin quantization, which the self-rotation model of the electron could never explain by itself.

Dirac also found that when an external force field is present, the procedure used to obtain Equation (2.47) from Equation (2.44) does not yield the Klein-Gordon equation (Equation (2.42)). The discrepancy can be explained by taking into account the interaction of the spin magnetic moment with the external field (recall the Zeeman interaction). Once this was demonstrated, the concept of spin was established on a firm footing.

2.4.1   Connection to Einstein's relativistic equation

If we substitute the results of Equation (2.4) in Equation (2.43), the Dirac
equation reduces to

$$i\hbar \frac{\partial}{\partial t}[\psi(x,y,z,t)] = \begin{bmatrix} \left(m_0 c^2 + V\right)\mathbf{I} & c\vec{\sigma} \cdot \left(\vec{p} + e\vec{A}\right) \\ c\vec{\sigma} \cdot \left(\vec{p} + e\vec{A}\right) & \left(-m_0 c^2 + V\right)\mathbf{I} \end{bmatrix} [\psi(x,y,z,t)]. \quad (2.53)$$

where $\mathbf{I}$ is, once again, the 2×2 identity matrix.

Interpreting the operator $i\hbar\partial/\partial t$ as the energy operator $\overline{E}_{op}$, the above
equation can be written as

$$\overline{E}_{op} = \begin{bmatrix} \left(m_0 c^2 + V\right)\mathbf{I} & c\vec{\sigma} \cdot \left(\vec{p} + e\vec{A}\right) \\ c\vec{\sigma} \cdot \left(\vec{p} + e\vec{A}\right) & \left(-m_0 c^2 + V\right)\mathbf{I} \end{bmatrix}. \quad (2.54)$$

When both vector and scalar potentials are absent, i.e., $V = A = 0$, we can
square both sides of the last equation to obtain

$$\overline{E}_{op}^2 = \begin{bmatrix} m_0 c^2 \mathbf{I} & c\vec{\sigma} \cdot \vec{p} \\ c\vec{\sigma} \cdot \vec{\sigma} & -m_0 c^2 \mathbf{I} \end{bmatrix}^2$$

$$= \left(p^2 c^2 + m_0^2 c^4\right) \begin{bmatrix} 1 & 0 & 0 & 0 \\ 0 & 1 & 0 & 0 \\ 0 & 0 & 1 & 0 \\ 0 & 0 & 0 & 1 \end{bmatrix}, \quad (2.55)$$

which matches exactly Equation (2.37). Therefore, indeed the Dirac equation
is the relativistic version of the Schrödinger equation.

2.5   Time-independent Dirac equation

From Equation (2.53), it is straightforward to show that the *time independent*
Dirac equation will be

$$\begin{bmatrix} A & 0 & C & D^* \\ 0 & A & D & -C \\ C & D^* & B & 0 \\ D & -C & 0 & B \end{bmatrix} \begin{pmatrix} \psi_1 \\ \psi_2 \\ \psi_3 \\ \psi_4 \end{pmatrix} = E \begin{pmatrix} \psi_1 \\ \psi_2 \\ \psi_3 \\ \psi_4 \end{pmatrix}, \quad (2.56)$$

where the asterisk denotes complex conjugate and

$$
\begin{aligned}
A &= m_0 c^2 + V \\
B &= -m_0 c^2 + V \\
C &= c(p_z + e A_z) \\
D &= c[(p_x + e A_x) + i(p_y + e A_y)].
\end{aligned}
\tag{2.57}
$$

Equation (2.56) can be written more compactly as

$$
\begin{bmatrix} (m_0 c^2 + V)[I] & c\vec{\sigma} \cdot \left[\vec{p} + e\vec{A}\right] \\ c\vec{\sigma} \cdot \left[\vec{p} + e\vec{A}\right] & (-m_0 c^2 + V)[I] \end{bmatrix} \begin{bmatrix} \{\psi\}(x,y,z) \\ \{\phi\}(x,y,z) \end{bmatrix} = \overline{E} \begin{bmatrix} \{\psi\}(x,y,z) \\ \{\phi\}(x,y,z) \end{bmatrix},
\tag{2.58}
$$

where V is the scalar potential energy and

$$
\{\psi\}(x,y,z) = \begin{pmatrix} \psi_1(x,y,z) \\ \psi_2(x,y,z) \end{pmatrix},
\tag{2.59}
$$

and

$$
\{\phi\}(x,y,z) = \begin{pmatrix} \psi_3(x,y,z) \\ \psi_4(x,y,z) \end{pmatrix}.
\tag{2.60}
$$

From Equation (2.58), we can show that

$$
\left\{ (m_0 c^2 + V)\,[I] + \left[c\vec{\sigma}\cdot(\vec{p}+e\vec{A})\right] \frac{1}{\overline{E}+m_0 c^2 - V}[I]\left[c\vec{\sigma}\cdot(\vec{p}+e\vec{A})\right] \right\} [\psi] = \overline{E}[\psi],
$$
$$
\left\{ (-m_0 c^2 + V)\,[I] + \left[c\vec{\sigma}\cdot(\vec{p}+e\vec{A})\right] \frac{1}{\overline{E}-m_0 c^2 - V}[I]\left[c\vec{\sigma}\cdot(\vec{p}+e\vec{A})\right] \right\} [\phi] = \overline{E}[\phi].
\tag{2.61}
$$

2.5.1 Non-relativistic approximation to the Dirac equation

Consider a non-relativistic electron moving at speeds much less than the speed of light in vacuum. For such a particle, Equation (2.37) yields $\overline{E} \approx m_0 c^2$. Using that result in the first of the two equations above yields

$$
\left\{ (m_0 c^2 + V)\,[I] + \frac{\left[\vec{\sigma}\cdot(\vec{p}+e\vec{A})\right]^2}{2 m_0} \right\} [\psi] = \overline{E}[\psi],
\tag{2.62}
$$

which reduces to

$$
\left(\overline{E} - m_0 c^2\right)[\psi] = E[\psi] = \left(\frac{(\vec{p}+e\vec{A})^2}{2 m_0}[I] + \mu_B \vec{B}\cdot\vec{\sigma} + V[I] \right) [\psi],
$$
$$
= \{[H_0] + [H_B]\}\,[\psi],
\tag{2.63}
$$

since $[\vec{\sigma} \cdot (\vec{p} + e\vec{A})]^2 = (\vec{p} + e\vec{A})^2[I] + 2m_0\mu_B\vec{B} \cdot \vec{\sigma}$ (see Problem 2.5). Here E is the total energy *minus* the rest energy m_0c^2.

The second of the two equations in Equation (2.61) will yield the same result as Equation (2.63) provided we make the transformation $m_0 \rightarrow -m_0$. This shows that the second equation applies to particles with negative mass, namely, *anti-matter*. The existence of anti-matter is a foregone conclusion from Equation (2.37). Noting that the De Broglie relation relates momentum p to wavevector k of a particle as $p = \hbar k$, Equation (2.37) gives *two* dispersion relations E versus k. They are shown in Fig. 2.1. One branch has a positive curvature and therefore positive mass. This corresponds to "matter." The other has a negative curvature and therefore negative mass. That corresponds to "anti-matter." The energy separation between the two curves is $2m_0c^2$, which is ~ 1 MeV for a free electron. These energy scales are seldom encountered in solid state physics, which is why anti-matter is usually of concern only in high energy physics. We will not have any occasion to worry about anti-matter anywhere in this textbook.

2.5.2 Relationship between the non-relativistic approximation to the Dirac equation and the Pauli equation

The reader will immediately recognize Equation (2.63) as the Pauli equation (Equation 2.33) without the spin-orbit interaction term. The latter term does not arise here since, strictly speaking, spin-orbit interaction is a relativistic effect and therefore cannot be captured within a non-relativistic picture. However, what is amazing is that the Zeeman interaction term appears automatically, without having to introduce it separately. Therefore, Dirac was able to explain the Zeeman interaction directly from his equation!

The spin-orbit interaction term is not beyond the Dirac equation, but it is beyond the non-relativistic approximation since spin-orbit interaction has a relativistic origin. If we make a binomial expansion of Equation (2.61) and retain only the lowest order terms, then we get an equation [6]

$$\left[\frac{|\vec{p} + e\vec{A}|^2}{2m_0} + V - \frac{|\vec{p} + e\vec{A}|^4}{8m_0^3c^2} + \frac{\hbar}{4im_0^2c^2} \left(\vec{\nabla}V \cdot (\vec{p} + e\vec{A}) \right) \right.$$
$$\left. + \frac{\hbar}{4m_0^2c^2} \left(\vec{\nabla}V \times (\vec{p} + e\vec{A}) \right) \cdot \vec{\sigma} \right] [\psi] = E[\psi], \quad (2.64)$$

where the last term in the left hand side represents the spin-orbit interaction term. Therefore, the Dirac equation incorporates the spin-orbit interaction physics as well.

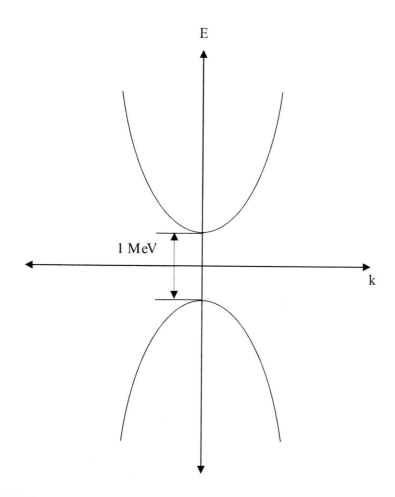

FIGURE 2.1
Dispersion relations for matter and anti-matter branches.

2.6   Problems

- **Problem 2.1**

Show that the operators for S_x, S_y and S_z satisfy Equations (2.7) and (2.17).

Solution

$$S_x S_y - S_y S_x = (\hbar^2/4) \left\{ \begin{bmatrix} 0 & 1 \\ 1 & 0 \end{bmatrix} \begin{bmatrix} 0 & -i \\ i & 0 \end{bmatrix} - \begin{bmatrix} 0 & -i \\ i & 0 \end{bmatrix} \begin{bmatrix} 0 & 1 \\ 1 & 0 \end{bmatrix} \right\},$$

$$= \left(\frac{\hbar}{2}\right)^2 \begin{bmatrix} 2i & 0 \\ 0 & -2i \end{bmatrix} = i\hbar \left[\frac{\hbar}{2}\sigma_z\right] = i\hbar S_z. \qquad (2.65)$$

Furthermore,

$$|S|^2 = S_x^2 + S_y^2 + S_z^2,$$

$$= (\hbar^2/4) \left\{ \begin{bmatrix} 0 & 1 \\ 1 & 0 \end{bmatrix}^2 + \begin{bmatrix} 0 & -i \\ i & 0 \end{bmatrix}^2 + \begin{bmatrix} 1 & 0 \\ 0 & -1 \end{bmatrix}^2 \right\},$$

$$= (3/4)\hbar^2 \begin{bmatrix} 1 & 0 \\ 0 & 1 \end{bmatrix} = (3/4)\hbar^2[I]. \qquad (2.66)$$

- **Problem 2.2**

Using Equation (2.32), show that if the 2-component spinor is the state $|+>_z$ given in Equation (2.20), then $< S_x > = < S_y > = 0$ and $< S_z > = \hbar/2$. Also show that if the 2-component spinor is the state $|->_z$ given in Equation (2.21), then $< S_x > = < S_y > = 0$ and $< S_z > = -\hbar/2$. Hence the state $|+>_z$ is referred to as the *+z-polarized state* and the state $|->_z$ is referred to as the *−z-polarized state*. This means that an electron in these states has its spin polarized along the +z and −z axes, respectively.

Solution

For the state $|+>_z$, $\phi_1 = 1$ and $\phi_2 = 0$ in Equation (2.30).

From Equation (2.32)

$$< S_x >= \hbar Re(\phi_1^* \phi_2) = 0,$$
$$< S_y >= \hbar Im(\phi_1^* \phi_2) = 0,$$
$$< S_z >= (\hbar/2)[|\phi_1|^2 - |\phi_2|^2] = \hbar/2. \tag{2.67}$$

For the state $|->_z$, $\phi_1 = 0$ and $\phi_2 = 1$, hence

$$< S_x >= \hbar Re(\phi_1^* \phi_2) = 0,$$
$$< S_y >= \hbar Im(\phi_1^* \phi_2) = 0,$$
$$< S_z >= (\hbar/2)[|\phi_1|^2 - |\phi_2|^2] = -\hbar/2. \tag{2.68}$$

Similarly, for the states $|\pm>_x$, $< S_x >= \pm\frac{\hbar}{2}$ and $< S_y >=< S_z >= 0$, so that these states are $\pm$x-polarized states. One can show that the states $|\pm>_y$ are $\pm$y-polarized states.

- **Problem 2.3**

  Show that if any 2×2 Hermitian matrix $[H]$ is used to perform a unitary transformation of the Pauli matrices defined above, i.e., if matrices $\sigma_x{}'$, $\sigma_y{}'$, $\sigma_z{}'$ are defined such that

  $$\sigma_n{}' = S\sigma_n S^{-1}, \tag{2.69}$$

  where $S = e^{iH}$, for $n = x$, y, or z, the commutation rules in Equation (2.7) are satisfied by the matrices $\sigma_n{}'$.

  Furthermore, the square of each of the matrices $\sigma_n{}'$ is equal to the 2×2 identity matrix.

  Solution

  The solution of this problem is left to the reader. Use the fact that the matrix S is

  $$[S] = [M]^{-1} \begin{bmatrix} e^{i\lambda_1} & 0 \\ 0 & e^{i\lambda_2} \end{bmatrix} [M], \tag{2.70}$$

  where $\lambda_{1,2}$ are the eigenvalues of the matrix $[H]$, and $[M]$ is a 2×2 matrix whose two columns are the eigenfunctions of $[H]$ corresponding to eigenvalues λ_1 and λ_2.

- **Problem 2.4**

  Show the following properties of the Pauli spin matrices by actual computation.

1. $\det(\sigma_j) = -1$; for $j = $ x, y, or z.
2. $\text{Tr}(\sigma_j) = 0$.
3. $\sigma_x^2 = \sigma_y^2 = \sigma_z^2 = I$.
4. $\sigma_x \sigma_y \sigma_z = iI$.
5. $\sigma_x \sigma_y = -\sigma_y \sigma_x = i\sigma_z$.
6. $\sigma_y \sigma_z = -\sigma_z \sigma_y = i\sigma_x$.
7. $\sigma_z \sigma_x = -\sigma_x \sigma_z = i\sigma_y$.
8. $\sigma_p \sigma_q + \sigma_q \sigma_p = 0, \quad (p \neq q; p, q = x, y, z)$.

- **Problem 2.5**

  Show that $\left[\vec{\sigma} \cdot (\vec{p} + e\vec{A})\right]^2 = (\vec{p} + e\vec{A})^2[I] + 2m_0 \mu_B \vec{B} \cdot \vec{\sigma}$.

  Solution

$$
\left[\vec{\sigma} \cdot (\vec{p} + e\vec{A})\right]^2 = \left[\sigma_x(p_x + eA_x) + \sigma_y(p_y + eA_y) + \sigma_z(p_z + eA_z)\right]
$$
$$
\times \left[\sigma_x(p_x + eA_x) + \sigma_y(p_y + eA_y) + \sigma_z(p_z + eA_z)\right],
$$
$$
= \sigma_x^2(p_x + eA_x)^2 + \sigma_y^2(p_y + eA_y)^2 + \sigma_z^2(p_z + eA_z)^2
$$
$$
+ \sigma_x\sigma_y(p_x + eA_x)(p_y + eA_y) + \sigma_y\sigma_x(p_y + eA_y)(p_x + eA_x)
$$
$$
\sigma_y\sigma_z(p_y + eA_y)(p_z + eA_z) + \sigma_z\sigma_y(p_z + eA_z)(p_y + eA_y)
$$
$$
\sigma_z\sigma_x(p_z + eA_z)(p_x + eA_x) + \sigma_x\sigma_z(p_x + eA_x)(p_z + eA_z).
$$
$$
\tag{2.71}
$$

Now the results of Problem 2.4 lead to

$$
\left[\vec{\sigma} \cdot (\vec{p} + e\vec{A})\right]^2 = (p_x + eA_x)^2 + (p_y + eA_y)^2 + (p_z + eA_z)^2
$$
$$
+ i\sigma_z\left(-ie\hbar\frac{\partial A_y}{\partial x} + ie\hbar\frac{\partial A_x}{\partial y}\right)
$$
$$
+ i\sigma_y\left(-ie\hbar\frac{\partial A_x}{\partial z} + ie\hbar\frac{\partial A_z}{\partial x}\right)
$$
$$
+ i\sigma_x\left(-ie\hbar\frac{\partial A_z}{\partial y} + ie\hbar\frac{\partial A_y}{\partial z}\right)
$$
$$
= (p_x + eA_x)^2 + (p_y + eA_y)^2 + (p_z + eA_z)^2
$$
$$
+ e\hbar\sigma_z\left(\frac{\partial A_y}{\partial x} - \frac{\partial A_x}{\partial y}\right)
$$
$$
+ e\hbar\sigma_y\left(\frac{\partial A_x}{\partial z} - \frac{\partial A_z}{\partial x}\right)
$$
$$
+ e\hbar\sigma_x\left(\frac{\partial A_z}{\partial y} - \frac{\partial A_y}{\partial z}\right)
$$
$$
= \left[\vec{p} + e\vec{A}\right]^2 + e\hbar\left(\vec{\nabla} \times \vec{A}\right) \cdot \vec{\sigma}, \tag{2.72}
$$

where we have used the fact that $\vec{p}_n = -i\hbar(\partial/\partial \vec{x}_n)$.

Since $\mu_B = e\hbar/2m_0$ and $\vec{B} = \vec{\nabla} \times \vec{A}$, this proves the equality stated in the problem.

- **Problem 2.6**

  Show that the 4×4 Dirac matrices are related to the 2×2 Pauli matrices as follows:

  $$-i\{a\}_0\{a\}_2\{a\}_3 = \begin{pmatrix} \sigma_x & 0 \\ 0 & -\sigma_x \end{pmatrix}$$

  $$-i\{a\}_0\{a\}_3\{a\}_1 = \begin{pmatrix} \sigma_y & 0 \\ 0 & -\sigma_y \end{pmatrix}$$

  $$-i\{a\}_0\{a\}_1\{a\}_2 = \begin{pmatrix} \sigma_z & 0 \\ 0 & -\sigma_z \end{pmatrix} \tag{2.73}$$

  Show also that

  $$\{a\}_0\{a\}_1 = \begin{pmatrix} 0 & \sigma_x \\ -\sigma_x & 0 \end{pmatrix}$$

  $$\{a\}_0\{a\}_2 = \begin{pmatrix} 0 & \sigma_y \\ -\sigma_y & 0 \end{pmatrix}$$

  $$\{a\}_0\{a\}_3 = \begin{pmatrix} 0 & \sigma_z \\ -\sigma_z & 0 \end{pmatrix} \tag{2.74}$$

- **Problem 2.7**

  Show that when an electron's or positron's momentum $\hbar k \ll m_0 c$, the dispersion relations in Fig. 2.1 are approximately parabolic.

2.7 Appendix

2.7.1 Working with spin operators

Applying the operators σ_x, σ_y and σ_z to the 2-component wavefunction in Equation (2.30) yields

$$\sigma_x \psi(\mathbf{x}) = \begin{bmatrix} 0 & 1 \\ 1 & 0 \end{bmatrix} \begin{bmatrix} \phi_1(\mathbf{x}) \\ \phi_2(\mathbf{x}) \end{bmatrix},$$

$$\sigma_y \psi(\mathbf{x}) = \begin{bmatrix} 0 & -i \\ i & 0 \end{bmatrix} \begin{bmatrix} \phi_1(\mathbf{x}) \\ \phi_2(\mathbf{x}) \end{bmatrix},$$

$$\sigma_z \psi(\mathbf{x}) = \begin{bmatrix} 1 & 0 \\ 0 & 1 \end{bmatrix} \begin{bmatrix} \phi_1(\mathbf{x}) \\ \phi_2(\mathbf{x}) \end{bmatrix}, \tag{2.75}$$

and therefore

$$S_x\psi = (\hbar/2)\sigma_x\psi = \frac{\hbar}{2}\begin{bmatrix} \phi_2(\mathbf{x}) \\ \phi_1(\mathbf{x}) \end{bmatrix},$$

$$S_y\psi = (\hbar/2)\sigma_x\psi = \frac{\hbar}{2}\begin{bmatrix} -i\phi_2(\mathbf{x}) \\ i\phi_1(\mathbf{x}) \end{bmatrix},$$

$$S_z\psi = (\hbar/2)\sigma_x\psi = \frac{\hbar}{2}\begin{bmatrix} \phi_1(\mathbf{x}) \\ -\phi_2(\mathbf{x}) \end{bmatrix}. \tag{2.76}$$

Thus, operating with S_x interchanges the two components of the spinor, operating with S_y interchanges the two components of the spinor while causing a phase shift of $-90°$ to the second component and a phase shift of $90°$ to the first component, and operating with S_z introduces a phase shift of $180°$ to the second component.

2.7.2 Two useful theorems

A trivial decomposition of any 2×2 matrix M is obviously

$$M = m_{11}\begin{pmatrix} 1 & 0 \\ 0 & 0 \end{pmatrix} + m_{12}\begin{pmatrix} 0 & 1 \\ 0 & 0 \end{pmatrix} + m_{21}\begin{pmatrix} 0 & 0 \\ 1 & 0 \end{pmatrix} + m_{22}\begin{pmatrix} 0 & 0 \\ 0 & 1 \end{pmatrix}, \tag{2.77}$$

since the four matrices on the right hand side form a complete basis for all 2×2 matrices. Now that we have introduced the Pauli spin matrices, a more subtle decomposition of any 2×2 complex matrix can be found, as discussed next.

Theorem I: Any 2×2 matrix M

$$M = \begin{pmatrix} m_{11} & m_{12} \\ m_{21} & m_{22} \end{pmatrix} \tag{2.78}$$

can be decomposed as

$$M = \frac{m_{11} + m_{22}}{2}I + \frac{m_{11} - m_{22}}{2}\sigma_z + \frac{m_{12} + m_{21}}{2}\sigma_x + i\frac{m_{12} - m_{21}}{2}\sigma_y. \tag{2.79}$$

·The proof is left as an exercise. In other words, the 4 matrices $(I, \sigma_x, \sigma_y, \sigma_z)$ form a complete set of bases in the space of 2×2 complex matrices.

The last equation can be written in the more condensed form

$$M = a_o I + \vec{a} \cdot \vec{\sigma}, \tag{2.80}$$

where

$$a_o = \frac{1}{2}Tr(M), \tag{2.81}$$

and

$$\vec{a} = \frac{1}{2}Tr(M\vec{\sigma}), \tag{2.82}$$

where $\vec{\sigma} = (\sigma_x, \sigma_y, \sigma_z)$, and Tr stands for the trace of the matrix.

A comparison of Equations (2.79) and (2.80) shows that M is Hermitian if a_0 and the three components of the vector $\vec{a}$ are real.

Exercise: Calculate the values of $(a_0, \vec{a})$ for the matrix M given by

$$M = \begin{pmatrix} 2 & \frac{i\sqrt{3}}{3} \\ \frac{-i\sqrt{3}}{3} & 4 \end{pmatrix}. \qquad (2.83)$$

Next, we prove an identity which will be used in the next chapter to interpret geometrically the Pauli matrices after the introduction of the Bloch sphere concept.

Theorem II: If θ is real and if the matrix A is such that $A^2 = I$, the following identity holds:

$$e^{i\theta A} = cos\theta I + isin\theta A. \qquad (2.84)$$

This is the generalization to operators of the well-known Euler relation for complex numbers, i.e., $e^{iz} = cosz + isinz$.

From the Taylor series expansion

$$e^x = \sum_{k=0}^{\infty} \frac{x^k}{k!}, \qquad (2.85)$$

and the definition of the function of an operator, we get

$$e^{i\theta A} = I + (i\theta)A + \frac{(i\theta)^2 A^2}{2!} + \frac{(i\theta)^3 A^3}{3!} + \frac{(i\theta)^4 A^4}{4!} + \dots \qquad (2.86)$$

or

$$e^{i\theta A} = (1 - \frac{\theta^2}{2!} + \frac{\theta^4}{4!} - \dots + (-1)^k \frac{\theta^{2k}}{(2k)!})I$$
$$+i(\theta - \frac{\theta^3}{3!} + \frac{\theta^5}{5!} - \dots + (-1)^k \frac{i\theta^{2k+1}}{(2k+1)!})A, \qquad (2.87)$$

which is indeed Equation (2.84) if we use the the Taylor expansions

$$sinx = \sum_{k=0}^{\infty} (-1)^k x^{2k+1}/(2k+1)!, \qquad (2.88)$$

and

$$cosx = \sum_{k=0}^{\infty} (-1)^k x^{2k}/(2k)!. \qquad (2.89)$$

2.7.3   Applications of the *Postulates of Quantum Mechanics* to a few spin problems

The Postulates of Quantum Mechanics are briefly reviewed in Chapter 18.

Example 1: If we measure the z-component of an electron's spin, apply the postulate of quantum projective measurement (Postulate 3) discussed in Chapter 18 to calculate the probability of the measurement to give the result $\frac{\hbar}{2}$ or $-\frac{\hbar}{2}$, respectively, if prior to the measurement, the state of the system $|\psi>$ is either

    (1) $|0>$,

    (2) $|1>$, or

    (3) $\frac{1}{\sqrt{2}}(|0>+|1>)$,

where $|0>$ is the +z-polarized state $|+>_z$ and $|1>$ is the -z-polarized state $|->_z$.

Solution

(1) The operator $S_z = \frac{\hbar}{2}\sigma_z$ has the spectral decomposition

$$S_z = \frac{\hbar}{2}|0><0| + (-\frac{\hbar}{2})|1><1| = \sum_m mP_m, \qquad (2.90)$$

where $P_m = |m><m|$, with $|m>$ being the eigenvectors of S_z.

Hence, if $|\psi> = |0>$, the probability of measuring $+\frac{\hbar}{2}$ is equal to

$$p(+\frac{\hbar}{2}) = <0|(|0><0|)|0> = 1, \qquad (2.91)$$

and the probability of measuring $-\frac{\hbar}{2}$ is equal to

$$p(-\frac{\hbar}{2}) = <0|(|1><1|)|0> = 0 \qquad (2.92)$$

and the sum of the probabilities is indeed equal to unity.

(2) If $|\psi> = |1>$, we get

$$p(\frac{\hbar}{2}) = <1|(|0><0|)|1> = 0, \qquad (2.93)$$

and

$$p(-\frac{\hbar}{2}) = <1|(|1><1|)|1> = 1. \qquad (2.94)$$

(3) Finally, if $|\psi> = \frac{1}{\sqrt{2}}(|0>+|1>)$, we find

$$p(+\frac{\hbar}{2}) = \frac{1}{\sqrt{2}}\{|0>+<1|\}(|0><0|)\frac{1}{\sqrt{2}}\{|0>+<1|\} = \frac{1}{2}\{1+0\}\cdot\{0+1\} = \frac{1}{2}, \qquad (2.95)$$

and similarly, $p(-\frac{\hbar}{2}) = \frac{1}{2}$.

According to *Postulate 3* discussed in Chapter 18, right after the measurement, the spinor collapses into the state

$$|\psi> \rightarrow |\psi^{new}> = \frac{(|0><0|)}{\sqrt{p(+\frac{\hbar}{2})}}|\psi> = (\frac{|0><0|}{\sqrt{\frac{1}{2}}})(\frac{(|0>+|1>)}{\sqrt{2}}) = |0>.$$

(2.96)

Exercise: Repeat the previous exercise if the component S_x is measured instead.

Example 2: Suppose an electron is prepared in the spinor $|0>$ eigenstate of S_z with eigenvalue $+\frac{\hbar}{2}$ and repeated measurements are made of the x-component of its intrinsic angular momentum; calculate the average value $< S_x >$ and the standard deviation $\Delta(S_x)$ of these measurements.

Solution

The spectral decomposition of the operator S_x is given by

$$S_x = \frac{\hbar}{2}\begin{pmatrix} 0 & 1 \\ 1 & 0 \end{pmatrix} = (+\frac{\hbar}{2})|+>_{x,x}< +| + (-\frac{\hbar}{2})|->_{x,x}< -|,$$

(2.97)

where $|+>_x$ is the $+x$-polarized state and $|->_x$ is the $-x$-polarized state. Hence,

$$< S_x > = \frac{\hbar}{2}\left[< 0|S_x|0 > = (1\ 0)\begin{pmatrix} 0 & 1 \\ 1 & 0 \end{pmatrix}\begin{pmatrix} 1 \\ 0 \end{pmatrix}\right] = 0.$$

(2.98)

Furthermore,

$$\Delta(S_x) = \sqrt{< S_x^2 > - < S_x >^2} = \sqrt{< S_x^2 >},$$

$$= \frac{\hbar}{2}\left[< 0|\begin{pmatrix} 0 & 1 \\ 1 & 0 \end{pmatrix}\begin{pmatrix} 0 & 1 \\ 1 & 0 \end{pmatrix}|0 >\right]^{1/2},$$

$$= \frac{\hbar}{2}\left[(1\ 0)\begin{pmatrix} 0 & 1 \\ 1 & 0 \end{pmatrix}\begin{pmatrix} 0 & 1 \\ 1 & 0 \end{pmatrix}\begin{pmatrix} 1 \\ 0 \end{pmatrix}\right]^{1/2} = \frac{\hbar}{2}.$$

(2.99)

Example 3: Suppose an electron is characterized by the spinor

$$|\psi> = \frac{1}{\sqrt{10}}\begin{pmatrix} 3 \\ -1 \end{pmatrix},$$

(2.100)

which is properly normalized, as easily checked. If we measure the y-component of the spin, what is the probability of finding that its value is $\frac{\hbar}{2}$?

Solution

The eigenvector of σ_y are $\frac{1}{\sqrt{2}}\begin{pmatrix}1\\i\end{pmatrix}$ with eigenvalue $+1$ and $\frac{1}{\sqrt{2}}\begin{pmatrix}1\\-i\end{pmatrix}$ with eigenvalue -1

Hence, if we write the spinor $|\psi>$ as a linear combination of $|+>_y$ and $|->_y$,

$$|\psi>= \frac{1}{\sqrt{10}}\begin{pmatrix}3\\-1\end{pmatrix} = \alpha\frac{1}{\sqrt{2}}\begin{pmatrix}1\\i\end{pmatrix} + \beta\frac{1}{\sqrt{2}}\begin{pmatrix}1\\-i\end{pmatrix}. \qquad (2.101)$$

The probability of finding $+\frac{\hbar}{2}$ when measuring S_y is given by $|\alpha|^2$ or

$$p(+\frac{\hbar}{2}) = |_y<+|\psi>|^2 = \left|\frac{1}{\sqrt{2}}(1, -i)\frac{1}{\sqrt{10}}\begin{pmatrix}3\\-1\end{pmatrix}\right|^2 = \frac{1}{20}|(3+i)|^2 = \frac{1}{2}. \qquad (2.102)$$

Example 4: Suppose an electron is characterized by the spinor

$$|\psi>= \frac{4}{5}|0> +\frac{3}{5}|1>. \qquad (2.103)$$

(1) What is the probability that a measurement of z-component of the spin will be $+\frac{\hbar}{2}$? and $-\frac{\hbar}{2}$?

Solution

$$p(+\frac{\hbar}{2}) = \left|< 0|(\frac{4}{5}|0> +\frac{3}{5}|1>)\right|^2 = \frac{16}{25}, \qquad (2.104)$$

and

$$p(\frac{-\hbar}{2}) = \left|< 1|(\frac{4}{5}|0> +\frac{3}{5}|1>)\right|^2 = \frac{9}{25}. \qquad (2.105)$$

(2) What is the expectation value of $S_z = \frac{\hbar}{2}\sigma_z$?

Solution

$$< \psi|S_z|\psi> = \left(\frac{4}{5}< 0| +\frac{3}{5}< 1|\right)\frac{\hbar}{2}\sigma_z\left(\frac{4}{5}|0> +\frac{3}{5}|1>\right),$$

$$= \frac{16}{25}< 0|\sigma_z|0>\frac{\hbar}{2} + \left(\frac{3}{5}\right)^2\frac{\hbar}{2}< 1|\sigma_z|1> = \frac{16}{25}\left(\frac{\hbar}{2}\right) - \frac{9}{25}\left(\frac{\hbar}{2}\right),$$

$$= \frac{7}{25}\left(\frac{\hbar}{2}\right). \qquad (2.106)$$

(3) What is the standard deviation of S_z if measurements are made on many electrons prepared in the state $|\psi>$ above?

Solution

By definition

$$\Delta(S_z) = \sqrt{<\psi|S_z{}^2|\psi> - <S_z>^2}. \tag{2.107}$$

Since

$$<S_z{}^2> = <\psi|S_z{}^2|\psi> = <\psi|\frac{\hbar^2}{4}\sigma_z{}^2|\psi> = \frac{\hbar^2}{4}, \tag{2.108}$$

and from the previous problem

$$\langle S_z \rangle = \frac{7}{25}\left(\frac{\hbar}{2}\right), \tag{2.109}$$

we get

$$\Delta(S_z) = \sqrt{\frac{\hbar^2}{4}(1 - \frac{49}{625})} = \frac{12}{25}\hbar. \tag{2.110}$$

Exercise: Proceeding as above, calculate the standard deviations $\Delta(S_x)$ and $\Delta(S_y)$.

Answer:

$$\Delta(S_x) = \frac{7}{25}\frac{\hbar}{2}, \tag{2.111}$$

and

$$\Delta(S_y) = \frac{\hbar}{2}. \tag{2.112}$$

2.7.4 The Heisenberg principle for spin components

In Chapter 18, we prove the general Heisenberg inequality

$$\Delta(C)\Delta(D) \geq \frac{|<\psi|[C,D]|\psi>|}{2}, \tag{2.113}$$

where $\Delta(C)$ and $\Delta(D)$ are the standard deviations associated with measurements of the observables C and D on an electron prepared many times in the same state $|\psi>$.

If we take as observables the two components of the intrinsic angular momentum of the electron σ_x and σ_y, the inequality above becomes

$$\Delta(S_x)\Delta(S_y) \geq \frac{\hbar}{2}|<\psi|S_z|\psi>| = \frac{\hbar^2}{4}|<\psi|\sigma_z|\psi>|. \tag{2.114}$$

Exercise: Prove that, in the last inequality, the equality signs holds when the state of the spinor is given by

$$|\psi> = \frac{3}{\sqrt{10}}|0> + \frac{-1}{\sqrt{10}}|1>, \tag{2.115}$$

considered in Example 3.

Solution
For an electron prepared in the state $|\psi >$ above, we use the results of Example 3 and find right-hand-side of Equation (2.114) to be equal to

$$\frac{\hbar^2}{4}| < \psi|\sigma_z|\psi > | = \frac{\hbar^2}{5}. \qquad (2.116)$$

The left hand side of the inequality is

$$\triangle(S_x)\triangle(S_y) = (\frac{2}{5}\hbar)(\frac{\hbar}{2}) = \frac{\hbar^2}{5}. \qquad (2.117)$$

Therefore, the inequality (2.114) reduces to an equality in this case.

Exercise: The following inequalities also hold

$$\triangle(S_y)\triangle(S_z) \geq \frac{\hbar}{2}| < \psi|S_x|\psi > | = \frac{\hbar^2}{4}| < \psi|\sigma_x|\psi > |, \qquad (2.118)$$

and

$$\triangle(S_x)\triangle(S_z) \geq \frac{\hbar}{2}| < \psi|S_y|\psi > | = \frac{\hbar^2}{4}| < \psi|\sigma_y|\psi > |. \qquad (2.119)$$

Prove that these inequalities are indeed satisfied for a spinor in the state (2.115) given above. The solution of this exercise is left for the reader.

2.8   References

[1] C. Ekart, "Operator calculus and the solution of the equations of dynamics", Phys. Rev., **28**, 711 (1926).

[2] W. Pauli, in Zur Quantenmechanik des Magnetischen Elektrons Zeitschrift für Physik, **43**, 601 (1927).

[3] R. P. Feynman, R. B. Leighton and M. Sands, *The Feynman Lectures on Physics*, Volume II, chapter 34, Addison-Wesley, Reading, Mass., (1965).

[4] P. A. M. Dirac, *The Principles of Quantum Mechanics*, Oxford University Press (1958).

[5] J. J. Sakurai, *Advanced Quantum Mechanics*, Addison-Wesley, Reading, Mass. (1967).

[6] E. U. Condon and G. H. Shortley, *Theory of Atomic Spectra*, Cambridge University Press, Cambridge (1935).

3

Bloch Sphere

The Bloch sphere concept is a useful tool to represent the actions of various quantum-mechanical operators on a spinor. Most important, it provides a link between the rather abstract concept of a spinor and the more intuitive way (although not rigorous) of thinking of the spin of an electron as being associated with a magnetic moment along the axis of self-rotation. The Bloch sphere is particularly useful when describing the action of a spatially uniform external (including time-dependent) magnetic field on the spin [1]. This will be illustrated in the next chapter through the derivation of the Rabi formula. Bloch sphere concepts are also frequently invoked in discussions of spin-based quantum computing, so that a sound understanding of this notion is imperative to comprehend the literature in that field.

3.1 Spinor and "qubit"

In Chapter 16, where we discuss spintronic implementations of quantum computers and logic gates, we will encounter the notion of "qubit." The term "qubit" is a short form for "quantum bit". Quantum computers do not process classical binary bits 0 and 1, but instead process *quantum bits*, which are coherent superpositions of both 0 and 1. A quantum computer derives its immense power from this superposition.

It was realized quite some time back that the spin polarization of an electron is the ideal physical implementation of a qubit since spin polarization can survive in a coherent superposition of two orthogonal states for a relatively long time. These two orthogonal states (i.e., states with anti-parallel spin polarizations) will represent the classical bits 0 and 1. We elucidate this below.

The 2-component wavefunction representing an arbitrary spin state can be written as

$$[\psi(\mathbf{x})] = \begin{bmatrix} \phi_1 \\ \phi_2 \end{bmatrix} = \phi_1 \begin{bmatrix} 1 \\ 0 \end{bmatrix} + \phi_2 \begin{bmatrix} 0 \\ 1 \end{bmatrix} = \phi_1 |+>_z + \phi_2 |->_z, \qquad (3.1)$$

where

$$|\phi_1|^2 + |\phi_2|^2 = 1 \qquad (3.2)$$

if $[\psi(\mathbf{x})]$ is properly normalized.

Therefore, any spin state can be expressed as a coherent superposition of the $+z$ and $-z$-polarized states. The two z-polarized states, being mutually orthogonal, can represent the classical bits 0 and 1. The quantities ϕ_1 and ϕ_2 are complex quantities with both a magnitude and a phase. To maintain a *coherent* superposition state, we have to maintain the phase relationship between ϕ_1 and ϕ_2 until the qubit is "read" and collapses to either the $+z$- or the $-z$-polarized state*.

With "spin", the coherent superposition can be maintained for a much longer duration than with "charge". Spin couples relatively weakly to its surroundings. As a result, the coupling to the environment does not destroy the phase coherence of spin (phase relationship between ϕ_1 and ϕ_2) as rapidly as it does in the case of charge. This is a major advantage of spin over charge and it has made "spin" the preferred vehicle for implementing qubits in solid state quantum computers.

Since the $+z$-polarized state was assumed to represent the classical bit 0, we will denote the corresponding spinor as $|0>$ using the Dirac ket notation. It is given by

$$|0>= \begin{bmatrix} 1 \\ 0 \end{bmatrix}. \tag{3.3}$$

Similarly, the spinor for the $-z$-polarized state will be denoted as the ket

$$|1>= \begin{bmatrix} 0 \\ 1 \end{bmatrix}. \tag{3.4}$$

We can rewrite Equation (3.1) as

$$|\chi>= \begin{bmatrix} \alpha \\ \beta \end{bmatrix} = \alpha|0> +\beta|1>, \tag{3.5}$$

where $|\chi>$ is an arbitrary spinor and α, β are complex numbers whose square magnitudes denote the probability that if a measurement of the spin component along the z-axis is performed on the electron, it will be found to be in state $|0>$ with probability $|\alpha|^2$ and in state $|1>$ with probability $|\beta|^2$. Therefore, we must have

$$|\alpha|^2 + |\beta|^2 = 1. \tag{3.6}$$

The above two equations immediately suggest that the spinor $|\chi>$ can be viewed as a qubit since it has the exact mathematical form of a qubit, namely, a coherent superposition of bits 0 and 1.

The quantities α, β are *continuous* variables whose magnitudes can take any value between 0 and 1 and whose phases can also take arbitrary values between 0 and 2π. On the other hand, the states $|0>$ and $|1>$ correspond

---

*The qubit will collapse to the $+z$-polarized state with probability $|\phi_1|^2$ and to $-z$-polarized state with probability $|\phi_2|^2$.

to digital (*discrete*) binary bits. Thus, quantum computation is neither quite analog computing, nor digital, but something in between.

Since complex numbers have a nice powerful geometrical (or Gauss) representation in the complex plane, one might ask if there is such a useful representation for the spinors or qubits in $\mathcal{C}^2$. The next sections will show that the *Bloch sphere* is precisely this representation.

3.2 Bloch sphere concept

3.2.1 Preliminaries

In Chapter 2, we introduced the concept that an electron's spin is described by the operator

$$\vec{S}_{op} = \frac{\hbar}{2}\vec{\sigma}, \tag{3.7}$$

whose three components (S_x, S_y, S_z) satisfy commutation rules similar to those for the orbital angular momentum $\vec{L}$ (see Equation (2.7)). A measurement of the spin component along an arbitrary direction characterized by a unit vector $\hat{n}$ will yield results given by the eigenvalues of the operator

$$\vec{S}.\hat{n}, \tag{3.8}$$

and these eigenvalues are $\pm\frac{\hbar}{2}$, irrespective of the direction of the unit vector $\hat{n}$.

We prove this last statement starting with the following equality for the Pauli spin matrices

$$(\vec{\sigma} \cdot \vec{a})(\vec{\sigma} \cdot \vec{b}) = i\vec{\sigma} \cdot (\vec{a} \times \vec{b}) + \vec{a} \cdot \vec{b}I, \tag{3.9}$$

where I is the 2×2 identity matrix and $\vec{a}$ and $\vec{b}$ are any arbitrary three-dimensional vectors in real space $\Re^3$. The above identity is very easy to verify and is left as an exercise for the reader.

If the vectors $\vec{a}$ and $\vec{b}$ are equal to a unit vector $\hat{n}$, then the equality above reduces to

$$(\vec{\sigma} \cdot \hat{n})^2 = I, \tag{3.10}$$

i.e., the square of any component of $\vec{\sigma}$ is equal to the unit 2×2 matrix. Hence, the eigenvalues of $\vec{\sigma}.\hat{n}$ are ± 1, and therefore (recall Equation (3.7)) the eigenvalues of the operator $\vec{S}.\hat{n}$ must be $\pm\hbar/2$, which proves the result we were after. In other words, the measurement of the spin angular momentum along any arbitrary axis always yields the values $\pm\hbar/2$.

Eigenvectors of $(\vec{\sigma} \cdot \hat{n})$

Let us derive the explicit analytical expressions for the eigenvectors of $\vec{\sigma} \cdot \hat{n}$ corresponding to the eigenvalues $+1$ and -1.

To achieve that goal, we consider the operators

$$\frac{1}{2}(I \pm \vec{\sigma} \cdot \hat{n}), \tag{3.11}$$

acting on an arbitrary spinor or qubit $|\chi>$. If we operate on that with the operator $(\vec{\sigma} \cdot \hat{n})$, we get

$$(\vec{\sigma} \cdot \hat{n}) \left[\frac{1}{2}(I \pm \vec{\sigma} \cdot \hat{n}) |\chi> \right] = \frac{1}{2} \vec{\sigma} \cdot \hat{n} |\chi> \pm \frac{1}{2} (\vec{\sigma} \cdot \hat{n})^2 |\chi> = \pm \left[\frac{1}{2}(I \pm \vec{\sigma} \cdot \hat{n}) |\chi> \right]. \tag{3.12}$$

This means that, for any $|\chi>$, $\frac{1}{2}(I \pm \vec{\sigma} \cdot \hat{n})|\chi>$ are eigenvectors of $\vec{\sigma}.\hat{n}$ with eigenvalues ± 1. Making use of the identity

$$\frac{1}{2}(I \pm \vec{\sigma}.\hat{n}) = \frac{1}{2}[I \pm \sigma_z n_z \pm \frac{1}{2}(\sigma_x + i\sigma_y)(n_x - in_y) \pm \frac{1}{2}(\sigma_x - i\sigma_y)(n_x + in_y)], \tag{3.13}$$

where n_x, n_y and n_z are the x-, y- and z-components of the vector $\hat{n}$, and using spherical coordinates with polar angle θ and azimuthal angle ϕ, so that

$$(n_x, n_y, n_z) = (sin\theta cos\phi, sin\theta sin\phi, cos\theta), \tag{3.14}$$

we get

$$n_x \pm in_y = sin\theta e^{\pm i\phi}, \tag{3.15}$$

which leads to

$$\frac{1}{2}(I \pm \vec{\sigma} \cdot \hat{n}) = \frac{1}{2} \left[I \pm cos\theta \sigma_z \pm \frac{1}{2}(sin\theta e^{-i\phi}\sigma_+ \pm sin\theta e^{i\phi}\sigma_-) \right], \tag{3.16}$$

where the operators σ_+ and σ_- are given by $\sigma_+ = \sigma_x + i\sigma_y$ and $\sigma_- = \sigma_x - i\sigma_y$, respectively.

Acting with these operators on the ket $|0>$, we get

$$\frac{1}{2}(I + \vec{\sigma} \cdot \hat{n})|0> = cos\frac{\theta}{2} \left[cos\frac{\theta}{2}|0> +sin\frac{\theta}{2}e^{i\phi}|1> \right], \tag{3.17}$$

and

$$\frac{1}{2}(I - \vec{\sigma} \cdot \hat{n})|0> = sin\frac{\theta}{2} \left[sin\frac{\theta}{2}|0> -cos\frac{\theta}{2}e^{i\phi}|1> \right]. \tag{3.18}$$

The last two spinors can be easily normalized by dividing the first by $cos\frac{\theta}{2}$ and the second by $sin\frac{\theta}{2}$. This leads to the spinors

$$|\xi_n^+> = cos\frac{\theta}{2}|0> +sin\frac{\theta}{2}e^{i\phi}|1>, \tag{3.19}$$

and

$$|\xi_n^- >= sin\frac{\theta}{2}|0 > -cos\frac{\theta}{2}e^{i\phi}|1 > . \tag{3.20}$$

Since we had proved that any spinor $(1/2)(1 \pm \vec{\sigma} \cdot \hat{n})|\chi >$ is an eigenvector of $\vec{\sigma} \cdot \hat{n}$ with eigenvalues ± 1, it is obvious that the spinors $|\xi_n^+ >$ and $|\xi_n^- >$ are eigenspinors of the operator $(\vec{\sigma} \cdot \hat{n})$ with eigenvalues $+1$ and -1, respectively.

Exercise: Prove that the two eigenspinors $|\xi_n^+ >$ and $|\xi_n^- >$ are indeed orthogonal, i.e., in Dirac's notations,

$$< \xi_n^+|\xi_n^- >= 0. \tag{3.21}$$

Solution

In their column format, these two spinors are given by

$$|\xi_n^+ >= cos\frac{\theta}{2}|0 > +sin\frac{\theta}{2}e^{i\phi}|1 >= \begin{bmatrix} cos\frac{\theta}{2} \\ sin\frac{\theta}{2}e^{i\phi} \end{bmatrix}, \tag{3.22}$$

which is the eigenstate of $\vec{\sigma}.\vec{n}$ with eigenvalue $+1$, and

$$|\xi_n^- >= sin\frac{\theta}{2}|0 > -cos\frac{\theta}{2}e^{i\phi}|1 >= \begin{bmatrix} sin\frac{\theta}{2} \\ -cos\frac{\theta}{2}e^{i\phi} \end{bmatrix}, \tag{3.23}$$

which is the eigenstate of $\vec{\sigma}.\vec{n}$ with eigenvalue -1. Therefore,

$$< \xi_n^+|\xi_n^- >= cos\frac{\theta}{2}sin\frac{\theta}{2} - cos\frac{\theta}{2}sin\frac{\theta}{2} = 0 \tag{3.24}$$

which proves the result.

We note the that spinor $|\xi_n^- >$ can actually be obtained from the spinor $|\xi_n^+ >$ as follows:

$$|\xi_n^-(\theta, \phi) >= |\xi_n^+(\theta \to \pi - \theta, \phi \to \phi + \pi) > . \tag{3.25}$$

Since varying the angle θ in the range $[0, \pi]$ and the azimuthal angle ϕ in the range $[0, 2\pi]$ allows the unit vector $\hat{n}$ to sweep the entire sphere of radius 1, we can think of the spinor $|\xi_n^+ >$ as the most general form of the spinorial part of the wavefunction characterizing a spin-1/2 particle.

Hereafter, we adopt the common convention of using $|\xi_n^+ >$ to represent the most general form of the spinor by including an additional overall phase factor $e^{i\gamma}$ which will not affect any measurement made on the spinor,

$$|\xi_n^+ >= e^{i\gamma} \left[cos\frac{\theta}{2}|0 > +sin\frac{\theta}{2}e^{i\phi}|1 > \right]. \tag{3.26}$$

The discussion above shows that there is a one-to-one relation between the unit vector $\hat{n}$ and the spinors $|\xi_n^+ >$ and $|\xi_n^- >$ since the latter are eigenspinors of $\vec{\sigma} \cdot \hat{n}$, which is the operator for the spin component along $\hat{n}$. This relation is represented geometrically using a sphere of radius 1 – the so-called *Bloch sphere* – where the unit vector $\hat{n}$ is the radius vector of the Bloch sphere and represents the spinor $|\xi_n^+ >$, namely, the eigenvector of $\vec{\sigma} \cdot \vec{n}$ with eigenvalue +1. This pictorial representation, shown in Fig. 3.1, is very useful when analyzing the action of 2×2 unitary matrices acting on the spinor $|\xi_n^+ >$ as actual rotations of the unit vector $\hat{n}$ on the Bloch sphere. This will be illustrated in more detail in the next sections.

3.2.2 Connection between the Bloch sphere concept and the classical interpretation of the spin of an electron

For an electron characterized by the spinor $|\xi_n^+ >$, it can be easily shown that the expectation values of the Pauli spin matrices are given by

$$< \xi_n^+ |\sigma_x|\xi_n^+ >= sin\theta cos\phi, \tag{3.27}$$

$$< \xi_n^+ |\sigma_y|\xi_n^+ >= sin\theta sin\phi, \tag{3.28}$$

$$< \xi_n^+ |\sigma_z|\xi_n^+ >= cos\theta, \tag{3.29}$$

which are exactly the Cartesian components of the unit vector $\hat{n}$.

So, we can think of the unit vector $\hat{n}$ characterized by the angles θ and ϕ as the representation of the spinor eigenstate of the operator $\vec{\sigma} \cdot \vec{n}$ with eigenvalue +1.

Similarly, starting with $|\xi_n^- >$, we easily obtain

$$< \xi_n^- |\sigma_x|\xi_n^- >= -sin\theta cos\phi, \tag{3.30}$$

$$< \xi_n^- |\sigma_y|\xi_n^- >= -sin\theta sin\phi, \tag{3.31}$$

$$< \xi_n^- |\sigma_z|\xi_n^- >= -cos\theta, \tag{3.32}$$

which are the Cartesian coordinates of a unit vector diametrically opposite to the previous unit vector $\hat{n}$.

Therefore, any two points on the surface of the Bloch sphere which intersect a straight line going through the center can be thought of as representations of eigenstates of the operator $\vec{\sigma} \cdot \hat{n}$ with eigenvalues ± 1.

Warning note:

At an early stage of exposure to the Bloch sphere concept, it is a common mistake to think of two orthogonal spinors on the Bloch sphere as unit vectors $\hat{n}_1$ and $\hat{n}_2$ whose vector product $\hat{n}_1 \cdot \hat{n}_2$ must be zero, i.e., the two unit vectors $\hat{n}_1$ and $\hat{n}_2$ must be perpendicular to one another. This is NOT correct! Orthogonal spinors on the Bloch sphere do not subtend an angle of 90°; instead, they subtend an angle of 180°. This may be an alien concept that requires getting used to.

Since the spinor $|\xi_n^+>$ given explicitly in Equation (3.26) can sweep the entire Bloch sphere by varying the angles θ and ϕ, and since the expectation values of the three components of the Pauli matrices coincide with the Cartesian components of the unit vector $\hat{n}$, we can indeed think of this unit vector as representing the magnetic moment of a small magnet attached to the electron with its main axis along the direction $\hat{n}$. This interpretation is intuitively attractive, but one should keep in mind that the spin of the electron is a purely quantum-mechanical concept. The unit vector $\hat{n}$ representing the general spinor $|\xi_n^+>$ on the Bloch sphere is referred to as the Bloch vector.

Since $|\xi_n^+>$ can be obtained by acting with the operator $\frac{1}{2}(1+\vec{\sigma}\cdot\hat{n})$ on an arbitrary qubit $|\chi>$, this operator is referred to as the *projection operator* in the direction $\hat{n}$. Similarly, the operator $\frac{1}{2}(1-\vec{\sigma}\cdot\hat{n})$ is the *projection operator* in the direction $-\hat{n}$.

Exercise: Prove that if a spinor is in the state $|\xi_{n_1}^+>$, its components along the two spinors $|\xi_{n_2}^+>$, $|\xi_{n_2}^->$, which are eigenstates of the projection operator along another unit vector $\hat{n}_2$, are such that

$$|<\xi_{n_2}^+|\xi_{n_1}^+>|^2 + |<\xi_{n_2}^-|\xi_{n_1}^+>|^2 = 1. \tag{3.33}$$

This is easy to prove and is left for the reader.

3.2.3 Relationship with qubit

The spinor $|\xi_n^+>$ can be easily related to a qubit $|\chi>$ defined as a column vector in $\mathcal{C}^2$:

$$|\chi>= \begin{bmatrix} \alpha \\ \beta \end{bmatrix}, \tag{3.34}$$

where α and β are complex numbers. This last expression can be rewritten as

$$|\chi>= \begin{bmatrix} Re(\alpha)+iIm(\alpha) \\ Re(\beta)+iIm(\beta) \end{bmatrix}, \tag{3.35}$$

or equivalently,

$$|\chi>= \begin{bmatrix} |\alpha|e^{i\phi_\alpha} \\ |\beta|e^{i\phi_\beta} \end{bmatrix}. \tag{3.36}$$

Hence,

$$|\chi>= e^{i\phi_\alpha} \begin{bmatrix} |\alpha| \\ |\beta|e^{i(\phi_\beta-\phi_\alpha)} \end{bmatrix}, \tag{3.37}$$

which has the general form of $|\xi_n^+>$ in Equation (3.26) for the following choice of parameters γ, θ, and ϕ:

$$\gamma = \phi_\alpha, \tag{3.38}$$

$$\theta = 2\ arctan\left(\frac{\sqrt{1-|\alpha|^2}}{|\alpha|}\right), \tag{3.39}$$

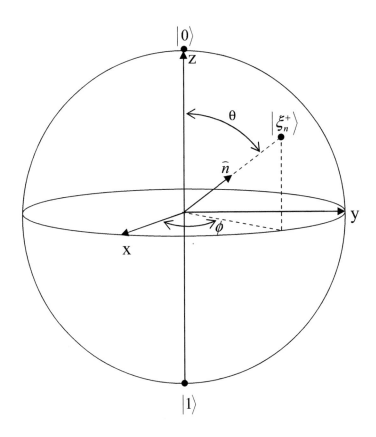

FIGURE 3.1
Bloch sphere representation of a qubit.

and

$$\phi = \phi_\beta - \phi_\alpha. \tag{3.40}$$

3.2.4 Special spinors

North and south poles Referring to Equation (3.26), it is obvious that the state $|0>$ can be written as

$$|0>= |\xi_n^+(\theta = 0, \phi, \gamma)>, \tag{3.41}$$

and the state $|1>$ as

$$|1>= |\xi_n^+(\theta = \pi, \phi, \gamma) > . \tag{3.42}$$

The fact that ϕ and γ are arbitrary is irrelevant since observables should not depend on the overall phase of the spinor. This interpretation is consistent with associating the spinor $|0>$ with the point representing the north pole on the Bloch sphere, whereas the spinor $|1>$, which is orthogonal to $|0>$, is associated with the south pole on the Bloch sphere. Since they are orthogonal, they are diametrically opposite through the sphere center.

Reaching the equator The Hadamard matrix is one of the most widely used in the design of quantum logic gates and is given by

$$H = \frac{1}{\sqrt{2}} \begin{pmatrix} 1 & 1 \\ 1 & -1 \end{pmatrix}. \tag{3.43}$$

One of the two eigenvectors of the Hadamard matrix is given by

$$|\chi >_{Hadamard} = \cos\left(\frac{\pi}{8}\right)|0> + \sin\left(\frac{\pi}{8}\right)|1> . \tag{3.44}$$

We wish to find the locations on the Bloch sphere of this eigenvector of the Hadamard matrix. If we express this eigenvector as $\alpha|0> +\beta|1>$, then, following the usual prescription:

$$\alpha = e^{i\gamma}\cos\frac{\theta}{2} = \cos\left(\frac{\pi}{8}\right), \tag{3.45}$$

$$\beta = e^{i\gamma}e^{i\phi}\sin\frac{\theta}{2} = \sin\left(\frac{\pi}{8}\right). \tag{3.46}$$

Hence, $\theta = \frac{\pi}{4}$, $\phi = 0$, and γ can be any integral multiple of 2π.

Hence the eigenvector of the Hadamard matrix given by Equation (3.44) is in the x-z plane, halfway between the north pole and the point where the positive x-axis pierces the Bloch sphere.

It is then easy to show that the other eigenvector of the Hadamard matrix is located diametrically opposite to the unit vector associated with the spinor in Equation (3.44).

3.2.5   Spin flip matrix

We now address the following question: What is the 2×2 matrix M which changes $|\xi_n^- >$ into $|\xi_n^+ >$? This would be defined by the relation

$$|\xi_n^+ >= e^{i\gamma} \begin{bmatrix} cos\frac{\theta}{2} \\ sin\frac{\theta}{2}e^{i\phi} \end{bmatrix} = M|\xi_n^- >= Me^{i\gamma} \begin{bmatrix} sin\frac{\theta}{2} \\ -cos\frac{\theta}{2}e^{i\phi} \end{bmatrix}. \qquad (3.47)$$

Evidently

$$M = \begin{bmatrix} 0 & -e^{-i\phi} \\ e^{i\phi} & 0 \end{bmatrix}, \qquad (3.48)$$

and indeed, we find

$$M|\xi_n^- >= \begin{bmatrix} cos\frac{\theta}{2} \\ sin\frac{\theta}{2}e^{i\phi} \end{bmatrix}, \qquad (3.49)$$

as required.

M can be written as a product of some well-known 2×2 matrices

$$M = \begin{bmatrix} 1 & 0 \\ 0 & e^{i\phi} \end{bmatrix} \begin{bmatrix} 0 & -1 \\ 1 & 0 \end{bmatrix} \begin{bmatrix} 1 & 0 \\ 0 & e^{-i\phi} \end{bmatrix} \qquad (3.50)$$

or, in a more compact form

$$M = e^{-i\frac{\pi}{2}} P(\phi)\sigma_y P(-\phi), \qquad (3.51)$$

where $P(\phi)$ is the phase shift matrix (see Problem 3.4).

3.2.6   Excursions on the Bloch sphere: Pauli matrices revisited

In Chapter 2, we proved the identity

$$e^{i\theta A} = cos\theta I + isin\theta A, \qquad (3.52)$$

where θ is real and the matrix A is such that $A^2 = I$. This can be looked at as a generalization to operators of the well-known Euler relation for complex numbers, i.e., $e^{iz} = cosz + isinz$.

We are now going to illustrate the usefulness of this relation for finding the unitary matrices which allow us to perform rotations on the Bloch sphere around the three axes x, y, and z. Since any general spinor in C^2 can be written as a linear superposition of the spinors $|0 >$ and $|1 >$ (see Equation (3.1)), we start with each of these two spinors and see how they are modified after being rotated through some angle θ, as indicated in Figure 3.2.

Consider first the spinor $|0 >$ and let us seek the matrix which would rotate that spinor an angle θ away from the z-axis as indicated in Figure 3.2. After rotation in the (y,z) plane (anticlockwise rotation, i.e., pushing the y-axis

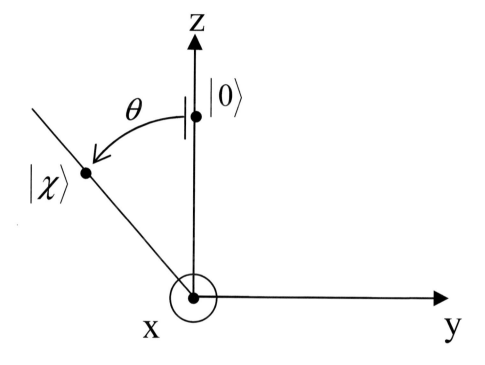

FIGURE 3.2

Illustration of the rotation of the spinor $|0>$ to the spinor $|\chi>$. The rotation is counterclockwise around the x-axis.

toward the z-axis using the shortest path), the new spinor $|\chi >_{new}$ is given by

$$|\chi >_{new}= cos\frac{\theta}{2}|0 > +sin\frac{\theta}{2}e^{-i\frac{\pi}{2}}|1 >, \tag{3.53}$$

or

$$|\chi >_{new}= \begin{bmatrix} cos\frac{\theta}{2} \\ sin\frac{\theta}{2}e^{-i\pi/2} \end{bmatrix} = \begin{bmatrix} cos\frac{\theta}{2} \\ -isin\frac{\theta}{2} \end{bmatrix}. \tag{3.54}$$

Since the rotation matrix is such that

$$|\chi >_{new}= R_x(\theta)|0 >, \tag{3.55}$$

the spinor in the last equation is the first column of the matrix $R_x(\theta)$.

Next, we apply $R_x(\theta)$ to the spinor $|1 >$ and find (see Figure 3.3)

$$|\chi' >_{new}= R_x(\theta)|1 >= cos(\frac{\pi - \theta}{2})|0 > +e^{i\pi/2}sin(\frac{\pi - \theta}{2})|1 >, \tag{3.56}$$

which has the explicit form

$$|\chi' >= \begin{bmatrix} sin\frac{\theta}{2} \\ icos\frac{\theta}{2} \end{bmatrix} = i \begin{bmatrix} -isin\frac{\theta}{2} \\ cos\frac{\theta}{2} \end{bmatrix}. \tag{3.57}$$

This second spinor represents the second column vector of the matrix $R_x(\theta)$. However, since the rotation matrix must leave the norm of the spinor unchanged, it must be a 2 × 2 unitary matrix. If we drop the overall phase factor i of the last column vector in Equation (3.57), an expression for the 2 × 2 matrix describing a rotation around the x-axis of a general qubit (which is a superposition, i.e., linear combination of the $|0 >$ and $|1 >$ bits) is given by

$$R_x(\theta) = \begin{bmatrix} cos\frac{\theta}{2} & -isin\frac{\theta}{2} \\ -isin\frac{\theta}{2} & cos\frac{\theta}{2} \end{bmatrix}, \tag{3.58}$$

which is unitary as easily checked.

Using the identity (3.52), the matrix $R_x(\theta)$ can be generated from the Pauli matrix σ_x as

$$R_x(\theta) = e^{-i\frac{\theta}{2}\sigma_x}. \tag{3.59}$$

Using similar arguments, it can be shown that, for rotation of a general spinor around the y-axis on the Bloch sphere, the rotation matrix is

$$R_y(\theta) = \begin{bmatrix} cos\frac{\theta}{2} & -sin\frac{\theta}{2} \\ sin\frac{\theta}{2} & cos\frac{\theta}{2} \end{bmatrix}, \tag{3.60}$$

and, for a rotation around the z-axis,

$$R_z(\theta) = \begin{bmatrix} e^{-i\frac{\theta}{2}} & 0 \\ 0 & e^{i\frac{\theta}{2}} \end{bmatrix}. \tag{3.61}$$

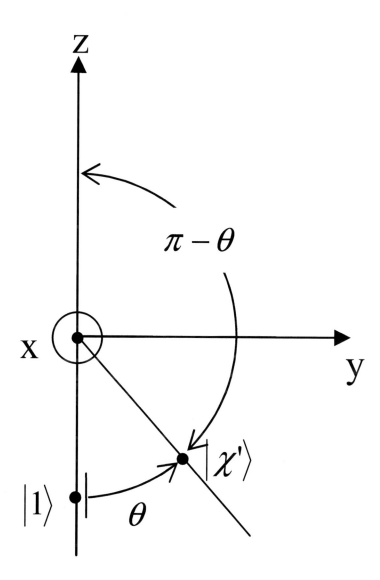

FIGURE 3.3

Illustration of the rotation of the spinor $|1>$ to the spinor $|\chi'>$ to find the second portion of the rotation matrix of a general spinor around the x-axis.

With the use of the identity (3.52), these last relations can be recast as

$$R_y(\theta) = e^{-i\frac{\theta}{2}\sigma_y},\tag{3.62}$$

and

$$R_z(\theta) = e^{-i\frac{\theta}{2}\sigma_z}.\tag{3.63}$$

Because the Pauli matrices constitute the building blocks to describe rotations on the Bloch sphere with the x-, y- or z-axis as the axis of rotation, they are sometimes referred to as the generators of the rotations in the Hilbert space of spinors associated with spin-1/2 particles [2].

Some sophisticated studies of the evolution of spinors on the Bloch sphere can be found in references [5, 6, 7]. The interested reader is referred to these works for further reading.

3.3   Problems

- **Problem 3.1**

   Consider the matrix

   $$\begin{pmatrix} 1 & 3 \\ 3 & 1 \end{pmatrix}.\tag{3.64}$$

   Calculate the eigenvalues and corresponding eigenvectors of this matrix. What are the angles θ and ϕ in the general expression of the spinor on the Bloch sphere, i.e.,

   $$|\xi_n^+> = \cos\frac{\theta}{2}|0> + \sin\frac{\theta}{2}e^{i\phi}|1>,\tag{3.65}$$

   associated with the two eigenvectors of the above matrix?

- **Problem 3.2**

   Consider the qubit

   $$|\chi> = \sqrt{\frac{3}{5}}|0> - \sqrt{\frac{2}{5}}|1>.\tag{3.66}$$

   Calculate the angles γ, θ, and ϕ associated with the corresponding spinor and show its location on the Bloch sphere.

- **Problem 3.3**

   Prove that Equation (3.44) is indeed the eigenvector of the Hadamard matrix H with eigenvalue $+1$. Find the other eigenvector of the Hadamard matrix and its corresponding eigenvalue.

- **Problem 3.4**

  Locate the points on the Bloch sphere which represent the eigenvectors of the Pauli matrices $(\sigma_x, \sigma_y, \sigma_z)$.

  Two other matrices frequently used in the design of quantum logic gates are the S matrix

  $$S = \begin{pmatrix} 1 & 0 \\ 0 & i \end{pmatrix}, \tag{3.67}$$

  which is a special case of the phase shift matrix P

  $$P(\phi) = \begin{pmatrix} 1 & 0 \\ 0 & e^{i\phi} \end{pmatrix}, \tag{3.68}$$

  and the $\pi/8$ or T-matrix

  $$T = \begin{pmatrix} 1 & 0 \\ 0 & e^{i\pi/4} \end{pmatrix}. \tag{3.69}$$

  Calculate the eigenspinors and eigenvalues of the S and T matrices. Locate the corresponding eigenvectors on the Bloch sphere.

- **Problem 3.5**

  Show that the spin flip matrix can also be written as a product of some phase shift matrices and the Pauli matrix σ_x.

- **Problem 3.6**

  The operator $\frac{1}{2}(1 + \vec{\sigma} \cdot \hat{n})$ is referred to as a *projection operator*. When applied to an arbitrary spinor $|\chi>$, the resulting state vector is an eigenstate of the operator $(\vec{\sigma} \cdot \hat{n})$, i.e., a spinor whose representation on the Bloch sphere is given by the unit vector $\hat{n}$. Using $|\chi>$ equal to $|0>$, $|1>$, and $\frac{1}{\sqrt{2}}(|0> +|1>)$, show that the three corresponding spinors $\frac{1}{2}(I + \vec{\sigma} \cdot \hat{n})|\chi>$ may differ by an overall phase factor but are all associated with a point on the Bloch sphere characterized by the Bloch vector $\hat{n}$.

  The projection operators $\frac{1}{2}(I \pm \vec{\sigma} \cdot \hat{n})$ are special cases of a very general decomposition of any 2×2 matrix in terms of the 2×2 identity and three Pauli matrices discussed in Chapter 2, i.e.,

  $$M = \begin{pmatrix} m_{11} & m_{12} \\ m_{21} & m_{22} \end{pmatrix} \tag{3.70}$$

  can be decomposed as

  $$M = \frac{m_{11} + m_{22}}{2}I + \frac{m_{11} - m_{22}}{2}\sigma_z + \frac{m_{12} + m_{21}}{2}\sigma_x + i\frac{m_{12} - m_{21}}{2}\sigma_y, \tag{3.71}$$

which can be written in a more condensed form

$$M = a_o I + \vec{a} \cdot \vec{\sigma}, \tag{3.72}$$

where

$$a_o = \frac{1}{2} Tr(M), \tag{3.73}$$

and

$$\vec{a} = \frac{1}{2} Tr(M\vec{\sigma}), \tag{3.74}$$

where Tr is the trace operator and $\vec{\sigma} = \sigma_x \hat{x} + \sigma_y \hat{y} + \sigma_z \hat{z}$.

Find the value of a_0 and $\vec{a}$ for the matrices H, T, and S defined in Problem 3.4.

- **Problem 3.7**

  Consider the two matrices M_1 and M_2 with the decomposition

  $$M_1 = a_o^{(1)} I + \vec{a_1} \cdot \vec{\sigma}, \tag{3.75}$$

  and

  $$M_2 = a_o^{(2)} I + \vec{a_2} \cdot \vec{\sigma}. \tag{3.76}$$

  What must be the relationship between the two vectors $\vec{a_1}$ and $\vec{a_2}$ for the matrices M_1 and M_2 to commute?

  Hint: Use the identity

  $$(\vec{\sigma} \cdot \vec{a_1})(\vec{\sigma} \cdot \vec{a_2}) = (\vec{a_1} \cdot \vec{a_2}) I + i\vec{\sigma} \cdot (\vec{a_1} \times \vec{a_2}), \tag{3.77}$$

  where I is the 2×2 identity matrix.

- **Problem 3.8**

  Prove the following identity for the Pauli spin matrices:

  $$\sigma_x \sigma_y \sigma_z = iI. \tag{3.78}$$

  Using the Bloch sphere concept, interpret this relation geometrically.

- **Problem 3.9**

  Generalize the previous problem to show that for any set of three vectors $(\vec{a}, \vec{b}, \vec{c})$ forming a right hand orthogonal basis, the following relation is true.

  $$(\vec{\sigma} \cdot \vec{a})(\vec{\sigma} \cdot \vec{b})(\vec{\sigma} \cdot \vec{c}) = iI. \tag{3.79}$$

  Hint: Use the identity

  $$(\vec{\sigma} \cdot \vec{a})(\vec{\sigma} \cdot \vec{b}) = \vec{a} \cdot \vec{b} I + i\vec{\sigma} \cdot (\vec{a} \times \vec{b}). \tag{3.80}$$

- **Problem 3.10**

Prove that the following is true.

$$R_y(\theta_1)R_y(\theta_2) = R_y(\theta_2)R_y(\theta_1) = R_y(\theta_1 + \theta_2). \qquad (3.81)$$

That is, rotations about a given axis are commutable. Prove similar relations with R_y replaced by R_x and R_z.

Hint: This can be shown easily by making use of the following trigonometric relations

$$sin(\alpha \pm \beta) = sin\alpha cos\beta \pm cos\alpha sin\beta, \qquad (3.82)$$

and

$$cos(\alpha \pm \beta) = cos\alpha cos\beta \mp sin\alpha sin\beta. \qquad (3.83)$$

- **Problem 3.11**

Show that for any orthogonal basis $(\vec{a}, \vec{b}, \vec{c})$, the following operator will bring any spinor back to its original location on the Bloch sphere (with an additional phase shift). What is the expression for that phase shift?

$$R_{\vec{c}}(\pi)R_{\vec{b}}(\pi)R_{\vec{a}}(\pi). \qquad (3.84)$$

Any cyclic permutation of the three vectors leads to a similar result.

- **Problem 3.12**

Equation (3.51) gives an expression for the *spin flip matrix* in terms of the phase shift matrix and the Pauli matrix σ_y. Interpret this equation geometrically as a succession of rotations on the Bloch sphere keeping in mind that the Pauli matrices were identified as the generators of spinor rotations on the Bloch sphere.

- **Problem 3.13**

Using the Bloch sphere concept, describe geometrically the action of the matrices S and T on a general Bloch vector $\hat{n}$ or qubit $|\xi_n^+ >$.

- **Problem 3.14**

What is the 2×2 matrix allowing transition from point 1 to 2, 2 to 3, and 3 to 1 on the Bloch sphere shown in Figure 3.4?

- **Problem 3.15**

Using the fact that

$$[\sigma_x, \sigma_y] = 2i\sigma_z, \qquad (3.85)$$

calculate the commutator $[R_x(\theta_1), R_y(\theta_2)]$ where the rotation matrices are given by

$$R_x(\theta_1) = e^{-i\frac{\theta_1}{2}\sigma_x}, \qquad (3.86)$$

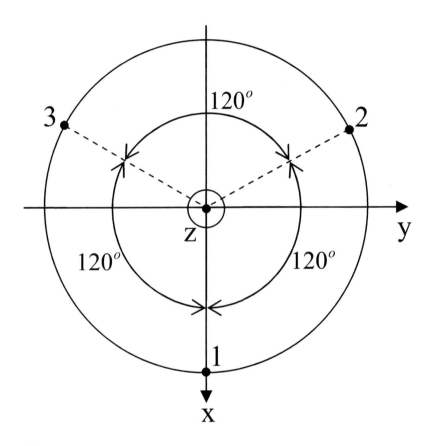

FIGURE 3.4
Locations of three spinors (1-2-3) located on the circle at the intersection of
the Bloch sphere and the $(x - y)$ plane. This figure is for Problem 3.14.

and
$$R_y(\theta_2) = e^{-i\frac{\theta_2}{2}\sigma_y}. \tag{3.87}$$

Hint: Use Equation (3.52).

- **Problem 3.16**

Consider the "Square-Root-of-Not" matrix defined as [3]

$$\sqrt{NOT} = \begin{pmatrix} \frac{1+i}{2} & \frac{1-i}{2} \\ \frac{1-i}{2} & \frac{1+i}{2} \end{pmatrix}. \tag{3.88}$$

(1) Show that this matrix is unitary.

(2) Find the values of a_o and $\vec{a}$ in the decomposition of the $\sqrt{NOT}$ matrix in the form
$$M = a_o I + \vec{a} \cdot \vec{\sigma} \tag{3.89}$$
discussed in a previous exercise.

(3) Show that the product of the $\sqrt{NOT}$ matrix with itself is σ_x, the bit flip matrix. (Hence $\sqrt{NOT} \cdot \sqrt{NOT} = NOT$, where NOT is the bit flip operation performed by a NOT-gate).

(4) Is the matrix $\sqrt{NOT}$ defined above unique?

(5) Calculate the eigenvalues and eigenvectors of the $\sqrt{NOT}$ matrix and show the locations of its corresponding eigenvectors on the Bloch sphere.

- **Problem 3.17**

It can be shown that any 2×2 unitary operator can be written in the form [4]
$$U = e^{i\alpha} R_{\hat{n}}(\theta), \tag{3.90}$$

for some real numbers α and θ and a real three dimensional vector $\hat{n}$.

Find the values of α, θ and $\hat{n}$, for the three Pauli matrices, Hadamard, Phase Shift, and $\pi/8$ gates.

3.4 References

[1] S-L Zhu, Z. D. Wang and Y-D Zhang, "Nonadiabatic noncyclic geometric phase of a spin-1/2 particle subject to an arbitrary magnetic field", Phys. Rev. B, **61**, 1142 (2000).

[2] A more formal discussion of the Pauli matrices, including their use as generators of infinitesimal rotations on spinors and their generalization to the Dirac matrices used in relativistic quantum mechanics, can be found in Chapter 4 of G. Arfken, *Mathematical Methods for Physicists*, 2nd Edition, (Academic Press, New York, 1970).

[3] C. P. Williams and S. H. Clearwater, *Exploration in Quantum Computing*, (Springer-Verlag, 1998). (Telos, the Electronic Library of Science).

[4] M. A. Nielsen and I. L. Chuang, *Quantum Computation and Quantum Information*, (Cambridge University Press, New York, 2000).

[5] N. D. Mermin, "From Cbits to Qbits: Teaching computer scientists quantum mechanics", Am. J. Phys., **71**, 23 (2002).

[6] D. M. Tong, J-L Chen, L.C. Kwek, C.H. Lai and C.H. Oh, "General formalism of Hamiltonians for realizing a prescribed evolution of a qubit", Phys. Rev. A, **68**, 062307 (2003).

[7] D. M. Sullivan and D.S. Citrin, "Time-domain simulation of quantum spin", J. Appl. Phys., **94**, 6518 (2003).

4

Evolution of a Spinor on the Bloch Sphere

In this chapter, we will consider the general approach used to calculate the time evolution of the probability of a spinor reaching a particular location on the Bloch sphere from another location, when subjected to the simultaneous actions of two spatially uniform magnetic fields. One of these fields is static (time-independent) and directed along a specific axis, while the other is time-dependent, has a different amplitude, and rotates in the plane perpendicular to the first magnetic field. This problem was originally considered by I. I. Rabi who calculated the probability of a "spin flip," i.e., the probability for a spinor, originally located at the north pole on the Bloch sphere, to ultimately reach the south pole [1, 2, 3]. Rabi calculated the time evolution of the probability in the case of a constant (dc) magnetic field along the z-axis and a circularly polarized time-varying (ac) magnetic field of different amplitude, with a specific frequency, rotating in the (x-y) plane. This led to the derivation of what is known as the *Rabi formula* which is one of the key ingredients in Nuclear Magnetic Resonance (NMR). In most textbooks, the Rabi formula is derived by solving the time-dependent Pauli equation [Equation (2.31)] with a judicious choice of the reference frame to describe the time-evolution of the 2-component spinor wavefunction. Here, we use a different approach based on the concept of the Bloch sphere described in the previous chapter. This approach is more instructive since it directly relates to NMR concepts and is more synergistic with concepts used in spin-based quantum computing.

4.1 Spin-1/2 particle in a constant magnetic field: Larmor precession

We use the well-known Ehrenfest theorem of quantum mechanics (see Chapter 18) to calculate the time evolution of the spin of an electron in a spatially uniform but time-dependent magnetic field $\vec{B}(t)$. Specifically, we will consider an electron's spin in a bulk piece of semiconductor for which (see Problem 4.1) the time evolution of the (spatial) average values of the Pauli spin matrices is governed by the following set of coupled equations:

$$\frac{d}{dt} < \sigma_x > = \frac{g\mu_B}{\hbar} (B_y < \sigma_z > - B_z < \sigma_y >)$$

$$\frac{d}{dt} < \sigma_y > = \frac{g\mu_B}{\hbar} (B_z < \sigma_x > - B_x < \sigma_z >)$$

$$\frac{d}{dt} < \sigma_z > = \frac{g\mu_B}{\hbar} (B_x < \sigma_y > - B_y < \sigma_x >), \tag{4.1}$$

where the brackets $< ... >$ denote an average of the operators over the ket $|\xi_n^+ >$ representing the spinor on the Bloch sphere, i.e., $< \sigma_n >=< \xi_n^+|\sigma_n|\xi_n^+ >$, and g is the effective Landé g-factor in the material.

Equation (4.1) can be written in matrix form as

$$\frac{d}{dt} \begin{bmatrix} < \sigma_x > \\ < \sigma_y > \\ < \sigma_z > \end{bmatrix} = \frac{g\mu_B}{\hbar} \begin{bmatrix} 0 & -B_z & B_y \\ B_z & 0 & -B_x \\ -B_y & B_x & 0 \end{bmatrix} \begin{bmatrix} < \sigma_x > \\ < \sigma_y > \\ < \sigma_z > \end{bmatrix}, \tag{4.2}$$

or, more compactly, in the vectorial form

$$\frac{d < \vec{\sigma} >}{dt} = \frac{g\mu_B}{\hbar} (\vec{B} \times < \vec{\sigma} >) = \vec{\Omega} \times < \vec{\sigma} >, \tag{4.3}$$

where the vector $< \vec{\sigma} > = (< \sigma_x >, < \sigma_y >, < \sigma_z >)$ and $\vec{\Omega} = \frac{g\mu_B}{\hbar} \vec{B}$, which is also equal to $\frac{e\vec{B}}{m_0}$ for a free electron since $g_{free-electron} = 2$ and $\mu_B = e\hbar/(2m_0)$. This is the well-known Larmor precession frequency.

Because the operator for any component of the spin angular momentum along a coordinate axis is given by $S_n = (\hbar/2)\sigma_n$, it is obvious that Equation (4.3) can be recast as

$$\frac{d < \vec{S} >}{dt} = \frac{g\mu_B}{\hbar} (\vec{B} \times < \vec{S} >) = \vec{\Omega} \times < \vec{S} >, \tag{4.4}$$

where $< \vec{S} >$ is, by definition, the expected value of the spin angular momentum. Equation (4.4) is the familiar Larmor equation describing the Larmor precession of a spin about a magnetic field.

Exercise: Consider the case of a spatially uniform and time-independent magnetic field along the z-axis. For that case, prove that the spin angular momentum vector $\vec{S}$ rotates around the z-axis with a constant angular frequency and find an expression for that angular frequency.

Solution

Equation (4.4) leads to

$$\vec{B} \cdot \frac{d\vec{S}}{dt} = 0. \tag{4.5}$$

Since $\vec{B}$ is time-independent, we obtain that $\frac{d(\vec{S} \cdot \vec{B})}{dt} = 0$, i.e., the angle θ between the two vectors $\vec{B}$ and $\vec{S}$ is time independent. In other words, the vector $\vec{S}$ moves on a cone with a fixed apex angle 2θ, and the axis of the cone coincides with the magnetic field $\vec{B}$.

Using spherical coordinates and taking into account the fact that the polar angle θ is time independent, Equation (4.4) immediately shows that the azimuthal angle ϕ (which the angle that the projection of $\vec{S}$ on the $(x - y)$ plane subtends with the x-axis) changes in time with a constant rate

$$\frac{d\phi}{dt} = \frac{g\mu_B B_z}{\hbar}. \tag{4.6}$$

Therefore, the spin angular momentum vector $\vec{S}$ rotates on a cone which makes a fixed angle with the magnetic field $\vec{B}$, and the rotation takes place at a constant angular frequency equal to the Larmor precession frequency.

4.1.1 Rotation on the Bloch sphere

Suppose we have a spinor initially located at the north pole on the Bloch sphere and we would like to flip it to the south pole using some combination of magnetic fields. What is the simplest combination of magnetic fields required to effect this transition?

In the previous section, we derived the equation of motion of a spinor in an arbitrary (including time-varying) spatially uniform magnetic field, and showed that the spinor will precess about the field in a cone with an angular frequency that is proportional to the strength of the magnetic field. So, if all we have is a constant magnetic field along the z-axis, nothing will happen to the spinor originally located at the north pole; it will remain at the north pole (polar angle $\theta = 0$). But what if we apply a magnetic field in the equatorial plane of the Bloch sphere (i.e., the $x - y$ plane) to force the spinor to move away from the north pole? If this field acts alone, the cone on which the spinor will rotate will actually be a circle on the Bloch sphere going through the north and south poles. If that new magnetic field is selected to be along the x-axis, the circle would be located in the $(y - z)$ plane. This does take the spinor to the south pole periodically and effect periodic spins flips, but only if the new field B_{new} acts alone and the original field B_{old} along the z-axis is removed.

If we leave B_{old} on, then as soon as the spinor starts to rotate in the $(y - z)$ plane under the influence of B_{new}, B_{old} will try to make it move in a cone around the z-axis. The spinor will then come out of the $(y - z)$ plane and the only way B_{new} can keep on exerting the right torque on the spinor to eventually bring it down to the south pole is to *chase* the spinor so that B_{new} always stays perpendicular to the direction of the spinor $\hat{n}$. This means that B_{new} needs to rotate in the equatorial plane at the right angular frequency to keep up with the spinor as it is trying to rotate on a cone around the z-axis. This situation is shown in Fig. 4.1. The magnetic field B_{new} in the equatorial

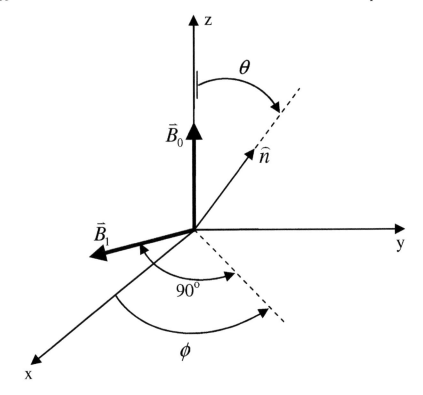

FIGURE 4.1
Spin flipping between north and south poles is achieved via the combination of a spatially uniform time-independent magnetic field along the z-axis and a rotating magnetic field in the $(x - y)$ plane ***chasing*** the unit vector $\hat{n}$ whose Larmor precession is controlled by the uniform magnetic field along the z-axis. Here $B_1 = B_{new}$ and $B_0 = B_{old}$.

plane should therefore be a constant magnetic field rotating on the equatorial plane at the angular frequency given by Equation (4.6).

How fast the spinor will be able to reach the south pole will depend on the strength of B_{new}. As B_{new} chases the spinor $\hat{n}$, the latter rotates in a circle on the plane defined by the z-axis and $\hat{n}$, with an angular frequency proportional to the strength of B_{new}. With both B_{old} and B_{new} present, the spinor $\hat{n}$ will periodically visit the north and south poles with a period inversely proportional to the strength of B_{new}. This is the basic physics underlying nuclear magnetic resonance (NMR) [4] which is routinely used in numerous applications and more recently for practical implementations of quantum computers [5].

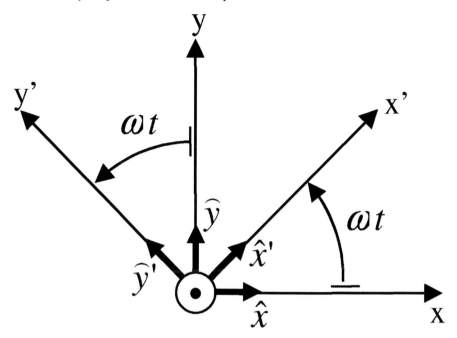

FIGURE 4.2

Illustration of the two reference frames used to described the temporal evolution of a spinor under the combined actions of B_{new} and B_{old}. The reference frame (x', y', z') rotates counterclockwise around the z-axis at a constant angular frequency ω.

An exact calculation of the probability of spin flip via the combined actions of B_{old} and B_{new} was originally performed by I. I. Rabi in the 1940s but here we present a different derivation based on the Bloch sphere concept discussed in the previous chapter and the results of the previous section.

4.2 Preparing to derive the Rabi formula

Consider a fixed (time-independent) reference frame with Cartesian coordinates (x, y, z) and a second time-varying reference frame with Cartesian coordinates $(x'(t), y'(t), z'(t))$ obtained from the initial reference frame through an anti-clockwise rotation around the z-axis at an angular frequency ω. This is shown in Fig. 4.2.

Referring to Fig. (4.2), it is easy to see that the relationship between the unit vector in the unprimed and primed reference frames is given by

$$\hat{x}' = (cos\omega t)\hat{x} + (sin\omega t)\hat{y}$$
$$\hat{y}' = -(sin\omega t)\hat{x} + (cos\omega t)\hat{y}$$
$$\hat{z}' = \hat{z}. \tag{4.7}$$

Using these relations, any three-dimensional vector $\hat{n}$ can be decomposed in the two reference frames as follows:

$$\hat{n} = n_{x'}\hat{x}' + n_{y'}\hat{y}' + n_{z'}\hat{z}' = n_x\hat{x} + n_y\hat{y} + n_z\hat{z}, \tag{4.8}$$

where (n_x, n_y, n_z) and $(n_{x'}, n_{y'}, n_{z'})$ are the components in the unprimed and primed reference frames, respectively. These components are related as

$$\hat{n}' = \begin{bmatrix} n_{x'} \\ n_{y'} \\ n_{z'} \end{bmatrix} = [A] \begin{bmatrix} n_x \\ n_y \\ n_z \end{bmatrix} = [A]\hat{n}, \tag{4.9}$$

where $[A]$ is given by

$$[A] = \begin{bmatrix} cos\omega t & sin\omega t & 0 \\ -sin\omega t & cos\omega t & 0 \\ 0 & 0 & 1 \end{bmatrix}. \tag{4.10}$$

Exercise: Prove that the transformation matrix $[A]$ is indeed given by Equation (4.10)

The reader can easily prove this result using Equations (4.7) - (4.9).

Earlier, we saw that when a spinor is subjected to a spatially uniform time dependent magnetic field, its time evolution is given by Equation (4.2). Therefore, in the unprimed reference frame, we have

$$\frac{d\hat{n}}{dt} = [X]\hat{n}, \tag{4.11}$$

where the matrix $[X]$ is given explicitly by the 3×3 matrix in Equation (4.2). Next, we ask what is the analytical form of the matrix $[X']$ such that

$$\frac{d\hat{n}'}{dt} = [X']\hat{n}', \tag{4.12}$$

which describes the time evolution of the spinor in the rotating frame.

From Equation (4.9), we get

$$\frac{d\hat{n}}{dt} = [A]^{-1}\frac{d\hat{n}'}{dt} + \frac{d[A]^{-1}}{dt}\hat{n}', \tag{4.13}$$

where (invert Equation (4.10))

$$[A]^{-1} = \begin{bmatrix} cos\omega t & -sin\omega t & 0 \\ sin\omega t & cos\omega t & 0 \\ 0 & 0 & 1 \end{bmatrix}. \tag{4.14}$$

Hence,

$$[A]^{-1}\frac{d\hat{n}'}{dt} + \frac{d[A]^{-1}}{dt}\hat{n}' = [X]\hat{n} = [X][A]^{-1}\hat{n}', \tag{4.15}$$

which can be rearranged as

$$\frac{d\hat{n}'}{dt} = [A]\left\{[X][A]^{-1} - \frac{d[A]^{-1}}{dt}\right\}\hat{n}' = \left\{[A][X][A]^{-1} - [A]\frac{d[A]^{-1}}{dt}\right\}\hat{n}' \tag{4.16}$$

where the expression between the last curly brackets is the matrix $[X']$.

Exercise: If a spin-1/2 particle is subjected to the simultaneous actions of B_{new} and B_{old} (see Figure 4.1), then the total magnetic field it experiences is given by

$$\vec{B} = (B_{new}cos\omega t, B_{new}sin\omega t, B_{old}). \tag{4.17}$$

Show that the matrix $[X']$ is

$$[X'] = \begin{bmatrix} 0 & \omega - \omega_0 & 0 \\ -(\omega - \omega_0) & 0 & -\omega_1 \\ 0 & \omega_1 & 0 \end{bmatrix}, \tag{4.18}$$

where

$$\omega_0 = \frac{g\mu_B B_{old}}{\hbar}, \tag{4.19}$$

and

$$\omega_1 = \frac{g\mu_B B_{new}}{\hbar}. \tag{4.20}$$

Note that ω_0 is the Larmor frequency associated with the z-directed static magnetic field and ω_1 is the Larmor frequency associated with the time varying magnetic field rotating in the $(x - y)$ plane.

Solution
Use the expressions for $[X]$ from Equation (4.2) and the expressions for $[A]$ and $[A]^{-1}$ given above. Then use the fact that

$$[X'] = \left\{[A][X][A]^{-1} - [A]\frac{d[A]^{-1}}{dt}\right\} \tag{4.21}$$

to prove the result.

Since the elements of the matrix $[X']$ are independent of time, the solution of Equation (4.12) is given by

$$\hat{n}'(t) = e^{[Q](t)}\hat{n}'(0), \tag{4.22}$$

where

$$[Q](t) = \int_0^t [X'](t')dt'. \tag{4.23}$$

The matrix $[Q](t)$ is therefore given by

$$[Q](t) = \begin{bmatrix} 0 & (\omega - \omega_0)t & 0 \\ -(\omega - \omega_0)t & 0 & -\omega_1 t \\ 0 & \omega_1 t & 0 \end{bmatrix}. \tag{4.24}$$

Suppose we have the matrix $[S](t)$ which diagonalizes $[Q](t)$, i.e.,

$$[\Lambda](t) = [S]^{-1}(t)[Q](t)[S](t) = diag(\lambda_1(t), \lambda_2(t), \lambda_3(t)), \tag{4.25}$$

where the λ_i (i=1,2,3) are the three eigenvalues of the matrix $[Q](t)$.
Then,

$$e^{[Q](t)} = e^{[S](t)[\Lambda](t)[S]^{-1}(t)} = [I] + [S][\Lambda][S]^{-1} + \frac{1}{2!}([S][\Lambda][S]^{-1})^2 + ...,$$

$$= [I] + [S]([\Lambda] + \frac{[\Lambda]^2}{2!} + \frac{[\Lambda]^3}{3!} + ...)[S]^{-1} = [S]e^{[\Lambda](t)}[S]^{-1}. \tag{4.26}$$

Hence, we get

$$\hat{n}'(t) = [S](t)e^{[\Lambda](t)}[S]^{-1}(t)\hat{n}'(0), \tag{4.27}$$

which leads to

$$\hat{n}(t) = [U]\hat{n}(0) = \left\{ [A]^{-1}(t) \left[[S](t)e^{[\Lambda](t)}[S]^{-1}(t) \right] A(0) \right\} \hat{n}(0). \tag{4.28}$$

The last equation describes the time evolution of the spinor on the Bloch sphere in the initial (fixed) frame of reference.

Next, we calculate the unitary matrix $[U] = [A]^{-1}(t) \left\{ [S](t)e^{[\Lambda](t)}[S]^{-1}(t) \right\} [A](0)$. The inverse $[A]^{-1}(t)$ was already calculated above and the matrix $[A](0)$ is simply the 3×3 identity matrix. We are therefore left with calculating the matrix $[S](t)e^{[\Lambda](t)}[S]^{-1}(t)$.

Hereafter, we use the shorthand notations

$$[Q](t) = \begin{bmatrix} 0 & \alpha & 0 \\ -\alpha & 0 & \beta \\ 0 & -\beta & 0 \end{bmatrix}, \tag{4.29}$$

where $\alpha = (\omega - \omega_0)t$ and $\beta = -\omega_1 t$.

The eigenvalues λ of the matrix $[Q](t)$ are found to be

$$\lambda_1 = 0, \tag{4.30}$$

$$\lambda_2 = i\sqrt{\alpha^2 + \beta^2}, \tag{4.31}$$

and

$$\lambda_3 = -i\sqrt{\alpha^2 + \beta^2}. \tag{4.32}$$

For $\lambda_1 = 0$, the corresponding eigenvector is

$$\vec{q_1} = [\beta/\alpha, 0, 1]. \tag{4.33}$$

For $\lambda_2 = i\sqrt{\alpha^2 + \beta^2}$, the corresponding eigenvector is

$$\vec{q_2} = [-\alpha/\beta, -i\sqrt{\alpha^2 + \beta^2}/\beta, 1], \tag{4.34}$$

and for $\lambda_3 = -i\sqrt{\alpha^2 + \beta^2}$, the eigenvector is

$$\vec{q_3} = [-\alpha/\beta, i\sqrt{\alpha^2 + \beta^2}/\beta, 1]. \tag{4.35}$$

Exercises:

- Prove that the eigenvalues of the matrix $[Q](t)$ are indeed given by the expressions λ_1, λ_2, and λ_3 given above.

- Derive the expressions of the corresponding eigenvectors, $\vec{q_1}$, $\vec{q_2}$ and $\vec{q_3}$.

These are easy to show and left as exercises for the reader.

It is trivial to show that if we form a matrix $[S]$ whose column vectors are equal to the three eigenvectors of the matrix $[Q]$ found above, then Equation (4.25) will be satisfied. Therefore, the explicit form of the matrix $[S]$ is

$$[S] = \begin{bmatrix} \beta/\alpha & -\alpha/\beta & -\alpha/\beta \\ 0 & -i\sqrt{\alpha^2 + \beta^2}/\beta & i\sqrt{\alpha^2 + \beta^2}/\beta \\ 1 & 1 & 1 \end{bmatrix}, \tag{4.36}$$

and its inverse is given by

$$[S]^{-1} = \begin{bmatrix} \alpha\beta/[\alpha^2 + \beta^2] & 0 & \alpha^2/[\alpha^2 + \beta^2] \\ -0.5\alpha\beta/[\alpha^2 + \beta^2] & 0.5i\beta/[\sqrt{\alpha^2 + \beta^2}] & 0.5\beta^2/[\alpha^2 + \beta^2] \\ -0.5\alpha\beta/[\alpha^2 + \beta^2] & -0.5i\beta/[\sqrt{\alpha^2 + \beta^2}] & 0.5\beta^2/[\alpha^2 + \beta^2] \end{bmatrix}. \tag{4.37}$$

Using the results obtained so far, the matrix
$[U] = [A]^{-1}(t)\left\{ [S](t)e^{[\Lambda](t)}[S]^{-1}(t) \right\}[A](0)$ appearing on the right hand side of Equation (4.28) is found to be

$$[U] = \begin{bmatrix} g(\delta,\chi)sin\omega t + h(\delta,\chi)cos\omega t & g(\delta,\chi)cos\omega t - cos\delta sin\omega t & [f(\delta)cos\omega t cos\chi - sin\delta sin\omega t]sin\chi \\ -g(\delta,\chi)cos\omega t + h(\delta,\chi)sin\omega t & g(\delta,\chi)sin\omega t + cos\delta cos\omega t & [f(\delta)sin\omega t cos\chi + sin\delta cos\omega t]sin\chi \\ f(\delta)cos\chi sin\chi & -sin\delta sin\chi & cos^2\chi + sin^2\chi cos\delta \end{bmatrix} \tag{4.38}$$

where

$$\delta = \sqrt{\alpha^2 + \beta^2} = \sqrt{(\omega - \omega_0)^2 + \omega_1^2}t$$
$$f(\delta) = 1 - cos\delta$$
$$h(\delta, \chi) = cos\delta cos^2\chi + sin^2\chi$$
$$g(\delta, \chi) = sin\delta cos\chi$$
$$\chi = atan[\omega_1/(\omega_0 - \omega)]$$
$$sin^2\chi = \frac{\omega_1^2}{\omega_1^2 + [\omega_0 - \omega]^2}. \tag{4.39}$$

4.3   Rabi formula

Suppose a spinor at time $t = 0$ is at the north pole on the Bloch sphere. We wish to calculate the probability for the spinor to flip to the south pole, as a function of time, when subjected to the combined actions of the static magnetic field B_{old} along the z-axis and the rotating time varying magnetic field B_{new} in the $(x - y)$ plane.

The probability for any arbitrary spinor to be eventually located at the south pole is given by (use Equation (3.26))

$$|<1|\xi_n^+>|^2 = sin^2(\frac{\theta(t)}{2}) = \frac{1 - cos\theta(t)}{2}, \tag{4.40}$$

which is equal to

$$sin^2(\frac{\theta(t)}{2}) = \frac{1 - n_z(t)}{2}. \tag{4.41}$$

Now, at $t = 0$, the spinor $\hat{n}(0) = ((n_x(0), n_y(0), n_z(0)) = (0, 0, 1)$. Next we can find $n_z(t)$ from Equation (4.28) and substitute that in the above equation to find

$$sin^2(\frac{\theta(t)}{2}) = \frac{sin^2\chi(t)}{2}[1 - cos\delta(t)], \tag{4.42}$$

which is the *Rabi formula*. The quantities χ and δ are defined in Equation (4.39).

The Rabi formula indicates that the probability of spin flipping reaches a maximum of $sin^2\chi$ and therefore can reach unity only if the following resonance condition is satisfied:

$$\omega = \omega_0, \tag{4.43}$$

which is mathematically equivalent to the physical discussion at the beginning of this chapter, i.e., the magnetic field B_{new} in the $(x-y)$ plane must rotate at a frequency equal to the Larmor frequency $g\mu_B B_{old}/\hbar$ in order for the spinor to reach the south pole.

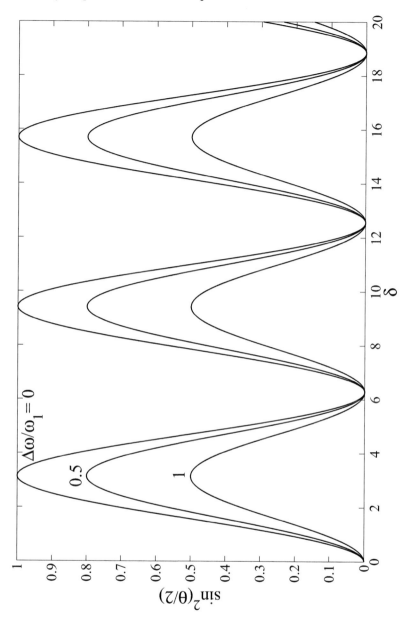

FIGURE 4.3

Plot of the probability for a spinor, initially located at the north pole of the Bloch sphere, to reach the south pole, as a function of δ for various values of the ratio $\Delta\omega/\omega_1$, where $\Delta\omega = \omega - \omega_0$. At resonance, i.e., when $\omega = \omega_0$, the probability of reaching the south pole reaches unity periodically. The shortest time to reach the south pole is π/ω_1.

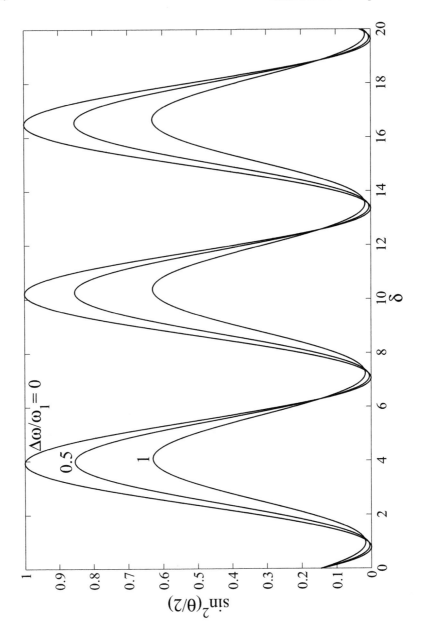

FIGURE 4.4

Time dependence of the probability of a spinor reaching the south pole if initially located in the $+x$ direction on the Bloch sphere. The probability reaches unity at resonance, i.e., when $\omega = \omega_0$, if the phase ϕ of the rotating magnetic field in the $(x - y)$ plane at time $t = 0$ is set equal to ϕ.

4.3.1 Spin flip time

The magnetic field $\vec{B}_{new}$ must rotate in the $(x - y)$ plane with an angular frequency ω_0, which is the Larmor frequency associated with the static field $\vec{B}_{old}$, in order to guarantee a spin flip.

The spinor $\hat{n}$ will periodically visit the north and south poles of the Bloch sphere with a frequency equal to the Larmor frequency of the field in the $(x - y)$ plane, which is ω_1. Therefore, the time that elapses between two consecutive visits to either pole is $T = 2\pi/\omega_1$ and the time required to switch from one pole to the other, which is the time required for the spin to flip, is

$$t_s = \frac{T}{2} = \frac{\pi}{\omega_1} = \frac{\pi\hbar}{g\mu_B B_{new}}. \tag{4.44}$$

This time is inversely proportional to the strength of the field $\vec{B}_{new}$.

The derivation of Rabi's formula was based on the general equation (4.28), which allows a direct calculation of the time dependence of the probability of a spin transfer between any two points of the Bloch sphere, starting at a location at time $t = 0$ other than the north pole, which is usually the case derived in most textbooks based on a direct solution of the time-dependent Schrödinger equation.

Example 1: Using the formulation above, calculate the time dependence of the probability (using the same configuration of constant and rotating magnetic fields) for a spinor originally in the eigenstate of the Hadamard matrix with eigenvalue +1 to reach the south pole on the Bloch sphere.

Solution

At $t = 0$, $((n_x(0), n_y(0), n_z(0)) = (\frac{1}{\sqrt{2}}, 0, \frac{1}{\sqrt{2}})$ is the unit vector or Bloch vector associated with the eigenstate of the Hadamard matrix with eigenvalue +1. Using Equation (4.40), we get

$$|<1|\xi_n^+>|^2 = sin^2(\frac{\theta(t)}{2}) = \frac{\sqrt{2} - 1 + sin\chi(sin\chi - cos\chi)(1 - cos\delta)}{2\sqrt{2}}. \tag{4.45}$$

A plot of that probability as a function of δ is shown in Figure 4.5. Compare it to the case when the initial spinor at time $t = 0$ is at the north pole.

In this case, the probability of reaching the south pole does not ever become unity even when the resonance condition $\omega = \omega_0$ is satisfied. The reason is that the phase of the magnetic field $\vec{B}_1$ in the $(x - y)$ plane must be such that, at time $t = 0$, $\vec{B}_1$ is orthogonal to the plane of the constant magnetic field $\vec{B}_0$ and the original spinor at $t = 0$. Therefore, the phase of the magnetic field $\vec{B}_1$

$$\vec{B}_1 = (B_1 cos(\omega t + \phi), B_1 sin(\omega t + \phi), 0) \tag{4.46}$$

must be set to $-\pi/2$ for the spinor to eventually reach the south pole (see Problem 4.7).

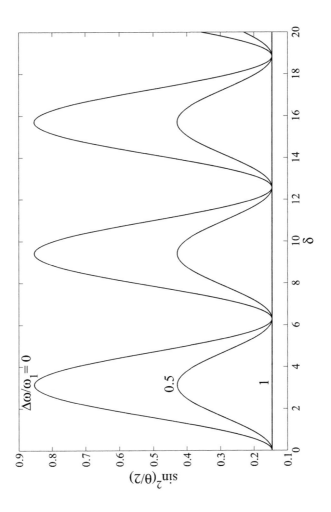

FIGURE 4.5

Time dependence of the probability of a spinor reaching the south pole if initially located on the Bloch sphere at $((n_x(0), n_y(0), n_z(0)) = (\frac{1}{\sqrt{2}}, 0, \frac{1}{\sqrt{2}})$, which is an eigenvector of the Hadamard matrix. The probability is maximized, but does **not** reach unity, at resonance when $\omega = \omega_0$. In order for the probability to reach unity, the phase ϕ at time $t = 0$ for the rotating magnetic field in the $(x - y)$ plane must be suitably adjusted (see Problem 4.7).

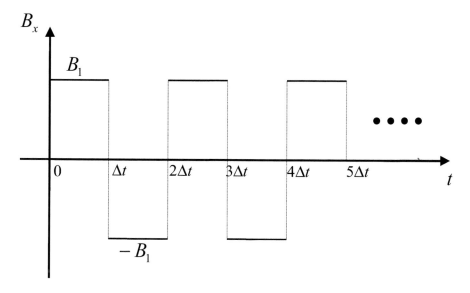

FIGURE 4.6

Waveform of the square wave magnetic field of intensity B_1 applied along the x-axis.

Example 2: Square magnetic pulse along the x-axis

In this section, we investigate the dynamics of spin flipping when the spinor is initially located at the north pole of the Bloch sphere and submitted to the action of a constant magnetic field B_0 along the z-axis, as well as a square wave magnetic field of amplitude B_1 along the x-axis. The waveform of the latter field is shown in Fig. 4.6.

Figs. 4.7 – 4.9 show the time dependence of the x-, y-, and z-components of the Bloch vector for the case where the intensities of the magnetic fields are such that $\omega_1/\omega_0 = 0.1$ and ω_1 is positive, i.e., the g-factor for the electron is assumed to be positive. In Figs. 4.7 – 4.9, the time interval Δt is one-half of the period associated with the Larmor frequency ω_0, i.e., the period of the square pulse is exactly the same as the period associated with the Larmor precession due to the magnetic field B_0 alone. In that case, there is a spin flip occurring in a time slightly less than 8 periods. There is also a beat pattern observed in the x and y-components of the Bloch vector corresponding to the qubit spiraling down the surface of the Bloch sphere until it finally reaches the south pole and then spiraling back up towards the north pole. This flipping back and forth between the north and south poles keeps occurring at longer time scales as in the case of a small rotating magnetic field in the $(x - y)$ plane examined above. This case corresponds to the resonant condition when the period of the magnetic field along the x-axis is the same as the inverse of the Larmor frequency due to B_0 alone.

Figs. 4.10 – 4.12 show the time evolution of the the x-, y-, and z-components of the Bloch vector for a qubit initially at the north pole. The period of the magnetic pulse along the x-axis is $1/10$ of the period associated with the Larmor frequency due to B_0 alone. This case corresponds to the $off-resonance$ case discussed earlier in the description of Rabi oscillations. The magnetic field along the x-axis is changing too frequently, preventing a flip of the spin. Figs. 4.10 – 4.12 show that the spin wobbles around the north pole and is never able to leave its proximity.

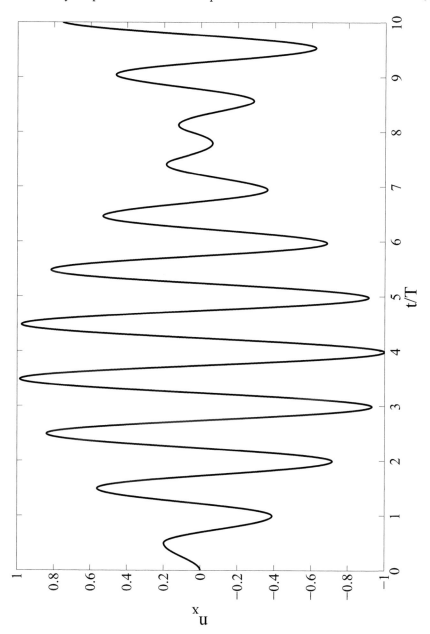

FIGURE 4.7

Time dependence of the x-component of the Bloch vector for a qubit originally at the north pole and submitted to the action of a constant magnetic field along the z-axis and a magnetic field square pulse along the x-axis where the intensities of the two magnetic fields are selected such that $\omega_1/\omega_0 = 0.1$ and ω_1 is positive.

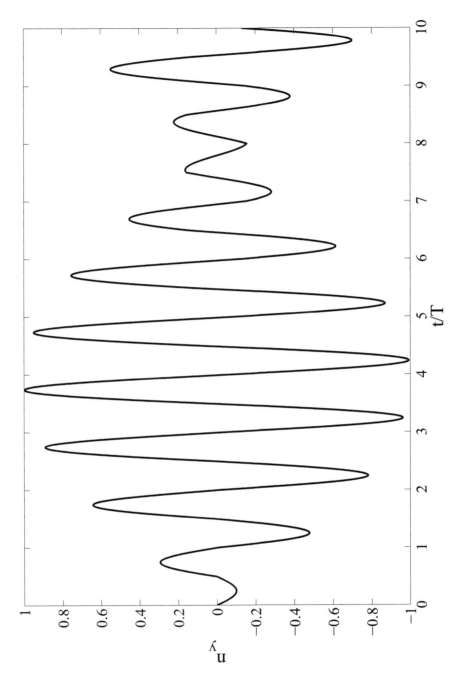

FIGURE 4.8
Same as the previous figure for the *y*-component of the Bloch vector.

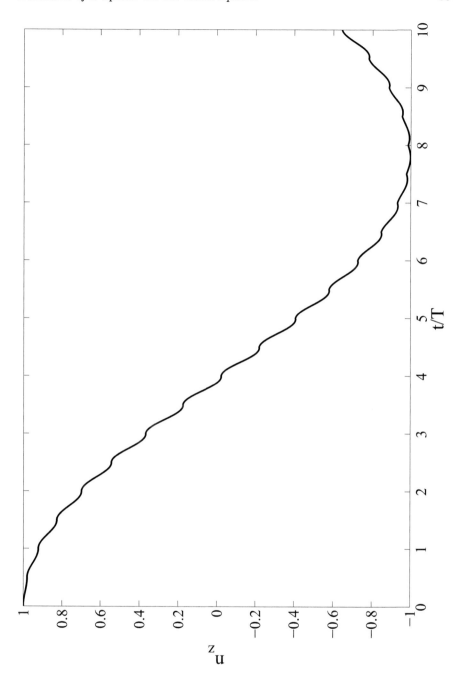

FIGURE 4.9
Same as the previous figure for the z-component of the Bloch vector.

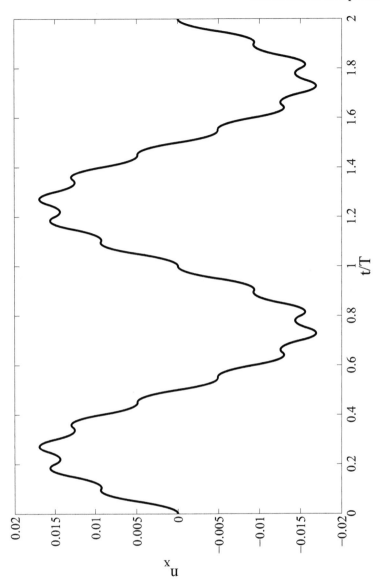

FIGURE 4.10

Time dependence of the x-component of the Bloch vector for a qubit originally at the north pole and submitted to the action of a constant magnetic field along the z-axis and a magnetic field square pulse along the x-axis where the intensities of the two magnetic fields are selected such that $\omega_1/\omega_0 = 0.1$ and ω_1 is positive. In this case, the period of the magnetic square pulse along the x-axis is equal to $1/10$ of the period associated with the Larmor frequency due to the magnetic field B_0 alone.

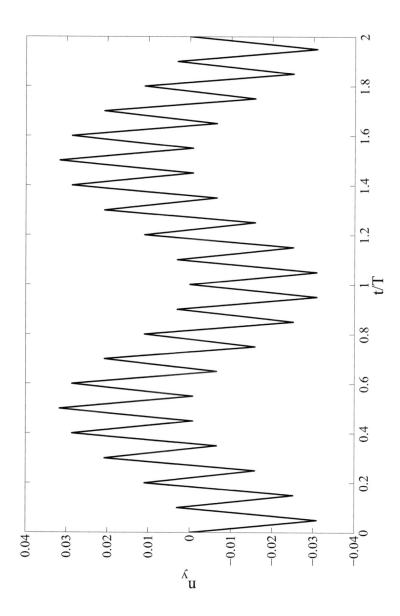

FIGURE 4.11
Same as the previous figure for the y-component of the Bloch vector.

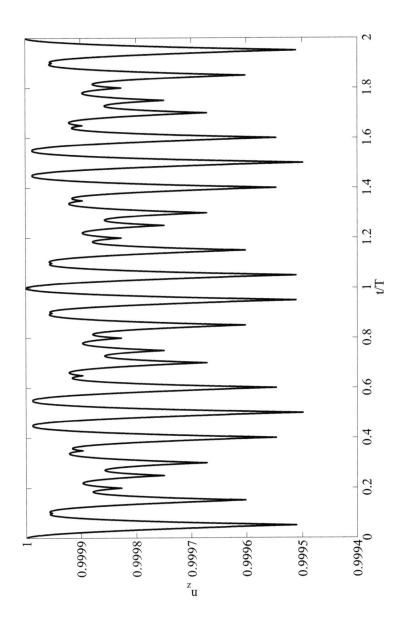

FIGURE 4.12
Same as the previous figure for the z-component of the Bloch vector.

4.4   Problems

- **Problem 4.1**

  Derive Equation (4.1) starting from the Ehrenfest theorem.

  Solution

  The Ehrenfest theorem states that the expectation value (averaged over space) of any time-dependent operator $< O >$ obeys the equation

  $$i\hbar \frac{d < O >}{dt} = < [HO - OH] >, \tag{4.47}$$

  where H is the Hamiltonian and the angular brackets denote spatial average.

  For our case, the Hamiltonian is

  $$H = H_0 - (g/2)\mu_B \vec{B} \cdot \vec{\sigma}. \tag{4.48}$$

  Therefore,

  $$i\hbar \frac{d < \sigma_x >}{dt} = -\frac{g\mu_B}{2} \left[B_y(\sigma_y \sigma_x - \sigma_x \sigma_y) + B_z(\sigma_z \sigma_x - \sigma_x \sigma_z) \right]. \tag{4.49}$$

  Next, we use the well-known commutation relations of the Pauli spin matrices (see Chapter 2) to obtain

  $$\frac{d < \sigma_x >}{dt} = \frac{g\mu_B}{\hbar} \left(B_y \sigma_z - B_z \sigma_y \right). \tag{4.50}$$

  One can similarly obtain equivalent relations for $d < \sigma_y > /dt$ and $d < \sigma_z > /dt$, to ultimately yield Equation (4.1).

- **Problem 4.2**

  Prove that the inverse of the matrix $[S]$ is indeed given by Equation (4.37).

- **Problem 4.3**

  Reproduce the plots shown in Figures (4.3) – (4.5).

- **Problem 4.4**

  Show that the matrix $[U]$ in Equation (4.38) is unitary.

- **Problem 4.5**

  Starting with the general expression for the spinor given in Equation (3.26) in the previous chapter, show that the probability for the spinor to be aligned along the $+x$-axis at time t is given by

  $$| < \xi_{+x} | \xi_n^+ > |^2 = \frac{1 + n_x(t)}{2}. \tag{4.51}$$

  Calculate the explicit time dependence of that probability when the original spinor is located at the north pole at $t = 0$. Repeat for the case when the original spinor is the eigenstate of the Hadamard matrix with eigenvalue $+1$. Find under what condition that probability is maximized.

- **Problem 4.6**

  Starting again with Equation (3.26), show that the probability for the spinor to be aligned along the $+y$-axis at time t is given by

  $$| < \xi_{+y} | \xi_n^+ > |^2 = \frac{1 + n_y(t)}{2}. \tag{4.52}$$

  Calculate the explicit time dependence of that probability when the original spinor is located along the $+x$-axis of the Bloch sphere.

- **Problem 4.7**

  In Example 1 above, is there a way to make the probability reach unity by appropriately choosing the intensity of the magnetic field and the phase of the magnetic field rotating in the $(x - y)$ plane? Note that

  $$\vec{B} = (B_1 cos(\omega t + \varphi), B_1 sin(\omega t + \varphi), B_0). \tag{4.53}$$

  Hint: The matrix elements of the unitary matrix in Equation (4.38) are given by

  $$\begin{aligned}
  u_{11} &= -g_1 sin(\omega t - \varphi) + h_1 cos(\omega t - \varphi) + P, \\
  u_{12} &= -g_2 sin(\omega t - \varphi) - h_2 cos(\omega t - \varphi) - Q, \\
  u_{13} &= [f(\delta) cos\chi cos(\omega t - \varphi) - sin\delta sin(\omega t - \varphi)] sin\chi, \\
  u_{21} &= h_1 sin(\omega t - \varphi) + g_1 cos(\omega t - \varphi) + Q, \\
  u_{22} &= -h_2 sin(\omega t - \varphi) + g_2 cos(\omega t - \varphi) + P, \\
  u_{23} &= [f(\delta) cos\chi sin(\omega t - \varphi) + sin\delta cos(\omega t - \varphi)] sin\chi, \\
  u_{31} &= [f(\delta) cos\chi cos\varphi - sin\delta sin\varphi] sin\chi, \\
  u_{32} &= -[f(\delta) cos\chi sin\varphi + sin\delta cos\varphi] sin\chi, \\
  u_{33} &= cos^2\chi + cos\delta sin^2\chi,
  \end{aligned} \tag{4.54}$$

where the following quantities were used:

$$g_1(\delta, \chi, \varphi) = sin\varphi cos\delta sin^2\chi, \tag{4.55}$$

$$g_2(\delta, \chi, \varphi) = cos\varphi cos\delta sin^2\chi, \tag{4.56}$$

$$h_1(\delta, \chi, \varphi) = cos\varphi sin^2\chi, \tag{4.57}$$

$$h_2(\delta, \chi, \varphi) = sin\varphi sin^2\chi, \tag{4.58}$$

$$P(\omega t, \delta, \chi) = cos\omega t cos\delta cos^2\chi + sin\omega t sin\delta cos\chi, \tag{4.59}$$

$$Q(\omega t, \delta, \chi) = -cos\omega t sin\delta cos\chi + sin\omega t cos\delta cos^2\chi. \tag{4.60}$$

4.5   References

[1] I. I. Rabi, "On the process of space quantization", Phys. Rev., **49**, 324 (1936).

[2] I. I. Rabi, "Space quantization in a gyrating magnetic field", Phys. Rev., **51**, 652 (1937).

[3] I. I. Rabi, N.F. Ramsey and J. Schwinger, "Use of rotating coordinates in magnetic resonance problems", Rev. Mod. Phys., **26**, 167 (1954).

[4] R. P. Feynman, R.N. Leighton and M. Sands, *The Feynman Lectures of Physics*, Volume II, Chapter IV, Sixth Edition, Addison-Wesley Publishing Company, Massachusetts (1977).

[5] T. F. Havel et al., "Quantum information processing by nuclear magnetic resonance spectroscopy", Am. J. Phys., **70**, 345 (2002).

5

The Density Matrix

The density matrix is a useful concept which will allow one to deal with a large collection of spinors with a statistical distribution at time $t = 0$ and follow their time evolution by tracking the time evolution of a new operator (the *density matrix*). This will require the extension of the *Bloch sphere* to the *Bloch ball* concept, in which the vector representing the collection of spinors will be able to reach points in the interior of the Bloch sphere as a function of time. This requires the inclusion of damping and scattering mechanisms representing interactions among spinors, and also their interactions with the environment. The phenomenological equations allowing a description of these relaxation mechanisms are known as **Bloch's equations**. They are introduced here through a connection to the Bloch sphere and Bloch ball concepts. The longitudinal and transverse relaxation times, T_1 and T_2, respectively, are also introduced and some numerical examples (results of Monte-Carlo simulations) are presented to illustrate their physical meaning.

5.1 Density matrix concept: Case of a pure state

The concept of density matrix can be introduced starting with the general ket $|\psi\rangle$ describing the quantum state of a system associated with a single particle or with an ensemble of particles, as prescribed in the general axioms of quantum-mechanics (see Chapter 18). To make the concept easier to grasp, we restrict ourselves to a general ket in the two-dimensional complex space $\mathcal{C}_2$, i.e.,

$$|\psi\rangle = \alpha|0> +\beta|1> . \tag{5.1}$$

This state is also referred to as a *pure state* for which the coefficients α and β must satisfy the normalization condition

$$|\alpha|^2 + |\beta|^2 = 1. \tag{5.2}$$

Starting with the ket $|\psi\rangle$, we define a 2×2 *density matrix* as

$$\rho = |\psi\rangle\langle\psi| = \begin{pmatrix} \alpha \\ \beta \end{pmatrix} (\alpha^*\beta^*) . \tag{5.3}$$

Hence,

$$\rho = \begin{pmatrix} \alpha\alpha^* & \alpha\beta^* \\ \beta\alpha^* & \beta\beta^* \end{pmatrix} = \begin{pmatrix} |\alpha|^2 & \alpha\beta^* \\ \beta\alpha^* & |\beta|^2 \end{pmatrix}. \tag{5.4}$$

5.2 Properties of the density matrix

The following properties of this matrix are easily derived:

1. It is definite positive, i.e., for any other ket in C_2,

$$< \psi' |\rho| \psi' > \geq 0. \tag{5.5}$$

2. It is Hermitian and, by virtue of property 1, its eigenvalues are therefore real and positive.

3. The trace of the matrix is unity (this results from the normalization of the spinor).

4. Furthermore, $\rho^2 = \rho$.

   Proof:

   $$\rho^2 = \begin{pmatrix} \alpha\alpha^* & \alpha\beta^* \\ \beta\alpha^* & \beta\beta^* \end{pmatrix} \begin{pmatrix} \alpha\alpha^* & \alpha\beta^* \\ \beta\alpha^* & \beta\beta^* \end{pmatrix}. \tag{5.6}$$

Hence

$$\rho^2 = \begin{pmatrix} |\alpha|^4 + |\alpha|^2|\beta|^2 & \alpha\beta^*|\alpha|^2 + \alpha\beta^*|\beta|^2 \\ \beta\alpha^*|\alpha|^2 + \beta\alpha^*|\beta|^2 & |\alpha|^2|\beta|^2 + |\beta|^4 \end{pmatrix}, \tag{5.7}$$

leading to

$$\rho^2 = \begin{pmatrix} |\alpha|^2 \left(|\alpha|^2 + |\beta|^2\right) & \alpha\beta^* \left(|\alpha|^2 + |\beta|^2\right) \\ \beta\alpha^* \left(|\alpha|^2 + |\beta|^2\right) & |\beta|^2 \left(|\alpha|^2 + |\beta|^2\right) \end{pmatrix}. \tag{5.8}$$

Taking into account the normalization condition (5.2) for the original spinor, we get

$$\rho^2 = \rho. \tag{5.9}$$

For the case of a spinor described by the ket in C_2 given by Equation (5.1), the average values of the individual spin components are given by

$$< \psi |S_x| \psi > = \frac{\hbar}{2} \left(\alpha\beta^* + \beta\alpha^*\right), \tag{5.10}$$

$$< \psi | S_y | \psi > = \frac{\hbar}{2i} \left(\beta \alpha^* - \alpha \beta^* \right), \tag{5.11}$$

and

$$< \psi | S_z | \psi > = \frac{\hbar}{2} \left(|\alpha|^2 - |\beta|^2 \right). \tag{5.12}$$

These results can also be obtained as follows using the definition of the density matrix and the explicit form of the Pauli matrices (where Tr stands for the trace operator)

$$\langle S_x \rangle = \text{Tr}(\rho S_x) = \frac{\hbar}{2} \text{Tr} \left[\begin{pmatrix} \alpha \alpha^* & \alpha \beta^* \\ \beta \alpha^* & \beta \beta^* \end{pmatrix} \begin{pmatrix} 0 & 1 \\ 1 & 0 \end{pmatrix} \right], \tag{5.13}$$

$$\langle S_x \rangle = \frac{\hbar}{2} \left(\alpha \beta^* + \beta \alpha^* \right), \tag{5.14}$$

$$\langle S_y \rangle = \text{Tr}(\rho S_y) = \frac{\hbar}{2} \text{Tr} \left[\begin{pmatrix} \alpha \alpha^* & \alpha \beta^* \\ \beta \alpha^* & \beta \beta^* \end{pmatrix} \begin{pmatrix} 0 & -i \\ i & 0 \end{pmatrix} \right], \tag{5.15}$$

$$\langle S_y \rangle = \frac{\hbar}{2i} \left(\beta \alpha^* - \alpha \beta^* \right), \tag{5.16}$$

and

$$\langle S_z \rangle = \text{Tr}(\rho S_z) = \frac{\hbar}{2} \text{Tr} \left[\begin{pmatrix} \alpha \alpha^* & \alpha \beta^* \\ \beta \alpha^* & \beta \beta^* \end{pmatrix} \begin{pmatrix} 1 & 0 \\ 0 & -1 \end{pmatrix} \right], \tag{5.17}$$

$$\langle S_z \rangle = \frac{\hbar}{2} \left(|\alpha|^2 - |\beta|^2 \right). \tag{5.18}$$

So, the calculation of the average value of any component of the spin operators can be readily obtained from a product of the density matrix and the corresponding Pauli matrix.

Using the above results, we get that for any pure state in $\mathcal{C}_2$

$$< S_x >^2 + < S_y >^2 + < S_z >^2 = \frac{\hbar^2}{4}, \tag{5.19}$$

at all time.

Exercises:

- Prove that the average value $< M > = < \psi | M | \psi >$ of any operator M operating in $\mathcal{C}_2$ can be calculated as:

$$< M > = Tr(\rho M). \tag{5.20}$$

The calculation of the average of an operator using Equation (5.20) is actually valid in any Hilbert space.

Solution

In $\mathcal{C}_2$, by definition,

$$\langle M \rangle = (\alpha^\star \beta^\star) \begin{bmatrix} M_{11} & M_{12} \\ M_{21} & M_{22} \end{bmatrix} \begin{pmatrix} \alpha \\ \beta \end{pmatrix},$$

$$= M_{11}|\alpha|^2 + M_{12}\alpha^\star\beta + M_{21}\beta^\star\alpha + M_{22}|\beta|^2. \tag{5.21}$$

On the other hand,

$$\text{Tr}(\rho M) = \text{Tr}\left[\begin{pmatrix} \alpha\alpha^\star & \alpha\beta^\star \\ \beta\alpha^\star & \beta\beta^\star \end{pmatrix} \begin{pmatrix} M_{11} & M_{12} \\ M_{21} & M_{22} \end{pmatrix} \right],$$

$$= M_{11}|\alpha|^2 + M_{12}\alpha^\star\beta + M_{21}\beta^\star\alpha + M_{22}|\beta|^2. \tag{5.22}$$

Hence,

$$\langle M \rangle = \langle \psi | M | \psi \rangle = \text{Tr}(\rho M), \tag{5.23}$$

for any 2×2 matrix.

- What is the form of the density matrix ρ associated with the most general form of the spinor on the Bloch sphere which was introduced in Chapter 3? Recall that the spinor is

$$|\psi\rangle = |\xi_n^\dagger\rangle = \cos\frac{\theta}{2}|0\rangle + \sin\frac{\theta}{2}e^{i\phi}|1\rangle. \tag{5.24}$$

By definition:

$$\rho(\theta,\phi) = |\psi\rangle\langle\psi| = \begin{pmatrix} \cos\frac{\theta}{2} \\ \sin\frac{\theta}{2}e^{i\phi} \end{pmatrix} \begin{pmatrix} \cos\frac{\theta}{2}, & \sin\frac{\theta}{2}e^{-i\phi} \end{pmatrix}. \tag{5.25}$$

Hence

$$\rho(\theta,\phi) = \begin{pmatrix} \cos^2\frac{\theta}{2} & \cos\frac{\theta}{2}\sin\frac{\theta}{2}e^{-i\phi} \\ \cos\frac{\theta}{2}\sin\frac{\theta}{2}e^{i\phi} & \sin^2\frac{\theta}{2} \end{pmatrix}. \tag{5.26}$$

- Show explicitly that $\rho(\theta,\phi)$ satisfy the four properties of the density matrix listed in Section 2.5.2.

- Is there a state of the form

$$|\psi\rangle = \alpha|+\rangle + \beta|-\rangle, \tag{5.27}$$

which represents a non-polarized spin, i.e., a state for which

$$\langle S_x \rangle = \langle S_y \rangle = \langle S_z \rangle = 0? \tag{5.28}$$

Solution

If we want $\langle S_x \rangle = 0$, we must select α and β such that $\alpha^*\beta$ is purely imaginary.

For $\langle S_y \rangle = 0$, we must have $\alpha^*\beta$ purely real. From these requirements, we must have

$$\alpha^*\beta = 0. \tag{5.29}$$

So, either $\alpha = 0$ and $|\beta| = 1$ and $\langle S_z \rangle = -\frac{\hbar}{2}$, or $\beta = 0$ and $|\alpha| = 1$ and $\langle S_z \rangle = +\frac{\hbar}{2}$.

Hence, $\langle S_z \rangle$ cannot be zero simultaneously with $\langle S_x \rangle$ and $\langle S_y \rangle$. The state $|\psi\rangle = \alpha|+\rangle + \beta|-\rangle$ cannot represent a non-polarized state. This observation will lead us in Section 5.3 to the generalization of the concept of the density matrix to represent a mixture of quantum states (including a non-polarized ensemble of spinors).

- The time evolution of the density matrix can be obtained using the definition of ρ and using the fact that the Schrödinger equation must be satisfied. This leads to

$$\frac{d\rho}{dt} = \frac{d}{dt} \left[|\psi\rangle\langle\psi| \right] = \left[\left(\frac{d}{dt}|\psi\rangle \right) \langle\psi| + |\psi\rangle \left(\frac{d}{dt}\langle\psi| \right) \right], \tag{5.30}$$

or

$$\frac{d\rho}{dt} = \frac{-i}{\hbar} \left[H|\psi\rangle\langle\psi| - |\psi\rangle\langle\psi|H \right] = \frac{-i}{\hbar} [H, \rho], \tag{5.31}$$

where the square braket [...] denotes the commutator. The above equation is called the **Liouville equation**.

- Starting with the equation for the time evolution of the density operator above and the calculation of the average of an operator using Equation (5.20), calculate the time evolution of the average value of an operator M and rederive Ehrenfest's equation which is proved in Chapter 18.

Solution

Taking the time derivative of $\text{Tr}(\rho M)$, we get

$$\frac{d\langle M\rangle}{dt} = \frac{d}{dt}\text{Tr}(\rho M),$$

$$= \text{Tr}\left(\frac{d\rho}{dt} M \right) + \text{Tr}\left(\rho \frac{dM}{dt} \right). \tag{5.32}$$

Using Equation (5.31) above, we get

$$\frac{d\langle M\rangle}{dt} = -\frac{i}{\hbar}\text{Tr}\left[[H,\rho]M\right] + \left\langle\frac{dM}{dt}\right\rangle,$$

$$= -\frac{i}{\hbar}\text{Tr}\left[H\rho M - \rho HM\right] + \left\langle\frac{dM}{dt}\right\rangle$$

$$-\frac{i}{\hbar}\text{Tr}\left[\rho MH - \rho HM\right] + \left\langle\frac{dM}{dt}\right\rangle. \tag{5.33}$$

Using the properties of the trace, $\text{Tr}(AB) = \text{Tr}(BA)$,

$$\frac{d<M>}{dt} = -\frac{i}{\hbar}\text{Tr}\left[\rho[M,H]\right] + \left\langle\frac{dM}{dt}\right\rangle, \tag{5.34}$$

or

$$\frac{d<M>}{dt} = -\frac{i}{\hbar}\langle[M,H]\rangle + \left\langle\frac{dM}{dt}\right\rangle, \tag{5.35}$$

which is Ehrenfest's theorem, proved in a different way in Chapter 18.

5.3 Pure versus mixed state

Example 1:

Starting with Equation (5.26) for the density matrix associated with a general spinor on the Bloch sphere and performing an average over the angles assuming a uniform distribution of the spinors on the Bloch sphere, we generate a new density matrix:

$$\rho = \frac{1}{4\pi}\int d\Omega\rho(\theta,\varphi) = \frac{1}{4\pi}\int_0^{2\pi}d\varphi\int_0^{\pi}\sin\theta d\theta\rho(\theta,\varphi). \tag{5.36}$$

Performing the integration, we get $\rho = \begin{pmatrix} \frac{1}{2} & 0 \\ 0 & \frac{1}{2} \end{pmatrix}$. It is easy to show that the four properties listed in Section 5.2 are satisfied, except for $\rho^2 = \rho$. In fact,

$$\rho^2 = \begin{pmatrix} \frac{1}{4} & 0 \\ 0 & \frac{1}{4} \end{pmatrix}, \tag{5.37}$$

and $\text{Tr}\left(\rho^2\right) = 0.5$.

Furthermore, if we compute the average values of the spin components using Equations (5.13), (5.15), and (5.18), we find that the three average values are $\langle S_i\rangle = \text{Tr}(\rho S_i) = 0$, for $i = x, y, z$.

So, the density matrix obtained by averaging over the angles seems to represent a collection of spinors uniformly distributed on the Bloch sphere. For that distribution, we do not expect an average magnetic moment for the collection of spinors. Indeed, we found that the ensemble averages of the three components of the spin operators were each equal to zero.

Example 2:

- Starting with the normalized kets

$$|a_1> = \sqrt{\frac{3}{5}}|0> + \sqrt{\frac{2}{5}}|1>, \tag{5.38}$$

and

$$|a_2> = \sqrt{\frac{3}{5}}|0> - \sqrt{\frac{2}{5}}|1> \tag{5.39}$$

(which are not orthogonal), consider the operator ρ defined as follows

$$\rho = \frac{1}{2}|a_1><a_1| + \frac{1}{2}|a_2><a_2|. \tag{5.40}$$

Show that

$$\rho = \frac{3}{5}|0><0| + \frac{2}{5}|1><1|. \tag{5.41}$$

Then, show that

$$Tr(\rho) = 1. \tag{5.42}$$

Calculate $Tr(\rho^2)$ and show that it is less than 1.

- Show that $< S_x > \; = \; < S_y > \; = \; 0$, and $< S_z > = \frac{\hbar}{4}$ and, therefore, $< S_x >^2 + < S_y >^2 + < S_z >^2 = \frac{\hbar^2}{16}$ which is less than the value for the case of a pure state.

Equation (5.41) looks like a superposition of the density matrices associated with the spinors $|0>$ and $|1>$ with the additional property that the sum of the coefficients in front of the individual matrices is unity.

We can generalize the results of the previous example by considering a density matrix defined using a set of normalized spinors on the Bloch sphere, i.e.,

$$\rho = \sum_j p_j |\psi_j\rangle\langle\psi_j| = \sum_j p_j \rho_j, \tag{5.43}$$

where the p_i's are the probabilities of finding a spinor in state $|\psi_j\rangle$ and ρ_j is the density matrix associated with state $|\psi_j\rangle$. If the sum contains more than two terms, the $|\psi_j\rangle$'s cannot be orthogonal to one another. Furthermore, since the p_j's are probabilities, the sum over the index j of these coefficients appearing in the general expression of the density matrix associated with a collection of spinors must be unity.

- For the density matrix ρ defined by Equation (5.43), show that the properties 1 through 3 in Section 5.2 are still satisfied.

- Using the general expression (5.43) for ρ, show that $(Tr\rho^2) < 1$.

Let us now compute $\langle\vec{V}\rangle = Tr[\rho\vec{V}]$, which is the generalized Bloch vector associated with a collection of spinors described by a density matrix of the form (5.43).

Solution

$$\langle\vec{V}\rangle = \text{Tr}[\rho\vec{V}] = \text{Tr}\left[\sum_j p_j|\psi_j\rangle\langle\psi_j|\vec{V}\right],$$

$$= \sum_j p_j\text{Tr}\left[|\psi_j\rangle\langle\psi_j|\vec{V}\right],$$

$$= \sum_j p_j Tr[\langle\psi_j|\vec{V}|\psi_j\rangle],$$

$$= \sum_j p_j\langle\psi_j|\vec{V}|\psi_j\rangle,$$

$$= \sum_j p_j\vec{v_j}, \tag{5.44}$$

where $\vec{v_j} = \langle\psi_j|\vec{V}|\psi_j\rangle$ and the summation is a vector sum. Hence, the generalized Bloch vector $\langle\vec{V}\rangle$ corresponding to an ensemble of pure states $|\psi_j\rangle$ is the weighted vector sum of the Bloch vectors representing the members of the ensemble.

From a physical point of view, the quantity $\langle\vec{V}\rangle$ is intuitively appealing since the weighted sum $\sum_j p_j\vec{v_j}$ is proportional to the average magnetic moment associated with the collection of spinors. The next question to address is to find out the location of the vector $\langle\vec{V}\rangle$ with respect to the Bloch sphere.

Exercises:

- Show that for any equally weighted two-member ensemble with $\langle\psi_i|\psi_j\rangle = 0$, the ensemble Bloch vector will be zero if $|\psi_1\rangle$ and $|\psi_2\rangle$ correspond to antipodal points on the Bloch sphere.

  Solution
  For such an ensemble, we have $\rho = \frac{1}{2}\left(|\psi_1\rangle\langle\psi_1| + |\psi_2\rangle\langle\psi_2|\right) = \frac{1}{2}\mathbf{1}$, and therefore $\langle\vec{V}\rangle = Tr(\rho\vec{V}) = 0$.

- Consider two ensembles of spinors, one in which the spinors are in the states below with the corresponding probabilities

$$|\psi_1\rangle, \ p_1 = 0.5,$$
$$|\psi_2\rangle, \ p_2 = 0.5. \tag{5.45}$$

The second ensemble is composed of spinors in the following states with the associated probabilities:

$$|\psi_+\rangle = \frac{|\psi_1\rangle + |\psi_1\rangle}{\sqrt{2}}, \ p_1 = 0.5,$$

$$|\psi_-\rangle = \frac{|\psi_1\rangle - |\psi_1\rangle}{\sqrt{2}}, \ p_2 = 0.5. \tag{5.46}$$

Show that the density matrices associated with the two mixed states described above are identical. Since the density matrix completely characterizes the state of a system, this last exercise shows that two seemingly different mixtures of states can be physically equivalent. In other words, a given density matrix can describe a large number of different mixed states.

- Which of the following matrices qualify as density matrices? Why?

$$\rho_1 = \begin{bmatrix} 1 & 2 \\ 2 & 1 \end{bmatrix}, \tag{5.47}$$

$$\rho_2 = \begin{bmatrix} .25 & 1 \\ 1 & .75 \end{bmatrix}, \tag{5.48}$$

$$\rho_3 = \begin{bmatrix} .25 & 0.2 + i0.1 \\ 0.2 - i0.1 & .75 \end{bmatrix}, \tag{5.49}$$

and

$$\rho_4 = \begin{bmatrix} 0.5 & .25 \\ .25 & 0.5 \end{bmatrix}. \tag{5.50}$$

For those which do not qualify, state the reason(s) why.

- In the previous problem, which of the density matrices are associated with a pure state? a mixed state? State the reason why.

5.4 Concept of the Bloch ball

In the previous sections and exercises, the density matrices associated with pure or mixed states were found to have the general form

$$\rho = \begin{bmatrix} a+b & c-id \\ c+id & a-b \end{bmatrix}, \tag{5.51}$$

where (a, b, c, d) are real.

This form is just another way to restate the theorem proven in Chapter 2 that any matrix M associated with a Hermitian operator can be decomposed in terms of the 2×2 identity and spin Pauli matrices. Indeed, since $Tr(\rho) = 1$ for density matrices associated with pure and mixed states, we must have $a = \frac{1}{2}$ and Equation (5.51) can be rewritten as

$$\rho = \frac{1}{2}\begin{bmatrix} 1 & 0 \\ 0 & 1 \end{bmatrix} + b\begin{bmatrix} 1 & 0 \\ 0 & -1 \end{bmatrix} + c\begin{bmatrix} 0 & 1 \\ 1 & 0 \end{bmatrix} + d\begin{bmatrix} 0 & -i \\ i & 0 \end{bmatrix}, \tag{5.52}$$

or, in a more compact form,

$$\rho = \frac{1}{2}[I + 2c\sigma_x + 2d\sigma_y + 2b\sigma_z]. \tag{5.53}$$

Using this last expression, the coefficients (b, c, d) can be easily related to the average components of the spin operators. In fact,

$$< \sigma_x >= Tr(\rho\sigma_x) = 2c, < \sigma_y >= Tr(\rho\sigma_y) = 2d, < \sigma_z >= Tr(\rho\sigma_z) = 2b. \tag{5.54}$$

Hence, we have

$$\rho = \frac{1}{2}[I + \vec{\nu} \cdot \vec{\sigma}], \tag{5.55}$$

where $\vec{\nu} = (\nu_x, \nu_y, \nu_z) = (< \sigma_x >, < \sigma_y >, < \sigma_z >)$.

A density matrix corresponds to a pure state when $Tr(\rho)^2 = 1$, i.e., when

$$Tr(\rho)^2 = \frac{1}{4}[(1 + \nu_z)^2 + (1 - \nu_z)^2 + 2\nu_x{}^2 + 2\nu_y{}^2] = \frac{1}{2}[1 + \vec{\nu} \cdot \vec{\nu}] = 1, \tag{5.56}$$

or whenever $|\vec{\nu}| = 1$.

Since ρ is a Hermitian operator which must be positive, all its eigenvalues are positive. Solving for the eigenvalues of the general density matrix (5.51), we find

$$\lambda = \frac{1}{2} \pm \sqrt{b^2 + c^2 + d^2}. \tag{5.57}$$

For $\lambda \geq 0$, we must have $\sqrt{b^2 + c^2 + d^2} \leq 1/2$. This implies that $|\vec{\nu}| \leq 1$.

This last property shows that unless the norm $|\vec{\nu}|$ is equal to unity, the generalized Bloch vector $\vec{\nu}$ associated with a mixed state will represent a point inside the Bloch sphere. In other words, these points can only be reached as representative of a state of a collection of particles whose density matrix corresponds to a mixed state. The ensemble of these points which are within the Bloch sphere makes up the *Bloch ball*.

Exercises

- Consider the matrix A

$$A = \alpha|0><0| + \beta|0><1| + \delta|1><0| + \gamma|1><1|, \tag{5.58}$$

where α, β, γ, and δ are complex numbers. What are the conditions on these parameters for the matrix A to be the density matrix associated with a pure state? mixed state?

- For each of the density matrices listed in the third exercise at the end of section 5.3, calculate the component of the vector $\vec{\nu}$ characterizing the vector representation of the pure or mixed state on the Bloch sphere or inside the Bloch ball.

- Consider the Bloch ball in Fig. 5.1. For the points A, B, C, D, E marked on or inside the Bloch ball, what is the density matrix operator associated with that point? The Cartesian coordinates of the points are A: $(1.0, 0.0, 0.0, 0.0)$, B: $(0.0, 1.0, 0.0, 0.0)$, C: $(0.0, 0.0, 0.0, 0.5)$, D: $(1/\sqrt{3}, 1/\sqrt{3}, 1/\sqrt{3})$, and E: $(1/\sqrt{2}, 1/\sqrt{2}, 0.0)$.

5.5 Time evolution of the density matrix: Case of mixed state

Assuming we have an ensemble of spinors represented by a density operator ρ given by Equation (5.43), we can derive its equation of motion in the following way:

$$
\begin{aligned}
\frac{d\rho}{dt} &= \frac{d}{dt} \left[\sum_j p_j |\psi_j\rangle \langle \psi_j| \right], \\
&= \sum_j p_j \left[\left(\frac{d}{dt} |\psi_j\rangle \right) \langle \psi_j| + |\psi_j\rangle \left(\frac{d}{dt} |\psi_j\rangle \right) \right], \\
&= \sum_j p_j \left[\left(\frac{-i}{\hbar} H |\psi_j\rangle \right) \langle \psi_j| + |\psi_j\rangle \left(\frac{i}{\hbar} H |\psi_j\rangle \right) \right], \\
&= \frac{-i}{\hbar} \sum_j p_j \left[H |\psi_j\rangle \langle \psi_j| - |\psi_j\rangle \langle \psi_j| H \right], \\
&= \frac{-i}{\hbar} [H, \rho], \quad\quad\quad (5.59)
\end{aligned}
$$

which is the **Liouville equation**.

In this derivation, the Hamiltonian H is the form of the Hamiltonian associated with a single spinor, and the time evolution of the density matrix for a mixed state is found to be equivalent to that for a pure state.

Similarly, we know that, from postulate 2 of quantum mechanics discussed in Chapter 18,

$$
|\psi_j(t)\rangle = U(t) |\psi_j(0)\rangle, \quad\quad\quad (5.60)
$$

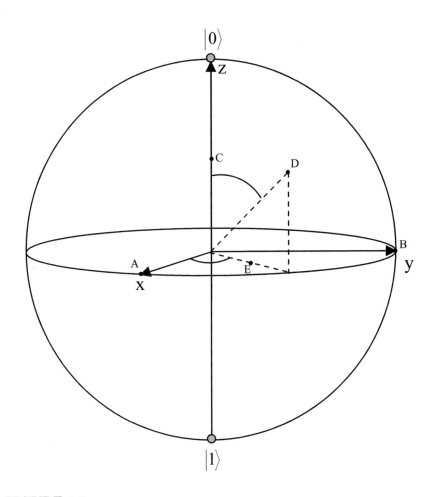

FIGURE 5.1
Bloch ball.

where $U(t)$ is a unitary operator and therefore,

$$\langle \psi_j(t)| = \langle \psi_j(0)|U^\dagger(t). \tag{5.61}$$

Therefore,

$$\rho(t) = \sum_j p_j U(t)|\psi_j(0)\rangle\langle\psi_j(0)|U^\dagger(t),$$

$$= U(t)\left[\sum_j p_j |\psi_j(0)\rangle\langle\psi_j(0)|\right]U^\dagger(t),$$

$$= U(t)\rho(0)U^\dagger(t). \tag{5.62}$$

Hence,

$$Tr\left[\rho^2(t)\right] = Tr\left[U(t)\rho^2(0)U^\dagger(t)\right] = Tr\left[U(t)\rho(0)\rho(0)U^\dagger(t)\right]. \tag{5.63}$$

Using the properties of the trace, we get

$$Tr\left[\rho^2(t)\right] = Tr\left[\rho(0)U^\dagger(t)U(t)\rho(0)\right] = Tr\left[\rho(0)\rho(0)\right], \tag{5.64}$$

and therefore,

$$Tr\left[\rho^2(t)\right] = Tr\left[\rho^2(0)\right]. \tag{5.65}$$

So, for a system characterized by a Hamiltonian (Hermitian matrix), the time evolution preserves the length of the generalized Bloch vector. As time evolves, the latter will move inside the Bloch ball but will stay at a fixed distance from its center.

We can use the above results to derive the equations of motion for the generalized Bloch vector components. Using Equation (5.55):

$$\frac{d\nu_j}{dt} = \frac{d}{dt}\left(Tr[\rho\sigma_j]\right),$$

$$= \mathrm{Tr}\left[\sigma_j\frac{d}{dt}\rho\right],$$

$$= \frac{-i}{\hbar}Tr\left[\sigma_j[H,\rho]\right],$$

$$= \frac{-i}{2\hbar}Tr\left[\sigma_j[H,(I+\vec{\nu}\cdot\vec{\sigma})]\right]. \tag{5.66}$$

Hence, the evolutions of the three components will generally be coupled.

Example: We seek the explicit form of the differential equations governing the time evolution of the components v_j associated with a spin-1/2 particle with gyromagnetic factor γ in a spatially independent magnetic field $\vec{B}$. In this case, the Hamiltonian for an individual spinor is given by

$$H = -\gamma \vec{S} \cdot \vec{B} = -\gamma \frac{\hbar}{2}(\vec{\sigma} \cdot \vec{B}). \tag{5.67}$$

Equation (5.66) becomes

$$\frac{dv_j}{dt} = \frac{-i}{2\hbar} Tr\left[<\sigma_j> \left[-\gamma\frac{\hbar}{2}\vec{\sigma}\cdot\vec{B},(1+\vec{v}\cdot\vec{\sigma})\right]\right], \tag{5.68}$$

where, by definition,

$$\vec{\sigma} \cdot \vec{B} = \sigma_x B_x + \sigma_y B_y + \sigma_z B_z, \tag{5.69}$$

and

$$\vec{v} \cdot \vec{\sigma} = v_x \sigma_x + v_y \sigma_y + v_z \sigma_z. \tag{5.70}$$

Equation (5.68) leads to

$$\frac{dv_j}{dt} = \frac{i\gamma}{4} Tr\left[\sigma_j(\vec{\sigma}\cdot\vec{B})(1+\vec{v}\cdot\vec{\sigma}) - \sigma_j(1+\vec{v}\cdot\vec{\sigma})(\vec{\sigma}\cdot\vec{B})\right]. \tag{5.71}$$

Evaluating the products on the right hand side and regrouping terms, we find

$$\frac{dv_j}{dt} = \frac{i\gamma}{4}Tr[\sigma_j\sigma_x\sigma_y(B_x v_y - v_x B_y) + \sigma_j\sigma_x\sigma_z(B_x v_z - v_x B_z)$$
$$+\sigma_j\sigma_y\sigma_x(B_y v_x - v_y B_x)]$$
$$+\frac{i\gamma}{4}Tr[\sigma_j\sigma_y\sigma_z(B_y v_z - v_y B_z) + \sigma_j\sigma_z\sigma_x(B_z v_x - v_z B_x)$$
$$+\sigma_j\sigma_z\sigma_y(B_z v_y - v_z B_y)]. \tag{5.72}$$

Focusing on the x-component first,

$$\frac{dv_x}{dt} = \frac{i\gamma}{4}Tr[\sigma_x\sigma_x\sigma_y(B_x v_y - v_x B_y) + \sigma_x\sigma_x\sigma_z(B_x v_z - v_x B_z)$$
$$+\sigma_x\sigma_y\sigma_x(B_y v_x - v_y B_x)]$$
$$+\frac{i\gamma}{4}Tr[\sigma_x\sigma_y\sigma_z(B_y v_z - v_y B_z) + \sigma_x\sigma_z\sigma_x(B_z v_x - v_z B_x)$$
$$+\sigma_x\sigma_z\sigma_y(B_z v_y - v_z B_y)], \tag{5.73}$$

and using the properties of the trace operator,

$$\frac{dv_x}{dt} = \frac{i\gamma}{4}Tr[\sigma_x\sigma_x\sigma_y(B_x v_y - v_x B_y) + \sigma_x\sigma_x\sigma_z(B_x v_z - v_x B_z)$$
$$+\sigma_x\sigma_x\sigma_y(B_y v_x - v_y B_x)]$$
$$+\frac{i\gamma}{4}Tr[\sigma_x\sigma_y\sigma_z(B_y v_z - v_y B_z) + \sigma_x\sigma_x\sigma_z(B_z v_x - v_z B_x)$$
$$+\sigma_x\sigma_z\sigma_y(B_z v_y - v_z B_y)]. \tag{5.74}$$

Since $\sigma_x{}^2 = 1$, we get

$$\frac{d\nu_x}{dt} = \frac{i\gamma}{4} Tr \left[\sigma_y(B_x\nu_y - \nu_x B_y) + \sigma_z(B_x\nu_z - \nu_x B_z) + \sigma_y(B_y\nu_x - \nu_y B_x)\right]$$

$$+\frac{i\gamma}{4} Tr [\sigma_x\sigma_y\sigma_z(B_y\nu_z - \nu_y B_z)$$

$$+\sigma_z(B_z\nu_x - \nu_z B_x) + \sigma_x\sigma_z\sigma_y(B_z\nu_y - \nu_z B_y)]. \tag{5.75}$$

In this last equation, the first trace is identically zero because $Tr\sigma_x = Tr\sigma_y = Tr\sigma_z = 0$. Hence,

$$\frac{d\nu_x}{dt} = \frac{i\gamma}{4} Tr \left[\sigma_x\sigma_y\sigma_z(B_y\nu_z - \nu_y B_z) + \sigma_x\sigma_z\sigma_y(B_z\nu_y - \nu_z B_y)\right]. \tag{5.76}$$

This leads to

$$\frac{d\nu_x}{dt} = \frac{i\gamma}{4}(B_y\nu_z - \nu_y B_z)Tr \left[\sigma_x\sigma_y\sigma_z - \sigma_x\sigma_z\sigma_y\right], \tag{5.77}$$

and, since $\sigma_x\sigma_y\sigma_z = iI$ and $\sigma_x\sigma_z\sigma_y = -iI$, we finally get

$$\frac{d\nu_x}{dt} = -\gamma(B_y\nu_z - \nu_y B_z). \tag{5.78}$$

Proceeding the same way for the other two components ν_y and ν_z, we get

$$\frac{d\nu_y}{dt} = \frac{i\gamma}{4}(B_x\nu_z - \nu_x B_z)Tr \left[\sigma_y\sigma_x\sigma_z - \sigma_y\sigma_z\sigma_x\right]$$

$$= -\gamma(B_z\nu_x - \nu_z B_x), \tag{5.79}$$

and

$$\frac{d\nu_z}{dt} = \frac{i\gamma}{4}(B_x\nu_y - \nu_x B_y)Tr \left[\sigma_z\sigma_x\sigma_y - \sigma_z\sigma_y\sigma_x\right]$$

$$= -\gamma(B_x\nu_y - \nu_x B_y). \tag{5.80}$$

In vector notation, the last three equations can be written in the very compact form

$$\frac{d\vec{\nu}}{dt} = \gamma\vec{\nu} \times \vec{B}. \tag{5.81}$$

5.6 Relaxation times T_1 and T_2 and the Bloch equations

If we consider a large ensemble of spinors located on the Bloch sphere at time $t = 0$ and submitted to the action of a constant magnetic field B_0 along the

z-axis, the solution of the three equations (5.78, 5.79, 5.80) can easily be found since the differential equations reduce to

$$\frac{d\nu_x}{dt} = -\gamma(B_y\nu_z - \nu_y B_z) = \gamma B_0 \nu_y, \tag{5.82}$$

$$\frac{d\nu_y}{dt} = -\gamma(B_z v_x - v_z B_x) = -\gamma B_0 \nu_y, \tag{5.83}$$

and

$$\frac{d\nu_z}{dt} = -\gamma(B_x v_y - v_x B_y) = 0. \tag{5.84}$$

From these equations, it is easy to show that

$$\nu_x \frac{d\nu_x}{dt} + \nu_y \frac{d\nu_y}{dt} = 0 \tag{5.85}$$

and

$$\frac{d\nu_z}{dt} = 0. \tag{5.86}$$

This means that the generalized Bloch vector representing the collection of spinors rotates on a cone centered around the z-axis with the Larmor frequency. This is expected since each individual spinor rotates on a cone with the Larmor frequency when subjected to the action of a constant magnetic field. Since the relative locations of the spinors stay unchanged, the generalized Bloch vector, which represents a weighted average of their locations, also rotates with the Larmor frequency around the z-axis and its magnitude stays unchanged. For an observer in a frame rotating with the Larmor frequency with respect to a fixed frame of reference of the enclosure containing the spinors, all spinors will appear frozen in time.

In a realistic situation, there will be some relaxation mechanisms which will have two major effects on the evolution of the ensemble of the spinors on the Bloch sphere. Imagine a gas of atoms in a finite enclosure in the presence of a constant magnetic field along the z-axis. Suppose all spinors are at the north pole at time $t = 0$, which corresponds to a high energy state for the Hamiltonian (5.67) when γ is negative. Eventually, some of the spinors will relax toward the south pole by losing energy via interaction with the environment until a distribution of spinors is reached agreeing with a thermodynamic description of the spinors. The population of the low and high energy Zeeman states must be in agreement with a Boltzmann distribution of the spinors characterized by a Bloch vector located inside the Bloch ball:

$$(\nu_x, \nu_y, \nu_z) = (0, 0, \nu_{z0}). \tag{5.87}$$

For the case of an ensemble of spin-1/2 particles in a constant external magnetic field, ν_{z0} is given by the well-known Langevin equation [3].

The timescale for relaxation from the initial distribution of spinors toward the ground state in agreement with the laws of thermodynamics is called the *longitudinal relaxation time* T_1 and is a measure of the time required for spins in the higher-energy eigenstate to decay back down to the ground state. The time T_1 is usually associated with energy dissipation. Hence, a long T_1 corresponds to a slow rate of energy loss due to interaction with the environment. For the case of gas atoms mentioned above, one possible mechanism contributing to T_1 is the interaction between individual spins as a result of collisions. As the spin of any given atom in the ensemble is randomly perturbed through collisions with other atoms (which carry their own magnetic moment), this atom will feel a time-dependent magnetic field, whose magnitude and direction will essentially be random owing to the randomness of the atomic spatial trajectories during collisions.

Whatever the time duration of the collisions is, the x and y components of the field pulse will have many Fourier components, one of which will be close in frequency to the Larmor frequency associated with the externally applied magnetic field. This component will trigger collision-induced spin flips, as was discussed in the description of Rabi oscillations in Chapter 4. If the collision time is short compared to the inverse of the Larmor frequency, many collisions will be necessary before a spin flip occurs, and therefore the time T_1 can be relatively large. Also, as was shown in the derivation of the Rabi formula, the time for spin flip is inversely proportional to the intensity of the magnetic field in the $(x - y)$ plane. T_1 not only depends on the magnitude of the pulses along the $(x - y)$ axes but also on their sign. Pulses of opposite signs will delay the time for spin flip events and increase the time scale T_1.

The short magnetic pulses due to collisional interactions can also lead to small variation in the z-component of the magnetic field felt by each spin. This leads to a spread in the Larmor frequency for each individual spin and a resulting spread in the x- and y-components of the spinors on the Bloch sphere. For the observer in a frame rotating at the Larmor frequency, the spinors will no longer appear frozen in time and the relative positions of the spinors will change rapidly. The timescale over which this dephasing between spinors takes place can be very short since the sign of the z-component of the magnetic pulses can be either positive or negative. The timescale T_2 for phase randomization in the (x,y) plane is called the *transverse relation time*. Contrary to the case of the timescale T_1, magnetic pulses along the z-axis with opposite signs will lead to a more rapid spread in the x and y components of the spinors, hence to a shorter timescale T_2.

The general arguments above suggest that the timescales T_1 and T_2 are typically such that T_2 will be much smaller than T_1*. There are many other scattering mechanisms which can contribute to both relaxation times, and the arguments above are not restricted to the spinors associated with atomic species of a gas in a cavity. In fact, the introduction of the timescales T_1 and

---

*This is actually not always true and there are situations when the reverse may be true. What is true has to be ascertained on a case-by-case basis.

T_2 was originally developed by Bloch to support the experimental effort of Rabi and his group in the earlier development of nuclear magnetic resonance experiments [1, 2, 3].

Bloch proposed that to describe phenomenologically the interaction of spinors with the environment, Equations (5.82 - 5.84) should be modified to include relaxation mechanisms which eventually help a collection of spinors reach their thermodynamic state of equilibrium. In this model, the new set of equations needed to characterize the relaxation of the generalized Bloch vector is given by the **Bloch equations**

$$\frac{d\nu_x}{dt} = -\gamma(B_y\nu_z - \nu_y B_z) - \nu_x T_2^{-1}, \tag{5.88}$$

$$\frac{d\nu_y}{dt} = -\gamma(B_z\nu_x - \nu_z B_x) - \nu_y T_2^{-1} \tag{5.89}$$

and

$$\frac{d\nu_z}{dt} = -\gamma(B_x\nu_y - \nu_x B_y) - \nu_z T_1^{-1}, \tag{5.90}$$

where the components (B_x, B_y, B_z) are the components of the externally applied magnetic field, whereas the last terms have been added to account phenomenologically for the rapidly varying magnetic field pulses arising from collisions among atoms or other sources representing additional interactions with the environment (like collisions of each individual atom with the walls of a cavity which could also be modeled as an effective magnetic field acting on each individual spin).

If we neglect the first terms on the right hand side of the Bloch equations, the solutions of the latter become

$$\nu_x(t) = \nu_x(0)\exp(-t/T_2), \tag{5.91}$$
$$\nu_y(t) = \nu_y(0)\exp(-t/T_2), \tag{5.92}$$
$$\nu_z(t) = (\nu_z(0) - \nu_{z0})\exp(-t/T_1) + \nu_{z0}, \tag{5.93}$$

which show that the *transverse* (x and y) components of the generalized Bloch vector in the equatorial plane decay exponentially to zero on a timescale T_2, whereas the *longitudinal* z-component decays exponentially to its equilibrium value ν_{z0} with time constant T_1. This is why T_1 is labeled the "longitudinal" relaxation time and T_2 is labeled the "transverse" relaxation time.

The Bloch equations have been used extensively to describe a variety of physical systems [1]. Typically, $T_2 < T_1$ when the effects of the environment can be described as a small perturbation to the effects of an externally applied magnetic field. In the next section, we use numerical simulations based on a Monte Carlo approach to describe the time evolution of an ensemble of spinors submitted to a macroscopic external magnetic field but also additional rapidly varying magnetic pulses with intensity which can be much larger than the externally applied field. Even in that case, the relaxation of an ensemble of spinors can be fitted to simple analytical functions that are solutions to the Bloch equations with phenomenological time constants for which the timescale T_1 can be smaller than T_2. This shows the generality of the Bloch equations in describing the effects of the environment on an ensemble of spins. This is a very valuable tool that underpins the importance of various (and sometimes competing) physical phenomena contributing to spin relaxation.

Numerical Examples

To illustrate the concept of density matrix for a mixed state and the corresponding time evolution of the generalized Bloch vector, a Monte Carlo approach is used to study the influence of a constant magnetic field B_0 applied along the z-axis and a random magnetic field pulse applied along the x-axis on an ensemble of spinors, each of which is characterized by a Bloch vector at $t = 0$ given by

$$(n_x(0), n_y(0), n_z(0)) = (\frac{1}{\sqrt{3}}, \frac{1}{\sqrt{3}}, \frac{1}{\sqrt{3}}). \qquad (5.94)$$

The random magnetic field along the x-axis is assumed to be different for each spin in the ensemble and represents the effective magnetic field acting on each spin due to its interaction with the environment (which could be due to the influence of an effective magnetic field caused by the presence of other spins during collisions, or the influence of a time-varying spin-orbit interaction, and any other type of interaction which can be modeled as a randomly fluctuating magnetic field). Since the spins are at different locations, the effective magnetic field they will feel will vary from spin to spin and, for simplicity, we assume that this effective magnetic field has only a component along the x-direction. Furthermore, we assume that the randomly fluctuating magnetic field changes over a typical time scale Δt whose value is comparable to the period of the Larmor precession of each spin due to the magnetic field B_0 along the z-axis. During each pulse, the field intensity along the x-axis is assumed to be constant but changed randomly between $\pm B_1{}^{max}$, as shown in Fig. 5.2.

Case 1: $B_1{}^{max}/B_0 = 10.0$ and $\omega_1 = -\gamma B_1{}^{max}$ is positive

Figs. 5.3 – 5.5 show the time dependence of the average (over an ensemble of 1,000 spins) of the x-, y- and z-components of the Bloch vector. The time interval Δt is selected such that $T/\Delta t$ is equal to 10, 50, and 100, respectively, where T is the period associated with the Larmor precession due to the magnetic field along the z-axis alone. All three components $(\bar{n}_x, \bar{n}_y, \bar{n}_z)$ rapidly decay toward 0, i.e., the Bloch vector representing the ensemble of spins moves quickly from the surface of the Bloch sphere toward the center of the Bloch ball. Since $B_1{}^{max} = 10B_0$, the precession period of the Bloch vector due to

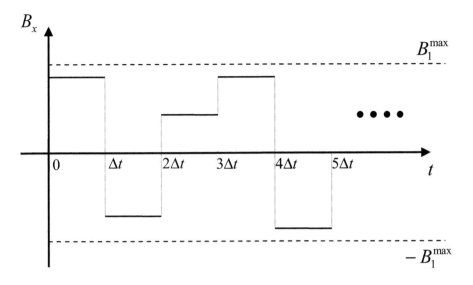

FIGURE 5.2
Random magnetic pulse sequence applied along the x-axis where the field
intensity is changed randomly and uniformly between $\pm B_1{}^{max}$ at every time
interval Δt.

B_1 alone around the x-axis is ten times smaller than the period for Larmor
precession around the z-axis. As a result, $\bar{n}_y$ and $\bar{n}_z$ decay towards 0 very
quickly. The decay is more rapid as the magnetic field pulse is changed less
frequently, i.e., for smaller values of $T/\Delta t$ since, when the Larmor precession
around the x-axis changes sign less frequently, $\bar{n}_y$ and $\bar{n}_z$ will average out to
zero faster.

Consider $\bar{n}_x$. If Δt is very large, i.e., $T/\Delta t$ converges toward 0, the component B_x will not change very frequently and the generalized Bloch vector will
rotate on a cone with principal axis along $(B_1, 0., B_0)$ with B_1 constant. The
average component $\bar{n}_x$ will therefore have mostly positive values. It is only
after flipping the sign of B_x more frequently in the negative direction that $\bar{n}_x$
can pick up negative components. As a result, $\bar{n}_x$ decays faster toward zero
as $T/\Delta t$ increases.

As the pulse rate increases, both $\bar{n}_x$ and $\bar{n}_y$ show some oscillatory behavior
with decaying amplitude, as illustrated in Figs. 5.6 and 5.7. The dashed lines
in Figs. 5.6 and 5.7 are fits to the Monte Carlo simulation results based on
analytical expressions of the form

$$\bar{n}_x = A_x cos(\omega t + \theta_x)e^{-\frac{\alpha_x t}{T}}, \tag{5.95}$$

and

$$\bar{n}_y = A_y cos(\omega t + \theta_y)e^{-\frac{\alpha_y t}{T}}. \tag{5.96}$$

The values of the parameters (A_x, A_y), (θ_x, θ_y), and (α_x, α_y) are listed in Tables 5.1 and 5.2 as a function of $N = T/\Delta t$. Within the statistical fluctuations of the Monte-Carlo results, the amplitudes A_x and A_y are basically equal to each other and the difference between θ_x and θ_y is equal to $\pi/2$. Furthermore, α_x and α_y are essentially identical. This implies that, in the $(x-y)$ plane, the $(\bar{n}_x, \bar{n}_y)$ vector rotates uniformly with an angular frequency ω and the amplitude of the vector decays progressively toward zero. As N increases, the value of ω (in units of T^{-1}) converges toward 2π, i.e., approaches the Larmor frequency due to B_0 alone.

On the other hand, the decay of $\bar{n}_z$ is monotonic and well approximated by an exponential, as shown in Figure 5.8, i.e.,

$$\bar{n}_z = \frac{1}{\sqrt{3}} e^{-\frac{\alpha_z t}{T}}. \tag{5.97}$$

The absence of oscillations in $\bar{n}_z$ is due to the fact that frequent changes in sign of B_x as $T/\Delta t$ increases prevent a completion of a full Larmor precession around the $\hat{x}$ axis as positive and negative pulses offset each other's influence on $\bar{n}_z$. This leads to the gradual decay of $\bar{n}_z$ toward zero.

The decay rates of the amplitudes of $\bar{n}_x$ and $\bar{n}_y$ are identical and lower than the decay rate of $\bar{n}_z$ which is listed in the last column of Table 5.2. This implies that

$$T_1 = \frac{T}{\alpha_z} < T_2 = T_x = T_y = \frac{T}{\alpha_x} = \frac{T}{\alpha_y}, \tag{5.98}$$

i.e., the longitudinal relaxation time is smaller than the transverse relaxation time in the numerical simulations described above.

TABLE 5.1
Fitting parameters for the
x-component of the generalized Bloch
vector

N $=$ T/Δt	A_x	ω	θ_x	α_x
100	0.621	5.766	0.433	3.279
200	0.692	6.106	0.638	1.609
300	0.790	6.089	0.722	1.249

Case 2: $B_1{}^{max}/B_0 = 1.0$ and $\omega_1 = -\gamma B_1{}^{max}$ is positive

Figs. 5.9 – 5.11 show the time dependence of the average (over an ensemble of 1000 spins) of the x-, y-, and z-components of the Bloch vector for the case where the ratio $B_1{}^{max}/B_0$ is equal to 1, i.e., ten times smaller than in case 1. In this case, the effect of the magnetic field along the z-axis is more prominent and the oscillations in the $\bar{n}_x$ and $\bar{n}_y$ components of the generalized

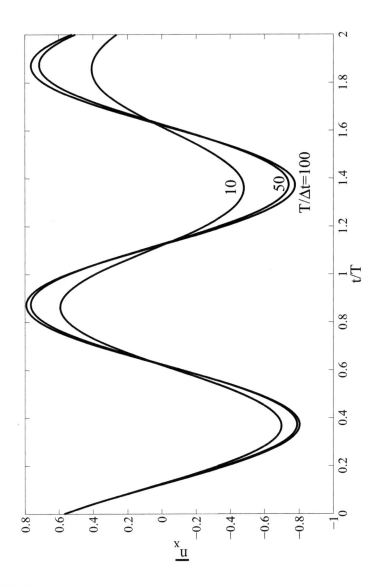

FIGURE 5.3

Time dependence of the average (over 1000 configurations) of the x-component of the generalized Bloch vector associated with an ensemble of qubits originally at $(n_x(0), n_y(0), n_z(0)) = (\frac{1}{\sqrt{3}}, \frac{1}{\sqrt{3}}, \frac{1}{\sqrt{3}})$ and submitted to the action of a constant magnetic field along the z-axis and a magnetic field square pulse along the x-axis where the intensities of the two magnetic fields are selected such that $B_1^{max}/B_0 = 10.0$ and $\omega_1 = -\gamma B_1^{max}$ is positive.

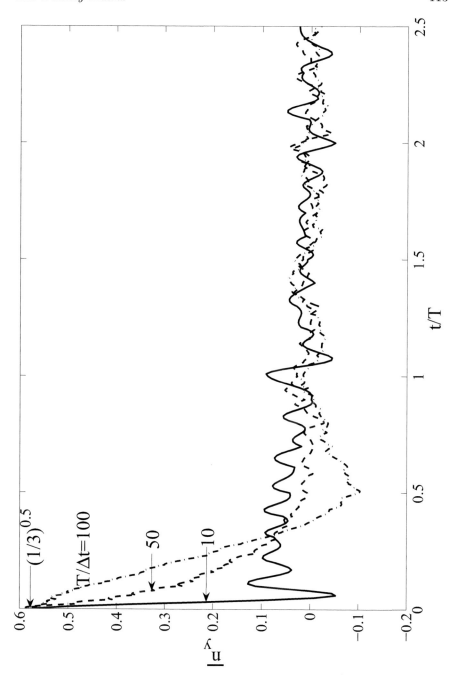

FIGURE 5.4

Same as Fig. 5.3 for the average component $\bar{n}_y(t)$ of the generalized Bloch vector associated with an ensemble of 1000 qubits located at $(n_x(0), n_y(0), n_z(0)) = (\frac{1}{\sqrt{3}}, \frac{1}{\sqrt{3}}, \frac{1}{\sqrt{3}})$ at time $t = 0$.

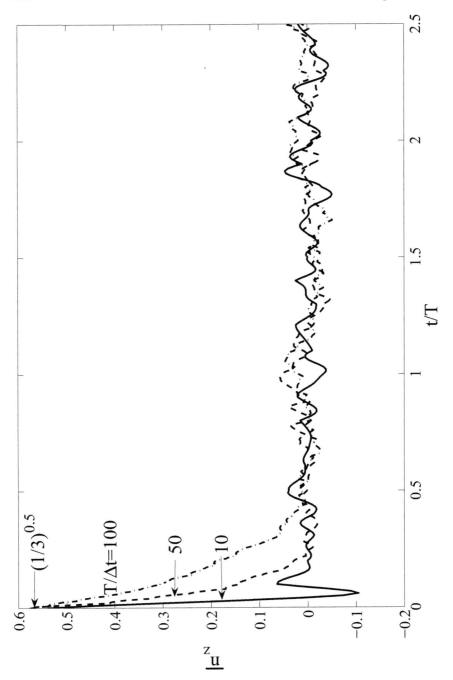

FIGURE 5.5

Same as Fig. 5.3 for the average component $\bar{n}_z(t)$ of the generalized Bloch vector associated with an ensemble of 1000 qubits located at $(n_x(0), n_y(0), n_z(0)) = (\frac{1}{\sqrt{3}}, \frac{1}{\sqrt{3}}, \frac{1}{\sqrt{3}})$ at time $t = 0$.

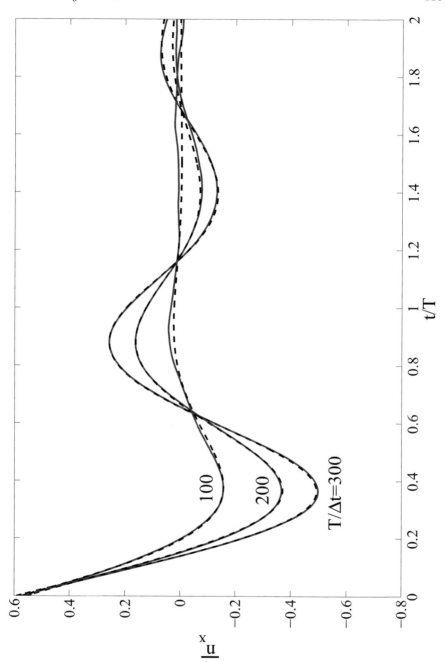

FIGURE 5.6
Same as Fig. 5.3 for a small pulse, i.e., for larger values of $T/\Delta t$. The dashed curve is a fit to the Monte Carlo simulations using the analytical expression given in Equation (5.95).

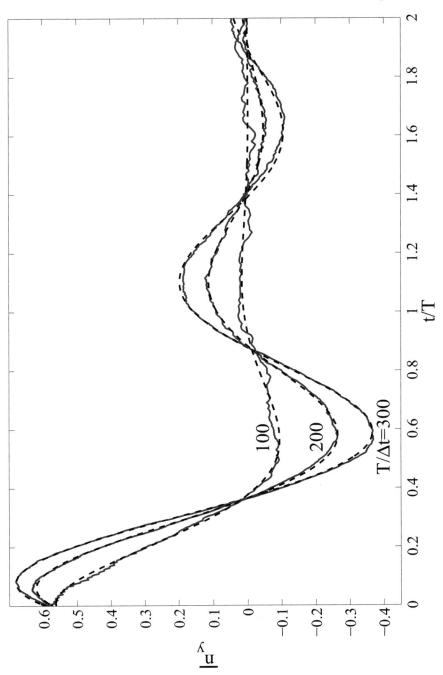

FIGURE 5.7

Same as Fig. 5.4 for a small pulse, i.e., for larger values of $T/\Delta t$. The dashed curve is a fit to the Monte Carlo simulations using the analytical expression given in Equation (5.96).

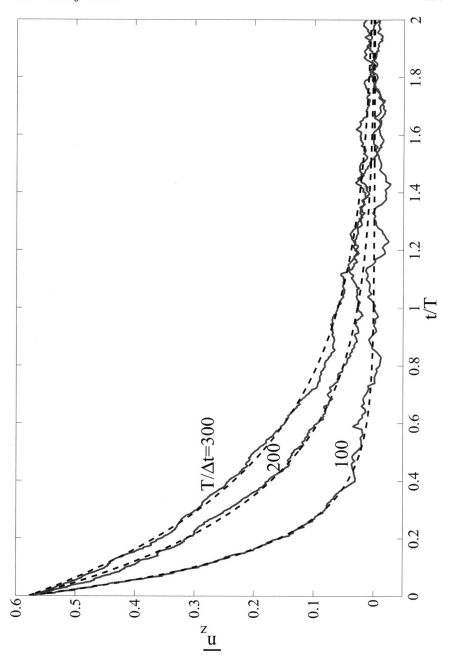

FIGURE 5.8

Same as Fig. 5.5 for a small pulse, i.e., for larger values of $T/\Delta t$. The dashed curve is a fit to the Monte Carlo simulations using the analytical expression given in Equation (5.97).

TABLE 5.2
Fitting parameters for the y- and z-components of the generalized Bloch vector

$N = T/\Delta t$	A_y	ω	θ_y	α_y	α_z
100	0.664	5.660	-0.563	3.226	6.49
200	0.711	6.045	-0.634	1.643	3.04
300	0.763	6.144	-0.682	1.213	2.26

Bloch vector take longer to decay, as can be seen by comparing Figs. 5.9 and 5.10, and 5.3 and 5.4, respectively. Also, Fig. 5.11 shows that $\bar{n}_z$ decays more slowly for the same value of $T/\Delta t$, when compared to the results displayed in Fig. 5.5.

The numerical results described above show that the time evolution of the generalized Bloch vector associated with a collection of spinors seem to be in agreement with the phenomenological approach based on the Bloch equations. The values of the fitting parameters (relaxation times T_1 and T_2) depend strongly on the signal (amplitude and time variation) modeling the effects of the environment. The time T_1 and T_2 can also be a strong function of the external magnetic field B_0. For the case of spin transport through semiconductors, one needs to clearly identify the possible sources of time relaxation mechanisms which can affect T_1 and T_2. This subject is revisited in Chapter 8.

5.7   Problems

- **Problem 5.1**

  Starting with Equation (5.68), prove Equations (5.79) and (5.80) for the time evolution of the y- and z-components of the generalized Bloch vector.

  Solution

  $$\frac{d\nu_y}{dt} = \frac{i\gamma}{4} Tr[\sigma_y \sigma_x \sigma_y (B_x \nu_y - \nu_x B_y)$$
  $$+\sigma_y \sigma_x \sigma_z (B_x \nu_z - \nu_x B_z) + \sigma_y \sigma_y \sigma_x (B_y \nu_x - \nu_y B_x)]$$
  $$+\frac{i\gamma}{4} Tr[\sigma_y \sigma_y \sigma_z (B_y \nu_z - \nu_y B_z)$$
  $$+\sigma_y \sigma_z \sigma_x (B_z \nu_x - \nu_z B_x) + \sigma_y \sigma_z \sigma_y (B_z \nu_y - \nu_z B_y)]. \qquad (5.99)$$

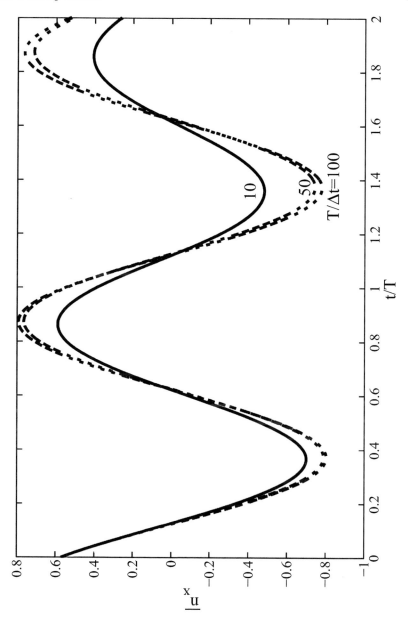

FIGURE 5.9

Time dependence of the average (over 1000 configurations) of the x-component of the generalized Bloch vector associated with an ensemble of qubits originally at $(n_x(0), n_y(0), n_z(0)) = (\frac{1}{\sqrt{3}}, \frac{1}{\sqrt{3}}, \frac{1}{\sqrt{3}})$ and submitted to the action of a constant magnetic field along the z-axis and a magnetic field square pulse along the x-axis where the intensities of the two magnetic fields are selected such that $B_1^{max}/B_0 = 1.0$ and $\omega_1 = -\gamma B_1^{max}$ is positive.

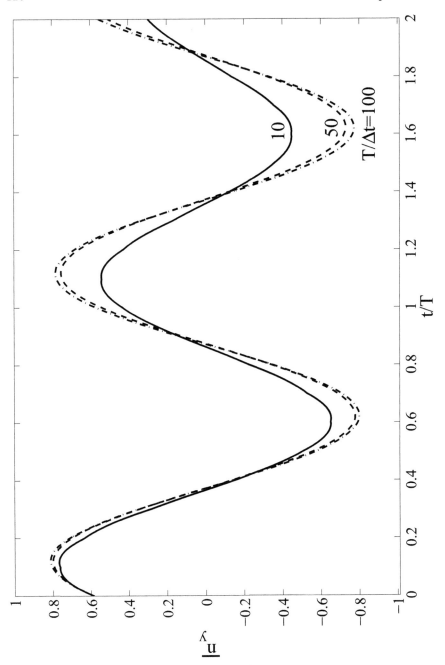

FIGURE 5.10

Same as Fig. 5.9 for the average component $\bar{n}_x(t)$ of the general-ized Bloch vector associated with an ensemble of 1000 qubits located at $(n_x(0), n_y(0), n_z(0)) = (\frac{1}{\sqrt{3}}, \frac{1}{\sqrt{3}}, \frac{1}{\sqrt{3}})$ at time t = 0.

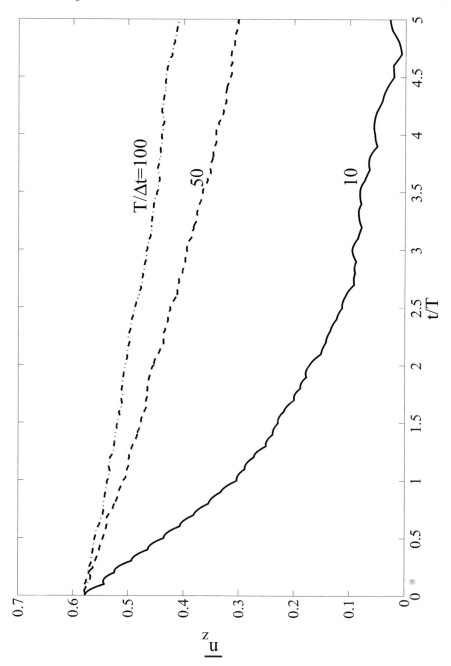

FIGURE 5.11
Same as Fig. 5.9 for the average component $\bar{n}_x(t)$ of the Bloch vector associated with a qubit located at $(n_x(0), n_y(0), n_z(0)) = (\frac{1}{\sqrt{3}}, \frac{1}{\sqrt{3}}, \frac{1}{\sqrt{3}})$ at time t = 0.

Using the cyclic property of the trace,

$$
\begin{aligned}
\frac{d\nu_y}{dt} &= \frac{i\gamma}{4} Tr[\sigma_y\sigma_y\sigma_x(B_x\nu_y - \nu_x B_y) \\
&\quad + \sigma_y\sigma_x\sigma_z(B_x\nu_z - \nu_x B_z) \\
&\quad + \sigma_y\sigma_y\sigma_x(B_y\nu_x - \nu_y B_x)] \\
&\quad + \frac{i\gamma}{4} Tr[\sigma_y\sigma_y\sigma_z(B_y\nu_z - \nu_y B_z) \\
&\quad + \sigma_y\sigma_z\sigma_x(B_z\nu_x - \nu_z B_x) \\
&\quad + \sigma_y\sigma_y\sigma_z(B_z\nu_y - \nu_z B_y)].
\end{aligned} \tag{5.100}
$$

Since $\sigma_y{}^2 = 1$, we get

$$
\begin{aligned}
\frac{d\nu_y}{dt} &= \frac{i\gamma}{4} Tr[\sigma_x(B_x\nu_y - \nu_x B_y) + \sigma_y\sigma_x\sigma_z(B_x\nu_z - \nu_x B_z) \\
&\qquad\qquad\qquad\qquad + \sigma_x(B_y\nu_x - \nu_y B_x)] \\
&\quad + \frac{i\gamma}{4} Tr[\sigma_z(B_y\nu_z - \nu_y B_z) + \sigma_y\sigma_z\sigma_x(B_z\nu_x - \nu_z B_x) \\
&\qquad\qquad\qquad\qquad + \sigma_z(B_z\nu_y - \nu_z B_y)].
\end{aligned} \tag{5.101}
$$

Hence,

$$
\begin{aligned}
\frac{d\nu_y}{dt} &= \frac{i\gamma}{4} Tr[\sigma_y\sigma_x\sigma_z(B_x\nu_z - \nu_x B_z)] \\
&\quad + \frac{i\gamma}{4} Tr[\sigma_y\sigma_z\sigma_x(B_z\nu_x - \nu_z B_x)].
\end{aligned} \tag{5.102}
$$

Pulling out the term $(B_z\nu_x - \nu_z B_x)$, we get

$$
\frac{d\nu_y}{dt} = \frac{i\gamma}{4}(B_z\nu_x - \nu_z B_x) Tr[\sigma_y\sigma_x\sigma_z - \sigma_y\sigma_z\sigma_x].
$$

Using the identities $\sigma_y\sigma_x\sigma_z = -iI$ and $\sigma_y\sigma_z\sigma_x = iI$, we get $\sigma_y\sigma_x\sigma_z - \sigma_y\sigma_z\sigma_x = -2iI$, and therefore

$$
\frac{d\nu_y}{dt} = -\gamma(B_z\nu_x - \nu_z B_x). \tag{5.103}
$$

Similarly,

$$
\begin{aligned}
\frac{d\nu_z}{dt} &= \frac{i\gamma}{4}Tr[\sigma_z\sigma_x\sigma_y(B_x\nu_y - \nu_x B_y) \\
&+ \sigma_z\sigma_x\sigma_z(B_x\nu_z - \nu_x B_z) \\
&+ \sigma_z\sigma_y\sigma_x(B_y\nu_x - \nu_y B_x)] \\
&+ \frac{i\gamma}{4}Tr[\sigma_z\sigma_y\sigma_z(B_y\nu_z - \nu_y B_z) \\
&+ \sigma_z\sigma_z\sigma_x(B_z\nu_x - \nu_z B_x) \\
&+ \sigma_z\sigma_z\sigma_y(B_z\nu_y - \nu_z B_y)].
\end{aligned}
\tag{5.104}
$$

Using the cyclic property of the trace once again,

$$
\begin{aligned}
\frac{d\nu_z}{dt} &= \frac{i\gamma}{4}Tr[\sigma_z\sigma_x\sigma_y(B_x\nu_y - \nu_x B_y) \\
&+ \sigma_z\sigma_z\sigma_x(B_x\nu_z - \nu_x B_z) \\
&+ \sigma_z\sigma_y\sigma_x(B_y\nu_x - \nu_y B_x)] \\
&+ \frac{i\gamma}{4}Tr[\sigma_z\sigma_z\sigma_y(B_y\nu_z - \nu_y B_z) \\
&+ \sigma_z\sigma_z\sigma_x(B_z\nu_x - \nu_z B_x) \\
&+ \sigma_z\sigma_z\sigma_y(B_z\nu_y - \nu_z B_y)].
\end{aligned}
\tag{5.105}
$$

Since $\sigma_z{}^2 = 1$, we get

$$
\begin{aligned}
\frac{d\nu_z}{dt} &= \frac{i\gamma}{4}Tr[\sigma_z\sigma_x\sigma_y(B_x\nu_y - \nu_x B_y) \\
&+ \sigma_x(B_x\nu_z - \nu_x B_z) \\
&+ \sigma_z\sigma_y\sigma_x(B_y\nu_x - \nu_y B_x)] \\
&+ \frac{i\gamma}{4}Tr[\sigma_y(B_y\nu_z - \nu_y B_z) \\
&+ \sigma_x(B_z\nu_x - \nu_z B_x) \\
&+ \sigma_y(B_z\nu_y - \nu_z B_y)].
\end{aligned}
\tag{5.106}
$$

Hence,

$$
\frac{d\nu_z}{dt} = \frac{i\gamma}{4}Tr\left[\sigma_z\sigma_x\sigma_y(B_x\nu_y - \nu_x B_y) + \sigma_z\sigma_y\sigma_x(B_y\nu_x - \nu_y B_x)\right].
\tag{5.107}
$$

Pulling out the term $(B_x\nu_y - \nu_x B_y)$, we get

$$
\frac{d\nu_z}{dt} = \frac{i\gamma}{4}(B_x\nu_y - \nu_x B_y)Tr\left[\sigma_z\sigma_x\sigma_y - \sigma_z\sigma_y\sigma_x\right].
\tag{5.108}
$$

Using the identities $\sigma_z\sigma_x\sigma_y = iI$ and $\sigma_z\sigma_y\sigma_x = -iI$, we get $\sigma_z\sigma_x\sigma_y - \sigma_z\sigma_y\sigma_x = 2iI$, leading to

$$\frac{d\nu_z}{dt} = -\gamma(B_x\nu_y - \nu_x B_y). \tag{5.109}$$

• **Problem 5.2**

In the ensuing problems, we show how the equation of motion for the density matrix can be used to re-derive Equation (4.1) in Chapter 4 which is the starting point of the discussion of the Rabi oscillations and the Rabi formula.

Starting with Equation (5.24), establish the following relationship between the general expression for a spinor in a frame of reference rotating at a constant angular velocity ω around the z-axis (counterclockwise) and the expression for the same spinor expressed with respect to a fixed frame of reference:

$$|\psi'\rangle = e^{i\gamma}e^{i\frac{\omega t}{2}\sigma_z}|\psi\rangle. \tag{5.110}$$

What is the expression for the phase γ?

Solution

In the fixed frame of reference,

$$|\psi\rangle = |\xi_n^{\dagger}\rangle = \cos\frac{\theta}{2}|0\rangle + \sin\frac{\theta}{2}e^{i\phi}|1\rangle. \tag{5.111}$$

In the rotating frame, the expression of the spinor is obtained from the previous one by replacing ϕ by $\phi - \omega t$. Hence,

$$|\psi'\rangle = e^{i\frac{\omega t}{2}}\begin{bmatrix} e^{i\frac{\omega t}{2}} & 0 \\ 0 & e^{-i\frac{\omega t}{2}} \end{bmatrix}|\psi\rangle \tag{5.112}$$

or, more succinctly,

$$|\psi'\rangle = e^{i\frac{\omega t}{2}}e^{i\frac{\omega t}{2}\sigma_z}|\psi\rangle. \tag{5.113}$$

Hence, the phase factor γ is Equation (5.110) is given by $\frac{\omega t}{2}$.

• **Problem 5.3**

Show that the expression for the density matrix ρ' in the rotating frame of reference, i.e., associated with the spinor $|\psi'\rangle$, is related to the density matrix ρ associated with the spinor $|\psi\rangle$ in the fixed reference frame as follows

$$\rho' = U_z\rho U_z^{\dagger}, \tag{5.114}$$

where U_z is the unitary matrix

$$U_z = e^{i\omega\sigma_z t/2}. \tag{5.115}$$

- **Problem 5.4**

Using the results of the previous problem and Equation (5.59) for the time evolution of the density matrix in the fixed frame of reference, show that the time evolution of the density matrix in the rotating frame of reference is given by

$$\frac{d\rho'}{dt} = \frac{-i}{\hbar} \left[H', \rho' \right],$$ (5.116)

where the Hamiltonian H' in the rotating frame is related to the Hamiltonian H in the fixed frame as follows:

$$H' = U_z H U_z^{-1} - i\hbar U_z \frac{dU_z^{-1}}{dt}.$$ (5.117)

Note that with the use of Equation (5.115), the last equation can be rewritten as follows

$$H' = U_z H U_z^{-1} - \frac{\hbar\omega}{2} \sigma_z.$$ (5.118)

- **Problem 5.5**

For the case where H is the Hamiltonian of a spin-1/2 particle in the presence of a constant magnetic field along the z-axis and a rotating magnetic field with intensity B_1 rotating with a constant angular velocity in the (x,y) plane and located along the x-axis at $t = 0$, we have

$$H = \omega_L S_z B_0 - \gamma B_1 (cos(\omega t) S_x + sin(\omega t) S_y).$$ (5.119)

Show that the explicit form of the Hamiltonian H' is given by

$$H' = \frac{1}{2} \hbar (\Delta \sigma_z - \gamma B_1 \sigma_x),$$ (5.120)

where $\Delta = \omega_L - \omega$ and $\omega_L = -\gamma B_0$.

- **Problem 5.6**

Since the equation governing the time evolution of the density matrix in the rotating frame is similar to the one derived in the fixed frame of reference, the equations of motion of the components of the Bloch vector in the rotating frame are the same as in the fixed frame and given by Equation (5.66), where all quantities (except for the Pauli matrices) must be primed.

Neglecting the primes for simplicity, the equations of evolution of each component of the Bloch vector in the rotating frame is therefore given by

$$\frac{d\nu_i}{dt} = -\frac{i}{4} Tr \left[\sigma_i \left[\Delta \sigma_z - \gamma B_1 \sigma_x, (1 + \nu_x \sigma_x + \nu_y \sigma_y + \nu_z \sigma_z) \right] \right]. (5.121)$$

Starting with the expression for the Hamiltonian in the rotating frame of reference, show that the equations of the time evolution of each component of the Bloch vector in the rotating frame is given in a vectorial form by

$$\frac{d}{dt}\vec{\nu} = \gamma\vec{\nu} \times \vec{B_{eff}}, \tag{5.122}$$

where

$$\vec{B_{eff}} = \left(B_0 + \frac{\omega}{\gamma}\right)\hat{z} + B_1\hat{x}. \tag{5.123}$$

Solution

Starting with Equation (5.121), we get

$$\frac{d\nu_i}{dt} = -\frac{i}{4}Tr\left[\sigma_i(\Delta\sigma_z(\nu_x\sigma_x + \nu_y\sigma_y) - \gamma B_1\sigma_x(\nu_y\sigma_y + \nu_z\sigma_z))\right]$$
$$+\frac{i}{4}Tr\left[\sigma_i(\Delta(\nu_x\sigma_x + \nu_y\sigma_y)\sigma_z - \gamma B_1(\nu_y\sigma_y + \nu_z\sigma_z)\sigma_x)\right]. \tag{5.124}$$

For the x-component, we obtain

$$\frac{d\nu_x}{dt} = -\frac{i}{4}Tr\left[\Delta\sigma_x\sigma_z(\nu_x\sigma_x + \nu_y\sigma_y) - \gamma B_1\sigma_x\sigma_x(\nu_y\sigma_y + \nu_z\sigma_z))\right]$$
$$+\frac{i}{4}Tr\left[\Delta(\nu_x\sigma_x\sigma_x + \nu_y\sigma_x\sigma_y)\sigma_z - \gamma B_1(\nu_y\sigma_x\sigma_y + \nu_z\sigma_x\sigma_z)\sigma_x)\right], \tag{5.125}$$

$$\frac{d\nu_x}{dt} = -\frac{i}{4}Tr\left[\Delta\nu_x\sigma_x\sigma_z\sigma_x + \Delta\nu_y\sigma_x\sigma_z\sigma_y - \gamma B_1(\nu_y\sigma_y + \nu_z\sigma_z))\right]$$
$$+\frac{i}{4}Tr\left[\Delta(\nu_x\sigma_x\sigma_x + \nu_y\sigma_x\sigma_y)\sigma_z - \gamma B_1(\nu_y\sigma_x\sigma_y + \nu_z\sigma_x\sigma_z)\sigma_x)\right], \tag{5.126}$$

$$\frac{d\nu_x}{dt} = -\frac{i}{4}Tr\left[-\Delta\nu_x\sigma_z - i\Delta\nu_y I - \gamma B_1(\nu_y\sigma_y + \nu_z\sigma_z))\right]$$
$$+\frac{i}{4}Tr\left[\Delta(\nu_x\sigma_z + i\nu_y I) - \gamma B_1(\nu_y\sigma_x\sigma_y\sigma_x + \nu_z\sigma_x\sigma_z\sigma_x))\right], \tag{5.127}$$

$$\frac{d\nu_x}{dt} = -\frac{i}{4}Tr\left[-\Delta\nu_x\sigma_z - i\Delta\nu_y I - \gamma B_1(\nu_y\sigma_y + \nu_z\sigma_z))\right]$$
$$+\frac{i}{4}Tr\left[\Delta(\nu_x\sigma_z + i\nu_y I) - \gamma B_1(-\nu_y\sigma_y - \nu_z\sigma_z))\right], \tag{5.128}$$

$$\frac{d\nu_x}{dt} = -\frac{i}{4}Tr\left[-\Delta\nu_x\sigma_z - i\Delta\nu_y I - \gamma B_1\nu_y\sigma_y - \gamma b_1\nu_z\sigma_z\right)]$$
$$+\frac{i}{4}Tr\left[\Delta\nu_x\sigma_z + i\Delta\nu_y I + \gamma B_1\nu_y\sigma_y + \gamma b_1\nu_z\sigma_z\right],$$

$$(5.129)$$

which reduces to

$$\frac{d\nu_x}{dt} = \frac{i}{4}Tr\left[i\Delta\nu_y I\right] + \frac{i}{4}Tr\left[i\Delta\nu_y I\right], \tag{5.130}$$

and hence,

$$\frac{d\nu_x}{dt} = -\Delta\nu_y. \tag{5.131}$$

Similarly, the y-component of Equation (5.121) leads to

$$\frac{d\nu_y}{dt} = -\frac{i}{4}Tr\left[\sigma_y(\Delta\sigma_z(\nu_x\sigma_x + \nu_y\sigma_y) - \gamma b_1\sigma_x(\nu_y\sigma_y + \nu_z\sigma_z))\right]$$
$$+\frac{i}{4}Tr\left[\sigma_y(\Delta(\nu_x\sigma_x + \nu_y\sigma_y)\sigma_z - \gamma B_1(\nu_y\sigma_y + \nu_z\sigma_z)\sigma_x)\right],$$

$$(5.132)$$

$$\frac{d\nu_y}{dt} = -\frac{i}{4}Tr\left[\Delta\sigma_y\sigma_z(\nu_x\sigma_x + \nu_y\sigma_y) - \gamma B_1\sigma_y\sigma_x(\nu_y\sigma_y + \nu_z\sigma_z))\right]$$
$$+\frac{i}{4}Tr\left[\Delta\sigma_y(\nu_x\sigma_x + \nu_y\sigma_y)\sigma_z - \gamma B_1\sigma_y(\nu_y\sigma_y + \nu_z\sigma_z)\sigma_x)\right],$$

$$(5.133)$$

$$\frac{d\nu_y}{dt} = -\frac{i}{4}Tr\left[\nu_x\Delta\sigma_y\sigma_z\sigma_x + \nu_y\Delta\sigma_y\sigma_z\sigma_y - \nu_y\gamma B_1\sigma_y\sigma_x\sigma_y + \nu_z\gamma B_1\sigma_y\sigma_x\sigma_z\right]$$
$$+\frac{i}{4}Tr\left[\nu_x\Delta\sigma_y\sigma_x\sigma_z + \nu_y\Delta\sigma_y\sigma_y\sigma_z - \gamma B_1\sigma_y\nu_y\sigma_y\sigma_x - \nu_z\gamma B_1\sigma_y\sigma_z\sigma_x\right],$$

$$(5.134)$$

$$\frac{d\nu_y}{dt} = -\frac{i}{4}Tr\left[\nu_x\Delta\sigma_y\sigma_z\sigma_x + \nu_y\Delta\sigma_y\sigma_y\sigma_z - \nu_y\gamma B_1\sigma_y\sigma_y\sigma_x + \nu_z\gamma B_1\sigma_y\sigma_x\sigma_z\right]$$
$$+\frac{i}{4}Tr\left[\nu_x\Delta\sigma_y\sigma_x\sigma_z + \nu_y\Delta\sigma_y\sigma_y\sigma_z - \gamma B_1\nu_y\sigma_y\sigma_y\sigma_x - \nu_z\gamma B_1\sigma_y\sigma_z\sigma_x\right],$$

$$(5.135)$$

$$\frac{d\nu_y}{dt} = -\frac{i}{4}Tr\left[\nu_x\Delta\sigma_y\sigma_z\sigma_x + \nu_y\Delta\sigma_z - \nu_y\gamma b_1\sigma_x + \nu_z\gamma B_1\sigma_y\sigma_x\sigma_z\right]$$
$$+\frac{i}{4}Tr\left[\nu_x\Delta\sigma_y\sigma_x\sigma_z + \nu_y\Delta\sigma_z - \gamma b_1\nu_y\sigma_x - \nu_z\gamma B_1\sigma_y\sigma_z\sigma_x\right],$$

$$(5.136)$$

$$\frac{d\nu_y}{dt} = -\frac{i}{4}Tr\left[\nu_x\Delta\sigma_y\sigma_z\sigma_x + \nu_z\gamma b_1\sigma_y\sigma_x\sigma_z\right]$$
$$+\frac{i}{4}Tr\left[\nu_x\Delta\sigma_y\sigma_x\sigma_z - \nu_z\gamma b_1\sigma_y\sigma_z\sigma_x\right],$$

$$(5.137)$$

and eventually,

$$\frac{d\nu_y}{dt} = \Delta\nu_x + \frac{i\gamma B_1\nu_z}{4}Tr\left[\sigma_y\sigma_x\sigma_z - \sigma_y\sigma_z\sigma_x\right] = \Delta\nu_x + \gamma B_1\nu_z \quad (5.138)$$

- Finally, the z-component of Equation (5.121) is given by

$$\frac{d\nu_z}{dt} = -\frac{i}{4}Tr\left[\sigma_z(\Delta\sigma_z(\nu_x\sigma_x + \nu_y\sigma_y) - \gamma B_1\sigma_x(\nu_y\sigma_y + \nu_z\sigma_z))\right]$$
$$+\frac{i}{4}Tr\left[\sigma_z(\Delta(\nu_x\sigma_x + \nu_y\sigma_y)\sigma_z - \gamma B_1(\nu_y\sigma_y + \nu_z\sigma_z)\sigma_x)\right],$$

$$(5.139)$$

$$\frac{d\nu_z}{dt} = -\frac{i}{4}Tr\left[\Delta\sigma_z\sigma_z(\nu_x\sigma_x + \nu_y\sigma_y) - \gamma B_1\sigma_z\sigma_x(\nu_y\sigma_y + \nu_z\sigma_z))\right]$$
$$+\frac{i}{4}Tr\left[\nu_x\Delta\sigma_z\sigma_x\sigma_z + \nu_y\Delta\sigma_z\sigma_y\sigma_z - \gamma B_1\sigma_z\nu_y\sigma_y\sigma_x - \gamma B_1\nu_z\sigma_z\sigma_z\sigma_x\right],$$

$$(5.140)$$

$$\frac{d\nu_z}{dt} = -\frac{i}{4}Tr\left[\Delta\sigma_z\sigma_z\nu_x\sigma_x + \nu_y\Delta\sigma_z\sigma_z\sigma_y - \gamma B_1\nu_y\sigma_z\sigma_x\sigma_y - \gamma B_1\nu_z\sigma_z\sigma_x\sigma_z\right]$$
$$+\frac{i}{4}Tr\left[\nu_x\Delta\sigma_z\sigma_x\sigma_z + \nu_y\Delta\sigma_z\sigma_y\sigma_z - \gamma B_1\nu_y\sigma_z\sigma_y\sigma_x - \gamma B_1\nu_z\sigma_z\sigma_z\sigma_x\right],$$

$$(5.141)$$

$$\frac{d\nu_z}{dt} = -\frac{i}{4}Tr\left[\Delta\nu_x\sigma_x + \nu_y\Delta\sigma_y - \gamma B_1\nu_y\sigma_z\sigma_x\sigma_y - \gamma B_1\nu_z\sigma_x\right]$$
$$+\frac{i}{4}Tr\left[\nu_x\Delta\sigma_x + \nu_y\Delta\sigma_y - \gamma B_1\nu_y\sigma_z\sigma_y\sigma_x - \gamma B_1\nu_z\sigma_x\right], \quad (5.142)$$

$$\frac{d\nu_z}{dt} = -\frac{i}{4} Tr\left[-\gamma B_1 \nu_y \sigma_z \sigma_x \sigma_y\right] + \frac{i}{4} Tr\left[-\gamma B_1 \nu_y \sigma_z \sigma_y \sigma_x\right] \quad (5.143)$$

leading to

$$\frac{d\nu_z}{dt} = -\frac{i}{4} Tr\left[\sigma_z(-\gamma B_1 \sigma_x \nu_y \sigma_y) - \sigma_z(-\gamma b_1 \nu_y \sigma_y \sigma_x)\right] = -\gamma B_1 \nu_y.$$

$$(5.144)$$

Equations (5.131),(5.138), and (5.144) can be written in the more compact vectorial form (5.122).

- Starting with Equation (5.122), show that, in the frame rotating at the Larmor frequency, the effective magnetic field experienced by the spinor is a constant magnetic field of amplitude B_1 along the x-axis.

Solution

Starting with Equations (5.122-5.123) which describes the time evolution of the generalized Bloch vector in the rotating frame:

$$\vec{B}_{eff} = \left(B_0 + \frac{\omega}{\gamma}\right)\hat{z} + B_1\hat{x}, \quad (5.145)$$

the z-component of the effective magnetic field will be equal to zero when

$$B_0 + \frac{\omega}{\gamma} = 0, \quad (5.146)$$

i.e., when $\omega = \omega_L = -\gamma B_0$ (which is positive since, in our case, γ is negative). In this case the effective magnetic field is a constant magnetic field of intensity B_1 along the $+x$-axis and all the spinors will rotate around the x-axis of the rotating frame with an angular frequency equal to the Larmor frequency corresponding to the intensity B_1. If all the spinors start at the north pole, they will all reach the south pole simultaneously if the effects of relaxation due to interactions with the environment are neglected.

5.8 References

[1] K. Blum, *Density Matrix Theory and Applications*, (Plenum, New York, 1996).

[2] R. P. Feynman, *Statistical Mechanics*, Frontiers in Physics Lecture Note Series, (Addison-Wesley, Redwood City, California, 1972)

[3] C. Kittel, *Introduction to Solid State Physics*, 2nd edition, (John Wiley & Sons, London, 1961).

6

Spin–Orbit Interaction

In the previous chapter, we introduced the concept of the longitudinal and transverse relaxation times T_1 and T_2. One of the major mechanisms that determine T_1 and T_2 for a single spin – or an ensemble of spins – in a solid is *spin-orbit interaction*. It is caused by the coupling of a moving electron's spin with an effective magnetic field due to an electric field in the solid. The electric field could be either microscopic (as in an atom due to the charged nucleus) or macroscopic (due to a global electric field caused by doping in a semiconductor or band structure modulation). Either type will make a moving electron experience an effective magnetic field. The magnetic field will not appear in the laboratory frame, but will appear in the rest frame of the electron due to Lorentz transformation of the electric field.

Spin-orbit interaction is important to understand since it not only affects spin relaxation, but is also at the heart of many spin-based devices that are discussed in Chapter 14. These devices operate by modulating the spin polarizations of charge carriers with an external electric field that controls spin-orbit interaction. A simple way of viewing this is that the electric field causes an effective magnetic field via Lorentz transformation, which, in turn, makes spins precess about it. By controlling the electric field (and hence the resulting magnetic field), one can alter the angular frequency of Larmor precession of spins and therefore the angle through which they precess in a given time. This affords control over the spin precession through an external electric field, which is the basis of many spintronic devices.

There are essentially two types of spin-orbit interaction that we need to discuss. One is *microscopic* or intrinsic (as in an atom) and the other is *macroscopic* or extrinsic (as in a solid). The latter is usually controllable by external agents and forms the basis of many spintronic devices, but we will start by discussing microscopic spin-orbit interaction first.

6.1 Microscopic or intrinsic spin–orbit interaction in an atom

While orbiting around the nucleus, a negatively charged electron in an atom feels the electric field due to the positively charged nucleus. As a result, a magnetic field that did not exist in the laboratory frame will appear in the rest frame of the electron (see Problem 6.1) through a Lorentz transformation. According to Einstein's relativity theory, the flux density associated with this magnetic field is

$$\vec{\mathcal{B}} = \frac{\vec{\mathcal{E}} \times \vec{v}}{c^2 \sqrt{1 - v^2/c^2}}, \tag{6.1}$$

where $\vec{\mathcal{E}}$ is the electric field seen by the electron, $\vec{v}$ is its orbital velocity, and c is the speed of light in vacuum. Note (very important) that the strength of this magnetic field depends on the electron's velocity.

The above equation is actually not entirely correct. Thomas, in a paper published in *Nature* in 1926 [1], had pointed out that the Lorentz transformation that we normally use to connect the electron's rest frame to the laboratory frame is inexact. If there is a component of the electric field in a direction perpendicular to the instantaneus velocity, the electron will be accelerating perpendicular to the velocity. Therefore, it is not enough to transform the laboratory frame to the rest frame using the electron's instantaneous velocity. An observer in the electron's rest frame finds that an additional rotation is required to align her/his coordinate axes with the ones obtained by Lorentz transforming the laboratory frame. If this fact is taken into account, it introduces an additional factor of 2 in the denominator of the expression above*. Thomas' derivation is exceedingly complex and omitted here, but the interested reader is referred to [4] which provides a physically appealing, albeit heuristic, explanation of how the factor of 2 comes about.

The correct expression for the magnetic flux density (in vacuum) is therefore

$$\vec{\mathcal{B}} = \frac{\vec{\mathcal{E}} \times \vec{v}}{2c^2 \sqrt{1 - v^2/c^2}}. \tag{6.2}$$

If we write the magnetic moment of the self rotating electron as $\vec{\mu}_e$, then the energy of its interaction with $\vec{\mathcal{B}}$ is (recall the discussion in Chapter 2):

$$E_{rel} = -\vec{\mu}_e \cdot \vec{\mathcal{B}}. \tag{6.3}$$

---

*This is actually only strictly true in vacuum. In an arbitrary material, the correction factor is $2/(g\text{-}1)$ where g is the Landé g-factor. Since the g-factor in vacuum is exactly 2, this factor becomes 2 in vaccum. See J. D. Jackson, Classical Electrodynamics, 3rd. edition, (John Wiley & Sons, New York, 1999).

We can call this interaction "spin-orbit interaction" since $\vec{\mu_e}$ arises from the "spin" and $\vec{B}$ arises from the orbital motion of a negatively charged electron around a positively charged nucleus.

Kronig, who had thought of spin as being associated with the self-rotation of an electron about its own axis (recall Chapter 1), had thought of this idea as well. He understood that the magnetic moment associated with the self-rotation of a charged particle $(\vec{\mu_e})$ will interact with the magnetic field $(\vec{B})$ due to Lorentz transformation of the electric field in an atom. This led to the formulation of spin-orbit interaction.

The ratio of the magnetic moment μ_e (in units of the Bohr magneton μ_B) to the angular momentum of self-rotation (in units of $\hbar$) is defined as the gyromagnetic ratio (or the Landé g-factor) g_0:

$$\frac{magnetic\ moment}{angular\ momentum} = g_0 . \tag{6.4}$$

Therefore,

$$\vec{\mu_e} = g_0 \mu_B \vec{s}, \tag{6.5}$$

since the angular momentum of self-rotation in units of $\hbar$ is equal to $\vec{s}$.

As a result, Equation (6.3) can be rewritten as

$$E_{rel} = -g_0 \mu_B \vec{s} \cdot \vec{B}. \tag{6.6}$$

Using Equation (6.2), the above equation can be re-cast as

$$E_{rel} = -g_0 \mu_B \frac{\vec{E} \times \vec{v}}{2c^2 \sqrt{1 - v^2/c^2}} \cdot \vec{s}, \tag{6.7}$$

.

In order to derive the quantum-mechanical operator for spin-orbit interaction, we simply have to replace $\vec{s}$ with the corresponding operator $(1/2)\vec{\sigma}$ to get

$$H_{so} = -\frac{g_0}{2} \frac{e\hbar}{2m} \frac{\vec{E} \times \vec{v}}{2c^2 \sqrt{1 - v^2/c^2}} \cdot \vec{\sigma}, \tag{6.8}$$

where we have used the relation $\mu_B = e\hbar/(2m)$.

Since Dirac had shown that $g_0 = 2$ for a free electron in vacuum, the spin-orbit interaction Hamiltonian for an atom in vacuum can also be written as:

$$H_{so} = \frac{e\hbar}{4m^2c^2 \sqrt{1 - v^2/c^2}} \left(\vec{\nabla} V \times \vec{p}\right) \cdot \vec{\sigma}$$

$$\approx \frac{e\hbar}{4m^2c^2} \left(\vec{\nabla} V \times \vec{p}\right) \cdot \vec{\sigma} [\text{if } v \ll c], \tag{6.9}$$

where the electric field is related to the electric potential V as $\vec{\mathcal{E}} = -\vec{\nabla} V$ and the velocity operator is $\vec{p}/m$ where $\vec{p} = -i\hbar\vec{\nabla}$ is the momentum operator.

As a side discussion, we point out that for a non-relativistic electron, i.e., an electron orbiting around a nucleus with an orbital velocity v_{orbit} much less than the speed of light in vacuum $(v_{orbit}^2/c^2 \ll 1)$, the non-relativistic version of Equation (6.2) can be derived in another way. An observer sitting on the electron and moving with it will think that the electron is at rest and the nucleus is revolving around it with a velocity $-\vec{v}_{orbit}$. To the observer, the radius of the orbit will be $-\vec{r}$ and the nuclear charge will be $+Ze$. Consequently, according to Biot Savart's law, the magnetic flux density at the position of the electron will be given by

$$\vec{B}_0 = Ze\frac{\vec{r} \times \vec{v_{orbit}}}{4\pi\epsilon_0 c^2 r^3}, \tag{6.10}$$

where ϵ_0 is the permittivity of free space.

Since the electric field $\mathcal{E}$ seen by the orbiting electron is the Coulomb field $Ze\vec{r}/(4\pi\epsilon_0 r^3)$, the above equation can be written as

$$B_0 = \frac{\vec{\mathcal{E}} \times \vec{v}}{c^2}, \tag{6.11}$$

which is the same as Equation (6.2) except for the Thomas factor of 2 in the denominator and the relativistic factor $\sqrt{1 - v^2/c^2}$. Therefore, for a non-relativistic electron, the magnetic field appearing at the orbiting electron can be derived from Biot Savart's law without invoking Lorentz transformation. Of course, we will have to then add the Thomas correction.

Note that Equation (6.10) can be written as

$$\vec{B}_0 = Ze\frac{\vec{K}}{4\pi\epsilon_0 mc^2 r^3}, \tag{6.12}$$

where $\vec{K}$ is the orbital angular momentum $(\vec{K} = m\vec{r} \times \vec{v}_{orbital})$ and m is the electron's mass. Since $\vec{K}$ is quantized in units of $\hbar$ $(\vec{K} = \vec{l}\hbar)$, we can rewrite the above equation as

$$\vec{B}_0 = \frac{Ze\vec{l}\hbar}{4\pi\epsilon_0 mc^2 r^3}. \tag{6.13}$$

Substituting the above expression for $\vec{B}$ in Equation (6.6), we get

$$E'_{rel} = -g_0\mu_B\hbar\frac{Ze}{4\pi\epsilon_0 mc^2 r^3}\vec{l} \cdot \vec{s}. \tag{6.14}$$

After we add the Thomas correction, the above equation becomes

$$E'_{rel} = -g_0\mu_B\hbar\frac{Ze}{8\pi\epsilon_0 mc^2 r^3}\vec{l} \cdot \vec{s}. \tag{6.15}$$

Finally, in the Bohr atomic model, the orbital radius of the n-th orbit is given by

$$r_n = \frac{n^2\hbar^2 4\pi\epsilon_0}{mZe^2}. \tag{6.16}$$

Substituting this in the preceding equation, we finally get

$$E'_{rel} = -\frac{g_0}{2} \frac{Z^4 e^8 m}{n^6 \hbar^4 c^2 (4\pi\epsilon_0)^4} \vec{l} \cdot \vec{s}. \tag{6.17}$$

The above energy depends on the scalar product $\vec{l} \cdot \vec{s}$ where $\vec{l}$ is the orbital quantum number and $\vec{s}$ is the spin quantum number. Hence the name "spin-orbit interaction." The extension of this type of coupling to multi-electron atoms was carried out by Russell and Saunders in 1925 [5]. The general form of the spin-orbit interaction energy is still given by Equation (6.7). Note that the intrinsic spin-orbit interaction strength is proportional to the *fourth power* of the atomic number Z. That means electrons in atoms of lighter elements will experience weaker spin-orbit interaction. Organic semiconductors, consisting mainly of hydrocarbons, are made of light elements, and hence have weak intrinsic spin-orbit interaction. As a result, they should have large T_1 and T_2 times. This has led to immense interest in "organic spintronics," which is the field of spin phenomena in organic semiconductors.

6.2 Macroscopic or extrinsic spin–orbit interaction

In a solid, a quasi-free electron does not experience the strong nuclear attraction that it would have experienced in an isolated atom because the nuclear electric field is strongly screened by other electrons. Therefore, the intrinsic (or microscopic) spin-orbit interaction of the type discussed in the preceding sections should be relatively weak in a solid. This is even more true of electrons in the conduction band of direct gap semiconductors. In direct-bandgap semiconductors (e.g., GaAs, InSb, etc.) the lowest conduction band valley is at the Brillouin zone center (Γ-valley) where the electron orbitals are nearly $|S>$-type for which $\vec{l} = 0$. Hence, the $\vec{l} \cdot \vec{s}$ coupling nearly vanishes and the intrinsic spin-orbit interaction will be very weak in the conduction band. The wavefunctions of holes in the valence band are, however, more $|P>$-type orbitals for which $\vec{l} \neq 0$. Hence, there is much stronger intrinsic spin-orbit interaction in the valence band of direct-bandgap semiconductors than in the conduction band.

Electrons in the conduction bands of indirect-bandgap semiconductors like silicon and germanium can experience stronger intrinsic spin-orbit interaction since the conduction band minima are at Brillouin zone edges and not at the zone center. Zone-edge electron orbital wavefunctions are *not* $|S>$-type ($\vec{l} \neq 0$), so $\vec{l} \cdot \vec{s} \neq 0$. Electrons in the conduction band of silicon, however, experience much weaker intrinsic spin-orbit interaction than those in the conduction band of germanium since silicon is the lighter element and the intrinsic spin-orbit interaction strength is proportional to the fourth power of the atomic number.

Although electrons in a solid do not experience the strong localized nuclear electric field owing to screening, they may still see a macroscopic (delocalized) electric field (or potential gradient) due to internal effects. Such an electric field could arise from an internal potential gradient (such as that associated with the conduction band discontinuity in a heterostructure or doping variations), or because of an externally applied electric field. This electric field will cause spin-orbit interaction.

This problem was examined by E. I. Rashba in 1961 [2]. The associated spin-orbit interaction bears his name (Rashba interaction) and is also sometimes referred to as spin-orbit interaction due to structural inversion asymmetry (SIA) since an external or internal electric field that causes this interaction breaks inversion asymmetry in the conduction and valence band profiles. Another type of internal electric field may arise in a solid due to crystallographic inversion asymmetry in a crystal, and that too will cause spin-orbit interaction. This latter problem was examined by G. Dresselhaus in 1954 [3] and the associated spin-orbit interaction bears his name (Dresselhaus interaction). It is sometimes referred to as spin-orbit interaction due to bulk inversion asymmetry (BIA) since it is caused by the lack of crystallographic inversion symmetry through the bulk of the crystal. Here, we will discuss these two effects briefly. They are sometimes called *symmetry-dependent* spin-orbit interactions since they require broken symmetry, while the intrinsic atomic type spin-orbit interaction is called *symmetry-independent* spin-orbit interaction since it does not require broken symmetry.

6.2.1 Rashba interaction

Since any three vectors $\vec{A}$, $\vec{B}$, and $\vec{C}$ obey the relation

$$\left(\vec{A} \times \vec{B}\right) \cdot \vec{C} = -\vec{A} \cdot \left(\vec{C} \times \vec{B}\right) , \tag{6.18}$$

we can write the spin-orbit interaction Hamiltonian in a solid as

$$H_{so}^{solid} = -\frac{g(g-1)e\hbar}{8(m^*)^2 c^2} \vec{\nabla} V \cdot (\vec{\sigma} \times \vec{p}), \tag{6.19}$$

where we have replaced the free electron mass m with the effective mass m^* since we are in a crystal, and omitted the relativistic factor $\sqrt{1 - v^2/c^2}$ since we are dealing with electrons that move with speeds much smaller than the speed of light in vacuum. We have also reintroduced the g-factor of the material.

We can rewrite the last equation compactly as

$$H_{so}^{solid} = \Gamma \left(\vec{\sigma} \times \vec{p}\right) \cdot \left[-\vec{\nabla} V\right], \tag{6.20}$$

where $\Gamma = \frac{g(g-1)e\hbar}{8(m^*)^2 c^2}$.

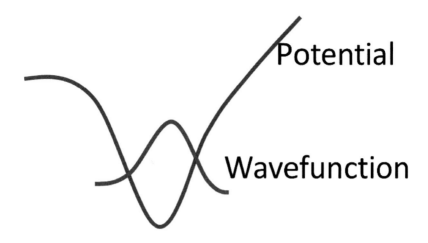

FIGURE 6.1
The confining potential profile and the wavefunction in an eigenstate of a quantum confined system.

Since we are dealing with delocalized electric fields in a solid, it is tempting to define a spin-orbit interaction Hamiltonian that depends on the spatially averaged value of the electric field or the potential gradient $\vec{\nabla}V$. Such a Hamiltonian would be written as

$$H_{so}^{av} = -\Gamma\left(\vec{\sigma}\times\vec{p}\right)\cdot\int d\vec{r}\psi^*(\vec{r})\vec{\nabla}V\psi(\vec{r}) = -\Gamma\left(\vec{\sigma}\times\vec{p}\right)\cdot\langle\vec{\nabla}V\rangle. \qquad (6.21)$$

Let us calculate the quantity $\langle\vec{\nabla}V\rangle$ in a quantum confined system where there is a spatially varying potential, as shown in Fig. 6.1. From Ehrenfest theorem, one can conclude that the time rate of change of the expectation value of an electron's momentum in this potential will obey

$$\frac{d\langle p\rangle}{dt} = \frac{-i}{\hbar}\langle[p, H]\rangle, \qquad (6.22)$$

where p is the momentum operator $(-i\hbar\vec{\nabla})$ and H is the Hamiltonian.

Since the only spatially varying term in the Hamiltonian is the potential V, it is easy to show that the right hand side of the preceding equation is $\langle-\vec{\nabla}V\rangle$. This of course yields Newton's law

$$\frac{d\langle p\rangle}{dt} = \langle-\vec{\nabla}V\rangle. \qquad (6.23)$$

However, in a quantum confined system, the expectation value of the momentum $\langle p\rangle = 0$ at all times since the electron wavefunction is a standing wave. Therefore, the last relation yields that $\langle -\vec{\nabla}V \rangle$, and hence H_{so}^{av} is exactly zero in any quantum confined system! This result is of course incorrect; the spin-orbit inetraction does *not* vanish in a quantum confined system and it exists even in quantum dots where an electron's motion is confined in all three dimensions.

If we neglect bandstructure effects, then the Rashba interaction Hamiltonian can be obtained from Equation (6.19) by direct analogy:

$$H_R = \frac{g(g-1)e\hbar}{8(m^*(\vec{r}))^2 c^2} \vec{\mathcal{E}}(\vec{r}) \cdot (\vec{\sigma} \times \vec{p}) = \vec{\eta}_R(\vec{r}) \cdot (\vec{\sigma} \times \vec{p}), \qquad (6.24)$$

where

$$\vec{\eta}_R(\vec{r}) = \frac{g(g-1)e\hbar}{8(m^*(\vec{r}))^2 c^2} \vec{\mathcal{E}}(\vec{r}). \qquad (6.25)$$

and $\vec{\mathcal{E}}(\vec{r})$ is the local electric field experienced by the electron at location $\vec{r}$.

However, band structure effects will inevitably play a role in a crystalline solid. Accounting for them results in more complex physics and the following expression for $\vec{\eta}_R(\vec{r})$ [6]:

$$\vec{\eta}_R(\vec{r}) = \frac{e\hbar}{m^*(\vec{r})} \frac{\pi \Delta_s (2E_g + \Delta_s)}{E_g(E_g + \Delta_s)(3E_g + 2\Delta_s)} \vec{\mathcal{E}}(\vec{r}), \qquad (6.26)$$

where E_g is the band gap of the semiconductor and Δ_s is the spin-orbit splitting in the valence band. Here, we have also made allowance for the fact that the effective mass of an electron in a heterostructure can vary spatially. There are some subtle issues regarding the spatial variation of $\vec{\eta}_R$ which we will not discuss. They are outside the scope of this textbook.

In the presence of a magnetic field, the momentum operator $\vec{p}$ must be replaced by $\vec{p} + e\vec{A}$, where $\vec{A}$ is the vector potential associated with the magnetic field. In that case, the general expression for the Hamiltonian describing the Rashba interaction in a magnetic field is

$$H_R = \vec{\eta}_R(\vec{r}) \cdot \left[\vec{\sigma} \times (\vec{p} + e\vec{A}) \right]. \qquad (6.27)$$

Another popular form for the Rashba interaction is

$$H_R = \frac{a_{46}}{\hbar} \vec{\mathcal{E}}(\vec{r}) \cdot \left[\vec{\sigma} \times (\vec{p} + e\vec{A}) \right], \qquad (6.28)$$

where a_{46} is a material constant. Obviously, $\frac{a_{46}}{\hbar} \vec{\mathcal{E}}(\vec{r}) = \vec{\eta}_R(\vec{r})$. Therefore,

$$a_{46} = \frac{\pi e \hbar^2}{m^*} \frac{\Delta_s (2E_g + \Delta_s)}{E_g(E_g + \Delta_s)(3E_g + 2\Delta_s)}, \qquad (6.29)$$

where E_g is the bandgap and Δ_s is the spin-orbit splitting energy in the valence band. In GaAs, the value of a_{46} has been reported to be of the order of 10^{-38} C-m^2 [7].

Using this type of spin-orbit interaction, Rashba and Bychkov [8] were able to explain many features observed in heterostructures such as the combined resonances (electric dipole resonance + cyclotron resonance) observed in two-dimensional electron gases formed at AlGaAs/GaAs hetero-interfaces. These experiments were reported in the early 1980s by Stein [9] and Störmer [10] and suggested the existence of a spin-splitting due to the Rashba spin-orbit interaction. Additionally, the observation of beating patterns in Shubnikov-deHaas oscillations in two-dimensional electron gases formed at InAs/InAlAs hetero-interfaces was explained by the spin-splitting of the lowest subband due to the Rashba interaction [11]. The Rashba interaction plays an extremely important role in spintronics since this interaction can be *tuned* by an external electric field $\vec{\mathcal{E}}$ (see Equation (6.25)). As such, it is the basis of many spintronic devices, both classical such as the Spin Field Effect Transistor [12], and quantum mechanical such as universal spin-based quantum gate [13].

6.2.2 Dresselhaus interaction

The Dresselhaus type of spin-orbit interaction arises due to crystallographic inversion asymmetry in a crystal that results in an effective electric field. This interaction will be absent in a centro-symmetric crystal like silicon or germanium, which do not have crystallographic inversion asymmetry in any direction. That is why the Dresselhaus interaction is sometimes referred to as spin-orbit interaction due to bulk inversion asymmetry (BIA).

Recently, it has been claimed that the Dresselhaus interaction also does not exist in a strictly one-dimensional structure (quantum wire with a single occupied subband) of a zinc-blende semiconductor if the wire axis is in the [100] or [111] crystallographic direction, even if there is inversion asymmetry [14]. This follows from group-theoretic arguments. Here, we will discuss the general principles underlying the Dresselhaus interaction in bulk crystals lacking inversion symmetry.

The Pauli equation describing a single electron in a crystal lattice in the absence of any external magnetic field but in the presence of spin-orbit interaction is

$$\left[\frac{|\vec{p}|^2}{2m_0} + V_{lattice} + \frac{e\hbar}{4m_0^2 c^2} \left(\vec{\nabla} V \times \vec{p} \right) \cdot \vec{\sigma} \right] [\psi] = E[\psi] \,, \qquad (6.30)$$

where $V_{lattice}$ is the (spatially periodic) electrostatic potential energy due to the ionized atoms in the crystal.

Since $V_{lattice}$ is periodic in space, the 2-component wavefunction should have the Bloch form

$$[\psi] = e^{i\vec{k} \cdot \vec{r}} \left[u_{\vec{k}} (\vec{r}) \right] = e^{i\vec{k} \cdot \vec{r}} \begin{bmatrix} v_{\vec{k}} (\vec{r}) \\ w_{\vec{k}} (\vec{r}) \end{bmatrix} \,. \qquad (6.31)$$

Substituting Equation (6.31) in Equation (6.30), we get

$$\left[\frac{|\vec{p}|^2}{2m_0} + V_{lattice} + \frac{e\hbar}{4m_0^2c^2} \left(\vec{\nabla}V \times \vec{p} \right) \cdot \vec{\sigma} \right] \left[u_{\vec{k}}(\vec{r}) \right]$$

$$+\hbar\vec{k} \cdot \left[\frac{\vec{p}}{m_0} + \frac{e\hbar}{4m_0^2c^2} \left(\vec{\sigma} \times \vec{\nabla}V \right) \right] \left[u_{\vec{k}}(\vec{r}) \right] = \left[E - \frac{\hbar^2 k^2}{2m_0} \right] \left[u_{\vec{k}}(\vec{r}) \right].$$

$$(6.32)$$

The equation for $\vec{k} + \vec{K}$ (where $\vec{K}$ is the reciprocal lattice vector) is

$$\left[\frac{|\vec{p}|^2}{2m_0} + V_{lattice} + \frac{e\hbar}{4m_0^2c^2} \left(\vec{\nabla}V \times \vec{p} \right) \cdot \vec{\sigma} \right] \left[u_{\vec{k}+\vec{K}}(\vec{r}) \right]$$

$$+\hbar\vec{k} \cdot \left[\frac{\vec{p}}{m_0} + \frac{e\hbar}{4m_0^2c^2} \left(\vec{\sigma} \times \vec{\nabla}V \right) \right] \left[u_{\vec{k}+\vec{K}}(\vec{r}) \right]$$

$$+\hbar\vec{K} \cdot \left[\frac{\vec{p}}{m_0} + \frac{e\hbar}{4m_0^2c^2} \left(\vec{\sigma} \times \vec{\nabla}V \right) \right] \left[u_{\vec{k}+\vec{K}}(\vec{r}) \right]$$

$$= \left[E - \frac{\hbar^2 \left(\vec{k} + \vec{K} \right)^2}{2m_0} \right] \left[u_{\vec{k}+\vec{K}}(\vec{r}) \right].$$

$$(6.33)$$

Treating the term $\hbar\vec{K} \cdot \left[\frac{\vec{p}}{m_0} + \frac{e\hbar}{4m_0^2c^2} \left(\vec{\sigma} \times \vec{\nabla}V \right) \right]$ as a perturbation and applying group theoretic results, Dresselhaus was able to derive the spin-orbit interaction Hamiltonian and spin-splitting energies in principal crystallographic directions (e.g., [100], [111], etc.) and show that they are non-zero if these crystallographic directions do not possess reflection symmetry [3]. Dresselhaus' complete derivation is too complex and involved to reproduce here and the reader is referred to [3] for more details.

For the [100] crystallographic direction, the Dresselhaus spin-orbit interaction Hamiltonian becomes

$$H_D = a_{42}\sigma \cdot \vec{\kappa} , \qquad (6.34)$$

where

$$\kappa_x = \frac{1}{2\hbar^3} \left[p_x \left(p_y^2 - p_z^2 \right) + \left(p_y^2 - p_z^2 \right) p_x \right]$$

$$\kappa_y = \frac{1}{2\hbar^3} \left[p_y \left(p_z^2 - p_x^2 \right) + \left(p_z^2 - p_x^2 \right) p_y \right]$$

$$\kappa_z = \frac{1}{2\hbar^3} \left[p_z \left(p_x^2 - p_y^2 \right) + \left(p_x^2 - p_y^2 \right) p_z \right] , \qquad (6.35)$$

and a_{42} is a material constant. If there is a magnetic field with an associated vector potential $\vec{A}$, then as usual the momentum operator in the above equation should be transformed according to (see Chapter 18)

$$\vec{p} \rightarrow \vec{p} + e\vec{A}. \qquad (6.36)$$

Looking back at Equation (6.19), we see that the spin-orbit interaction Hamiltonian involves the gradient of the potential. In a crystal with no structural or bulk inversion asymmetry, the gradient need not vanish *locally*, even if it vanishes globally. This is reminiscent of the microscopic spin-orbit interaction. Recently, it has been shown that the global symmmetry (the bulk space group) is not important, but it is the local symmetry (the site point group) that matters. As a result, even centrosymmetric crystals with no inversion asymmetry can have a Dresselhaus effect if the site point group is non-centrosymmetric [15]. This is apparently the case with hexagonal boron nitride [16]. Similarly, the Rashba effect can exist in a crystal with no external electric field to break the global structural inversion symmetry, as is the case with BiTeI [17]. These effects are only now coming to light and challenge age-old wisdom about spin-orbit interactions in solids.

6.3 Problems

- **Problem 6.1**

  Consider an electron orbiting a proton in a circular loop. Show, using Biot-Savart's law, that the orbiting electron experiences a magnetic flux density of $\vec{B} = \frac{1}{2}\frac{\vec{E}\times\vec{v}}{c^2}$, where we have inserted the Thomas correction of a factor of 2 in the denominator.

  Solution

  If we sit on the electron, the proton will appear to rotate around us and create a circular loop of current which is given by

  $$I = \frac{e}{T} = \frac{ev}{2\pi R},\tag{6.37}$$

  where v is the velocity of the proton moving on the circular orbit, R is the radius of the circular orbit, and T is the period of rotation, which is identical to the period of rotation of the electron around the heavy proton in the rest frame of the latter.

  According to Biot-Savart's law, the magnetic flux density at a distance x along a line passing through the center of the proton orbit and perpendicular to the plane of the orbit is given by

  $$B = \frac{\mu_0 I R^2}{2\sqrt{R^2 + x^2}^{3/2}}.\tag{6.38}$$

  So, in the plane of the orbit, where $x = 0$, we have

  $$B = \frac{\mu_0 I}{2R} = \frac{\mu_0 e}{2RT} = \frac{\mu_0 ev}{4\pi R^2}.\tag{6.39}$$

Furthermore, since $c^2 = \frac{1}{\mu_0 \epsilon_0}$, we get

$$B = \frac{Ev}{c^2}, \tag{6.40}$$

which, in the rest frame of the electron, can be written in a vectorial form as

$$\vec{B} = \frac{1}{2} \frac{\vec{E} \times \vec{v}}{c^2}, \tag{6.41}$$

where we have inserted the Thomas "factor of 2 correction" by hand.

- **Problem 6.2**

Consider a two-dimensional electron gas in the $(x - z)$ plane and a symmetry breaking electric field $\mathcal{E}_y$ in the y-direction, which induces Rashba spin-orbit interaction. Show that the Rashba Hamiltonian is

$$[H_R] = \frac{a_{46}}{\hbar} \mathcal{E}_y \left(p_x \sigma_z - p_z \sigma_x \right). \tag{6.42}$$

Assume now that there is a magnetic flux density B in the y-direction as well. In that case, use the Landau gauge $\vec{A} = (Bz, 0, 0)$ to show that the Rashba Hamiltonian is

$$[H_R] = \frac{a_{46}}{\hbar} \mathcal{E}_y \left[(p_x + eBz)\sigma_z - p_z \sigma_x \right]$$

$$= \frac{a_{46}}{\hbar} \mathcal{E}_y \begin{bmatrix} p_x + eBz & -p_z \\ -p_z & -p_x - eBz \end{bmatrix}. \tag{6.43}$$

- **Problem 6.3**

Derive Equations (6.32) and (6.33). Hint: $\vec{p} = -i\hbar\nabla$ and $\nabla \left[e^{i\vec{k}\cdot\vec{r}} u_{\vec{k}}(\vec{r}) \right] = i\vec{k} \left[e^{i\vec{k}\cdot\vec{r}} u_{\vec{k}}(\vec{r}) \right] + e^{i\vec{k}\cdot\vec{r}} \nabla u_{\vec{k}}(\vec{r})$.

- **Problem 6.4**

Silicon and germanium are centro-symmetric crystals that have no inverssion asymmetry and hence should not have Dresselhaus interaction. In the absence of internal electric fields, they should not have Rashba interaction either. Thus, spin-orbit interaction experienced by electrons in these materials will be very weak and the T_1, T_2 times should be very long. Is this true?

Solution

Not really. First of all, spin-orbit interaction is not the only cause for electron spin relaxation and there are other channels for spin relaxation as well. However, even if they were absent, electrons in the conduction bands of Si and Ge can experience intrinsic spin-orbit interaction. The lowest conduction band valley is at the X-point in silicon and L-point in germanium. Since the electron wavefunctions are not $|S>$-type orbitals in the conduction band, there is non-zero $\vec{l} \cdot \vec{s}$ coupling in the conduction bands of Si and Ge, resulting in non-vanishing intrinsic spin-orbit interaction. Fortunately for Si, it is a relatively light element and hence the intrinsic spin-orbit interaction (which is proportional to the fourth power of the atomic number) is small; however, Ge is a much heavier element and it has quite strong intrinsic spin-orbit interaction in the conduction band.

- **Problem 6.5**

Standard Lorentz transformation predicts that an electron moving with a velocity $\vec{v}$ in an electric field $\vec{E}$ will see a magnetic flux density $\vec{B}$ in its own reference frame given by

$$\vec{B} = \frac{\vec{E} \times \vec{v}/c^2}{\sqrt{1 - (v/c)^2}} \approx \frac{\vec{E} \times \vec{v}}{c^2} \quad (v \ll c). \tag{6.44}$$

This does not quite apply to the case of a curved trajectory where the magnetic field will be roughly one-half of what is given by the above equation, as shown by Thomas. Prove this result semi-empirically by forcing the electron to move along a straight line (when standard Lorentz transformation will be valid) by adding a magnetic field $\vec{B}'$ in the rest frame so that the resulting Lorentz force will balance the force due to the electric field, i.e., $\vec{E} = -\vec{v} \times \vec{B}'$. Choose this magnetic field to be perpendicular to the velocity.

Solution

The solution is taken from [4].

The electric field $\vec{E}$, the added magnetic flux density $\vec{B}'$, and the velocity $\vec{v}$ are all mutually perpendicular. Without loss of generality, we can choose the velocity to be in the x-direction, the electric field in the y-direction and the added magnetic flux density in the z-direction in Cartesian coordinates. Clearly, this results in

$$E_y = v_x B'_z. \tag{6.45}$$

If the combination of the added magnetic field and the electric field are Lorentz transformed into the (non-rotating) frame of the electron, it results in a magnetic flux density

$$B_z = \frac{B'_z - E_y v_x/c^2}{\sqrt{1 - (v_x/c)^2}}. \tag{6.46}$$

Expanding the above result in a binomial series, we get

$$B_z = B'_z + B_z \left(\frac{v_x}{c}\right)^2 \left[\frac{1}{2} + \frac{3}{8}\left(\frac{v_x}{c}\right)^2\right] - E_y \frac{v_x}{c^2}\left[1 + \frac{1}{2}\frac{v_x}{c^2}\right] + \cdots. \tag{6.47}$$

Using the first equation in the third, we immediately get

$$B_z = B'_z - E_y \frac{v_x}{c^2} \left[\frac{1}{2} + \frac{1}{8} \left(\frac{v_x}{c} \right)^2 \right] + \cdots. \tag{6.48}$$

In the limit $v_x \ll c$, we obtain

$$B_z = B'_z - \frac{1}{2} E_y \frac{v_x}{c^2}. \tag{6.49}$$

Generalizing to the vector form, the preceding result can be written as

$$\vec{B} = \vec{B}' + \frac{1}{2} \frac{\vec{E} \times \vec{v}}{c^2}. \tag{6.50}$$

The above magnetic flux density is the one experienced by the moving electron. If we then remove the added flux density needed to convert the rotating frame to a non-rotating frame, what remains is the flux density experienced in a rotating frame. Therefore, we get

$$B_{\text{rot}} = \frac{1}{2} \frac{\vec{E} \times \vec{v}}{c^2}, \tag{6.51}$$

where the Thomas's "factor of 2" correction has appeared.

6.4   References

[1] L. H. Thomas, "The motion of the spinning electron", Nature, **117**, 514 (1926). For a slightly easier derivation, see H. Kroemer, "The Thomas precession factor in spin orbit interaction", Am. J. Phys., **72**, 51 (2004) and J. A. Rhodes and M. D. Semon, "Relativistic velocity space, Wigner rotation, and Thomas precession", Am. J. Phys., **72**, 943 (2004).

[2] E. I. Rashba and V. I. Sheka, "Combinational resonance of zonal electrons in crystals having a zinc-blende lattice", Sov. Phys. Solid State, **3**, 1257 (1961); "Combined resonance of electrons in InSb", ibid **3**, 1357 (1961).

[3] G. Dresselhaus, "Spin orbit coupling effects in zinc blende structures", Phys. Rev., **100**, 580 (1955).

[4] H. Kroemer, "The Thomas precession factor in spin-orbit interaction", Am. J. Phys., **72**, 51 (2004).

[5] H. N. Russell and F. A. Saunders, "New regularities in the spectra of the alkaline earths", Astrophysical Journal, **61**, 38 (1925).

[6] F. G. Pikus and G. E. Pikus, "Conduction band spin splitting and negative magnetoresistance in A_3B_5 heterostructures", Phys. Rev. B, **51**, 16928 (1995).

[7] A. Bournel, P. Dollfus, S. Galdin, F. X. Musalem and P. Hesto, "Modelling of gate induced spin precession in a striped channel high electron mobility transistor", Solid St. Commun., **104**, 85 (1997).

[8] Yu A. Bychkov and E. I. Rashba, "Oscillatory effects and the magnetic susceptibility of carriers in inversion layers", J. Phys. C, **17**, 6039 (1984).

[9] D. Stein, K. v. Klitzing and G. Weimann, "Electron spin resonance on $GaAs-Al_xGa_{1-x}As$ heterostructures", Phys. Rev. Lett., **51**, 130 (1983).

[10] H. L. Störmer, Z. Schlesinger, A. Chang, D. C. Tsui, A. C. Gossard and W. Weigmann, "Energy structure and quantized Hall effect of two dimensional holes", Phys. Rev. Lett., **51**, 126 (1983).

[11] B. Das, D. C. Miller, S. Datta, R. Reifenberger, W. P. Hong, P. K. Bhattacharyya, J. Singh and M. Jaffe, "Zero field spin splitting in a two-dimensional electron gas", Phys. Rev. B, **39**, 1411 (1989).

[12] S. Datta and B. Das, "Electronic analog of the electro-optic modulator", Appl. Phys. Lett., **56**, 665 (1990).

[13] S. Bandyopadhyay, "Self assembled nanoelectronic quantum computer based on the Rashba effect in quantum dots", Phys. Rev. B, **61**, 13813 (2000).

[14] J-W Luo, L. Zhang and A. Zunger, "Absence of intrinsic spin splitting in one-dimensional quantum wires of tetrahedral semiconductors", Phys. Rev. B., **84**, 121303(R), (2011).

[15] X. Zhang, Q. Liu, J-W Luo, A. J. Freeman and A. Zunger, "Hidden spin polarization in inversion symmetric bulk crystals" Nature Phys., **10**, 387 (2014).

[16] B. Partoens, "Hide and seek", Nature Phys., **10**, 333 (2014).

[17] K. Ishizaka, et al., "Giant Rashba type spin splitting in bulk BiTeI", Nature Mater., **10**, 521 (2011).

7

---

*Magneto-Electric Subbands in Quantum
Confined Structures in the Presence of
Spin–Orbit Interaction*

When an electron is confined in a structure of restricted dimensionality (such as a quantum well or a quantum wire) and is subjected to a magnetic field, the electron experiences both *electrostatic confinement* due to structure and *magnetostatic confinement* due to the magnetic field. As a result, the allowed electronic states are *hybrid magneto-electric states* which form *magneto-electric subbands.* In this chapter, we will derive the dispersion relations (energy versus wavevector) of electrons in magneto-electric subbands in the presence of *spin-orbit interaction.* The two types of spin-orbit interactions that we will consider are the common ones in a solid, the Rashba and the Dresselhaus interactions.

7.1 Dispersion relations of spin resolved magneto-electric subbands and eigenspinors in a two-dimensional electron gas in the presence of spin–orbit interaction

Consider a two-dimensional electron gas (2-DEG) such as the one encountered at the interface of a heterostructure or a quantum well. Figure 7.1 shows such a system. We will assume that there is a symmetry-breaking electric field $\mathcal{E}_y$ along the y-axis that induces a Rashba spin-orbit interaction. This electric field could come about because of the conduction band discontinuity at the hetero-interface between two different materials, or could be applied from outside by attaching a gate electrode and connecting it to the floating terminal of a grounded voltage source (see the discussion of the Spin Field Effect Transistor in Chapter 14). In addition to the Rashba interaction, there is also the Dresselhaus interaction accruing from bulk (or crystallographic) inversion asymmetry. Finally, there is also an external magnetic field inducing the Zeeman interaction that we encountered in Chapter 2. Our intention is to derive the dispersion relation (relation between electron kinetic energy and wavevector) in this two-dimensional system. We will also derive the

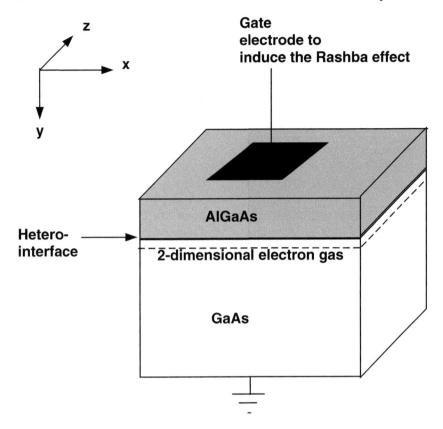

FIGURE 7.1

A heterostructure containing a two-dimensional electron gas.

eigenspinors (or spin eigenstates) at any given wavevector in a given subband. Since the boundaries of the quantum confined structure restrict motion in the direction perpendicular to the boundary, and the external magnetic field also constrains the electron's motion by forcing it into closed Landau orbits, the electron experiences both electrostatic and magnetostatic confinement. This discretizes the allowed energies, breaking it up into discrete subbands. They are called *magneto-electric subbands* since they are due to the combined effect of electrostatic and magnetostatic confinements.

In this chapter, we will find the energy dispersion relations (electron energy E as a function of electron wavevector k) in the magneto-electric subbands when Rashba and Dresselhaus spin-orbit interactions are present. We will find that generally the eigenspinors (and therefore the spin orientation) will be wavevector dependent. We will find the eigenspinors as a function of the wavevector in different subbands.

Because of the wavevector dependence of spin orientation, whenever an electron changes momentum (or wavevector) owing to scattering, its spin orientation also changes. Consequently, momentum relaxation simultaneously causes spin relaxation. This is the basis of a spin relaxation mechanism discussed in [1] which arises in the presence of both spin-orbit interaction and an external magnetic field. It is similar to the Elliott-Yafet mechanism of spin relaxation discussed in Chapter 8.

Consider the quantum well containing a two-dimensional electron gas (2-DEG) as shown in Fig. 7.1. Assume that both Rashba and Dresselhaus spin-orbit interactions are present. The Rashba interaction is caused solely by a structural symmetry-breaking electric field directed perpendicular to the plane of the 2-DEG. Additionally, there is an external magnetic field lying either in the plane of the 2-DEG, or directed perpendicular to it.

The Rashba spin-orbit interaction has the Hamiltonian

$$H_R = \vec{\eta}_R(\mathcal{E}) \cdot \left[\vec{\sigma} \times (\vec{p} + e\vec{A}) \right], \tag{7.1}$$

where $\vec{A}$ is the vector potential associated with the magnetic field, $\vec{\eta}_R(\mathcal{E})$ is a vector whose magnitude is the strength of the Rashba interaction. This strength depends on the magnitude of the symmetry breaking electric field $\vec{\mathcal{E}}$ and the direction of $\vec{\eta}_R(\mathcal{E})$ is along $\vec{\mathcal{E}}$. Finally, $\vec{\sigma}$ is the Pauli spin matrix and $\vec{p}$ is the momentum operator.

The Hamiltonian for the Dresselhaus interaction can be complicated and depends on the crystallographic orientation of the plane, but for the (100) plane, it has the form

$$H_D^{100} = a_{42}\vec{\sigma} \cdot \vec{\kappa}_{100}, \tag{7.2}$$

where $a_{42} = \nu_D$ is a material constant, $\vec{\kappa}_{100} = \kappa_x\hat{x} + \kappa_y\hat{y} + \kappa_z\hat{z}$, and

$$\begin{aligned}
\kappa_x &= \frac{1}{2\hbar^3} \big[(p_x + eA_x)\left\{ (p_y + eA_y)^2 - (p_z + eA_z)^2 \right\} \\
&\quad + \left\{ (p_y + eA_y)^2 - (p_z + eA_z)^2 \right\} (p_x + eA_x) \big] \\
\kappa_y &= \frac{1}{2\hbar^3} \big[(p_y + eA_y)\left\{ (p_z + eA_z)^2 - (p_x + eA_x)^2 \right\} \\
&\quad + \left\{ (p_z + eA_z)^2 - (p_x + eA_x)^2 \right\} (p_y + eA_y) \big] \\
\kappa_z &= \frac{1}{2\hbar^3} \big[(p_z + eA_z)\left\{ (p_x + eA_x)^2 - (p_y + eA_y)^2 \right\} \\
&\quad + \left\{ (p_x + eA_x)^2 - (p_y + eA_y)^2 \right\} (p_z + eA_z) \big].
\end{aligned} \tag{7.3}$$

Note that by using the symmetric combination for the κ-s, we ensure that the Hamiltonian H_D is Hermitian. Note also that the operators $(p_i + eA_i)[(p_j + eA_j)^2 - (p_k + eA_k)^2]$ and $[(p_j + eA_j)^2 - (p_k + eA_k)^2](p_i + eA_i)$ do not necessarily commute; hence, it is imperative to use the symmetric combination to generate a Hermitian operator.

The expression for the Dresselhaus interaction in a two-dimensional electron gas in the (110) plane is [2]

$$H_D^{110} = b_{42}\vec{\sigma} \cdot \vec{\kappa}_{110}, \tag{7.4}$$

where b_{42} is a material constant, $\kappa_{110} = \kappa'_x \hat{x} + \kappa'_y \hat{y} + \kappa'_z \hat{z}$, the x-, y- and z-directions correspond to $[\bar{1}10]$, $[001]$ and $[110]$ crystallographic directions, and

$$\kappa'_x = -\frac{1}{4\hbar^3} \left[(p_z + eA_z) \left\{ (p_x + eA_x)^2 - (p_z + eA_z)^2 + 2(p_y + eA_y)^2 \right\} \right.$$
$$\left. + \left\{ (p_x + eA_x)^2 - (p_z + eA_z)^2 + 2(p_y + eA_y)^2 \right\} (p_z + eA_z) \right]$$
$$\kappa'_y = \frac{2}{\hbar^3} (p_x + eA_z)(p_y + eA_z)(p_z + eA_z)$$
$$\kappa'_z = \frac{1}{4\hbar^3} \left[(p_x + eA_x) \left\{ (p_x + eA_x)^2 - (p_z + eA_z)^2 - 2(p_y + eA_y)^2 \right\} \right.$$
$$\left. + \left\{ (p_x + eA_x)^2 - (p_z + eA_z)^2 - 2(p_y + eA_y)^2 \right\} (p_x + eA_x) \right]. \tag{7.5}$$

The expression for the Dresselhaus interaction in a two dimensional electron gas in the (111) plane is

$$H_D^{111} = b_{42}\vec{\sigma} \cdot \vec{\kappa}_{111}, \tag{7.6}$$

where $\kappa_{111} = \kappa''_x \hat{x} + \kappa''_y \hat{y} + \kappa''_z \hat{z}$, the x-, y- and z-directions correspond to the $[11\bar{2}]$, $[\bar{1}10]$ and $[111]$ crystallographic directions, and

$$\kappa''_x = \frac{1}{\hbar^3} \left[-\sqrt{2/3}(p_x + eA_x)(p_y + eA_y)(p_z + eA_z) - 1/(2\sqrt{3})(p_y + eA_y)^3 \right.$$
$$-1/(4\sqrt{3}) \left\{ (p_y + eA_y)(p_x + eA_x)^2 + (p_x + eA_x)^2(p_y + eA_y) \right\}$$
$$+1/(\sqrt{3}) \left\{ (p_y + eA_y)(p_z + eA_z)^2 + (p_z + eA_z)^2(p_y + eA_y) \right\}$$
$$\left. -\sqrt{2}/6 \left\{ (p_y + eA_y)^2(p_z + eA_z) + (p_z + eA_z)(p_y + eA_y)^2 \right\} \right]$$
$$\kappa''_y = \frac{1}{\hbar^3} \left[1/(4\sqrt{3}) \left\{ (p_x + eA_x)(p_y + eA_y)^2 + (p_y + eA_y)^2(p_x + eA_x) \right\} \right.$$
$$+1/(2\sqrt{3})(p_x + eA_x)^3$$
$$-1/(2\sqrt{6}) \left\{ (p_x + eA_x)^2(p_z + eA_z) + (p_z + eA_z)(p_x + eA_x)^2 \right\}$$
$$-1/(2\sqrt{6}) \left\{ (p_z + eA_z)[(p_x + eA_x)^2 + (p_y + eA_y)^2] \right.$$
$$\left. + [(p_x + eA_x)^2 + (p_y + eA_y)^2](p_z + eA_z) \right\} \right]$$
$$\kappa''_z = \frac{1}{\hbar^3} \left[\sqrt{3}/(2\sqrt{2}) \left\{ (p_x + eA_x)^2(p_y + eA_y) + (p_y + eA_y)(p_x + eA_x)^2 \right\} \right.$$
$$-1/\sqrt{6}(p_y + eA_y)^3$$
$$\left. +(1/3) \left\{ (p_z + eA_z)(p_y + eA_y)^2 + (p_y + eA_y)^2(p_z + eA_z) \right\} \right]. \tag{7.7}$$

Since the Dresselhaus interaction is so much more complicated for the non-(100) plane than the (100) plane, in this textbook, we will always assume the two-dimensional electron gas to reside in the (100) plane. Similarly, when we discuss quantum wires, we will assume that the axis of the wire is in the [100] direction, which will simplify the Dresselhaus interaction term.

The general Hamiltonian describing the 2-DEG system in a magnetic field of flux density $\vec{B}$ is

$$H_{2-DEG} = \frac{\left|\vec{p} + e\vec{A}\right|^2}{2m^*} + V(y), + H_Z + H_R + H_D \qquad (7.8)$$

where the electrons are confined along the y-direction by the electrostatic confining potential $V(y)$, H_Z is the Zeeman interaction Hamiltonian and $\vec{A}$ is the vector potential associated with the magnetic field such that

$$\vec{\nabla} \times \vec{A} = \vec{B}. \qquad (7.9)$$

7.1.1   Magnetic field in the plane of the 2-DEG

We are now ready to derive the energy dispersion $(E - k)$ relation for the 2-DEG. First consider the situation when there is a magnetic field in the plane of the 2-DEG. In this case, the magnetic flux density is given by $\vec{B} = B_x \hat{x} + B_z \hat{z}$ since the 2-DEG is in the $(x - z)$ plane.

We choose the gauge $\vec{A} = -B_z y \hat{x} + B_x y \hat{z}$. If the 2-DEG has vanishing thickness, then the spatial average $< p_y^2 > >> < p_x^2 >, < p_z^2 >$, and the effective mass Hamiltonian describing the 2-DEG can be written in the form

$$H^{||}_{2-DEG} \approx \frac{1}{2m^*} \left[(p_x - eB_z y)^2 + p_y^2 + (p_z + eB_x y)^2 \right] + V(y)$$
$$- \frac{g}{2} \mu_B \left[B_x \sigma_x + B_z \sigma_z \right] - \frac{\eta}{\hbar} \left[(p_x - eB_z y) \sigma_z - (p_z + eB_x y) \sigma_x \right]$$
$$\frac{\nu_D}{\hbar} \frac{p_y^2}{\hbar^2} \left[(p_x - eB_z y) \sigma_x - (p_z + eB_x y) \sigma_z \right], \qquad (7.10)$$

where $V(y)$ is the confining potential in the y-direction (including the effect of the y-directed electric field causing the Rashba interaction) and $\eta = \hbar |\vec{\eta}_R(\mathcal{E})|$. We have assumed the crystallographic plane to be [100]. The quantity η is related to the material constant a_{46} as $\eta = -a_{46}|\mathcal{E}|$ where $\mathcal{E}$ is the symmetry-breaking electric field inducing the Rashba effect. In InAs heterostructures with built-in electric fields of the order of 100 kV/cm, η is of the order of $10^{-12} - 10^{-11}$eV-m.

The terms in the second and third line of the last equation are spin-dependent terms since they involve the Pauli spin matrix. Note that the first of these spin-dependent terms is the Zeeman interaction, the second is the Rashba interaction, and the last is the Dresselhaus interaction.

To find the energy dispersion relations in this 2-DEG and the 2-component wavefunction for the spin eigenstates, we have to solve the non-relativistic

Pauli equation (recall Chapter 2):

$$\left[H^{||}_{2-DEG}\right]\left[\psi^{||}_{2-DEG}(x,y,z)\right] = E\left[\psi^{||}_{2-DEG}(x,y,z)\right], \qquad (7.11)$$

where $\left[H^{||}_{2-DEG}\right]$ is a 2×2 matrix and $\left[\psi^{||}_{2-DEG}(x,y,z)\right]$ is the 2-component wavefunction which is a 2×1 column vector. The superscript $||$ reminds the reader that there is a magnetic field parallel to the plane of the 2-DEG.

Since the Hamiltonian is invariant in x and z, the wavevectors k_x and k_z are good quantum numbers and we can write the 2-component wavefunction as

$$\left[\psi^{||}_{2-DEG}(x,y,z)\right] = \frac{1}{\sqrt{L_x L_z}}e^{ik_x x}e^{ik_z z}[\lambda(y)], \qquad (7.12)$$

where L_x and L_z are normalizing lengths ($\frac{1}{L_x L_y}\int\int dx dz e^{-ik_x x}e^{-ik_z z}e^{ik_x x}e^{ik_z z} = \frac{1}{L_x L_y}\int\int dx dz = 1$), and $[\lambda(y)]$ is a 2×1 column vector representing the spinor in the Pauli equation. It is very important to note that in this case, the spinor depends only on the y-coordinate and hence the spin polarization in the 2-DEG can vary only along the thickness (y-direction), but not in the plane of the 2-DEG.

The Pauli equation yields

$$\left[H^{||}_{2-DEG}\right]\frac{1}{\sqrt{L_x L_z}}e^{ik_x x}e^{ik_z z} = E\frac{1}{\sqrt{L_x L_z}}e^{ik_x x}e^{ik_z z}. \qquad (7.13)$$

Integrating the above over x- and z-coordinates, after multiplying with the complex conjugates of the x- and z-components of the wavefunction, yields

$$\frac{1}{L_x L_y}\int\int dx dz e^{-ik_x x}e^{-ik_z z}\left[H^{||}_{2-DEG}\right]e^{ik_x x}e^{ik_z z}[\lambda(y)]$$

$$= E\frac{1}{L_x L_y}\int\int dx dz e^{-ik_x x}e^{-ik_z z}e^{ik_x x}e^{ik_z z}[\lambda(y)]$$

$$= E[\lambda(y)]. \quad (7.14)$$

We have to evaluate the double integral on the left hand side of the above equation using Equation (7.10) to replace $\left[H^{||}_{2-DEG}\right]$. Once we have done that, the last equation will reduce to

$$\left\{ \frac{\hbar^2 k_x^2}{2m^*} + \frac{\hbar^2 k_z^2}{2m^*} - \frac{\hbar^2}{2m^*} \frac{\partial^2}{\partial y^2} + V(y) \right.$$

$$-eB_z y \frac{\hbar k_x}{m^*} + eB_x y \frac{\hbar k_z}{m^*} + e^2 y^2 (B_x^2 + B_z^2)/(2m^*)$$

$$-\frac{g}{2} \mu_B [B_x \sigma_x + B_z \sigma_z]$$

$$-\frac{\eta}{\hbar} [(\hbar k_x - eB_z y) \sigma_z - (\hbar k_z + eB_x y) \sigma_x]$$

$$\left. +\frac{\nu_D}{\hbar} \frac{p_y^2}{\hbar^2} [(\hbar k_x - eB_z y) \sigma_x - (\hbar k_z + eB_x y) \sigma_z] \} [\lambda(y)] \right\} = E[\lambda(y)], \ (7.15)$$

and the boundary conditions are

$$[\lambda(y)](y = d) = [\lambda(y)](y = -d) = [0], \tag{7.16}$$

where the width of the well in the y-direction is assumed to be $2d$, and $[0]$ is the null vector. Here we have assumed a rectangular potential well of infinite barrier for the sake of mathematical simplicity. For a triangular potential well that is usually found at the interface of two semiconductors of different bandgaps, as in Fig. 7.1, the boundary conditions will be, of course, different. If the barriers are finite, then the problem is solved by enforcing the continuity of the wavefunction and its first derivative across the interface between the well and the barrier materials.

In order to find the dispersion relation, we will have to find the values of the wavevector k_x that satisfy the above two equations for a given value of k_z and E. We then repeat this for various values of E and the same value of k_z to find the dispersion relation (E versus k_x) for a fixed value of k_z. This procedure can be continued to find E versus k_x for different values of k_z, which will finally yield E versus $\vec{k}_t = k_x \hat{x} + k_z \hat{z}$.

Unfortunately, this is not straightforward since Equation (7.15) is not an eigenequation in k_x because it is non-linear in k_x. We therefore have to convert Equation (7.15) to an eigenequation in k_x using the following procedure.

Let

$$[\zeta(y)] = k_x [\lambda(y)]. \tag{7.17}$$

Equation (7.15) can be written as

$$\{ \mathbf{B} k_x + \mathbf{C} \} [\lambda(y)] = \mathbf{A} k_x^2 [\lambda(y)], \tag{7.18}$$

where the 2×2 matrices $\mathbf{A}$, $\mathbf{B}$, and $\mathbf{C}$ are given by

$$\mathbf{A} = -\frac{\hbar^2}{2m^*} \mathbf{I}, \tag{7.19}$$

$$\mathbf{B} = -eB_z y \frac{\hbar}{m^*} \mathbf{I} - \eta[\sigma_{\mathbf{z}}] - \nu[\sigma_{\mathbf{x}}], \tag{7.20}$$

$$\mathbf{C} = \frac{\hbar^2 k_z^2}{2m^*}\mathbf{I} - \frac{\hbar^2}{2m^*}\frac{\partial^2}{\partial y^2}\mathbf{I} + eB_x y\frac{\hbar k_z}{m^*}\mathbf{I} + V(y)\mathbf{I}$$

$$+\frac{e^2 y^2}{2m^*}\left(B_x^2 + B_z^2\right)\mathbf{I} - \frac{g}{2}\mu_B\left\{B_x[\sigma_\mathbf{x}] + B_z[\sigma_\mathbf{z}]\right\} - E\mathbf{I}$$

$$+\frac{eB_z y}{\hbar}\left[\eta[\sigma_\mathbf{z}] + \nu_D\frac{\partial^2}{\partial y^2}[\sigma_\mathbf{x}]\right]$$

$$+\left(k_z + \frac{eB_x y}{\hbar}\right)\left[\eta[\sigma_\mathbf{x}] + \nu_D\frac{\partial^2}{\partial y^2}[\sigma_\mathbf{z}]\right]. \tag{7.21}$$

Using the transformation in Equation (7.17), Equation (7.18) can be recast as an eigenequation in k_x

$$\left[\begin{matrix} \mathbf{0} & \mathbf{I} \\ \{\mathbf{A}^{-1}\mathbf{C}\} & \{\mathbf{A}^{-1}\mathbf{B}\} \end{matrix}\right]\left[\begin{matrix} \lambda(y) \\ \zeta(y) \end{matrix}\right] = k_x\left[\begin{matrix} \lambda(y) \\ \zeta(y) \end{matrix}\right]. \tag{7.22}$$

To find the eigenvalues k_x for a given E and k_z, we will have to use a numerical solution of Equation (7.22) subject to the boundary condition in Equation (7.16). If we actually carry this out, we will not only find the dispersion relation E_n versus k_x (in the n-th magneto-electric subband) for any given k_z, but we will also find the 2-component wavefunction $[\lambda(y)] = [\lambda_{+,n,k_x,k_z}(y), \lambda_{-,n,k_x,k_z}(y)]^\dagger$ in the n-th magneto-electric subband at wavevectors k_x and k_z. It is entirely possible that $\lambda_{+,n,k_x,k_z}(y) \neq \lambda_{-,n,k_x,k_z}(y)$, i.e., the spatial parts of the wavefunctions of the two spin eigenstates at given wavevectors k_x, k_z are *different*. In other words, the spatial wavefunction is spin-dependent.

While the above procedure for finding the dispersion relation is exact, it is also painstaking, cumbersome and does not yield an analytical result. In order to obtain an analytical result, we will make the following approximation.

First, we will assume that

$$[\lambda(y)] = \left[\begin{matrix} \lambda_1(y) \\ \lambda_2(y) \end{matrix}\right] = \lambda_0(y)\left[\begin{matrix} \alpha \\ \beta \end{matrix}\right]. \tag{7.23}$$

We call this the "variable separation" approximation, since we have written the eigenspinor as the *product* of a space-dependent part $\lambda_0(y)$ and a spin-dependent part $[\alpha\ \beta]^\dagger$. The latter does not depend on the y-coordinate, which means that we are assuming (heuristically) that the spin orientation of an electron does not depend on the y-coordinate, or conversely, the spatial part of the wavefunction is not spin dependent. A consequence of this approximation is that we can miss some interesting and important physics, such as "spin texturing", whereby the spin orientation (and therefore the spin density) can vary along the y-direction [7, 8]. The second approximation that we make is that the width of the 2-DEG is so narrow that the subbands are well separated in energy. As a result, we can ignore subband mixing. With these two approximations, we are able to obtain an analytical result.

Note that with the variable separation approximation, the 2-component wavefunction can be written as

$$[\psi_{2-DEG}^{||}] = e^{ik_x x} e^{ik_z z} \lambda_0(y) \begin{bmatrix} \alpha \\ \beta \end{bmatrix}. \tag{7.24}$$

Let $< y > = \int_0^\infty \lambda_0^*(y) y \lambda_0(y) dy = y_0$, and without loss of generality, we can assume that $y_0 = 0$. In that case, the spatial average of the above Hamiltonian (spatial average is obtained by averaging over the x-, y- and z-coordinates) is

$$
\begin{aligned}
E_n &\approx \left\langle H_{2-DEG}^{||} \right\rangle = \left\langle H^{||} \right\rangle = \left\langle \psi_{2-DEG}^{||} | H_{2-DEG} | \psi_{2-DEG}^{||} \right\rangle \\
&= \epsilon_n + \frac{\hbar^2 \left(k_x^2 + k_z^2 \right)}{2m^*} - \frac{g}{2} \mu_B \left[B_x \sigma_x + B_z \sigma_z \right] \\
&\quad - \eta \left[k_x \sigma_z - k_z \sigma_x \right] - \nu \left[k_x \sigma_x - k_z \sigma_z \right],
\end{aligned} \tag{7.25}
$$

where $\nu = -(\nu_D/\hbar^2) < p_y^2 > = \nu_D < \partial^2/\partial y^2 >$ and

$$\left[-\frac{\hbar^2}{2m^*} \frac{\partial^2}{\partial y^2} + \frac{e^2 y^2 (B_x^2 + B_z^2)}{2m^*} \right] \lambda_0^n(y) = \epsilon_n \lambda_0^n(y). \tag{7.26}$$

Writing the above equation explicitly in terms of the Pauli spin matrices, we get that

$$
\begin{aligned}
E_n[\mathbf{I}] &= \frac{\hbar^2 \left(k_x^2 + k_z^2 \right)}{2m^*} [\mathbf{I}] + \epsilon_n[\mathbf{I}] - \frac{g}{2} \mu_B \begin{bmatrix} B_z & B_x \\ B_x & -B_z \end{bmatrix} \\
&\quad - \eta \begin{bmatrix} k_x & 0 \\ 0 & -k_x \end{bmatrix} + \eta \begin{bmatrix} 0 & k_z \\ k_z & 0 \end{bmatrix} - \nu \begin{bmatrix} 0 & k_x \\ k_x & 0 \end{bmatrix} + \nu \begin{bmatrix} k_z & 0 \\ 0 & -k_z \end{bmatrix} \\
&= \begin{bmatrix} \overline{E_n} - (g/2)\mu_B B_z - \eta k_x + \nu k_z & -(g/2)\mu_B B_x + \eta k_z - \nu k_x \\ -(g/2)\mu_B B_x + \eta k_z - \nu k_x & \overline{E_n} + (g/2)\mu_B B_z + \eta k_x - \nu k_z \end{bmatrix},
\end{aligned} \tag{7.27}
$$

where

$$\overline{E_n} = \frac{\hbar^2 \left(k_x^2 + k_z^2 \right)}{2m^*} + \epsilon_n. \tag{7.28}$$

In order to find the spin-dependent energy eigenstates E_n (as a function of k_x and k_z) for the n-th magneto-electric subband, as well as the corresponding eigenspinors, we have to "diagonalize the Hamiltonian," meaning that we must find the eigenvalues and eigenfunctions of the matrix on the right hand side of Equation (7.27). Since we have a 2×2 matrix, there are two eigenvalues and two corresponding eigenfunctions. The two eigenvalues are the spin-dependent eigenenergies, and the two eigenfunctions are the eigenspinors. It is easy to evaluate the eigenvalues and eigenfunctions of the 2×2 matrix in Equation (7.27). We find that the eigenvalues in the n-th subband are

$$E_{\pm,n}^{||} = \frac{\hbar^2 \left(k_x^2 + k_z^2 \right)}{2m^*} + \epsilon_n \pm \sqrt{\left(\frac{g}{2}\mu_B B_z + \eta k_x - \nu k_z \right)^2 + \left(\frac{g}{2}\mu_B B_x - \eta k_z + \nu k_x \right)^2}. \tag{7.29}$$

The above equation gives the E versus (k_x, k_z) relations for the two spin-split levels in any subband n. Again the superscript '||' reminds the reader that there is an external magnetic field parallel to the plane of the 2-DEG.

The corresponding eigenspinors are:

$$\psi_+^{||}(B_x, B_z, k_x, k_z) = \begin{bmatrix} -sin\theta_k \\ cos\theta_k \end{bmatrix} \tag{7.30}$$

and

$$\psi_-^{||}(B_x, B_z, k_x, k_z) = \begin{bmatrix} cos\theta_k \\ sin\theta_k \end{bmatrix}, \tag{7.31}$$

where

$$\theta_k = \frac{1}{2}arctan\left[\frac{(g/2)\mu_B B_z - \eta k_z + \nu k_x}{(g/2)\mu_B B_x + \eta k_x - \nu k_z}\right]. \tag{7.32}$$

Note that at any given wavevector state (k_x, k_z), the two eigenspinors are orthogonal, which means that the corresponding spins are anti-parallel. The angle θ_k, however, depends on the wavevector components k_x and k_z, and is independent of the subband index n. This means that: (1) in any given subband, spins in the two spin split levels are *not* anti-parallel in *different* wavevector states but are anti-parallel in the same wavevector state, and (2) in any given wavevector state (k_x, k_z), spins in all the lower spin-split levels, in all subbands, are mutually parallel, as are the spins in all the upper spin-split levels.

Because of the wavevector dependence of the eigenspinors, neither spin-split level whose energy dispersion relation is given by Equation (7.29) has a fixed spin quantization axis. The spin quantization axis, or the spin polarization in any level, changes with changing wavevector. In Fig. 7.2, we plot the energy versus wavevector relations (E vs. k_x, k_z) given by Equation (7.29) for the lowest magneto-electric subband ($n = 1$). Each magneto-electric subband is spin-split into two levels. The lower one has inflections along lines $k_x = 0$ and $k_z = 0$, whereas the upper one is nearly a paraboloid.

Velocity versus wavevector relation

The velocity operator along the $\vec{q}$-direction is given by

$$v_q = \frac{\partial H_{2-DEG}^{||}}{\partial p_q}, \tag{7.33}$$

where $H_{2-DEG}^{||}$ is the Hamiltonian and p_q is the momentum along the $\vec{q}$-direction.

Using the Hamiltonian in Equation (7.10), we get that the expectation values of the velocity components along the x- and z-directions are

$$<v_x> = \frac{<p_x>}{2m^*} - \frac{eB<y>}{m^*} - \frac{\eta}{\hbar}<\sigma_z> - \frac{\nu}{\hbar}<\sigma_x>,$$

$$<v_z> = \frac{<p_z>}{2m^*} + \frac{eB<y>}{m^*} + \frac{\eta}{\hbar}<\sigma_x> + \frac{\nu}{\hbar}<\sigma_z>, \tag{7.34}$$

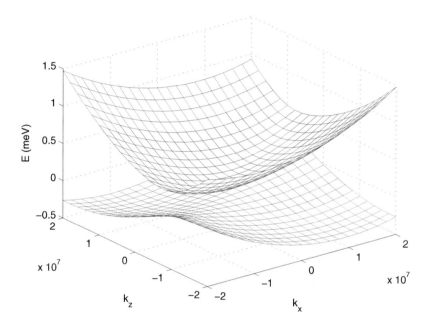

FIGURE 7.2
Energy dispersion relation of spin-split levels in the lowest subband in a two-dimensional electron gas subjected to a magnetic field in the plane of the gas. We assumed $m^* = 0.05m_0$, $g = 4$, $B = 1$ Tesla, $\eta = 10^{-11}$ eV-m, $\nu = 2\eta$. The wavevectors are in units of cm^{-1}.

where angular brakets $< ... >$ denote average over both the spatial and the spin parts of the wavefunctions. Using the eigenspinors in Equations (7.30) and (7.31) to calculate the expectation values of the velocity operators and dropping the angular brakets (they will henceforth be implied), we obtain that the velocities in the two spin-split levels in any wavevector state (k_x, k_z), in any subband, are different and given by

$$v_x^\pm = \frac{\hbar k_x}{m^*} \pm \frac{\eta}{\hbar} cos(2\theta_k) \pm \frac{\nu}{\hbar} sin(2\theta_k),$$

$$v_z^\pm = \frac{\hbar k_z}{m^*} \mp \frac{\nu}{\hbar} cos(2\theta_k) \mp \frac{\eta}{\hbar} sin(2\theta_k) \ , \tag{7.35}$$

$$\tag{7.36}$$

where we have assumed, for convenience, that $< y > = 0$ since the origin of the y-coordinate is arbitrary.

Note that v_q is *not* proportional to k_q because of the additional terms that depend on the spin-orbit interaction strengths. Moreover, since θ_k depends on k_x and k_z, v_x depends on both k_x and k_z, as does v_z. This is expected since the dispersion relations are not parabolic.

Special case

Consider a situation when there is no external magnetic field and no Dresselhaus interaction. This is a situation that arises in the discussion of the Intrinsic Spin Hall Effect visited later. For this case, $tan(2\theta_k) = -k_z/k_x$, so that $sin(2\theta_k) = -k_z/k$ and $cos(2\theta_k) = k_x/k$, where $k = \sqrt{k_x^2 + k_z^2}$. Therefore, we obtain that

$$v_x^\pm = \frac{\hbar k_x}{m^*} \pm \frac{\eta}{\hbar} \frac{k_x}{k}$$

$$v_z^\pm = \frac{\hbar k_z}{m^*} \pm \frac{\eta}{\hbar} \frac{k_z}{k}. \tag{7.37}$$

Therefore,

$$\left(v^\pm\right)^2 = \left(v_x^\pm\right)^2 + \left(v_z^\pm\right)^2 = \left(\frac{\hbar k}{m^*} \pm \frac{\eta}{\hbar}\right)^2$$

$$or \quad v^\pm = \pm \left(\frac{\hbar k}{m^*} \pm \frac{\eta}{\hbar}\right). \tag{7.38}$$

Another special case

There is one other special case of interest. Consider the situation if two conditions are fulfilled:

$$B_x = B_z$$

$$\eta = \nu \tag{7.39}$$

meaning that the magnetic field subtends an angle of 45° with the x-axis, and the strengths of the Rashba and Dresselhaus interactions are equal. We can always choose our x and y axes orientation in a 2-DEG plane such that the magnetic field subtends an angle of 45° with either axes. All we need then is that the two spin-orbit interactions have equal strength. In that case we find from Equation (7.32) that

$$\theta_k = 22.5° \tag{7.40}$$

and is independent of wavevector*. Therefore, the eigenspinors in this case are

$$\psi_+^{||}(B_x, B_z, k_x, k_z) = \begin{bmatrix} -sin(\pi/8) \\ cos(\pi/8) \end{bmatrix} \tag{7.41}$$

and

$$\psi_-^{||}(B_x, B_z, k_x, k_z) = \begin{bmatrix} cos(\pi/8) \\ sin(\pi/8) \end{bmatrix}, \tag{7.42}$$

---

*Note that θ_k becomes independent of wavevector also if $B_x = -B_z$ and $\eta = -\nu$.

In the above special case, each subband has a fixed quantization axis. We can find this spin quantization axis (spin polarization) easily. Let S_n be the component of spin polarization along the n-axis. Then

$$S_x^{\pm} = (\hbar/2)[\psi_{\pm}]^T[\sigma_x][\psi_{\pm}]$$
$$S_y^{\pm} = (\hbar/2)[\psi_{\pm}]^T[\sigma_y][\psi_{\pm}]$$
$$S_z^{\pm} = (\hbar/2)[\psi_{\pm}]^T[\sigma_z][\psi_{\pm}] \tag{7.43}$$

where $\psi_{\pm}$ is given by Equations (7.41) and (7.42). We thus obtain that $S_x^- = (\hbar/2)sin(45°) = \hbar/(2\sqrt{2})$, $S_y^- = 0$ and $S_z^- = (\hbar/2)cos(45°) = \hbar/(2\sqrt{2})$. Therefore, the spin polarization lies in the $(x-z)$ plane and subtends an angle of $45°$ with either the x- or z-axis. In other words, it is directed along the magnetic field. The reader can easily verify that the spin polarization in the other spin-split level is *anti-parallel* to the spin polarization in the first level. Thus, the two spin polarizations are parallel and anti-parallel, respectively, to the magnetic field, as we expect from the bare Zeeman effect.

In this special case,

$$E_{\pm}^{\parallel} = \frac{\hbar^2}{2m^*}\left(k_x \pm (m^*\eta)/\hbar^2\right)^2 + \frac{\hbar^2}{2m^*}\left(k_z \mp (m^*\eta)/\hbar^2\right)^2$$
$$- \frac{m^*\eta^2}{\hbar^2} \pm (g/2)\mu_B B + \epsilon_n$$
$$= \frac{\hbar^2}{2m^*}\left(k_x \pm (m^*\nu)/\hbar^2\right)^2 + \frac{\hbar^2}{2m^*}\left(k_z \mp (m^*\nu)/\hbar^2\right)^2$$
$$- \frac{m^*\nu^2}{\hbar^2} \pm (g/2)\mu_B B + \epsilon_n. \tag{7.44}$$

The velocity versus wavevector relations are

$$v_x^{\pm} = \frac{\hbar k_x}{m^*} \pm \sqrt{2}\frac{\eta}{\hbar} = \frac{\hbar k_x}{m^*} \pm \sqrt{2}\frac{\nu}{\hbar}$$
$$v_z^{\pm} = \frac{\hbar k_z}{m^*} \mp \sqrt{2}\frac{\eta}{\hbar} = \frac{\hbar k_z}{m^*} \mp \sqrt{2}\frac{\nu}{\hbar}. \tag{7.45}$$

The velocity is linearly related to the wavevector. Moreover, the x-component of the velocities in the two spin-split levels (in any given subband) differ by a constant $(= 2\sqrt{2}\eta/\hbar = 2\sqrt{2}\nu/\hbar)$ at the same value of the wavevector k_x. The same is true of the z-component of the velocity at the same value of the wavevector k_z.

This situation (when the Rashba and Dresselhaus spin-orbit interactions are equal in strength) is sometimes referred to as the *persistent spin-helix* state [3] since it leads to invariance with respect to rotation of the electron's spin or SU(2) symmetry. Such a state would conserve the amplitude and phase of a helical spin density wave. It has been observed experimentally [4, 5, 6].

7.1.2 Magnetic field perpendicular to the plane of the 2-DEG

When the magnetic field is perpendicular to the plane of the 2-DEG (in the y-direction), we can write the vector potential $\vec{A} = Bz\hat{x}$, where B is the flux density. In that case, the effective mass Hamiltonian is approximately given by

$$H^{\perp}_{2-DEG} = \frac{(p_x + eBz)^2 + p_y^2 + p_z^2}{2m^*} + V(y) - \frac{g}{2}\mu_B B \sigma_y$$

$$- \frac{\eta}{\hbar}\left[(p_x + eBz)\sigma_z - p_z\sigma_x\right]$$

$$+ \frac{\nu_D}{\hbar^3}p_y^2\left[(p_x + eBz)\sigma_x - p_z\sigma_z\right]. \tag{7.46}$$

Since the Hamiltonian is invariant in the coordinate x, the wavevector k_x is a good quantum number and we can write the wavefunction as

$$\psi^{\perp}_{2-DEG} = \frac{1}{\sqrt{L_x}}e^{ik_x x}[\phi(y)\zeta(z)] = \frac{1}{\sqrt{L_x}}e^{ik_x x}[\varphi(y, z)], \tag{7.47}$$

where $[\varphi(y, z)] = [\phi(y)\zeta(z)]$. This allows for varation of the spin polarization along the y- and z-coordinates, but not along the x-coordinate.

We then write the Pauli equation

$$H^{\perp}_{2-DEG}\psi^{\perp}_{2-DEG} = E\psi^{\perp}_{2-DEG}, \tag{7.48}$$

multiply both sides of this equation by the quantity $\frac{1}{\sqrt{L_x}}e^{-ik_x x}$, and integrate over the x-coordinate to obtain

$$\frac{1}{L_x}\int dx e^{-ik_x x}H^{\perp}_{2-DEG}e^{ik_x x}[\varphi(y, z)] = E\frac{1}{L_x}\int dx e^{-ik_x x}e^{ik_x x}[\varepsilon(y, z)]$$

$$= E[\varphi(y, z)]. \tag{7.49}$$

Next, we evaluate the integral on the left-hand-side using Equation (7.46). This will reduce the last equation to

$$\left\{\frac{p_y^2}{2m^*} + V(y) + \frac{p_z^2}{2m^*} + \frac{m^*}{2}\omega_c^2(z + z_0)^2 - \frac{g}{2}\mu_B B \sigma_y\right.$$

$$- \frac{\eta}{\hbar}\left[m^*\omega_c(z + z_0)\sigma_z - p_z\sigma_x\right]$$

$$\left. - \frac{\nu}{\hbar}\left[m^*\omega_c(z + z_0)\sigma_x - p_z\sigma_z\right]\right\}[\varphi(y, z)]\right\} = E[\varphi(y, z)] \tag{7.50}$$

subject to the boundary conditions $[\varphi(y, z)]$ vanishes when $y = \pm\infty$ or $z = \pm\infty$.

Here $z_0 = \hbar k_x/eB$ and $\omega_c = eB/m^*$.

Equation (7.50) has to be solved numerically to find the 2-component wavefunction (spinor) $[\varphi(y, z)]$ and the energy dispersion relation E versus k_x. Note that in the above equation $z_0 = \hbar k_x / eB$. However, since this equation is non-linear in z_0 (and hence k_x), a transformation of the type in Equation (7.17) will be required.

7.2 Dispersion relations of spin resolved magneto-electric subbands and eigenspinors in a one-dimensional electron gas in the presence of spin–orbit interaction

Consider a quantum wire, or one dimensional electron gas, grown in the [100] crystallographic direction, as shown in Fig. 7.3. In contrast with the two-dimensional case, motions along both the y- and z-direction are constrained by quantum confinement and only motion along the x-direction is permitted. We will assume that the x-direction corresponds to the [100] crystallographic direction. There is a symmetry breaking electric field along the y-direction which induces the Rashba spin-orbit interaction. Furthermore, the quantum wire has crystallographic inversion asymmetry along the x-direction which induces the Dresselhaus spin-orbit interaction.

We will derive the dispersion relations and the eigenspinors in three different cases corresponding to an externally applied magnetic field being directed along three different coordinate axes - x, y and z.

7.2.1 Magnetic field directed along the wire axis (x-axis)

The effective mass Hamiltonian for the wire, in the Landau gauge $\vec{A} = (0, -Bz, 0)$, can be written as

$$H_{1-DEG}^x = (p_x^2 + p_y^2 + p_z^2)/(2m^*) - (eBzp_y)/m^* + (e^2B^2z^2)/(2m^*)$$
$$-(g/2)\mu_B B\sigma_x + V(y) + V(z)$$
$$+\nu_D[\sigma_x\kappa_x + \sigma_y\kappa_y + \sigma_z\kappa_z] - \eta[(p_x/\hbar)\sigma_z - (p_z/\hbar)\sigma_x], \quad (7.51)$$

where $V(y)$ and $V(z)$ are the confining potentials along the y- and z-directions, ν is the strength of the Dresselhaus spin-orbit interaction, and η is the strength of the Rashba spin-orbit interaction, as before.

Since the Hamiltonian is invariant in the x-coordinate, the wavevector k_x is a good quantum number and the eigenstates are plane waves traveling in the x-direction. Therefore, the 2-component wavefunction can be written as

$$[\psi_{1-DEG}] = e^{ik_x x}\phi_m(y)[\lambda(z)], \quad (7.52)$$

where $[\lambda(z)]$ is a 2×1 column vector.

The Pauli equation for this system is

$$H_{1-DEG}^x[\psi_{1-DEG}] = E[\psi_{1-DEG}]. \quad (7.53)$$

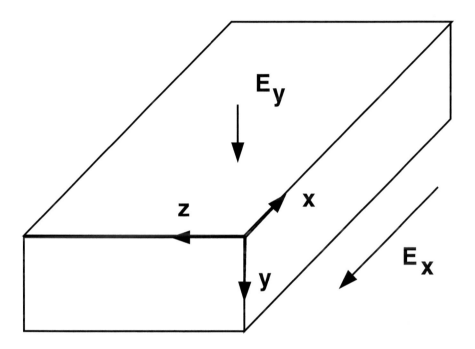

FIGURE 7.3
Schematic of a quantum wire showing the assignment of coordinate axes.

Spatially averaging both sides of the above equation over the x- and y-coordinates, we get

$$[\{(\hbar^2 k_x^2)/(2m^*) + E_m - \frac{\hbar^2}{2m^*}\frac{\partial^2}{\partial z^2} + \frac{e^2 B^2 z^2}{2m^*} + V(z)\}[I]$$
$$-(\nu k_x + \beta - i\eta\partial/\partial z)\sigma_x - (\eta k_x + i\nu\partial/\partial z)\sigma_z][\lambda(z)] = E[\lambda(z)],$$
$$(7.54)$$

where $\beta = g\mu_B B/2$, $\nu = -(\nu_D/\hbar^2) <p_y^2> = \nu_D < \partial^2/\partial y^2 >$ and

$$\left[\frac{p_y^2}{2m^*} + V(y)\right]\phi_m(y) = E_m\phi_m(y).\qquad(7.55)$$

In deriving the above, we assumed that the thickness of the quantum wire along the y-direction is much smaller than the width along the z-direction, so that $<p_y^2> \;>> \; <p_z^2>$ or $<p_x^2>$. Note that, since the y-component of the wavefunction is bounded, i.e., $\phi_m(y = d) = \phi_m(y = -d) = 0$, therefore $<p_y> = 0$.

In order to obtain the dispersion relation E versus k_x and the 2-component wavefunction $[\lambda(z)]$, we have to solve Equation (7.54) numerically, but only after transforming it following the trick of Equation (7.17) so that we first obtain an eigenequation in k_x. As always, this is the correct procedure, but it is also cumbersome and requires quite a bit of numerics. The approximate technique detailed below allows us to obtain an analytical expression for the energy dispersion relation.

First, we will make the "variable separation" approximation, i.e., use the approximation in Equation (7.23) where we write the eigenspinor $[\lambda(z)]$ as the product of a space dependent part and a spin dependent part. We will assume that the spin-dependent part is independent of the z-coordinate. As a result, we will miss spin texturing effects where the spin density can vary along the z-coordinate. We have to pay this price in order to obtain analytical solutions. Second, we will assume that the energy separation between subbands (caused by confinement along the y- and z-directions) is so large that any mixing between them caused by spin-orbit interaction is negligible. This mixing can have interesting effects such as "anti-crossings" in the energy dispersion relations [7], but we will miss those effects as well. Finally, we will assume that the confinement along the z-direction is parabolic (this will indeed be the case if the confinement along this direction is imposed by "split-gates", as is usually done). In that case $V(z) = (1/2)m^*\omega_0^2 z^2$. Therefore, the unperturbed z-component of the wavefunction $\lambda_0(z)$ (in the absence of spin-orbit interaction) obeys the equation:

$$\left[\frac{p_z^2}{2m^*} + \frac{1}{2}m^*\omega^2 z^2\right]\lambda_0(z) = \Xi\lambda_0(z),\qquad(7.56)$$

where $\omega^2 = \omega_0^2 + \omega_c^2$ and ω_c is the cyclotron frequency ($\omega_c = eB/m^*$).

In this case, the eigenenergies are $\Xi = (n + 1/2)\hbar\omega$, and $< p_z > = 0$. If we use the *unperturbed* wavefunction $\lambda_0(z)$ in Equation (7.54), and spatially average over the z-coordinate, then we obtain

$$E[\mathbf{I}] = \left[(\hbar^2 k_x^2)/(2m^*) + E_m + (n + 1/2)\hbar\omega\right][\mathbf{I}] - (\nu k_x + \beta)\sigma_x - (\eta k_x)\sigma_z. \tag{7.57}$$

The energy E is found from the two eigenvalues of the 2×2 matrix making up the right hand side of the above equation. This is the familiar "diagonalization" procedure and the two eigenvalues are

$$E_{\pm}^{(1)} = \frac{\hbar^2 k_x^2}{2m^*} + E_0 \pm \sqrt{(\eta^2 + \nu^2)\left(k_x + \frac{\nu\beta}{\eta^2 + \nu^2}\right)^2 + \frac{\eta^2}{\eta^2 + \nu^2}\beta^2}, \tag{7.58}$$

which are the energy dispersion (E vs. k_x) relations for the two spin-split subbands. Here, $E_0 = E_m + (n + 1/2)\hbar\omega$ and we have used the superscript '(1)' in $E_{\pm}^{(1)}$ to remind the reader that this is the first of the three cases we will consider - where the magnetic field is along the x-axis (axis of the quantum wire).

The eigenspinors (2-component wavefunctions) are the eigenvectors of the matrix making up the right-hand side of Equation (7.57) corresponding to the eigenvalues given in Equation (7.58). Therefore, these eigenspinors are

$$\Psi^{(1)}_-(B, k_x) = \begin{bmatrix} cos(\theta_{k_x}) \\ sin(\theta_{k_x}) \end{bmatrix} \tag{7.59}$$

and

$$\Psi^{(1)}_+(B, k_x) = \begin{bmatrix} sin(\theta_{k_x}) \\ -cos(\theta_{k_x}) \end{bmatrix}, \tag{7.60}$$

where $\theta_{k_x} = (1/2)arctan[(\nu k_x + \beta)/\eta k_x]$.

The dispersion relations given by Equation (7.58) are plotted in Fig. 7.4 (a) for the case $\eta = \nu$. Note that the dispersions are clearly nonparabolic and *asymmetric* about the energy-axis. More important, note that the eigenspinors given in Equation (7.60) are functions of k_x because θ_{k_x} depends on k_x. Therefore, the eigenspinors are not fixed in any spin-split level, in any subband, but change with k_x. In other words, neither level has a definite spin quantization axis and the spin of an electron in either level depends on the wavevector. We are familiar with this situation since we have seen it before in the case of the 2-DEG.

7.2.2 Spin components

Let us call the spin components along the x, y and z directions $S_x^{(1)\pm}$, $S_y^{(1)\pm}$ and $S_z^{(1)\pm}$, respectively. The $\pm$ sign refers to the two spin resolved levels. We will have:

$$S_m^{(1)\pm} = (\hbar/2)[\Psi_{\pm}^{(1)}]^\dagger[\sigma_m][\Psi_{\pm}^{(1)}] \quad m = x, y, z. \tag{7.61}$$

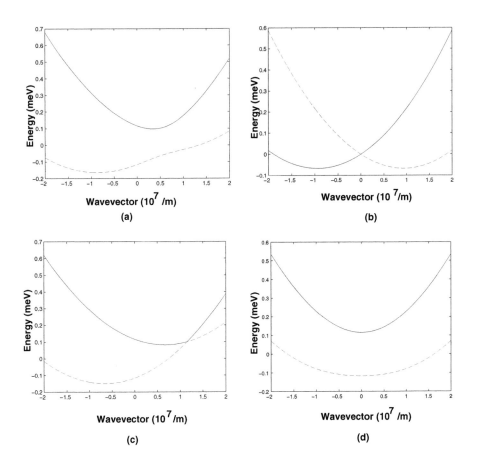

FIGURE 7.4

(a) Dispersion relations of spin-split subbands in a quantum wire with a magnetic field directed along the axis of the wire. We have plotted (assuming E_0 = 0) the dispersion relations for the cases: (a) $\eta = \nu = 10^{-11}$ eV-m, $m^* = 0.05m_0$, $B = 1$ Tesla, $g = $ -2, (b) $\eta = \nu = 10^{-11}$ eV-m, $m^* = 0.05m_0$, $B = 0$ Tesla, (c) $\eta = 0$, $\nu = 10^{-11}$ eV-m, $m^* = 0.05m_0$, $B = 1$ Tesla, $g = $ -2, and (d) $\eta = 10^{-11}$ eV-m, $\nu = 0$, $m^* = 0.05m_0$, $B = 1$ Tesla, and $g = $ -2.

This yields $S_x^{(1)\pm}(k_x) = \mp\frac{\hbar}{2}sin(2\theta_{k_x})$, $S_y^{(1)\pm} = 0$ and $S_z^{(1)\pm}(k_x) = \mp\frac{\hbar}{2}cos(2\theta_{k_x})$.

It is easy to see that, in general, $S_x^{(1)}(k_x) \neq S_x^{(1)}(-k_x)$ and $S_z^{(1)}(k_x) \neq S_z^{(1)}(-k_x)$. However, if the Dresselhaus interaction vanishes ($\nu = 0$), then $S_x^{(1)}(k_x) = -S_x^{(1)}(-k_x)$ and $S_z^{(1)}(k_x) = S_z^{(1)}(-k_x)$. If the Rashba interaction vanishes, then $S_x^{(1)}(k_x) = \hbar/2$ and $S_y^{(1)}(k_x) = S_z^{(1)}(k_x) = 0$.

Velocity versus wavevector relation

There is no velocity along y- and z-directions because of quantum confinement. The velocity along the x-direction is once again $\frac{\partial H}{\partial p_x}$:

$$v_x^\pm = \frac{\hbar k_x}{m^*} - \frac{\eta}{\hbar}<\sigma_z> -\frac{\nu}{\hbar}<\sigma_x>$$

$$= \frac{\hbar k_x}{m^*} \pm \frac{\eta}{\hbar}cos(2\theta_{k_x}) \pm \frac{\nu}{\hbar}sin(2\theta_{k_x}). \tag{7.62}$$

As expected, the velocity is not proportional to k_x because of the spin-orbit interaction. Note that θ_{k_x} depends on k_x.

Special case

A special case arises when there is no magnetic field. In that case, the energy dispersion relation is

$$E_\pm = \frac{\hbar^2}{2m^*}\left(k_x \pm \frac{m^*\sqrt{\eta^2+\nu^2}}{\hbar^2}\right)^2 + E_0 - \frac{m^*(\eta^2+\nu^2)}{2\hbar^2}. \tag{7.63}$$

The above dispersion relation is definitely parabolic, albeit the origin of k_x has shifted by $\pm m^*\sqrt{\eta^2+\nu^2}/\hbar^2$ for the two spin-split bands. These dispersion relations are plotted in Fig. 7.4 (b). The role of spin-orbit interactions is to translate the dispersion parabolas (in any subband) horizontally from each other and shift them vertically downwards by a fixed amount. Note that the two parabolas are parallel to each other in the sense that at any energy, the difference between the wavevectors in the two subbands is $2m^*\sqrt{\eta^2+\nu^2}/\hbar^2$, *which is independent of the energy.* This feature plays a major role in the Spin Field Effect Transistor to be discussed in Chapter 14. In this case, θ_{k_x} is independent of k_x, so that the velocity is linearly related to the wavevector:

$$v_x^\pm = \frac{\hbar(k_x \pm \text{constant})}{2m^*}. \tag{7.64}$$

Note that the velocity is zero when $k_x = \mp$ constant, but not when $k_x = 0$.

More important, if there is no magnetic field, then $\theta_{k_x} = (1/2)arctan[\nu/\eta]$, *independent* of k_x. In that case, the eigenspinors are wavevector-independent and each spin-split level in every subband has a definite spin quantization axis. The x-, y- and z-components of this spin quantization axis will be $S_x^{(1)\pm} = \mp \nu \hbar/(2\sqrt{\nu^2 + \eta^2})$, $S_y^{(1)\pm} = 0$, and $S_z^{(1)\pm} = \mp \eta \hbar/(2\sqrt{\nu^2 + \eta^2})$. The orientation of the spin quantization axis is determined by the relative strengths of the Dresselhaus and Rashba interactions alone since $S_x^{(1)\pm}/S_z^{(1)\pm} = \nu/\eta$. It does not depend on the wavevector.

Note that there is a crucial difference with the 2-DEG case. In the case of a 2-DEG, we needed that the strengths of the Rashba and Dresselhaus interactions be equal in order to make the eigenspinors independent of the wavevector. In the case of 1-DEG, we do *not* need the two strengths to be equal. All that we need to make the eigenspinors independent of the wavevector is that there should be no magnetic field Absence of a magnetic field guarantees that each spin-split level has a fixed spin quantization axis.

An ultra-sensitive magnetometer

The above feature has an important device application. Whenever the eigenspinors are wavevector dependent, any momentum relaxing scattering event that changes the electron's wavevector – even if it is caused by a non-magnetic scatterer – will alter spin polarization and therefore cause spin relaxation. Now consider a device consisting of a quantum wire with two ideal ferromagnetic contacts that inject and detect spin with 100% efficiency. These two ferromagnetic contacts are magnetized *anti-parallel* to each other. One contact injects electrons with a spin polarization that lies in the x-z plane and subtends an angle $arctan(\nu/\eta)$ with the z-axis. If there is no external magnetic field, then this spin is an eigenstate in the quantum wire. Therefore, unless there are magnetic scatterers to flip the spin, this spin will travel through the wire without flipping, even if there is plenty of momentum relaxing collisions that frequently change the electrons wavevector[†]. When the electron arrives at the other contact, it is completely blocked since the other contact is magnetized anti-parallel to the first contact, i.e., anti-parallel to the electrons spin. As a result, the current through the quantum wire will be ideally zero at any applied voltage. In other words, the wire resistance is infinite.

Now place this wire in a magnetic field along the wire axis. Immediately, the injected spins are no longer eigenstates in the wires since the eigenspinors become wavevector dependent. Any momentum relaxing scattering that changes the electron's wavevector will now also change the spin polarization. Consequently, the spins that arrive at the second contact will no longer necessarily be aligned anti-parallel to the magnetization of this contact. Therefore, there will be some current flow and the resistance of the wire will be finite. Consequently, this device can be used as a spintronic magnetic field sensor. It has infinite resistance in the absence of a magnetic field and a finite resistance in a

---

[†]Basically, the two primary spin relaxation mechanisms - the Elliott-Yafet and Dyakonov-Perel (both discussed in the next chapter) – are absent.

magnetic field. A large array of such sensors (e.g., an array of $10^6 \times 10^6$ wires) can be extremely sensitive and with proper electronics detect magnetic fields with a sensitivity of a few tens of femto-Tesla/$\sqrt{Hz}$ at room temperature [9].

Obviously such a sensor can be realized *only with a quantum wire*. It cannot be realized with a quantum well (2-DEG) since the eigenspinors in a 2-DEG are not necessarily wavevector independent even if there is no magnetic field.

Another special case

Another special case arises if the Rashba interaction is absent. Then, $\theta_{k_x} = 45°$, independent of k_x, even if there is a magnetic field. Each subband has a fixed spin quantization axis such that $S_y^{(1)\pm} = S_z^{(1)\pm} = 0$, and $S_x^{(1)\pm} = \pm\hbar/2$. The spin aligns itself parallel or anti-parallel to the direction of the *net* magnetic field which is the vector sum of the external magnetic field and the pseudo-magnetic field caused by the Dresselhaus interaction. Note that both these two magnetic fields are along the x-direction and therefore the spin lines up along the x-axis.

The reader can easily verify that the dispersion relation is again parabolic in this case and given by

$$E_\pm^{(1)} = \frac{\hbar^2 \left(k_x \pm \frac{m^*\nu}{\hbar^2} \right)^2}{2m^*} + E_0 - \frac{m^*\nu^2}{2\hbar^2} \mp \beta. \qquad (7.65)$$

The two parabolas representing the $E^{(1)}$ versus k_x relations of the two spin-split levels are now not only horizontally displaced from each other, but also vertically displaced. Again, the velocity is linearly related to the wavevector:

$$v_x^\pm = \frac{\hbar k_x}{m^*} \pm \frac{\nu}{\hbar}. \qquad (7.66)$$

The dispersion relations for this case are shown in Fig. 7.4 (c).

Yet another special case

The last special case that we will consider is when the Dresselhaus interaction is absent. In this case, θ_{k_x} is wavevector dependent if there is a magnetic field; otherwise, $\theta_{k_x} = 0$, regardless of k_x. It is easy to show that in the latter case, $S_x^{(1)\pm} = S_y^{(1)\pm} = 0$ and $S_z^{(1)\pm} = \pm\hbar/2$. Therefore, the spin aligns itself parallel or anti-parallel to the direction of the pseudo magnetic field caused by the Rashba interaction.

7.2.3 Magnetic field perpendicular to wire axis and along the electric field causing Rashba effect (i.e., along y-axis)

Using the Landau gauge $\vec{A} = (Bz, 0, 0)$, the Hamiltonian for the wire can be written as

$$
\begin{aligned}
H^y_{1-DEG} = {}& (p_x^2 + p_y^2 + p_z^2)/(2m^*) + (eBzp_x)/m^* + (e^2B^2z^2)/(2m^*) \\
& -(g/2)\mu_B B\sigma_y + V(y) + V(z) \\
& +\nu_D[\sigma_x\kappa_x + \sigma_y\kappa_y + \sigma_z\kappa_z] \\
& -(\eta/\hbar)[(p_x + eBz)\sigma_z - p_z\sigma_x].
\end{aligned}
\tag{7.67}
$$

Since the Hamiltonian is invariant in the coordinate x, the wavefunctions are plane waves traveling in the x-direction, and can be written as

$$
[\psi(x, y, z)] = e^{ik_x x}\phi_m(y)[\lambda(z)],
\tag{7.68}
$$

where $\phi_m(y)$ obeys the equation

$$
\left[\frac{p_y^2}{2m^*} + V(y)\right]\phi_m(y) = E_m\phi_m(y).
\tag{7.69}
$$

By now, the reader should be able to determine that, if we assume that the thickness of the quantum wire along the y-direction is much smaller than the effective thickness along the z-direction, the Pauli equation will yield

$$
\begin{aligned}
E[\lambda(z)] = {}& \{[(\hbar^2 k_x^2 - \hbar^2(\partial^2/\partial z^2))/(2m^*) + (eBz\hbar k_x)/m^* + (e^2B^2z^2)/(2m^*) \\
& E_m + V(z)][I] - (g/2)\mu_B B\sigma_y \\
& -\nu[(k_x + eBz/\hbar)\sigma_x + i(\partial/\partial z)\sigma_z] \\
& -\eta[k_x + eBz/\hbar)\sigma_z + i(\partial/\partial z)\sigma_x]\}[\lambda(z)]
\end{aligned}
\tag{7.70}
$$

subject to the boundary condition

$$
[\lambda(z = \infty)] = [\lambda(z = -\infty)] = [0].
\tag{7.71}
$$

We have to solve the above two equations, using the transformation trick of Equation (7.17), to find the energy dispersion relation E versus k_x, as well as the two-component wavefunction $[\lambda(z)]$. In Fig. 7.5, we show the dispersion relations calculated in this way for a quantum wire for two different values of the magnetic field. Then, in Fig. 7.6, we show the real and imaginary parts of the 2-component wavefunction $[\lambda(z)]$ $(= [\xi_\alpha(z), \xi_\beta(z)])$ as a function of the z-coordinate at a high magnetic field in two spin-split bands for a specific electron energy E_0. Note that the wavefunctions are skewed to one or the other edge of the wire depending on whether their velocities are positive or negative. This happens because the magnetic field exerts a Lorentz force $e\vec{v} \times \vec{B}$ on the electron, which pushes them toward one or the other edge of the wire depending on the direction of the velocity. Consequently, the wavefunctions are skewed toward one or the other edge of the quantum wire. These wavefunctions are therefore called "edge states" and play a critical role in the celebrated integer Quantum Hall Effect (as well as many other phenomena).

Since the Rashba and Dresselhaus spin-orbit interaction strengths vanish at $k_x = 0$, we would normally expect that the energy splitting between spin split bands at $k_x = 0$ should be the bare Zeeman splitting $g\mu_B B$. However, it turns out that this is *not* the case. Note that in Fig. 7.5, the spin-splitting energy at $k_x = 0$ clearly increases with increasing subband index, which would not have happened if it were pure Zeeman splitting since $g\mu_B B$ does not depend on the subband index. Although not obvious from the figure, it also turns out that the spin-splitting energy within any subband is not even linear in the magnetic field, showing once again that it is not simply $g\mu_B B$. It would have been *approximately* $g\mu_B B$ had we made the variable separation approximation. The reason why this happens is because the spin-orbit interaction has an indirect influence on the so-called zero-k spin splitting. This has been discussed in [10].

In order to obtain analytical results, we can make the variable separation approximation (recall Equation (7.23)) and neglect mixing between subbands caused by spin-orbit interaction. As mentioned earlier, we will then miss spin texturing effects and anti-crossing effects in the dispersion relation, but that is the price we have to pay in order to produce analytical results.

The analytical results are obtained under two different limiting cases: (i) the magnetostatic confinement in the z-direction is weaker than the electrostatic confinement, and (ii) the opposite scenario, whereby the electrostatic confinement in the z-direction is weaker than the magnetostatic confinement. We assume that the electrostatic confinement in the z-direction is parabolic so that the confining potential is expressed as

$$V(z) = \frac{1}{2}m^*\omega_0^2 z^2. \tag{7.72}$$

In that case, condition (i) above implies

$$\omega_0 >> \omega_c = \frac{eB}{m^*}, \tag{7.73}$$

while condition (ii) implies

$$\omega_0 << \omega_c = \frac{eB}{m^*}. \tag{7.74}$$

We will consider these two cases separately.

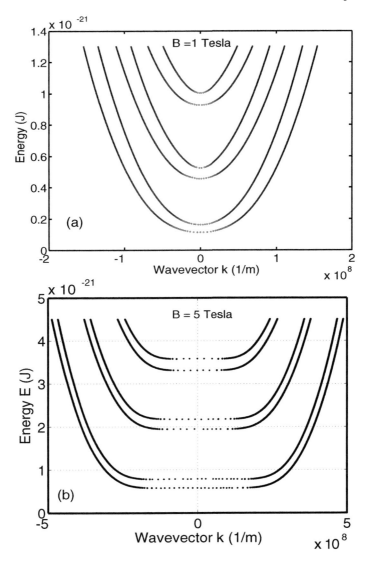

FIGURE 7.5

Energy dispersion relation ($E - k$ plot) in a quantum wire of width 100 nm in the presence of Rashba and Dresselhaus spin-orbit interaction and an external magnetic field in the y-direction. The effective mass is assumed to be 0.067 times the free electron mass and the Landé g-factor is assumed to be 4. The Rashba and Dresselhaus interaction strengths are $\eta = 2\nu = 10^{-11}$ eV-m. The magnetic flux density B is: (a) 1 Tesla, and (b) 5 Tesla. Reprinted with permission from Sandipan Pramanik, et al., Phys. Rev. B, **76**, 155325 (2007). Copyright (2007) American Physical Society. http://link.aps.org/abstract/PRB/v76/p155325.

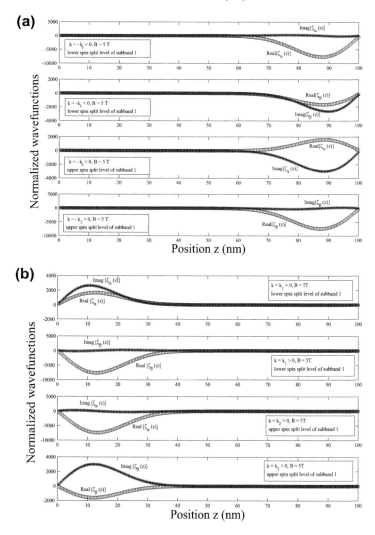

FIGURE 7.6

Real and imaginary parts of the two components of the wavefunction $[\zeta_\alpha(z) \; \zeta_\beta(z)]^T$ in a quantum wire as a function of the z-coordinate. The wavefunctions are plotted for electron energy $E_0 = 9.3$ meV and magnetic flux density $B = 5$ Tesla. The wavevectors corresponding to this energy in the two spin-split bands are $k_1^\pm = \pm\, 3.56 \times 10^8$ m^{-1} (lower band), and $k_2^\pm = \pm\, 3.3 \times 10^8$ m^{-1} (upper band). In Fig. (a), we show the wavefunctions for k_1^- and k_2^-, while in Fig. (b), we show the wavefunctions for k_1^+ and k_2^+. The wavefunctions are skewed to the left or right edge of the quantum wire, depending on whether the electrons are forward or backward traveling (i.e., whether k is positive or negative), because of the Lorentz force that pushes electrons toward one or the other edge.

Case I: Magnetostatic confinement is weaker

In this case, we can approximate Equation (7.70) as

$$E[\lambda(z)] \approx \{[(\hbar^2 k_x^2 - \hbar^2(\partial^2/\partial z^2))/(2m^*)$$
$$E_m + \frac{eBz\hbar k_x}{m^*} + (1/2)m^*\omega^2 z^2][I] - (g/2)\mu_B B\sigma_y$$
$$-\nu\left[(k_x + eBz/\hbar)\sigma_x + i(\partial/\partial z)\sigma_z\right]$$
$$-\eta\left[k_x + eBz/\hbar)\sigma_z + i(\partial/\partial z)\sigma_x\right]\}[\lambda(z)], \qquad (7.75)$$

where $\omega^2 = \omega_0^2 + \omega_c^2$.

Following the usual procedure (i.e., spatially averaging the Hamiltonian over x-, y- and z-coordinates) and then carrying out the "diagonalization,', we obtain the energy dispersion relations of the two spin-split levels as

$$E_\pm^{(2)} = \frac{\hbar^2 k_x^2}{2m^*} + E_m + \left(n + \frac{1}{2}\right)\hbar\omega + \frac{eB <z> \hbar k_x}{m^*} \pm \sqrt{P^2 - Q^2 + (g\mu_B B/2)^2},$$
$$(7.76)$$

where

$$P^2 = \left(\nu^2 + \eta^2\right)\left[(k_x + eB <z> /\hbar)^2 + <p_z/\hbar>^2\right]$$
$$Q^2 = 4(\nu\eta/\hbar^2)eB <z><p_z>, \qquad (7.77)$$

and n is an integer. Here we have used the superscript '(2)' in $E_\pm^{(2)}$ to remind the reader that we are now dealing with the second of the three cases, namely that the magnetic field is parallel to the y-axis.

Here, $<z> = \int_0^\infty dz\lambda_0^*(z)z\lambda_0(z)$, $<p_z> = \int_0^\infty dz\lambda_0^*(z)(--i\hbar(\partial/\partial z)\lambda_0(z)$ and E_m is the quantized energy in the n-th subband due to confinement along the y-direction. Because of confinement along the z-direction, $<p_z> = 0$. Furthermore, without loss of generality, we can choose the origin of the z-coordinate such that $<z> = 0$. Therefore, the above equation simplifies to

$$E_\pm^{(2)} = \frac{\hbar^2 k_x^2}{2m^*} + E_m + \left(n + \frac{1}{2}\right)\hbar\omega \pm \sqrt{(\nu^2 + \eta^2)k_x^2 + (g\mu_B B/2)^2}. \quad (7.78)$$

The eigenspinors are

$$\Psi_-^{(2)}(B, k_x) = \begin{bmatrix} cos(\theta'_{k_x}) \\ sin(\theta'_{k_x})e^{i\phi_{k_x}} \end{bmatrix} \qquad (7.79)$$

$$\Psi_+^{(2)}(B, k_x) = \begin{bmatrix} sin(\theta'_{k_x}) \\ -cos(\theta'_{k_x})e^{i\phi_{k_x}} \end{bmatrix}, \qquad (7.80)$$

where

$$\theta'_{k_x} = (1/2)arctan[\sqrt{(g\mu_B B/2)^2 + \nu^2 k_x^2}/\eta k_x] \qquad (7.81)$$

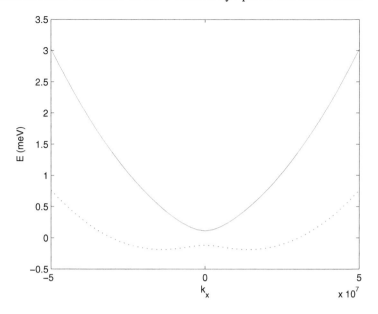

FIGURE 7.7

Dispersion relations of spin-split subbands in a quantum wire with a magnetic field directed perpendicular to the axis of the wire along the direction in which there is a symmetry breaking electric field causing a Rashba interaction. We have assumed that $< z > = 0$, $m^* = 0.05m_0$, $\eta = \nu = 10^{-11}$ eV-m, $g = 4$, and $B = 1$ Tesla.

and

$$\phi_{k_x} = \frac{g\mu_B B}{2\nu k_x}. \tag{7.82}$$

In Fig. 7.7, we plot the dispersion relations (E versus k_x) for the case $\eta = \nu$. Note that the dispersions are again clearly nonparabolic. The eigenspinors are functions of k_x because θ'_{k_x} and ϕ_{k_x} depend on k_x. Therefore, the eigenspinors are again not fixed in any subband, but change with k_x. Neither subband has a definite spin quantization axis, and the spin of an electron in either subband depends on the wavevector.

7.2.4 Spin components

The spin components along the x, y and z directions are given by

$$S_m^\pm = (\hbar/2)[\Psi_\pm^*]^\dagger[\sigma_m][\Psi_\pm] \quad m = x, y, z. \tag{7.83}$$

This yields

$$S_x^\pm(k_x) = \mp\frac{\hbar}{2}sin\theta_{k_x}cos\phi_{k_x}$$

$$S_y^\pm(k_x) = \mp\frac{\hbar}{2}sin\theta_{k_x}sin\phi_{k_x},$$

$$S_z^\pm(k_x) = \mp cos\theta_{k_x}. \tag{7.84}$$

Case II: When the magnetostatic confinement is stronger

In this case, we can approximate Equation (7.70) as

$$E[\lambda(z)] \approx \{[(\hbar^2 k_x^2 - \hbar^2(\partial^2/\partial z^2))/(2m^*) + (eBz\hbar k_x) + (e^2 B^2 z^2)/(2m^*) + E_m][I]$$
$$-(g/2)\mu_B B\sigma_y - \nu\,[(k_x + eBz/\hbar)\sigma_x + i(\partial/\partial z)\sigma_z]$$
$$-\eta\,[k_x + eBz/\hbar)\sigma_z + i(\partial/\partial z)\sigma_x]\}[\lambda(z)]. \tag{7.85}$$

The unperturbed states (i.e., the states in the absence of Zeeman, Dresselhaus and Rashba interactions) are Landau orbital states of energy $E_m + (n + 1/2)\hbar\omega_c$ whose wavefunctions are centered at $< z >= -\hbar k_x/(eB)$. The expectation value $< k_x + eBz/\hbar >$ calculated with the unperturbed states is exactly zero. Therefore, the eigenenergies are approximately expressed as

$$E_\pm^{(2)} = E_m + \left(n + \frac{1}{2}\right)\hbar\omega_c \pm \frac{g\mu_B B}{2} \tag{7.86}$$

and the eigenspinors are approximately

$$\Psi_\pm^{(2)}(B, k_x) = \frac{1}{\sqrt{2}}\begin{bmatrix}1\\\pm i\end{bmatrix}. \tag{7.87}$$

It makes perfect sense that, when the magnetic field is strong enough to make the magnetostatic confinement in the z-direction stronger than the electrostatic confinement, the spin-orbit interaction (Rashba and Dresselhaus) has virtually no effect. The puny effective magnetic fields caused by the spin-orbit interactions are completely overwhelmed by the much stronger external magnetic field, so that we see no effect of the spin-orbit interaction in this case.

7.3 Magnetic field perpendicular to the wire axis and the electric field causing the Rashba effect (i.e., along the z-axis)

Using the Landau gauge $\vec{A} = (-By, 0, 0)$, the Hamiltonian for the wire can be written as

$$
\begin{aligned}
H_{1-DEG}^z = {} & (p_x^2 + p_y^2 + p_z^2)/(2m^*) - (eByp_x)/m^* + (e^2B^2y^2)/(2m^*) \\
& - (g/2)\mu_B B\sigma_z + V(y) + V(z) \\
& + \nu_D[\sigma_x\kappa_x + \sigma_y\kappa_y + \sigma_z\kappa_z] \\
& - (\eta/\hbar)[(p_x - eBy)\sigma_z - p_z\sigma_x].
\end{aligned}
\tag{7.88}
$$

The 2-component wavefunction which is a solution of the above Hamiltonian can be written as

$$
[\psi(x, y, z)] = e^{ik_x x}\lambda(z)[\phi(y)],
\tag{7.89}
$$

where the z-component of the unperturbed wavefunction (in the absence of spin-orbit interaction) obeys the equation

$$
\left[\frac{p_z^2}{2m^*} + V(z)\right]\lambda_0(z) = \varepsilon\lambda_0(z).
\tag{7.90}
$$

If the confinement along the z-direction is parabolic, i.e., $V(z) = (1/2)m^*\omega_0^2 z^2$, then the wavefunctions $\lambda_0^n(z)$ for the n-th subband are simple harmonic oscillator wavefunctions:

$$
\lambda_0^n(z) = \left(\frac{\Gamma'}{\sqrt{\pi}2^n n!}\right)^{1/2} \mathcal{H}_n(\Gamma' z)e^{-\frac{1}{2}(\Gamma')^2 z^2},
\tag{7.91}
$$

where $\Gamma' = \{(m^*\omega_0)^2/(2\hbar^2)\}^{1/4}$ and $\mathcal{H}_n$ is the Hermite polynomial of the n-th order.

As before, we will consider two cases: (i) the magnetic field is so strong that the magnetostatic confinement along the y-direction is much stronger than the electrostatic confinement $V(y)$, and (ii) the opposite case, namely that the electrostatic confinement is stronger than the magnetostatic confinement.

Case I: Magnetostatic confinement is stronger

In this case, the y-component of the unperturbed wavefunction (in the absence of spin-orbit interaction) is given by the simple harmonic oscillator states

$$
\phi_0^m(y) = \left(\frac{\Gamma}{\sqrt{\pi}2^m m!}\right)^{1/2} \mathcal{H}_m(\Gamma[y - y_0])e^{-\frac{1}{2}\Gamma^2(y-y_0)^2},
\tag{7.92}
$$

where $\Gamma = \{(m^*\omega_c)^2/(2\hbar^2)\}^{1/4}$ and $\mathcal{H}_m$ is the Hermite polynomial of the m-th order, and $y_0 = \hbar k_x/eB$.

It is easy to show that the expectation value of the Hamiltonian in Equation (7.88) (evaluated using the unperturbed wavefunction) is

$$\langle \lambda_0^n(z)\phi_0^m(y)e^{ik_x x}|H_{1-DEG}^z|\lambda_0^n(z)\phi_0^m(y)e^{ik_x x}\rangle = E\mathbf{I}$$
$$= [(n+1/2)\hbar\omega_0 + (m+1/2)\hbar\omega_c][\mathbf{I}] - (g\mu_B B/2)\,\sigma_z. \qquad (7.93)$$

In deriving the above, we have used the fact that $< p_x - eBy > = 0$.

As usual, the energy eigenstates are found by diagonalization, i.e., they are the eigenvalues of the 2×2 matrix making up the right hand side of the above equation. They are

$$E_{\pm}^{(3)} = (n+1/2)\hbar\omega_0 + (m+1/2)\hbar\omega_c \pm g\mu_B B/2 \qquad (7.94)$$

where the superscript '(3)' in $E_{\pm}^{(3)}$ is intended to remind the reader that we are dealing with the last of the three cases, when the magnetic field is in the z-direction. The last equation gives the energies of the two spin split levels (independent of k_x) and the corresponding eigenspinors are

$$\Psi_+^{(3)}(B, k_x, x) = \begin{bmatrix} 1 \\ 0 \end{bmatrix} \qquad (7.95)$$

$$\Psi_-^{(3)}(B, k_x, x) = \begin{bmatrix} 0 \\ 1 \end{bmatrix}. \qquad (7.96)$$

Note that there is no influence of spin-orbit interaction in this case. Only the Zeeman interaction splits the spin eigenstates. This is a familiar situation. The spin-orbit interactions will cause effective magnetic fields which are completely overwhelmed by the externally applied magnetic field because it is much stronger. In that case, we see no effect of the spin-orbit interaction.

Case II: Electrostatic confinement is stronger

In this case, the expectation value of the Hamiltonian in Equation (7.88) is approximately

$$\langle H_{1-DEG}^z \rangle = E[\mathbf{I}] \approx \left[\frac{\hbar^2 k_x^2}{2m^*} + \epsilon_m - \frac{eB <y> \hbar k_x}{m^*} + (n+1/2)\hbar\omega \right][\mathbf{I}]$$
$$- (g\mu_B B/2)\sigma_z - \nu\,(k_x eB <y> /\hbar)\,\sigma_x$$
$$- \eta\,(k_x - eB <y> /\hbar)\,\sigma_z$$
$$= \left[\frac{\hbar^2 k_x^2}{2m^*} + \varepsilon_{m,n} \right][\mathbf{I}] - \frac{g\mu_B B}{2}\sigma_z - \nu k_x \sigma_x - \eta k_x \sigma_z,$$

$$(7.97)$$

where ϵ_m are the energies of the subbands caused by the confinement in the y-direction, $\varepsilon_{m,n} = \epsilon_m + (n + 1/2)\hbar\omega$. In deriving the above, we used the fact that the spatial averages $< p_y > = < p_z > = 0$. We have also assumed, without loss of generality, that the y-coordinate origin is chosen such that $< y > = 0$.

The energy eigenstates are the eigenvalues of the 2×2 matrix making up the right hand side of the above equation. These eigenenergies are

$$E_{\pm}^{(3)} = \frac{\hbar^2 k_x^2}{2m^*} + \varepsilon_{m,n} \pm \sqrt{(\eta^2 + \nu^2) k_x^2 + \eta g\mu_B B k_x + (g\mu_B B/2)^2},$$

(7.98)

and the corresponding eigenspinors are

$$\Psi^{(3)}{}_-(B, k_x, x) = \begin{bmatrix} cos(\theta_{k_x}'') \\ sin(\theta_{k_x}'') \end{bmatrix}$$

(7.99)

$$\Psi^{(3)}{}_+(B, k_x, x) = \begin{bmatrix} sin(\theta_{k_x}'') \\ -cos(\theta_{k_x}'') \end{bmatrix},$$

(7.100)

where $\theta_{k_x}'' = (1/2)arctan[(\nu k_x)/(\eta k_x + g\mu_B B/2)]$.

The eigenspinors are functions of k_x because θ_{k_x}'' depends on k_x. Therefore, the eigenspinors are again not fixed in any subband, but change with k_x. Neither subband has a definite spin quantization axis and the spin of an electron in either subband depends on the wavevector.

7.3.1 Spin components

The spin components along the x-, y- and z-directions are given by

$$S_m^{(3)\pm} = (\hbar/2)[\Psi_{\pm}^{(3)}]^\dagger [\sigma_m][\Psi_{\pm}^{(3)}] \quad m = x, y, z.$$

(7.101)

This yields $S_x^{(3)\pm}(k_x) = \mp(\hbar/2)sin(2\theta_{k_x}'')$, $S_y^{(3)\pm} = 0$ and $S_z^{(3)\pm}(k_x) = \mp(\hbar/2)cos(2\theta_{k_x}'')$.

In this case, $S_x^{(3)}(k_x) \neq S_x^{(3)}(-k_x)$ and $S_z^{(3)}(k_x) \neq S_z^{(3)}(-k_x)$.

7.3.2 Special case

A special case arises if the Dresselhaus interaction is absent. In this case, it is easy to show that

$$E_{\pm}^{(3)} = \frac{\hbar^2}{2m^*}\left[k_x \pm \frac{m^*\eta}{\hbar^2}\right]^2 + \varepsilon_{m,n} - \frac{\hbar^2}{2m^*}\left(\frac{m^*\eta}{\hbar^2}\right)^2 \pm \frac{g\mu_B B}{2}.$$

(7.102)

In this case, the dispersion relations are parabolic. The two spin resolved parabolas are horizontally displaced from each other by a fixed wavevector $2\frac{m^*\eta}{\hbar^2}$ and vertically displaced from each other by the Zeeman splitting energy $g\mu_B B$.

7.4   Eigenenergies of spin resolved subbands and eigenspinors in a quantum dot in the presence of spin–orbit interaction

Consider a quantum dot, or quasi zero-dimensional electron gas, as shown in Fig. 7.8. Here, motion along every direction is constrained by quantum confinement. There is a symmetry breaking electric field along the y-direction which, under proper conditions that we will enumerate later, induces the Rashba spin-orbit interaction. Furthermore, the quantum dot has crystallographic inversion asymmetry which induces the Dresselhaus spin-orbit interaction.

As always, we will consider scenarios when there is an external magnetic field. Let us assume that the magnetic field is directed along the x-direction. We choose the Landau gauge $\vec{A} = -Bz\hat{y}$.

The effective mass Hamiltonian describing this system is

$$
\begin{aligned}
H_{dot} =\ & \frac{p_x^2 + (p_y - eBz)^2 + p_z^2}{2m^*} + V_x(x) + V_y(y) + V_z(z) \\
& -(g/2)\mu_B B\sigma_x - \frac{\eta}{\hbar}\left[p_x\sigma_z - p_z\sigma_x\right] \\
& +\frac{\nu_D}{2\hbar^3}\left[p_x\{(p_y - eBz)^2 - p_z^2\} + \{(p_y - eBz)^2 - p_z^2\}p_x\right]\sigma_x \\
& +\frac{\nu_D}{2\hbar^3}\left[(p_y - eBz)\{p_z^2 - p_x^2\} + \{p_z^2 - p_x^2\}(p_y - eBz)\right]\sigma_y \\
& +\frac{\nu_D}{2\hbar^3}\left[p_z\{p_x^2 - (p_y - eBz)^2\} + \{p_x^2 - (p_y - eBz)^2\}p_z\right]\sigma_z
\end{aligned}
\tag{7.103}
$$

where $V_x(x)$, $V_y(y)$ and $V_z(z)$ are the confining potentials along the three coordinate axes, and $V_y(y)$ includes the potential due to the electric field inducing the Rashba interaction.

In a quantum dot, spin-orbit interaction is weak. Normally, the strengths of the spin-orbit interactions are proportional to the momentum. In any spin resolved subband in a quantum dot, where the spin degeneracy has been lifted by some agent (e.g., an external magnetic field), the expectation value of the momentum in the $\vec{q}$ direction $< \phi_m^\uparrow|p_{\vec{q}}|\phi_m^\uparrow > = < \phi_m^\downarrow|p_{\vec{q}}|\phi_m^\downarrow > = 0$ always, but $< \phi_m^\uparrow|p_{\vec{q}}|\phi_m^\downarrow > = 0$ only if $|\phi_m^\uparrow >= |\phi_m^\downarrow >$. Otherwise, $< \phi_m^\uparrow|p_{\vec{q}}|\phi_m^\downarrow > \neq 0$, and the momentum operator can couple the two spin states, with *different* wavefunctions in space, and cause weak spin-orbit interaction. Thus, what is needed for spin-orbit interaction is that the spatial parts of the wavefunctions of the two spin states in any subband be different. This can happen if the quantum dot has *finite barriers*. In that case, the spin state with higher energy will have a more "leaky" wavefunction that spreads out more than that of the lower energy spin state. This makes $|\phi_m^\uparrow >\neq |\phi_m^\downarrow >$, and results in weak spin-orbit interaction. Therefore, two ingredients are needed for spin-orbit

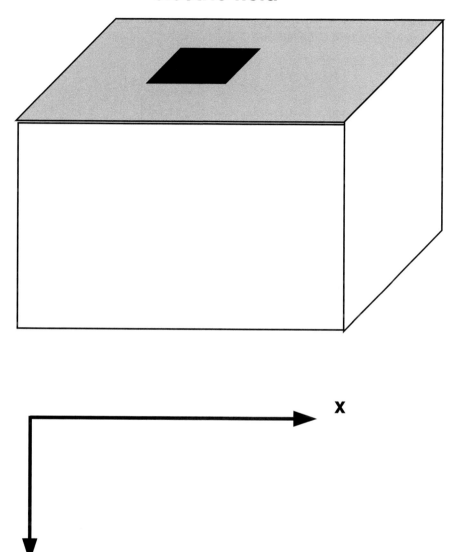

FIGURE 7.8

Schematic representation of a rectangular quantum dot with spin orbit interaction placed in a magnetic field.

interaction in a quantum dot: (1) the presence of some agent that lifts the spin degeneracy in a subband (e.g., an external magnetic field), and (2) finite potential barriers.

In order to find the eigenstates of the Hamiltonian in Equation (7.103), we will apply perturbation theory. We will treat the Rashba and Dresselhaus interactions as weak perturbations. The unperturbed wavefunctions (in the lowest Zeeman split subband) are

$$\{\phi_\uparrow(x,y,z)\} = \frac{1}{\sqrt{2}}\phi_x(x)\phi_y(y)\phi_z(z)\begin{bmatrix}1\\1\end{bmatrix}$$

$$\{\phi_\downarrow(x,y,z)\} = \frac{1}{\sqrt{2}}\phi'_x(x)\phi'_y(y)\phi'_z(z)\begin{bmatrix}1\\-1\end{bmatrix},$$

$$(7.104)$$

where we have assumed that the spatial parts of the wavefunctions of the two spin resolved states (which are resolved by the Zeeman effect) are *different*, i.e., the primed and unprimed wavefunctions are different. This will happen as long as the quantum dot has finite potential barriers. In that case, the upper spin level's wavefunction will be less confined than the lower spins level's wavefunction and therefore will have a larger spatial spread, making $\phi \neq \phi'$. Note that the magnetic field is essential for this inequality and without this inequality, there will be no Rashba or Dresselhaus spin-orbit interaction in the dot, at least to first order.

Note also that the unperturbed eigenspinors (unperturbed by spin-orbit interaction) are +x and -x-polarized spin states $(1/\sqrt{2})[1\ 1]^\dagger$ and $(1/\sqrt{2})[1\ -1]^\dagger$ since the external magnetic field is in the x-direction.

The Pauli equation is

$$[H_{dot}]\,[\psi] = E[\psi], \qquad (7.105)$$

where the 2-component wavefunction $[\psi]$ is a linear superposition of "up-spin" and "down-spin" states, i.e., +x- and −x-polarized states, that are mutually orthogonal:

$$[\psi] = a_\uparrow\{\phi^\uparrow\} + a_\downarrow\{\phi^\downarrow\}. \qquad (7.106)$$

Equation (7.105) can be rewritten as

$$[H_0 + H_{SO}]\,[\psi] = E[\psi], \qquad (7.107)$$

where H_0 is the unperturbed Hamiltonian and H_{SO} is the spin-orbit interaction Hamiltonian due to the Rashba and Dresselhaus interactions.

Substitution of Equation (7.106) into Equation (7.107) yields

$$\begin{bmatrix} <H_1> + <H_{SO}>_{11} & <H_{SO}>_{12} \\ <H_{SO}>_{21} & <H_2> + <H_{SO}>_{22} \end{bmatrix}\begin{pmatrix}a_\uparrow\\a_\downarrow\end{pmatrix} = E\begin{pmatrix}a_\uparrow\\a_\downarrow\end{pmatrix},$$

$$(7.108)$$

where $< H_1 > = < \phi_s^\uparrow|H_0|\phi_s^\uparrow >$, $< H_2 > = < \phi_s^\downarrow|H_0|\phi_s^\downarrow >$, $< H_{SO} >_{11} = < \phi_s^\uparrow|H_{SO}|\phi_s^\uparrow >$, $< H_{SO} >_{22} = < \phi_s^\downarrow|H_{SO}|\phi_s^\downarrow >$, $< H_{SO} >_{12} = < \phi_s^\uparrow|H_{SO}|\phi_s^\downarrow >$ and $< H_{SO} >_{21} = < \phi_s^\downarrow|H_{SO}|\phi_s^\uparrow >$. Here, $\phi_s^\uparrow = \phi_x(x)\phi_y(y)\phi_z(z)$ and $\phi_s^\downarrow = \phi_x'(x)\phi_y'(y)\phi_z'(z)$.

It is easy to show that $< H_{SO} >_{11} = < H_{SO} >_{22} = 0$. Thereafter, diagonalizing the above Hamiltonian (in the basis of the states given in Equation (7.104)), we get

$$E = \frac{< H_1 > + < H_2 >}{2} \pm \sqrt{\left(\frac{g\mu_B B}{2}\right)^2 + < H_{SO} >_{12}< H_{SO} >_{21}}, \quad (7.109)$$

where we have used the fact that $< H_1 > - < H_2 > = g\mu_B B$.

Let us first find the contribution to $< H_{SO} >_{12}$ due to Rashba interaction. The Rashba Hamiltonian is

$$H_{SO}^{Rashba} = -(\eta/\hbar)[p_x\sigma_z - p_z\sigma_x]. \quad (7.110)$$

Therefore,

$$\begin{aligned}
< H_{SO}^{Rashba} >_{12} &= \frac{-\eta}{2\hbar} \begin{bmatrix} 1 & 1 \end{bmatrix} \begin{bmatrix} -\overline{p_x} & \overline{p_z} \\ \overline{p_z} & \overline{p_x} \end{bmatrix} \begin{bmatrix} 1 \\ -1 \end{bmatrix} \\
&= \frac{\eta}{\hbar}\overline{p_x} \\
&= D, \quad (7.111)
\end{aligned}$$

where $\overline{p_x} = < \phi_x(x)\phi_y(y)\phi_z(z)| - i\hbar(\partial/\partial x)|\phi_x'(x)\phi_y'(y)\phi_z'(z) >$. Note that $\overline{p_x} \neq 0$, only because the spatial wavefunctions of the Zeeman split states are *different* $(\phi_x(x) \neq \phi_x'(x))$. It is easy to show (following the same prescription as above) that

$$< H_{SO}^{Rashba} >_{21} = \frac{\eta}{\hbar}\overline{p_x'} = D^*, \quad (7.112)$$

where $\overline{p_x'} = < \phi_x'(x)\phi_y'(y)\phi_z'(z)| - i\hbar(\partial/\partial x)|\phi_x(x)\phi_y(y)\phi_z(z) >$. Since the operator $-i\hbar(\partial/\partial x)$ is Hermitian, one can shown that $\overline{p_x'}$ is the complex conjugate of $\overline{p_x}$, so that D^* is the complex conjugate of D. Furthermore, D is completely imaginary since the wavefunction ϕ_x is *real*. Therefore, $D^* = -D$.

The Dresselhaus Hamiltonian is

$$\begin{aligned}
H_{SO}^{Dresselhaus} =\ & \frac{\nu_D}{2\hbar^3} \left[p_x\{(p_y - eBz)^2 - p_z^2\} + \{(p_y - eBz)^2 - p_z^2\}p_x \right] \sigma_x \\
& + \frac{\nu_D}{2\hbar^3} \left[(p_y - eBz)\{p_z^2 - p_x^2\} + \{p_z^2 - p_x^2\}(p_y - eBz) \right] \sigma_y \\
& + \frac{\nu_D}{2\hbar^3} \left[p_z\{p_x^2 - (p_y - eBz)^2\} + \{p_x^2 - (p_y - eBz)^2\}p_z \right] \sigma_z.
\end{aligned}$$
$$(7.113)$$

Therefore,

$$< H_{SO}^{Dresselhaus} >_{12} = \frac{\nu_D}{2\hbar^3}[[1\ 1]]\begin{bmatrix} \overline{A} & \overline{B} \\ \overline{C} & -\overline{A} \end{bmatrix}\begin{bmatrix} 1 \\ -1 \end{bmatrix}$$
$$= A - (B - C)/2, \qquad (7.114)$$

where A, B, and C are complex constants. The reader can evaluate these constants by evaluating the expectation values of the operators in Equation (7.113). It is easy to show that

$$< H_{SO}^{Dresselhaus} >_{21} = A' - (B' - C')/2, \qquad (7.115)$$

where, once again, the reader can evaluate the complex constants A', B' and C' from the expectation values of the operators in the Hamiltonian of Equation (7.113).

It should be obvious that $< H_{SO}^{Dresselhaus} >_{12} \neq 0$ and $< H_{SO}^{Dresselhaus} >_{21} \neq 0$, only because the spatial wavefunctions of the Zeeman split states are *different*. Thus, without the magnetic field, there will be no spin-orbit interaction effect on a quantum dot's energy dispersion relation or eigenspinors.

The reader can also show that $< H_{SO}^{Dresselhaus} >_{12} = < H_{SO}^{Dresselhaus} >_{21}^{*}$ where the asterisk denotes complex conjugate. This is a consequence of the fact that the spin-orbit Hamiltonian must be Hermitian like all quantum mechanical operators. Therefore, $< H_{SO} >_{12} < H_{SO} >_{21} = | < H_{SO} >_{12} |^2$.

Finally, substituting all this in Equation (7.109), we get the eigenenergies in a quantum dot in the presence of spin-orbit interaction:

$$E_{dot} = \frac{< H_1 > + < H_2 >}{2} \pm \sqrt{\left(\frac{g\mu_B B}{2}\right)^2 + |\delta|^2}, \qquad (7.116)$$

where

$$\delta = < H_{SO} >_{12} = [D + A - (B - C)/2]. \qquad (7.117)$$

We can now proceed to find the coefficients $a_\uparrow$ and $a_\downarrow$, which are the eigenfunction of the matrix in Equation (7.108). For the lower spin-split state, corresponding to the negative sign in Equation (7.116),

$$a_\uparrow^- = cos\zeta$$
$$a_\downarrow^- = sin\zeta e^{i\xi}, \qquad (7.118)$$

where $\zeta = (1/2)arctan[|\delta|/(g\mu_B B)]$ and ξ is the phase of δ, i.e., $\delta = |\delta|e^{i\xi}$. For the higher spin-split state,

$$a_\uparrow^+ = -sin\zeta$$
$$a_\downarrow^+ = cos\zeta e^{i\xi}. \qquad (7.119)$$

Therefore, the eigenspinor for the lower spin-split state is

$$
\begin{aligned}
\Psi_- &= \frac{a_\uparrow^-}{\sqrt{2}} \begin{bmatrix} 1 \\ 1 \end{bmatrix} + \frac{a_\downarrow^-}{\sqrt{2}} \begin{bmatrix} 1 \\ -1 \end{bmatrix} \\
&= \frac{1}{\sqrt{2}} \begin{bmatrix} cos\zeta + sin\zeta e^{i\xi} \\ cos\zeta - sin\zeta e^{i\xi} \end{bmatrix},
\end{aligned} \tag{7.120}
$$

and the eigenspinor in the higher spin-split state is

$$
\begin{aligned}
\Psi_+ &= \frac{a_\uparrow^+}{\sqrt{2}} \begin{bmatrix} 1 \\ 1 \end{bmatrix} + \frac{a_\downarrow^+}{\sqrt{2}} \begin{bmatrix} 1 \\ -1 \end{bmatrix} \\
&= \frac{1}{\sqrt{2}} \begin{bmatrix} sin\zeta - cos\zeta e^{i\xi} \\ sin\zeta + cos\zeta e^{i\xi} \end{bmatrix}.
\end{aligned} \tag{7.121}
$$

The spin components in the lower and upper spin-split states are:

$$
S_x^- = (\hbar/2)cos(2\zeta); \quad S_y^- = -(\hbar/2)sin(2\zeta)sin(\xi); \quad S_z^- = (\hbar/2)sin(2\zeta)cos(\xi) \tag{7.122}
$$

and

$$
S_x^+ = -(\hbar/2)cos(2\zeta); \quad S_y^+ = (\hbar/2)sin(2\zeta)sin(\xi); \quad S_z^+ = -(\hbar/2)sin(2\zeta)cos(\xi). \tag{7.123}
$$

Thus, the spins in the upper and lower energy states are anti-parallel, as they should be.

One can see from the above that it is possible to "rotate" the ground state electron's spin polarization using the applied electric field in the y-direction which induces the Rashba interaction. By changing this electric field, we can change η and therefore D (see Equation (7.111)) and δ (see Equation (7.117)) and finally ζ. This causes the spin polarization vector to rotate. However, we cannot rotate the spin by arbitrary angle since δ can only be varied between 0 and ∞, so that ζ can be varied between 0 and $\pi/4$ radians only.

Wavefunctions and energy-dispersion relations in quantum dots of other shapes, where both Rashba spin-orbit coupling and an external magnetic field are present, have been calculated by others [13].

7.5   Why are the dispersion relations important?

After all is said and done, readers may wonder what use the dispersion relations have. We have led them through all this complicated algebra, enough to leave even the most patient and trained minds reeling, but do these dispersion relations have any practical use? The answer is an emphatic "yes." Optical properties of semiconductors, such as absorption and emission of polarized

light, are determined by the spin resolved energy dispersion relations. Therefore, these dispersion relations are extremely important in opto-spintronics.
They are also important in spin transport since spin current under a given set
of circumstances may depend critically on the band structure. One can use the
band structure properties to measure the spin-orbit interaction strengths, such
as the Rashba interaction strength η, using optical absorption and emission
studies. Normally, these strengths are measured using transport experiments
[11] but optical studies are always more convenient and "user-friendly" than
transport studies, particularly in the case of nanostructures, since one does not
have to make contacts to the structure to study its optical properties. Finally,
many new spin devices have been proposed based on the peculiarities of the
band structure. These include switching transistors [12] and ultra-sensitive
magnetic field sensors [9]. Therefore, the dispersion relations are vital in more
than one way.

The dispersion relations are also important in understanding certain intriguing phenomena. One of them is the *Intrinsic Spin Hall Effect* and the
other is the *Spin Galvanic Effect* – both of which are discussed in the next
chapter.

7.6   Problems

- ## Problem 7.1

  Derive the second equality in Equation (7.50).

  Solution

  We have

  $$\left[\frac{(p_x + eBz)^2 + p_z^2}{2m^*}\right] e^{ik_x x}\zeta_n(z) = \varepsilon_n e^{ik_x x}\zeta_n(z). \qquad (7.124)$$

  This can be rewritten as

  $$\left[\frac{p_x^2}{2m^*} + \frac{p_x eBz}{m^*} + \frac{e^2 B^2 z^2}{2m^*} + \frac{p_z^2}{2m^*}\right] e^{ik_x x}\zeta_n(z) = \varepsilon_n e^{ik_x x}\zeta_n(z) \qquad (7.125)$$

  since the operators p_x and z commute. Next, multiplying by the "bra"
  $e^{-ik_x x}$ and integrating over the coordinate x, we get

  $$\left[\frac{\hbar^2 k_x^2}{2m^*} + \frac{\hbar k_x eBz}{2m^*} + \frac{e^2 B^2 z^2}{2m^*} + \frac{p_z^2}{2m^*}\right]\zeta_n(z) = \varepsilon_n \zeta_n(z). \qquad (7.126)$$

  Using the definition of z_0, we can rewrite the above equation as

  $$\frac{e^2 B^2 z_0^2}{2m^*} + \frac{e^2 B^2 2z z_0}{2m^*} + \frac{e^2 B^2 z^2}{2m^*} + \frac{p_z^2}{2m^*} = \varepsilon_n \zeta_n(z). \qquad (7.127)$$

Finally, using the definition of cyclotron frequency ω_c, we get from the above

$$\left[\frac{p_z^2}{2m^*} + \frac{m^*}{2}\omega_c^2 (z + z_0)^2\right] \zeta_n(z) = \varepsilon_n \zeta_n(z). \qquad (7.128)$$

- **Problem 7.2**

Show that if the thickness of a 2-DEG is vanishingly small, then the Dresselhaus interaction term can be written as in Equation (7.10).

Solution

Because there are no extended states in the direction of confinement in a 2-DEG (y-direction in our case), the expected value of the y-component of momentum $< p_y > = 0$. This makes $< \kappa_y > = 0$ (see Equation (7.3)) where the spatial averaging is carried out over the y-coordinate only. If the 2-DEG is vanishingly thin, then $< p_y^2 > \to \infty$. Therefore, $< \kappa_x > (p_x + eA_x) \times constant$, and $< \kappa_z > = -(p_z + eA_z) \times constant$. Defining ν such that the *constant* is equal to $-\nu/\hbar$ allows us to write the Dresselhaus interaction in the form given in Equation (7.10).

- **Problem 7.3**

Consider the case of a quantum wire with spin-orbit interaction and a magnetic field directed along the axis of the wire. Show that the dispersion relation is symmetric about the energy-axis ($k_x = 0$) only if the Dresselhaus interaction is absent. Show that when the magnetic field is along the direction of the electric field causing the Rashba interaction, the dispersion relation is always symmetric about the energy-axis and when the magnetic field is along the direction mutually perpendicular to the axis of the quantum wire and the electric field causing the Rashba interaction, the dispersion relation is symmetric about the energy-axis only if the Rashba interaction is absent.

Solution

Case I: When the magnetic field is along the wire axis.

In this case, if the Dresselhaus interaction is absent, then $\nu = 0$ and the dispersion relation in Equation (7.58) simplifies to

$$E_\pm^{(1)} = \frac{\hbar^2 k_x^2}{2m^*} + E_0 \pm \sqrt{(\eta k_x)^2 + \beta^2}, \qquad (7.129)$$

where $\beta = g\mu_B B/2$. It is obvious from the above that $E_\pm^{(1)}(k_x) = E_\pm^{(1)}(-k_x)$ which proves that the dispersion relation is symmetric about the energy-axis.

Case II: When the magnetic field is along the direction of the electric field causing the Rashba interaction.

The dispersion relation is

$$E_{\pm}^{(2)} = \frac{\hbar^2 k_x^2}{2m^*} + E_0' \pm \sqrt{(\eta^2 + \nu^2) k_x^2 + \beta^2}. \qquad (7.130)$$

In this case, $E_{\pm}^{(2)}(k_x) = E_{\pm}^{(2)}(-k_x)$, always.

Case III: When the magnetic field is along the direction mutually perpendicular to the axis of the quantum wire and the electric field causing the Rashba interaction, and the Rashba interaction is absent.

The dispersion relation is

$$E_{\pm}^{(3)} = \frac{\hbar^2 k_x^2}{2m^*} + E_0' \pm \sqrt{\nu^2 k_x^2 + \beta^2}. \qquad (7.131)$$

Here too, $E_{\pm}^{(2)}(k_x) = E_{\pm}^{(2)}(-k_x)$, which makes the dispersion relations symmetric about the energy-axis.

- **Problem 7.4**

Consider the case when a magnetic field is directed along the axis of a quantum wire and the Dresselhaus interaction is absent. Show that the minima of the upper spin resolved subband occur at $k_x = 0$, but the minima of the lower subband occur at $k_x = \pm\sqrt{\frac{m^{*2}\eta^2}{\hbar^4} - \frac{\beta^2}{\eta^2}}$. The lower subband also has a local maximum at $k_x = 0$. This gives rise to the "camel-back" feature in the lower subband seen in Fig. 7.4. Show that the "camel-back" feature disappears if the magnetic flux density $B > B_c$ where $B_c = \frac{2m^*\eta^2}{\hbar^2 g\mu_B}$. Show that the same is true in the case when the magnetic field is along the direction of the electric field causing the Rashba interaction. In the latter case, all of the above is true even if there is a Dresselhaus interaction. The value of the B_c then is given by $B_c = \frac{2m^*(\eta^2 + \alpha_2^2)}{\hbar^2 g\mu_B}$.

Solution

To find the extrema, we differentiate the energy with respect to k_x, set the result equal to zero, and solve for k_x.

In this case,

$$\frac{1}{\hbar}\frac{\partial E_{\pm}}{\partial k_x} = \frac{\hbar k_x}{m^*} \pm \frac{\eta^2 k_x}{\hbar\sqrt{(\eta k_x)^2 + \beta^2}}. \qquad (7.132)$$

Setting this equal to zero and solving for k_x, we get the following:

Upper band: In the case of the $+$ sign, there is only one solution for k_x which is $k_x = 0$. This shows that the upper subband has a minimum at $k_x = 0$.

Lower band: In this case, there are two solutions for k_x. One is $k_x = 0$, which corresponds to a local maximum. The other solution is

$$k_x^2 = \frac{m^{*2}\eta^2}{\hbar^4} - \frac{\beta^2}{\eta^2}. \qquad (7.133)$$

We note first that there is no *real* solution for k_x if $\beta > \eta^2 m^*/\hbar^2$, or $B > \frac{2m^*\eta^2}{\hbar^2 g\mu_B}$. Therefore, there are no other extrema if $B > B_c$ where $B_c = \frac{2m^*\eta^2}{\hbar^2 g\mu_B}$. In that case, there will be no "camel-back" feature. If, on the other hand, $B < B_c$, then from Equation(7.133), we get that there are minima located at

$$k_x = \pm\sqrt{\frac{m^{*2}\eta^2}{\hbar^4} - \frac{\beta^2}{\eta^2}}. \qquad (7.134)$$

The reader can easily extend this analysis to show that in the case when the magnetic field is along the direction of the electric field inducing the Rashba effect, the minima occur at $k_x = \pm\sqrt{\frac{m^*}{\hbar^2} - \frac{\beta^2}{\eta^2 + \alpha_2^2}}$ and $B_c = \frac{2m^*(\eta^2 + \alpha_2^2)}{\hbar^2 g\mu_B}$.

- **Problem 7.5**

Consider a quantum dot in which there is no Dresselhaus interaction, but there is a Rashba interaction due to a symmetry breaking electric field in the y-direction. There is also a magnetic field in the x-direction. In this case, show that the spin of an electron in the ground state lies in the x-y plane, i.e., $S_z = 0$.

Solution

If only the Rashba interaction is present, then from Equation (7.117), we see that $\delta = D$. However, since D is purely imaginary, we get that $\delta = i|\delta|$, so that the angle $\xi = \pi/2$. Using this result in Equation (7.122), we immediately find that $S_z = 0$. Note that when the Rashba interaction strength is extremely strong so that $|\delta| >> g\mu_B B$, then $\zeta = \pi/4$ and we have $S_x^- = S_z^- = 0$, and $S_y^- = $ -1, while $S_x^+ = S_z^+ = 0$, and $S_y^+ = 1$. In that case, the ground state spin points against the electric field inducing the Rashba effect and the excited state spin points in the direction of the electric field. Of course, when the Rashba interaction strength is very weak so that $|\delta| << g\mu_B B$, then $S_y^\pm = S_z^\pm = 0$, and $S_x^\pm = \mp 1$. In that case, the ground state spin points along the magnetic field and the excited state spin points against the magnetic field, as we expect.

- **Problem 7.6**

In a semiconductor, absorption of light is caused by a photon exciting an electron from a lower energy state to a higher energy state and in the process getting absorbed. Emission is the reverse process where an electron drops down from a higher energy state to a lower energy one, giving off a photon.

Consider a semiconductor quantum wire with its axis along the x-direction. For light polarized along the width of the wire (y-direction), the strength of optical absorption and the intensity of emission due to intrasubband transitions between the two spin-split levels are proportional to the squared matrix element $|M|^2 = K \left| \int_{-\infty}^{\infty} dy \psi_{final}^*(y) \left(\frac{\partial}{\partial y} \right) psi_{initial}(y) \right|^2$, where K is a constant, $\psi_{initial}$ and ψ_{final} are the electron's initial and final state envelope wavefunction components in the y-direction, and the asterisk denotes complex conjugate.

Show that if we make the variable separation approximation and ignore spin texturing effect, then absorption and emission of light between the two spin-split levels will be forbidden. However, if we do not make this approximation, then these transitions are allowed. Thus, the variable separation approximation has very non-trivial effects.

Solution

If we make the variable separation approximation, then

$$\left[\psi_{final}^*(y) \right] = \phi_{final}^*(y) \left[\alpha^* \ \beta^* \right]$$

$$\left[\psi_{initial}(y) \right] = \phi_{initial}(y) \begin{bmatrix} \alpha' \\ \beta' \end{bmatrix},$$

where spin orthogonality mandates

$$\left[\alpha^* \ \beta^* \right] \begin{bmatrix} \alpha' \\ \beta' \end{bmatrix} = \alpha^* \alpha' + \beta^* \beta' = 0. \tag{7.135}$$

In this case, the squared matrix element is

$$|M|^2 = K \left| \alpha^* \alpha' + \beta^* \beta' \right|^2 \left| \int_0^{\infty} dy \phi_{final}^*(y) \left(\frac{\partial}{\partial y} \right) \phi_{initial}(y) \right|^2 = 0. \tag{7.136}$$

The point to note is that since the α-s and β-s do *not* depend on space, we could pull them outside the integral. Spin orthogonality given by Equation (7.135) then immediately shows that the matrix element is zero, which implies that absorption and emission between spin-orthogonal states are forbidden.

But now consider the situation when we do not make the variable separation approximation. Here,

$$\left[\psi_{final}^*(y) \right] = \begin{bmatrix} \Phi_{final}^*(y) \\ \Psi_{final}^*(y) \end{bmatrix}$$

$$\left[\psi_{initial}(y) \right] = \begin{bmatrix} \Phi_{initial}^*(y) \\ \Psi_{initial}^*(y) \end{bmatrix}$$

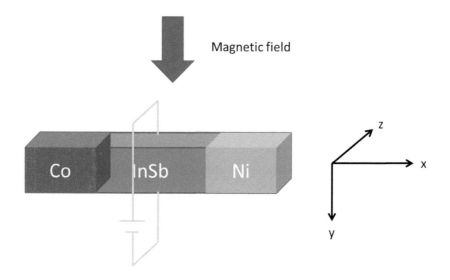

FIGURE 7.9
A one-dimensional spin valve of InSb placed in a magnetic field.

Spin orthogonality now mandates

$$\int_0^\infty dy \left[\Phi^*_{final}(y)\Phi_{initial}(y) + \Psi^*_{final}(y)\Psi_{initial}(y) \right] = 0. \qquad (7.137)$$

However, the squared matrix element is

$$|M|^2 = K \left| \int_0^\infty dy \left[\Phi^*_{final}(y)\frac{\partial \Phi_{initial}(y)}{\partial y} + \Psi^*_{final}(y)\frac{\partial \Psi_{initial}(y)}{\partial y} \right] \right|^2 \neq 0. \qquad (7.138)$$

The last equation clearly shows that optical transitions between spin orthogonal levels are not forbidden if we take spin texturing effects into account and do not make the variable separation approximation.

- **Problem 7.7**

  Consider a one-dimensional structure of InSb, as shown in Fig. 7.9, that is placed in a magnetic field of flux density B. The wavefunction can be written as $[\Psi(x,y,z)] = e^{ik_x x}\phi(y)[\psi(z)]$. Assume that there is no Dresselhaus interaction, but there is a very strong Rashba interaction due to the battery shown. Show that the the energy eigenvalues and the corresponding spinors at $k_x = 0$ can be found by solving the equation

$$
\begin{bmatrix}
-\frac{\hbar^2}{2m^*}\frac{d^2}{dz^2} + \frac{e^2 B^2 z^2}{2m^*} + \eta\frac{eBz}{\hbar} - E & i\left(\eta\frac{d}{dz} + \frac{g\mu_B B}{2}\right) \\
i\left(\eta\frac{d}{dz} - \frac{g\mu_B B}{2}\right) & -\frac{\hbar^2}{2m^*}\frac{d^2}{dz^2} + \frac{e^2 B^2 z^2}{2m^*} - \eta\frac{eBz}{\hbar} - E
\end{bmatrix}
\begin{bmatrix}
\psi_1(z) \\
\psi_2(z)
\end{bmatrix}
$$
$$= 0. \tag{7.139}$$

- **Problem 7.8**

  Assume that the eigenspinors are spatially invariant. Find the spin-splitting energy at $k_x = 0$.

7.7   References

[1] M. Cahay and S. Bandyopadhyay, "Phase coherent quantum mechanical spin transport in a weakly disordered quasi one-dimensional channel", Phys. Rev. B, **69**, 045303 (2004).

[2] C. Lü, H. C. Schneider and M. W. Wu, "Electron spin relaxation in n-type InAs quantum wires", J. Appl. Phys., **106**, 073703 (2009).

[3] B. A. Bernevig, J. Orenstein and S-C Zhang, "Exact SU(2) symmetry and persistent spin helix in a spin-orbit coupled system", Phys. Rev. Lett., **97**, 236601 (2006).

[4] J. D. Koralek, et al., "Emergence of a persistent spin helix in semiconductor quantum wells", Nature, **458**, 610 (2009).

[5] M. P. Walser, C. Reichl, W. Wegscheider and G. Salis, "Direct mapping of the formation of a persistent spin helix", Nature Phys., **8**, 757 (2012).

[6] A. Sasaki, et al., "Direct determination of spin-orbit interaction coefficients and realization of the persistent spin helix symmetry", Nature Nanotechnol., **9**, 703 (2014).

[7] U. Zülicke and M. Governale, "Spin accumulation in quantum wire with strong Rashba spin orbit coupling", Phys. Rev. B, **66**, 073311 (2002).

[8] S. Pramanik, S. Bandyopadhyay and M. Cahay, Energy dispersion relations of spin-split subbands in a quantum wire and electrostatic modulation of carrier spin polarization, Phys. Rev. B., **76**, 155325 (2007).

[9] S. Bandyopadhyay and M. Cahay, "Proposal for a spintronic femto-Tesla magnetic field sensor", Physica E, **27**, 98 (2005).

[10] S. Pramanik, S. Bandyopadhyay and M. Cahay, Phys. Rev. B, **76**, 155325 (2007).

[11] B. Das, D. C. Miller, S. Datta, R. Reifenberger, W. P. Hong, P. K. Bhattacharyya, J. Singh and M. Jaffe, "Zero field spin splitting in a two-dimensional electron gas", Phys. Rev. B, **39**, 1411 (1989).

[12] J. Schliemann, J. C. Egues and D. Loss, "Non ballistic spin field effect transistor", Phys. Rev. Lett., **90**, 146801 (2003); X. Cartoixá, D. Z-Y Ting and Y-C Chang, "A resonant spin lifetime transistor", Appl. Phys. Lett., **83**, 1462 (2003).

[13] E. Tsitsishvili, G. S. Lozno and A. O. Gogolin, "Rashba coupling in quantum dots: An exact solution", Phys. Rev. B., **70**, 115316 (2004).

8

Spin Relaxation

In Chapter 5, we were introduced to the concept of spin relaxation and the T_1 and T_2 times. Spin relaxation is of paramount importance in spintronics. Here, one is often concerned with using the spin polarization of either a single charge carrier (or the net spin polarization of an ensemble of charge carriers) to encode information. An extreme example of this strategy is the Single Spin Logic (SSL) paradigm discussed in Chapter 13. If spin is to host information reliably, it must be protected against random and spontaneous depolarization, caused by "spin relaxation." In this chapter, we discuss the various mechanisms for spin relaxation in a solid.

When an electron is introduced in a solid, interaction with the environment can affect its spin orientation. The environment may give rise to an *effective magnetic field* which interacts with the spin and causes it to alter its state, thereby changing the orientation. The effective magnetic field can arise from a multitude of sources, e.g., the spins of other electrons and holes in the solid, nuclear spins, phonons (vibrating atoms) that give rise to a time varying magnetic field in some circumstances, and, Most important, spin-orbit interactions in the solid that act like an effective magnetic field, as we saw in the previous chapter.

Any effective magnetic field interacts with the magnetic moment of the electron's spin. The interaction energy is given by Equation (6.6). This interaction can alter the electron's spin state. To understand how this occurs, first consider the time-independent Pauli equation

$$[H_0+H_{so}][\psi] = [H_0-g\mu_B\vec{B}_{eff}\cdot\vec{s}][\psi] = [H_0+(g/2)\mu_B\vec{B}_{eff}\cdot\vec{\sigma}][\psi] = E[\psi] \quad (8.1)$$

describing this electron. The spin-orbit interaction Hamiltonian is written as a Zeeman interaction term, $(g/2)\mu_B\vec{B}_{eff}\cdot\vec{\sigma}$, due to the effective magnetic field arising from the interaction. Assume, for simplicity, that the effective magnetic field is directed along the z-direction. In that case, the time independent Pauli equation reduces to

$$[H_0 - (g/2)\mu_B B_{eff}\sigma_z][\psi] = \begin{bmatrix} H_O - (g/2)\mu_B B_{eff} & 0 \\ 0 & H_O + (g/2)\mu_B B_{eff} \end{bmatrix}[\psi]$$
$$= E[\psi], \quad (8.2)$$

where $[\psi]$ is the 2-component wavefunction.

Diagonalization of the Hamiltonian in the basis of unperturbed states ϕ_0 (unperturbed states are eigenstates of H_0) yields the eigenenergies

$$E_\pm = < H_O > \pm (g/2)\mu_B B \tag{8.3}$$

where $H_0 >=< \phi_0|H_0|\phi_0 >$. The corresponding eigenspinors are

$$|1> = \begin{bmatrix} 0 \\ 1 \end{bmatrix} \tag{8.4}$$

$$|0> = \begin{bmatrix} 1 \\ 0 \end{bmatrix}, \tag{8.5}$$

which are the $-z$-polarized and $+z$-polarized spin states. In other words, the stable and metastable (time-invariant) spin polarizations in the presence of the effective magnetic field are those that are parallel and anti-parallel to the magnetic field. No other spin polarization is an eigenstate and stable (or metastable). Note that a magnetic field makes the spin polarization "bistable" (i.e., only the $+z$- and $-z$-polarized states are stable or metastable) and this feature is exploited in the realization of the Single Spin Logic (SSL) family discussed in Chapter 15.

Now consider an electron that finds itself in a different effective magnetic field caused by a magnetic scatterer, or other spins, or a different spin-orbit interaction. If the electron's spin polarization is already parallel or anti-parallel to the new effective magnetic field, then the spin will not change since the spin polarization is already an eigenstate and therefore stable. However, this will be a rare situation. Most of the time, the electron's initial spin will not be parallel or anti-parallel to the new effective magnetic field and therefore the initial spin will *not* be stable (or time invariant) and begin to change with time as soon as the electron encounters the new effective magnetic field. We can actually show, starting from the time dependent Pauli equation (see Problem 8.1 and also recall Chapter 4), that the spin will *precess* about the new effective magnetic field B_{eff} with an angular frequency $\Omega_{eff} = g\mu_B B_{eff}/\hbar$, where g is the Landé g-factor of the medium where the electron resides. This will happen only if there is no dissipation (becasue the Ehrenfest theorem from which we derived Larmor precession is valid only in the absence of dissipation) and the time evolution of the spin will be described by the Larmor precession equation

$$\frac{d\vec{S}}{dt} = \vec{\Omega}_{eff} \times \vec{S}, \tag{8.6}$$

which we derived in Chapter 4.

Now, since the flux density B_{eff}, caused by spin orbit interaction, depends on the electron's velocity $\vec{v}$ or wavevector k, therefore, if an electron's velocity or wavevector changes randomly owing to scattering, then two different things happen. First, the axis about which the spin precesses changes direction because the effective magnetic field changes direction, and second, the frequency of Larmor precession changes. These changes occur randomly in time because scattering events are random. As a result, the spin orientation of an electron changes randomly in time. This leads to spin relaxation. We will talk about this more when we discuss the D'yakonov-Perel' mechanism of spin relaxation [1].

If spin changes gradually with time, we normally call it "spin relaxation," while if the spin changes suddenly (discretely in time) then we tend to call it a "spin flip," A complete spin-flip is one where an "up-spin" state becomes a "down-spin" state and vice versa, i.e., the spin rotates by a full 180°. But two such anti-parallel spin states are mutually *orthogonal* and therefore we normally expect no coupling between them since the matrix element connecting these two states should be zero. So, what can couple two orthogonal states and cause transitions between them, resulting in a spin flip? The "scatterer" that couples them and causes transitions must have an internal magnetic field of some sort. In that case, the "up-spin" state acquires a little admixture of the "down-spin" state, and vice versa, so that the two states are no longer completely orthogonal. This will then result in some coupling between them.

A scatterer with an internal magnetic field can also make an electron originally in an eigenspinor state become unstable and change spin orientation. This happens because, when we add the scatterer's magnetic field (vectorially) to the original magnetic field that the electron was residing in, we get a new magnetic field. The electron's spin polarization is no longer parallel or anti-parallel to this new magnetic field. Therefore, the electron is no longer in a spin eigenstate. Thus, the scatterer makes the spin unstable and causes it to change.

In the following sections, we discuss the major spin flip mechanisms in a solid. But first we will revisit the issue of the spin-orbit field B_{eff} (or B_{so}) being proportional to either velocity v or wavevector k. We showed in the previous chapter that the B_{so} which is proportional to k is spin-dependent. Now, we will show that we can define a *spin-independent B_{so}* for either the Rashba or the Dresselhaus interaction which will be proportional to the *average* wavevector k_{av} in the two spin-split bands. This average wavevector is the arithmetic mean of the wavevectors in the two spin-split bands and is therefore spin-independent.

8.1   Spin-independent spin–orbit magnetic field

In Chapter 7 (Section 7.6), we applied the Ehrenfest theorem to deduce the spin-orbit field B_{so} from the spin-orbit splitting energy and came up with Equation (9.15). This approach yields that B_{so} is proportional to k, not v. However, that also immediately makes this B_{so} *spin-dependent* since an

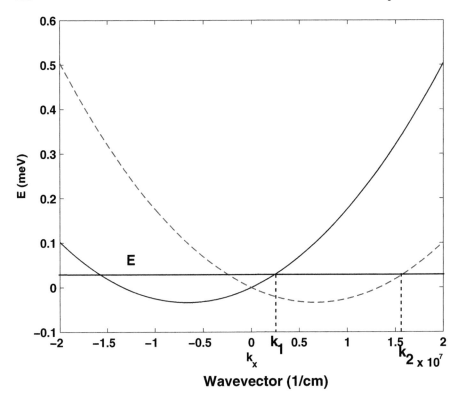

FIGURE 8.1

Dispersion relations of spin-split subbands in a quantum wire with just Rashba interaction. In this plot, we assumed that $\eta = 10^{-11}$ eV-m and $m^* = 0.05$ m_0.

electron with energy E has two different wavevectors k_1 and k_2 in the two spin-split bands in the presence of spin-orbit interaction. This is shown in Fig. 8.1.

To define a *spin-independent* or *spin-averaged* B_{so}, we proceed as follows. The dispersion relations in the presence of either the Rashba or the Dresselhaus interaction in a one- or two-dimensional electron gas are (neglecting band structure non-parabolicity and any external magnetic field):

$$E^{\pm}(k) = \frac{\hbar^2 k^2}{2m^*} \pm \eta k \quad (RASHBA)$$

$$E^{\pm}(k) = \frac{\hbar^2 k^2}{2m^*} \pm \nu k \quad (DRESSELHAUS). \tag{8.7}$$

These are the dispersion relations shown in Fig. 8.1.

The velocities of an electron with energy E in the two spin-split bands are

$$v_1(E) = \frac{1}{\hbar}\frac{\partial E^+}{\partial k}\bigg|_{E=E^+} = \frac{\hbar k_1(E)}{m^*} + \frac{\eta}{\hbar} \quad (RASHBA)$$

$$= \frac{1}{\hbar}\frac{\partial E^+}{\partial k}\bigg|_{E=E^+} = \frac{\hbar k_1(E)}{m^*} + \frac{\nu}{\hbar} \quad (DRESSELHAUS),$$

$$v_2(E) = \frac{1}{\hbar}\frac{\partial E^-}{\partial k}\bigg|_{E=E^-} = \frac{\hbar k_2(E)}{m^*} - \frac{\eta}{\hbar} \quad (RASHBA)$$

$$= \frac{1}{\hbar}\frac{\partial E^-}{\partial k}\bigg|_{E=E^+} = \frac{\hbar k_2(E)}{m^*} - \frac{\nu}{\hbar} \quad (DRESSELHAUS). \quad (8.8)$$

From Equation (8.7), we get

$$k_2(E) - k_1(E) = \frac{2m*\eta}{\hbar^2} \quad (RASHBA)$$

$$= \frac{2m*\nu}{\hbar^2} \quad (DRESSELHAUS). \quad (8.9)$$

Using the last equation in Equation (8.8), we get

$$v_1(E) = v_2(E) = v(E), \quad (8.10)$$

which shows that the velocity of an electron with energy E is the same in both spin-split bands, so that the velocity is *spin-independent*. Moreover, from Equation (8.8), we obtain

$$v(E) = \frac{v_1(E) + v_2(E)}{2} = \frac{\hbar}{m^*}\frac{k_1(E) + k_2(E)}{2} = \frac{\hbar k_{av}}{m^*}. \quad (8.11)$$

Since the *spin-averaged* B_{so} should be proportional to spin-averaged k, i.e., k_{av}, we can posit that the spin-averaged B_{so} should be proportional to v since v and k_{av} have turned out to be proportional to each other. Since v is spin-independent, the spin averaged B_{so} will also be spin-independent, as it should be.

It is easy to show (an exercise left for the reader) that the velocity $v(E)$ *in either spin-split band* at any energy E is given by

$$v(E) = \pm\sqrt{\frac{\eta^2}{\hbar^2} + \frac{2E}{m^*}} \quad (RASHBA)$$

$$v(E) = \pm\sqrt{\frac{\nu^2}{\hbar^2} + \frac{2E}{m^*}} \quad (DRESSELHAUS). \quad (8.12)$$

Therefore, to find an expression for the spin-independent B_{Rashba}, we should replace k in Equation (9.15) with k_{av} or $m^*v(E)/\hbar$. This will yield

$$|B_{Rashba}| = \frac{2\eta m*}{g\mu_B\hbar}v(E) = \frac{2\eta m*}{g\mu_B\hbar}\sqrt{\frac{\eta^2}{\hbar^2} + \frac{2E}{m^*}}. \quad (8.13)$$

Similarly,

$$|B_{Dresselhaus}| = \frac{2\nu m^*}{g\mu_B \hbar} v(E) = \frac{2\nu m*}{g\mu_B \hbar} \sqrt{\frac{\nu^2}{\hbar^2} + \frac{2E}{m^*}}. \tag{8.14}$$

Note that the above equations clearly show that the spin-independent magnetic fields depend only on the electron energy regardless of the spin orientation.

In the rest of this book, we will use the spin-independent magnetic fields in Equations (8.13) and (8.14) whenever we talk about spin precession about an effective magnetic field caused by spin-orbit interaction. In that case, we will not have to worry about the effective magnetic field itself changing with the precessing spin. The notion of the spin-independent effective magnetic field associated with spin-orbit interaction allows us to present a completely classical description of the Spin Field Effect Transistor's operation in Chapter 14.

8.2   Spin relaxation mechanisms

The four major spin relaxation mechanisms of electrons in the conduction band of a semiconductor are (i) the D'yakonov-Perel' mechanism, (ii) Elliott-Yafet mechanism, (iii) the Bir-Aronov-Pikus mechanism, and (iv) hyperfine interactions with nuclear spins. Physical pictures to aid in the understanding of the first three mechanisms were presented in a paper by Fabian and Das Sarma [2].

8.2.1   Elliott–Yafet mechanism

The Elliott-Yafet mechanism is a result of the fact that in a crystal, the Bloch states (which are solutions of the Schrödinger equation in the periodic lattice potential) are not spin eigenstates. To see this clearly, note that the Pauli equation governing an electron in a crystal is

$$\left[-\frac{\hbar^2}{2m}\nabla^2 + V_L\left(\vec{r}\right) + H_{so} \right] [\Psi] H[\Psi] = E[\Psi], \tag{8.15}$$

where $V_L\left(\vec{r}\right)$ is the periodic lattice potential. The solution of the unperturbed Hamiltonian (in the absence of spin-orbit interaction) is

$$\left[-\frac{\hbar^2}{2m}\nabla^2 + V_L\left(\vec{r}\right) \right] [\Psi_0] = H_0\left[\Psi_0\right] = E_0\left[\Psi_0\right], \tag{8.16}$$

where $[\Psi_0]$ is the wavefunction in a crystal given by the Bloch theorem, i.e., $[\Psi_0] = e^{i\vec{k}\cdot\vec{r}}\left[u_{\vec{k}}\left(\vec{r}\right)\right]$. Since $H \neq H_0$, $[\Psi] \neq e^{i\vec{k}\cdot\vec{r}}\left[u_{\vec{k}}\left(\vec{r}\right)\right]$. The spin-orbit interaction mixes up the up-spin and down-spin Bloch functions so that

$$[\Psi] = e^{i\vec{k}\cdot\vec{r}} \left\{ a_{\vec{k}} \left[u_{\vec{k}}^{\uparrow}(\vec{r}) \right] + b_{\vec{k}} \left[u_{\vec{k}}^{\uparrow}(\vec{r}) \right] \right\}, \tag{8.17}$$

where the coefficients $a_{\vec{k}}$ and $b_{\vec{k}}$ depend on the wavevector $\vec{k}$.

Assume that the spin quantization axis is the z-axis so that "up-spin" states are $+z$-polarized states and "down-spin" states are the $-z$-polarized states. In that case,

$$[\Psi] = e^{i\vec{k}\cdot\vec{r}} \left\{ a_{\vec{k}} \begin{bmatrix} 1 \\ 0 \end{bmatrix} + b_{\vec{k}} \begin{bmatrix} 0 \\ 1 \end{bmatrix} \right\} = e^{i\vec{k}\cdot\vec{r}} \begin{bmatrix} a_{\vec{k}} \\ b_{\vec{k}} \end{bmatrix}. \tag{8.18}$$

The above result clearly shows that *in the presence of spin-orbit interaction*, an electron's spin in a crystal does not really have one of two fixed polarizations, "up" or "down", contrary to what we normally assume. The actual polarization depends on what the electron's wavevector is. This is schematically depicted in Fig. 8.2, where we plot the energy-wavevector relation in an arbitrary band and show that the spin orientation of an electron (depicted by the arrow) is slightly different in different wavevector states. Each wavevector state still has two possible spin orientations that are mutually anti-parallel, but spin orientations associated with different wavevector states can have arbitrary angles between them. This is a situation that we encountered in the previous chapter, where it was caused by an external magnetic field. Here it is caused by the fact that the Bloch eigenstates are not spin eigenstates in a crystal that lacks inversion symmetry.

If a collision event with a non-magnetic scatterer (such as an impurity, device boundary, or phonon) changes the momentum or wavevector of an electron, then it will change the spin as well since the spin orientations associated with the initial and final wavevector states are never mutually parallel. This process is also shown in Fig. 8.2. By how much the spin orientation changes naturally depends on how much the wavevector changes. Collisions with charged impurities that do not usually change the wavevector by a lot* will not be very effective in changing spin orientation this way, but collisions with certain types of acoustic phonons that prefer scattering through large angles (and therefore change the wavevector by a lot) will be very effective in relaxing spin.

Elliott [3] had considered an even more drastic situation. The spin state of a nearly up-spin electron at a wavevector state k_1 and that of a nearly down-spin electron at a *different* wavevector state k_2 are *not* strictly orthogonal, since they are not strictly anti-parallel. Thus, any collision with a non-magnetic scatterer that changes the wavevector of an electron from an initial state k_1 to a state k_2 can also couple the nearly down-spin state at k_1 to the nearly

---

*The matrix element for electron-impurity interaction typically has an inverse dependence on the net momentum change, meaning that small angle scatterings are preferred over large angle scatterings.

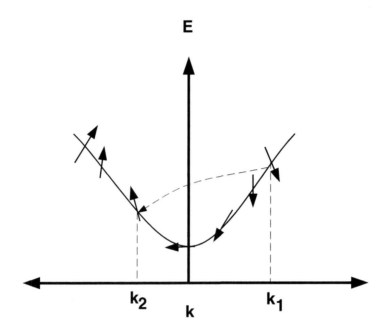

FIGURE 8.2
Energy dispersion relation showing the spin polarizations at different wavevector states. Each wavevector state has two possible mutually anti-parallel spin polarizations, one of which is shown by the arrowheads in the above figure. A momentum changing collision event, which changes the wavevector state from k_1 to k_2, also changes the spin orientation since the eigenspin polarizations at k_1 and k_2 are generally not parallel.

up-spin state at k_2. This coupling can flip an electron's spin from nearly up to nearly down. Note that this kind of spin relaxation is always accompanied by some degree of momentum relaxation since the wavevector must change in order to change the spin.

Pikus and Titkov [4] have calculated an approximate expression for the Elliott-Yafet spin relaxation rate in bulk semiconductors assuming that the carriers are in a non-degenerate population, in quasi equilibrium, and are described by a thermal distribution with temperature T. This expression is

$$\frac{1}{\tau_{EY}} = A \left(\frac{k_B T}{E_g}\right)^2 \mathcal{R}^2 \left(\frac{1 - \mathcal{R}/2}{1 - \mathcal{R}/3}\right)^2 \frac{1}{\tau_m}, \tag{8.19}$$

where E_g is the bandgap, $\mathcal{R} = \Delta/(E_g + \Delta)$, Δ is the spin-orbit splitting of the valence band, τ_m is the momentum relaxation time, k_B is the Boltzmann constant, and A is a constant varying from 2 to 6 depending on the dominant scattering mechanism for momentum relaxation.

This expression shows that the Elliott-Yafet spin relaxation rate $1/\tau_{EY}$ is directly proportional to the momentum relaxation rate $1/\tau_m$. This is to be expected since Elliot-Yafet relaxation is *caused* by momentum relaxation. However, what this expression also shows is that the spin and momentum relaxation rates have *different* temperature dependences. For example, if the momentum relaxation rate is independent of temperature (as expected for mostly elastic scattering events), then the spin relaxation rate should roughly increase quadratically with temperature since the temperature dependence of bandgap and spin-orbit interaction strengths is negligible. However, in a metal or strongly degenerate electron gas, the Elliott-Yafet spin relaxation rate and the momentum relaxation rate tend to have the same temperature dependence. Elliott derived a relation between the two rates as [5]

$$\frac{\tau_m}{\tau_{EY}} \propto \frac{\Delta}{E_g}, \tag{8.20}$$

which Yafet has shown to be temperature-independent [6]. According to the above equation, the two relaxation rates must have the same temperature dependence if we ignore the weak temperature dependences of Δ and E_g. There is also some experimental evidence that indeed the two rates have the same temperature dependence in metals [7].

The Elliott-Yafet mechanism is often the primary spin relaxation mechanism in low mobility materials, such as organics, where momentum relaxing scattering events are frequent. In high mobility semiconductors, the primary spin relaxation mechanism is usually the D'yakonov-Perel' mechanism, which we discuss next.

8.2.2 D'yakonov Perel' mechanism

If a solid lacks inversion symmetry because of its crystalline structure (such as in compound semiconductors) or because of an externally applied electric field (or a built-in electric field), then an electron in the solid will experience strong spin-orbit interaction. As shown in Chapter 6, the crystallographic inversion asymmetry (also known as bulk inversion asymmetry) gives rise to the Dresselhaus spin-orbit interaction, while the structural inversion asymmetry caused by an external (or built-in) electric field gives rise to the Rashba spin-orbit interaction. These interactions lift the degeneracy between up-spin and down-spin states at any non-zero wavevector (recall Chapters 6 and 7), so that these two states have different energies in the same wavevector state. In this respect, Dresselhaus and Rashba spin-orbit interactions truly behave like effective magnetic fields, since a magnetic field also lifts the degeneracy between up-spin and down-spin states at any given wavevector because of the Zeeman interaction. The effective magnetic field $\vec{B}(\vec{v})$, arising from either the Rashba or the Dresselhaus interaction, will be dependent on the electron's velocity $\vec{v}$ if it is spin-independent[†].

Earlier, we mentioned that the effective magnetic field causes an electron's spin to undergo Larmor precession. The precession axis is collinear with the magnetic field and the precession frequency is $\vec{\Omega}(\vec{v}) = g\mu_B \vec{B}(\vec{v})/\hbar$. Now consider an ensemble of electrons drifting and diffusing in a solid. If every electron's velocity $\vec{v}$ is the same and does not change with time, then the field $\vec{B}(\vec{v})$ is the same for all electrons and every electron precesses about this constant field with a fixed frequency $|\vec{\Omega}(\vec{v})|$. This does *not* cause any spin relaxation at all. If all electrons started with the same spin polarization, then after any arbitrary time, they all have precessed by exactly the same angle, and therefore they all have again the same spin polarization. The direction of this spin polarization may be different from the initial one, but that does not matter. The *magnitude* of the ensemble averaged spin does not change with time. Therefore, there is no spin relaxation.

However, if $\vec{v}$ changes *randomly* with time because of scattering, then different electrons would have precessed by different angles after a certain time because they have different scattering histories. Thus, if all electrons are injected with the same spin polarization, their spin polarizations gradually go out of phase with each other. After a sufficiently long time, the magnitude of the ensemble averaged spin will decay to zero. This is the basis of D'yakonov-Perel' (D-P) spin relaxation [8]. The relaxation of the ensemble averaged spin in a quantum wire, as a function of time and distance, is shown in Fig. 8.3.

It is obvious that $\vec{B}(\vec{v})$ – the velocity dependent effective magnetic field arising from Dresselhaus and Rashba spin-orbit interactions – is the cause of D'yakonov-Perel' relaxation. Therefore, anything that reduces $\vec{B}(\vec{v})$ will suppress D-P relaxation. Slower moving electrons experience a smaller $\vec{B}(\vec{v})$ since the magnitude of $\vec{B}(\vec{v})$ is proportional to the magnitude of $\vec{v}$. Thus, they are less susceptible to D-P relaxation. One would therefore expect that everything else being equal, high carrier mobility, or high saturation velocity

---

[†]See the discussion in Section 8.1.

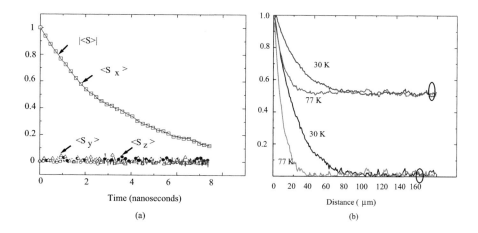

FIGURE 8.3

Decay of ensemble averaged spin components $< S_x >$, $< S_y >$ and $< S_z >$, as well as $| < S > | = \sqrt{< S_x >^2 + < S_y >^2 + < S_z >^2}$, in a GaAs quantum wire of width 30 nm and thickness 4 nm due to D'yakonov-Perel' relaxation. Electrons are intially injected into the quantum wire with all their spins polarized in the x-direction (axis of the wire). Relaxation-in-time plot (at a temperature of 30 K) (a) is reprinted with permission from S. Pramanik, et al., Phys. Rev. B, **68**, 075313 (2003). Copyright (2003) American Physical Society. Relaxation-in-space (distance along the length of the quantum wire) plot at two different temperatures (b) is reprinted with permission from S. Pramanik, et al., Appl. Phys. Lett., **84**, 266 (2004). Copyright (2004) American Institute of Physics. In (b), we show the relaxation of spin component $< S_x > (x)$ and the spin-filtered current $I_s(x)$. The spin-filtered current probed at any location x tells us the relative abundance of $+x$- and $-x$-polarized spins at that location. This current will decay by 50% when the spin has been completely randomized, i.e., there are equal number of $+x$- and $-x$-polarized spins.

of carriers, *reduces* spin lifetimes. That may indeed be true in most cases. Therefore, frequent momentum relaxing collisions may actually be beneficial to spin longevity. These collisions will tend to slow down electrons, reduce their average momentum, and therefore suppress D-P relaxation by reducing $\vec{B}(\vec{v})$, as originally noted by D'yakonov and Perel'. This phenomenon has come to be known as "motional narrowing." On the other hand, very strong momentum relaxing mechanisms also tend to put the spin polarizations of different electrons quickly out of phase with one another by causing a larger spread in the velocity of the carriers. Thus, it is not clear *a priori* whether momentum relaxing collisions are ultimately beneficial or detrimental to spin lifetimes. The answer may be different in different systems[‡].

The existence of the motional narrowing effect has been demonstrated indirectly in a remarkable experiment by Sandhu et al. [10]. They studied spin relaxation in a two-dimensional electron gas (a quantum well) which had a metal gate placed on top of the well. By applying an electrostatic potential to the gate, they could change the carrier concentration in the well and study the spin relaxation time as a function of carrier concentration. At high carrier concentration, the impurities in the well are screened by the electrons. As a result, the momentum relaxation rate due to impurity scattering is suppressed. Therefore, the D'yakonov-Perel' mode dominates. At lower carrier concentration, the screening is weakened and the momentum relaxation rate goes up. This reduces the D'yakonov-Perel' rate and the dominant spin relaxation mode becomes the Bir-Aronov-Pikus mode discussed in the next section. Sandhu's experiment was presumably carried out at small biases where hot carrier effects are unimportant and the motional narrowing effect is clearly manifested.

Pikus and Titkov [4] have derived an approximate expression for the spin relaxation rate due to the D'yakonov-Perel' mechanism in bulk *non-degenerate* semiconductors where the carriers are in quasi equilibrium described by Boltzmann statistics. Furthermore, the derivation assumes that the primary source of $\vec{B}(\vec{v})$ is the Dresselhaus interaction. Pikus' expression is

$$\frac{1}{\tau_{DP}} = Q\alpha^2 \frac{(kT)^3}{\hbar^2 E_g}\tau_m, \tag{8.21}$$

[‡]Many researchers believe that more frequent momentum relaxing scattering leads to longer D-P spin relaxation times. They think that the mean of the velocity distribution, which determines the average $\vec{B}(\vec{v})$, is determined by the frequency of momentum relaxation events, whereas the standard deviation of the velocity distribution is determined solely by temperature and has nothing to do with the momentum relaxation rate. This may be correct if the electron system is in quasi equilibrium and described by Boltzmann statistics. However, if the system is driven far out of equilibrium and experiences hot carrier effects, then this argument is invalid. Our simulations [9] have shown that there are two regimes of transport: the "motional narrowing regime" where frequent momentum relaxation events suppress D-P relaxation and the "inhomogeneous broadening regime" where frequent momentum relaxing scattering enhances D-P relaxation. The inhomogeneous broadening regime seems to appear only in high field transport where hot carrier effects are important.

where Q is a dimensionless quantity ranging from 0.8 to 2.7 depending on the dominant momentum relaxation process, E_g is the bandgap, τ_m is the momentum relaxation time, and α is a measure of the Dresselhaus interaction strength given by

$$\alpha \approx \frac{4\mathcal{R}}{\sqrt{3-\mathcal{R}}}\frac{m^*}{m_0}, \qquad (8.22)$$

where m_0 is the free electron mass, m^* is the electron's effective mass and $\mathcal{R}$ was defined previously in connection with the Elliott-Yafet mechanism. Note that this expression shows that the spin relaxation rate is inversely proportional to the momentum relaxation rate, which is true only in the motional narrowing regime. Furthermore, it shows that the spin relaxation rate will have a strong temperature dependence even if the momentum relaxation rate is temperature independent.

The temperature dependence of the D-P relaxation is important. In a strongly degenerate semiconductor, or metal, with large carrier concentration, the Fermi energy E_F is much larger than the thermal energy kT. Therefore, the velocity of electrons carrying current is largely determined by E_F and not kT. In that case, the D-P relaxation rate is more or less independent of temperature. By the same consideration, if the drift velocity of electrons is much larger than the thermal velocity, then the D-P relaxation will be again more or less temperature independent. Otherwise, the D-P relaxation will depend on the temperature. The exact dependence will depend on the temperature dependence of the momentum relaxing collisions which affect the carrier momentum. Generally, the D-P relaxation rate tends to increase slowly with increasing temperature [11].

Before we conclude our discussion of the D-P relaxation, we should stress that it is an "ensemble phenomenon" that can only be understood in a many-particle picture, as opposed to single-particle picture[§]. A single electron cannot experience D-P relaxation. Its spin will coherently precess about the effective (velocity-dependent) magnetic field caused by spin-orbit interaction, even when that field is changing randomly in time and space because of velocity randomizing collisions. The single electron's spin could always come back to the original state after some time, and, in fact, does come back to the original state if it completes a precession cycle before a scattering event changes the effective magnetic field. Hence, it cannot, by itself, experience D-P relaxation. However, in an ensemble of many electrons, different electrons are precessing about different magnetic fields simultaneously. Since all these precessions are not synchronous (the scattering histories of the different electrons are independent), the precessions go out of phase with each other. Hence the "ensemble average" spin decays in time and space, which is the basis of D-P

[§]Another well-known example of a phenomenon that can only be understood in a many-particle picture (with no single-particle analog) is "diffusion". Diffusion requires a concentration gradient which necesarily requires many particles forming a "concentration". A single particle does not form a concentration.

relaxation. Because the D-P relaxation is a result of ensemble averaging, it cannot be modeled by single-particle dynamics; it always requires many-body dynamics to be modeled correctly.

Elliott-Yafet relaxation, on the other hand, is easier to model since it is a single-particle phenomenon. A single electron's spin can change orientation if its wavevector changes, which is the basis of Elliott-Yafet relaxation. Hence, multi-particle dynamics is not necessary. A proper treatment of the E-Y mechanism, however, requires knowledge of the spin eigenstate at every wavevector state, which is not easily acquired.

D-P relaxation in a quantum wire

A very interesting question is whether there will be D-P relaxation in quantum wires? In a quantum wire (quasi one-dimensional system), $\vec{v}$ is always along the axis of the wire and therefore cannot change *direction* because of scattering or acceleration due to electric field (although it can change magnitude). Now, the spin-independent field $\vec{B}(\vec{v})$ depends on $\vec{v}$, but is *not* collinear with $\vec{v}$. In a quantum wire, the contribution to $\vec{B}(\vec{v})$ from the Dresselhaus interaction is indeed collinear with $\vec{v}$, but the contribution from the Rashba interaction is perpendicular to $\vec{v}$. Consider a quantum wire with rectangular cross-section shown in Fig. 8.4. The wire axis is along the x-direction and the symmetry breaking electric field which induces the Rashba spin-orbit interaction is along the y-direction. Current flows along the x-direction which is also assumed to be the crystallographic [100] direction. In this system, the Dresselhaus and Rashba contributions to $\vec{B}(\vec{v})$ will be given by (see Problem 8.2)

$$\vec{B}_{Dresselhaus}(\vec{v}) = \frac{2m^* a_{42}}{g\mu_B \hbar} \left[\left(\frac{m\pi}{W_z}\right)^2 - \left(\frac{n\pi}{W_y}\right)^2 \right] v_x \hat{x}$$

$$\vec{B}_{Rashba}(\vec{v}) = \frac{2m^* a_{46}}{g\mu_B \hbar} E_y v_x \hat{z}, \tag{8.23}$$

where W_z, W_y are the transverse dimensions of the wire, E_y is the symmetry-breaking electric field that induces the Rashba effect, a_{42}, a_{46} are material constants, v_x is the electron's velocity along the wire axis, and m, n are the transverse subband indices for the subband in which the electron resides. Note that the effective magnetic field due to the Dresselhaus interaction is along the wire axis, but that due to the Rashba interaction is mutually perpendicular to the wire axis and the symmetry-breaking electric field. The total magnetic field $\vec{B}(\vec{v})$ $(= \vec{B}_{Rashba}(\vec{v}) + \vec{B}_{Dresselhaus}(\vec{v}))$ is therefore in the $(x - z)$ plane

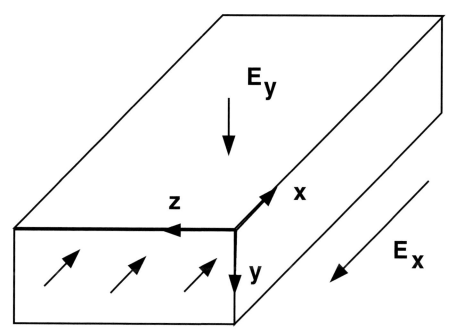

FIGURE 8.4
A quantum wire with rectangular cross-section. Electrons are injected at $x = 0$, with 100% spin injection efficiency.

and subtends an angle α with the wire axis given by

$$\alpha = arctan\left(\frac{B_{Rashba(\vec{v})}}{B_{Dresselhaus}(\vec{v})}\right) = arctan\left(\frac{a_{46}E_y}{a_{42}\left[\left(\frac{m\pi}{W_z}\right)^2 + \left(\frac{n\pi}{W_y}\right)^2\right]}\right).$$
(8.24)

Note that the angle α is independent of the velocity. Now, if the electron is always in a fixed subband defined by a fixed set of indices m, n, then the angle α (and therefore the axis of spin precession) will be *fixed* and invariant in time. In other words, the *direction* of the effective magnetic field $\vec{B}(\vec{v})$ does not change with time, although the *magnitude* does. Therefore, every electron precesses about the *same* axis, albeit with different frequencies. So, will there be any D-P spin relaxation in a quantum wire grown in the [100] crystallographic direction? The correct answer is extremely intriguing. There will be no D-P relaxation in *space* [12], but there will be D-P relaxation in *time*! We show this in Problem 8.3. Thus, there is a fundamental *difference* between spin relaxation in space versus spin relaxation in time. There is no analog of this in momentum or energy relaxation. Spin is indeed unique in this sense.

If there is inter-subband scattering and the electrons change subbands, then the angle α will change every time a subband transition occurs. In that case, both the magnitude *and* the direction of $\vec{B}(\vec{v})$ change randomly in time. In this case, there will be D-P relaxation in both time and space. Note also, that neither Rashba nor Dresselhaus interaction *alone* can cause spatial D-P relaxation in a quantum wire even if inter-subband scattering occurs. Both interactions have to be present. If either one is absent, then α is either 0 or $\pi/2$, i.e., α is fixed and D-P relaxation in space does not occur. Thus, the two necessary ingredients for *spatial* D-P relaxation in a quantum wire are (i) simultaneous presence of both Dresselhaus and Rashba interactions, and (ii) multichanneled transport. This is shown again in Problem 8.3.

Even if transport is not strictly single channeled, the D-P relaxation is progressively subdued in a quantum wire as its transverse dimensions shrink. This has been theorized in a number of publications [13, 14]. There is also some experimental evidence to support this postulate [15]. Thus, quantum wires have an advantage over quantum wells when it comes to spin relaxation. Spin relaxation via the D-P mode can be dramatically suppressed in quantum wires. As a result, many of the spin transistors described in Chapter 14, which are adversely affected by D-P spin relaxation, are best implemented with quantum wire channels as opposed to quantum well channels.

Another important issue that has lately received some attention is the anisotropy of spin relaxation. Because the effective magnetic fields due to the Dresselhaus and Rashba interactions point in different directions, as noted above, the interplay between these two interactions will make the spin relaxation rate depend on the initial polarization of the spin, i.e., what angle the initial spin polarization subtends with the net effective magnetic field caused by Rashba and Dresselhaus interactions. This has been noted in both quantum wires [16] and quantum wells. The Dresselhaus interaction also depends strongly on the crystallographic orientation (Miller indices) associated with the direction of spin transport. This can give rise to anisotropy of spin relaxation rates of a different kind, meaning that spin relaxation rate will depend strongly on the direction of spin transport within a crystal. This has been studied theoretically in the context of quantum wells [17, 18]. The anisotropy of spin relaxation rates can be quite strong. The relaxation rate for spins originally polarized along one particular crystallographic direction may differ from that originally polarized along another crystallographic direction by a few orders of magnitude! A dramatic experimental demonstration of the influence of crystallographic orientation was shown by Ohno, et al. [19]. They showed that the measured spin relaxation time in (110) GaAs quantum wells is an order of magnitude larger than in (100) GaAs quantum wells. Therefore, when it comes to implementing spintronic devices where spin relaxation is a major impediment to good performance, it pays to choose the crystallographic orientation of the device structure judiciously in order to improve device characteristics.

8.2.3 Bir–Aronov–Pikus mechanism

The Bir-Aronov-Pikus mechanism is a source of spin relaxation of electrons in semiconductors where there is a significant concentration of both electrons and holes. In that case, an electron will usually be in close proximity to a hole, so that their wavefunctions will overlap, which will cause an exchange interaction between them. Exchange interaction is the subject of the next chapter. If the hole's spin flips, then exchange interaction will cause the electron's spin to flip as well, leading to spin relaxation.

Pikus and Titkov [4] have calculated approximate expressions for the spin relaxation rate of electrons due to the Bir-Aronov-Pikus mechanism in bulk semiconductors. For a non-degenerate semiconductor, Pikus' expression is

$$\frac{1}{\tau_{BAP}} = \frac{2a_B^3}{\tau_0 v_B} \left(\frac{2E}{m^*}\right)^{1/2} \left[n_f|\psi(0)|^4 + \frac{5}{3}n_b\right],\tag{8.25}$$

where n_a (n_b) is the concentration of free (bound) holes, E is the electron energy in the conduction band, m^* is the electron effective mass, and

$$\tau_0 = \frac{64 E_B \hbar}{3\pi \Delta_x^2},$$

$$a_B = \left(\frac{m_0}{\mu}\right)\epsilon_r a_0 \quad [effective \; Bohr \; radius],$$

$$v_B = \frac{\hbar}{\mu a_B},$$

$$E_B = \frac{\hbar^2}{2\mu a_B^2} \quad [exciton \; binding \; energy],$$

$$|\psi(0)|^2 = \frac{2\pi}{\kappa}\left(1 - e^{-2\pi/\kappa}\right)^{-1},$$

$$\kappa = \sqrt{E/E_B},\tag{8.26}$$

where a_o is the Bohr radius of hydrogen atom (0.529 Å), ϵ_r is the relative dielectric constant of the material, μ is the electron-hole reduced mass, and Δ_x is the exchange splitting of the exciton ground state. For a semiconductor with degenerate concentration of holes, the result is

$$\frac{1}{\tau_{BAP}} = \frac{2a_B^3}{\tau_0 v_B}\left(\frac{E}{E_F}\right)n_f|\psi(0)|^4(2E/m^*)^{1/2}; \quad if \; E_F < E(m_v/m^*)$$

$$= \frac{2a_B^3}{\tau_0 v_B}\left(\frac{E}{E_F}\right)n_f|\psi(0)|^4(2E_F/m^*)^{1/2}; \quad if \; E_F > E(m_v/m^*)$$

$$\tag{8.27}$$

where E_F is the Fermi energy and m_v is the relevant hole effective mass (light- or heavy-hole, whichever is participating in the Bir-Aronov-Pikus mechanism).

8.2.4   Hyperfine interactions with nuclear spins

In a solid, the nuclear spins generate a magnetic field which interacts with the electron spins via hyperfine contact interaction and can cause spin relaxation. The Hamiltonian describing this interaction is given by

$$H_{nuclear} = \vec{S} \cdot \sum_i A_i \vec{I}_i, \tag{8.28}$$

where $\vec{I}_i$ is the spin of the i-th nucleus, A is a constant and A_i is the corresponding coupling coefficient. The quantity A_i is given by

$$A_i = AV|\psi(\vec{r}_i)|^2, \tag{8.29}$$

where $\psi(\vec{r}_i)$ is the electron envelope wavefunction at the nuclear site $\vec{r}_i$ and V is the volume of the crystal cell.

In order to study how the electron spin decays in time, one needs to evaluate the correlator [20]

$$C_n(t) =< n|[S_z(t) - S_z(0)]S_z(0)|n >, \tag{8.30}$$

where $|n>$ is the coupled electron and nuclear state, $S_z(0)$ is the initial spin eigenstate at time $t = 0$ and $S_z(t)$ is the eigenstate at some later time t given by

$$e^{iH_{nuclear}t}S_z(0)e^{-iH_{nuclear}t}. \tag{8.31}$$

From the behavior of $C_n(t)$, one can deduce the temporal behavior of $S_z(t)$.

Khaetski, et al. [20] have studied this problem for both unpolarized and fully polarized nuclei in a quantum dot. In the former case, the ensemble averaged spin eigenstate $< S_z(t) >$ oscillates between $S_z(0)$ and $-S_z(0)$ with a period $\sqrt{N}/A$ where N is the number of nuclear spins interacting with the electron. This behavior persists up to a time $\sim N/A$. Then $< S_z(t) >$ decays non-exponentially with time. In the case of a fully polarized nucleus, $< S_z(t) >$ initially oscillates with a period $1/A$ up to a time $\sim N/A$ and then it decays non-exponentially with time. The decay is much faster for the fully polarized case, but the steady state value $< S_z(\infty) >$ is much closer to $S_z(0)$ in the fully polarized case than in the unpolarized case. It is interesting to note that in neither case $< S_z(\infty) > = 0$, meaning that hyperfine interaction does not cause complete loss of electron spin polarization.

8.3   Spin relaxation in a quantum dot

We will pay special attention to the spin relaxation of an electron in a quantum dot since this system forms the basis of two important technologies: single spin logic (SSL) and spin-based quantum logic gates. These are discussed in Chapters 15 and 16. In a quantum dot hosting a single electron, there are *five* major sources of spin relaxation. These are (i) spin-orbit interaction arising from any local electric field that causes Rashba interaction, (ii) spin-orbit interaction associated with crystallographic inversion asymmetry in the quantum dot material causing Dresselhaus interaction, (iii) spin orbit interaction due to strain field in the solid generated by phonons, (iv) spin-phonon coupling in the presence of an external magnetic field which mixes the valence band and conduction band states, (v) hyperfine interactions with the nuclear spin. All of these effects give rise to essentially effective magnetic fields that can interact with the spin of an electron in the quantum dot and change its state.

The Hamiltonians describing these perturbations are [21]

$$H_1 = (a_{46}/\hbar)\,\vec{\mathcal{E}}\cdot[\vec{\sigma}\times(\vec{p}+e\vec{A})]$$
$$H_2 = a_{42}\vec{\sigma}\cdot\vec{\kappa}$$
$$H_3 = \gamma\vec{\sigma}\cdot\vec{\varsigma}$$
$$H_4 = \hat{g}\mu_B(u_{xz}\sigma_x + u_{yz}\sigma_y)B_z \tag{8.32}$$
$$H_5 = \mathcal{A}\sum_{i,k}\mathbf{S}_i\cdot\mathbf{I}_k\delta\left(\mathbf{r}_i - \mathbf{R}_k\right), \tag{8.33}$$

where a_{46} and a_{42} are the familiar material constants associated with Rashba and Dresselhaus interactions, respectively; $\vec{\mathcal{E}}$ is a local electric field caused by either a charged impurity or confinement potential; $\vec{\sigma}$ is the Pauli spin matrix; $\vec{p}$ is the momentum operator; $\vec{A}$ is the vector potential associated with any external magnetic field; $\vec{\kappa}$ is a vector whose x-component is given by $\kappa_x = (p_x + eA_x)\{(p_y + eA_y)^2 - (p_z + eA_z)^2)\}$, and the other components are obtained by cyclic permutation of the indices, with x, y, z being the principal crystallographic axes; γ is a material constant that depends on the spin-orbit splitting energy and bandgap of the material; $\vec{\varsigma}$ is a vector whose x-component is given by $\varsigma_x = (1/2)[u_{xy}, p_y]_+ - (1/2)[u_{xz}, p_z]_+$, where $[,]_+$ denotes the anti-commutator (the other components are obtained by cyclic permutations of the indices); u_{ij} is the lattice strain tensor; $\hat{g}$ is another material constant (not the electron Landé g-factor) that depends on effective mass, spin-orbit splitting energy in the valence band, bandgap and one of three deformation potential constants describing the strain effect on the hole band splitting; μ_B is the Bohr magneton; B_z is a z-directed magnetic flux density associated with any external magnetic field present (its direction is taken to be the z-direction); $\mathcal{A}$ is the hyperfine constant; while $\vec{S}_i(\vec{I}_k)$ and $\vec{r}_i(\vec{R}_k)$ denote the spin and position of the i-th electron (k-th nuclei).

The presence of the Pauli spin matrix $\vec{\sigma}$ in H_n ($n = 1,2,3,4$) indicates that, unless the electron is initially in an eigenspinor state of this Hamiltonian, it will be coupled to its orthogonal eigenspinor and therefore the Hamiltonian

$\sum_{n=1}^{4} H_n$ will cause a spin flip event. Within the Born approximation, this spin flip rate $\Gamma_{\uparrow\downarrow}$ can be found from Fermi's Golden Rule

$$\Gamma_{\uparrow\downarrow} = \frac{2\pi}{\hbar} | < \psi_\downarrow | \sum_{n=1}^{5} H_n | \psi_\uparrow > |^2 \delta(E_\downarrow - E_\uparrow \pm \hbar\omega), \qquad (8.34)$$

where $\psi_\downarrow$ and $\psi_\uparrow$ are the wavefunctions (inclusive of the spatial and spinor parts) of the final and initial states of the electron, $E_\downarrow$ and $E_\uparrow$ are the energies of the final and initial states, and $\hbar\omega$ is the energy of phonon emitted or absorbed in the transition process. In the equation above, the plus sign within the delta-function corresponds to absorption and the minus sign to emission of a phonon.

The spin flip rates have been calculated in different types of quantum dots at various temperatures (see, for example, [21]). These rates are very small, typically less than 10^5/sec in GaAs quantum dots under reasonable conditions. In fact, at very low temperatures, the spin-flip rate in GaAs quantum dots has been measured to be smaller than 1/sec [22]. Therefore, quantum dots are ideal hosts for spin since spin is a relatively long lived entity (lifetime > 10 μs) in these systems.

Many research groups have engaged in experimentally measuring spin flip times in quantum dots. One of the earlier experiments measured the spin flip rates in GaAs quantum dots each populated with a single electron [23]. A magnetic field was applied to split the lowest subband state into two Zeeman sublevels. A short voltage pulse that affects the confining potential of the quantum dot was used to tune the sublevel energies with respect to the chemical potentials in the leads and thus selectively populate the higher sublevel. Relaxation to the lower sublevel was then measured. At a temperature of 20 mK and under an applied magnetic field of 7.5 Tesla, the spin relaxation time in the GaAs quantum dots was found to be larger than 50 μsec. In InP quantum dots, the spin relaxation time exceeds 100 μsec at a temperature of 2 K [24]. More recent measurements have shown relaxation times can reach several milliseconds in GaAs quantum dots [25], 20 milliseconds in InGaAs quantum dots at a temperature of 1 K [26], and even 170 milliseconds in a GaAs quantum dot [27] at a magnetic field of $\sim$ 1.7 Tesla (later extended to 1 second [22]). But all these pale in comparison with organic nanostructures. The spin relaxation time in 50-nm diameter nanowires of the π-conjugated organic tris(8-hydroxyquinolinolato-aluminum), popularly known as *Alq3*[¶], was found to be at least several msec, and perhaps even 1 second, at the relatively high temperature of 100 K [28]. This is caused by the extremely weak spin-orbit interactions in organics [29]. InAs quantum dots have much stronger spin-orbit coupling and therefore reported relaxation time in these

---

[¶]This organic is one of the most popular materials used as the electron transport layer in organic light emitting diodes.

dots is considerably smaller [30]. There are now theoretical predications that the spin relaxation time in carbon nanotube quantum dots can exceed several tens of seconds at finite magnetic fields because of an interference effect [31].

The spin relaxation time in various materials is summarized in the Table 8.1 [29].

TABLE 8.1

Spin relaxation time in various materials

Material	Spin relaxation time (sec)
Copper (bulk)	5×10^{-13}
Cobalt (bulk)	4×10^{-14}
GaAs (bulk)	10^{-9} - 10^{-7}
GaAs (quantum dot)	10^{-1} - 1
Alq$_3$ (bulk or nanowire)	10^{-2} - 1

8.3.1 Longitudinal and transverse spin relaxation times in a quantum dot

In Chapter 5, we introduced the notion of longitudinal and transverse spin relaxation times T_1 and T_2. Since spins in quantum dots are often used as vehicles to host qubits, what is often important is not the longitudinal spin relaxation time T_1, but rather the spin coherence time, or the transverse relaxation time T_2. The T_1 time essentially refers to the mean time that elapses before an excited spin flips to the ground state. This time is of great importance in classical spin-based logic gates, which we will discuss in Chapter 15, but less relevant to quantum computing, where what matters is how rapidly a single spin loses its quantum mechanical phase coherence. That time is determined by the transverse relaxation time, or T_2 time. The definition of the T_2 time invariably involves the concept of quantum superposition. As we saw in Chapter 3, a qubit is a coherent superposition of up-spin and down-spin states:

$$qubit = \alpha| \uparrow> +\beta| \downarrow> . \tag{8.35}$$

In order to preserve the coherence between the up-spin and down-spin states, one needs to maintain the precise phase relationship between the complex coefficients α and β. The time constant associated with the temporal decay of this phase relation is the T_2 time. In fact, we can associate a density matrix with the qubit (recall Chapter 5):

$$\rho_{qubit} = \begin{pmatrix} |\alpha|^2 & \alpha^*\beta \\ \beta^*\alpha & |\beta|^2 \end{pmatrix} . \tag{8.36}$$

The off-diagonal terms represent the phase relationship between α and β. The quantity T_2 is the time constant associated with the temporal decay of the off-diagonal terms, and therefore provides a measure of the robustness of a qubit.

Whether T_1 will be larger than T_2, or the other way around, depends on the primary mode of spin relaxation. Golovach et al. [32] have argued that if the spin relaxation is mediated by Dresselhaus and Rashba spin-orbit interactions, then $T_2 = 2T_1$. However, for other modes of spin relaxation, it is very possible that $T_2 < T_1$ (see Equation (8.37) later).

It is difficult to measure the T_2 time directly since we have to probe decoherence of a single electron. In most experiments, one probes a large number of spins simultaneously. Even if these spins do not interact with each other, inhomogeneities in local magnetic fields caused by nuclear spins having random polarization, or spatially varying spin-orbit interaction, will increase the decoherence rate substantially. Therefore, what we measure will always be less than the actual T_2 time associated with spin-spin interaction encountered by a single electron. We will call the measured time the T_2^* time. It has been shown that T_2^* can be several times smaller than T_2 [33].

There are many ways of measuring T_2^*. One of the most direct methods is *electron spin resonance* (ESR). In ESR experiments, we apply a dc magnetic field to lift the degeneracy between up-spin and down-spin energies via the Zeeman effect. An ac magnetic field, produced by a microwave source, is then turned on. It induces Rabi oscillations between the Zeeman-split levels when the microwave frequency is resonant with the spin-splitting energy (recall Chapter 4). Consequently, microwave is absorbed when resonance occurs. The Rabi oscillation amplitude decays in time roughly exponentially, with a time constant T_2^*. The Fourier transform of the damped oscillation is the absorption spectrum (absorption intensity versus microwave frequency). The linewidth of the absorption peak (which should have a Lorentzian shape) is $1/T_2^*$. Consequently, from the absorption peak linewidth, we can deduce the T_2^* time. Although this is a straightforward technique, it has been seldom applied to quantum dots since the absorption intensity depends on the number of participating spins and that number has to be large in order to produce a strong enough signal. With quantum dots, the number of spins involved is typically small so that normally ESR is not a reliable method. However, if the density of quantum dots is sufficiently high, then it may be possible to obtain a strong enough ESR signal (particularly at low temperatures) to yield a reliable estimate of the T_2^* time. These days, it is possible to produce an extremely dense array of quantum dots (density $> 10^{11}$ cm^{-2}) using electrochemical self-assembly [34]. With such dense arrays, ESR can be applied successfully to measure the T_2^* time.

There are several reports of the measurement of T_1 times in quantum dots [23, 30, 26, 27], but few reports of the measurement of T_2 times (single spin decoherence time). That would require a specific $\pi/2$-π spin echo sequence [35], which is difficult to implement. Such measurements have been carried out in some systems, notably nitrogen vacancy color centers in diamond. The measured T_2 time of spin dephasing in the color center was amazingly long – 58 μs at room temperature (300 K) [36]. Coherent coupling between single spins in diamond was demonstrated at room temperature by Gaebel et al. [37], while Dutt and co-workers have now demonstrated a quantum register based on individual electronic and nuclear spin-based qubits in this system [38].

The *ensemble averaged* T_2^* time (decoherence time associated with many spins) has been measured in 2-8 nm CdSe quantum dots using femtosecond resolved Faraday rotation [39]. Dzhioev, et al. have measured the T_2^* time associated with the decoherence of several interacting spins in an n-GaAs heterostructure using polarized photoluminescence measurement [40]. They found that the T_2^* time is determined by hyperfine interaction with nuclear spins and is of the order of few hundred nanoseconds. In our laboratory, we too have measured the T_2^* time in electrochemically self assembled cylindrical CdS quantum dots, of 10 nm diameter, using direct ESR spectroscopy [41]. The measured T_2^* time was a few nanoseconds. Surprisingly, we found that the T_2^* time increases with temperature, as shown in Fig. 8.5. This can be explained by invoking dephasing due to hyperfine interaction with nuclear spins.

It is widely believed that the primary cause of spin dephasing in an ensemble of quantum dots is the spatially varying hyperfine interaction caused by nuclear spins [20, 33, 42, 43]. The spatial variation comes about from two sources: First, the electron wavefunction varies in space and the strength of the hyperfine interaction depends on this wavefunction. Second, the magnetic field due to nuclear spins is also inhomogeneous. This inhomogeneity is the major contributor to the dephasing rate $1/T_2^*$. However, it is possible in singly charged quantum dots to synchronize the electron spin precession about a magnetic field into well-defined modes with a laser pulse sequence. In that case, the light stimulated fluctuations of the hyperfine nuclear field acting on the electron are suppressed. The information about the electron spin precession can be imprinted on the nuclei and stored for tens of minutes in darkness at a temperature of 6 K. This was recently demonstrated by Greilich, et al [44].

In quantum wells, on the other hand, the major spin dephasing mechanism is believed to be Rashba spin-orbit interaction [45]. That interaction, however, is extremely weak in quantum dots since the electron motion is restricted in all three dimensions [46].

Organics typically exhibit a relatively short T_2 time of a few to a few tens of nanoseconds, but in the organic molecule *Alq3*, it is nearly *temperature independent* up to room temperature. Curiously, this time is sufficient to meet the requirements of fault tolerant quantum computing, which raises hopes of organic quantum computers utilizing spin based qubits at room temperature. Organic molecules also allow elegant schemes for qubit read out that have no analog in inorganic semiconductors. At this time, however, the longest T_2 times, particularly at room temperature, are reported to be associated with

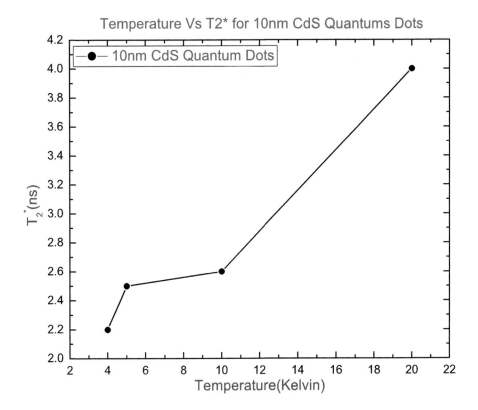

FIGURE 8.5

Measured T_2^* time as a function of temperature in CdS quantum dots of 50-nm diameter. The T_2 time increases with temperature since the primary relaxation mechanism is hyperfine interaction with nuclear spins. The strength of this interaction decreases with increasing temperature because of increasing depolarization of nuclear spins.

nitrogen vacancies in diamond. These T_2 times can exceed 100 μs at room temperature [36], which appears extremely promising.

Khaetski et al. [20] give an equation for the transverse relaxation rate due to hyperfine interactions with nuclear spins in the special case of electrons hopping between dots:

$$\frac{1}{T_2} = \frac{1}{2T_1} + \frac{(g\mu_B)^2}{2\hbar^2} \int_{-\infty}^{\infty} d\tau < H_{NZ}(0)H_{NZ}(\tau) >, \qquad (8.37)$$

where the bracket $< .. >$ denotes ensemble average and H_{NZ} is the effective magnetic field due to nuclear spins. Obviously, the dephasing rate increases with increasing autocorrelation given by the integral in Equation (8.37). With increasing temperature, the nuclear spin polarizations are randomized leading to a decrease in the autocorrelation function. That explains why the dephasing rate decreases with increasing temperature in the experiment in [41] (see also Fig. 8.5).

Thermal depolarization also plays a direct role in spin dephasing, as mentioned in [43]. For an unpolarized configuration of nuclear spins, the net nuclear magnetic field is $A/(\sqrt{N}g\mu_B)$ (where A is the hyperfine constant and N is the number of nuclei in a quantum dot, typically about 10^5). For a fully polarized configuration, the net field is $A/(g\mu_B)$. Therefore, thermally induced depolarization of nuclear spins will tend to reduce the dephasing rate by reducing the net nuclear magnetic field by a factor of $\sqrt{N} \sim 300$ in typical quantum dots.

Using optical pump and probe techniques, spin states of carriers and nuclei in *individual* semiconductor quantum dots have been studied experimentally [47, 48]. These measurements have revealed that the primary decoherence mechanism limiting T_2 is the hyperfine interaction of the conduction electron spins with nuclear spins. The measured T_2 time in individual GaAs quantum dots is on the order of 10 nanoseconds, which is $\sim \hbar\sqrt{N}/A$. Thus, the experimentally measured T_2 times in semiconductor dots are much smaller (by about six orders of magnitude) than the T_1 times, which clearly demonstrates that realizing *quantum* logic processors based on single electron spins in quantum dots is a much more difficult challenge than realizing *classical* logic processors (see Chapter 15). The latter does not require phase coherence and merely requires a long T_1 time, whereas the former requires phase coherence and therefore a long T_2 time, as well as a long T_1 time.

8.4   Problems

- **Problem 8.1**

  Consider an electron with x-polarized spin placed in a magnetic field directed along the z-coordinate. Show using quantum mechanics that, if there are no dissipative processes, then the spin will precess about the magnetic field according to Equation (8.6).

  Solution

  First note that the x-polarized spin is not parallel or anti-parallel to the z-directed magnetic field. Therefore at time $t = 0$, the spin is *not* in an eigenstate of the Hamiltonian containing the z-directed magnetic field. Consequently, the spin polarization must change with time and we have to show that the spin will precess about the magnetic field.

  If there are no dissipative processes, then the eigenspinor $[\phi]$ will obey the *time dependent* Pauli equation

  $$\left[i\hbar \frac{\partial}{\partial t} + <H_0> + \frac{g}{2}\mu_B \vec{B} \cdot \vec{\sigma} \right] [\phi] = 0, \qquad (8.38)$$

  which has the solution

  $$[\phi](t) = \exp[-i(<H_0> + (g/2)\mu_B \vec{B} \cdot \vec{\sigma})t/\hbar][\phi](0)$$

  $$= \begin{bmatrix} 1 & 0 \\ 0 & 1 \end{bmatrix}^{-1} \begin{bmatrix} e^{-iE_+ t/\hbar} & 0 \\ 0 & e^{-iE_- t/\hbar} \end{bmatrix} \begin{bmatrix} 1 & 0 \\ 0 & 1 \end{bmatrix} [\phi](0), \qquad (8.39)$$

  where E_{mp} are the eigenvalues of the 2×2 matrix $\left[<H_0> + \frac{g}{2}\mu_B \vec{B} \cdot \vec{\sigma} \right]$. Since the magnetic field is in the z-direction, the energy eigenstates and eigenspinors of $\left[<H_0> + \frac{g}{2}\mu_B \vec{B} \cdot \vec{\sigma} \right]$ will be given by Equations (8.3) and (8.5).

  The last equation tells us what the eigenspinor will be at any instant of time t.

  Since the initial spin polarization was along the x-direction, $[\phi](0)$ is the x-polarized state $(1/\sqrt{2})[1 \ -1]^\dagger$. As a result, the last equation reduces to

  $$[\phi](t) = \begin{bmatrix} e^{-iE_+ t/\hbar} \\ -e^{-iE_- t/\hbar} \end{bmatrix}. \qquad (8.40)$$

  Now the spin components will be given by

  $$S_x(t) = [\phi]^\dagger(t)[\sigma_x][\phi]^\dagger(t)$$

  $$= \begin{bmatrix} e^{+iE_+ t/\hbar}/\sqrt{2} & -e^{+iE_- t/\hbar}/\sqrt{2} \end{bmatrix} \begin{bmatrix} 0 & 1 \\ 1 & 0 \end{bmatrix} \begin{bmatrix} e^{-iE_+ t/\hbar}/\sqrt{2} \\ -e^{-iE_- t/\hbar}/\sqrt{2} \end{bmatrix}$$

  $$= -cos\theta, \qquad (8.41)$$

where $\theta = (E_+ - E_-)t/\hbar = g\mu_B Bt/\hbar$ where B is the flux density associated with the magnetic field.

Similarly,

$$S_y(t) = [\phi]^\dagger(t)[\sigma_y][\phi]^\dagger(t)$$

$$= \left[e^{+iE_+t/\hbar}/\sqrt{2} \;-e^{+iE_-t/\hbar}/\sqrt{2} \right] \begin{bmatrix} 0 & -i \\ i & 0 \end{bmatrix} \begin{bmatrix} e^{-iE_+t/\hbar}/\sqrt{2} \\ -e^{-iE_-t/\hbar}/\sqrt{2} \end{bmatrix}$$

$$= -sin\theta \tag{8.42}$$

and

$$S_z(t) = [\phi]^\dagger(t)[\sigma_y][\phi]^\dagger(t)$$

$$= \left[e^{+iE_+t/\hbar}/\sqrt{2} \;-e^{+iE_-t/\hbar}/\sqrt{2} \right] \begin{bmatrix} 1 & 0 \\ 0 & -1 \end{bmatrix} \begin{bmatrix} e^{-iE_+t/\hbar}/\sqrt{2} \\ -e^{-iE_-t/\hbar}/\sqrt{2} \end{bmatrix}$$

$$= 0. \tag{8.43}$$

Now

$$\frac{d\vec{S}}{dt} = \hat{x}\frac{dS_x}{dt} + \hat{y}\frac{dS_y}{dt} + \hat{z}\frac{dS_z}{dt}$$

$$= -\hat{x}\frac{d(cos\theta)}{dt} + -\hat{y}\frac{d(sin\theta)}{dt}$$

$$= \frac{d\theta}{dt}[\hat{x}sin\theta - \hat{y}cos\theta]$$

$$= \frac{g}{2}\frac{eB}{m}[\hat{x}sin\theta - \hat{y}cos\theta]$$

$$= \frac{eB}{m*}[\hat{x}sin\theta - \hat{y}cos\theta], \tag{8.44}$$

where we have used the fact that normally $g = 2m/m^*$ with m being the free electron mass and m^* the effective mass of the electron.

Now, since the B-field is directed along the z-direction,

$$\vec{\Omega} \times \vec{S} = \frac{e\vec{B}}{m*} \times \vec{S}$$

$$= \frac{eB}{m*}[-\hat{x}S_y + \hat{y}S_x]$$

$$= \frac{eB}{m*}[\hat{x}sin\theta - \hat{y}cos\theta]. \tag{8.45}$$

The last two equations show that

$$\frac{d\vec{S}}{dt} = \vec{\Omega} \times \vec{S}, \tag{8.46}$$

which is the result to be proved.

- **Problem 8.2**

  Derive Equation (8.23).

  Hint: Set the expectation values of the Rashba and Dresselhaus Hamiltonians in Equations (7.1) and (7.2) [for a one-dimensional system whose axis is along the x-direction] equal to the effective Zeeman splitting energies $g\mu_B B_{Rashba}$ and $g\mu_B B_{Dresselhaus}$ and replace $\hbar k_x$ with $m^* v_x$.

- **Problem 8.3**

  Show that in a quantum wire whose axis is in the [100] crystallographic direction, three ingredients must be simultaneously present to cause D'yakonov-Perel' spin relaxation in *space*: (a) Rashba interaction, (b) Dresselhaus interaction and (c) multi-channeled transport. How is the situation different for spin relaxation in *time*?

  Solution

  Consider the quantum wire of Fig. 8.4. We will assume that electrons are injected from the left (at $x = 0$) with their spins polarized along the wire axis, so that

  $$S_x = 1$$
  $$S_y = 0$$
  $$S_z = 0. \tag{8.47}$$

  In this system, the spin precession frequency $\vec{\Omega}(\vec{v})$ is given by

  $$\vec{\Omega}(\vec{v}) = \Omega_D(\vec{v})\hat{x} + \Omega_R(\vec{v})\hat{z}, \tag{8.48}$$

  where $\hat{n}$ is the unit vector along the n-axis. In the above, the first term is the contribution due to the Dresselhaus interaction and the second term is the contribution due to the Rashba interaction (recall Equation(8.23)). These quantities are given by

  $$\Omega_D(\vec{v}) = \frac{2a_{42}m^*}{\hbar^2}\left[\left(\frac{m\pi}{W_z}\right)^2 - \left(\frac{n\pi}{W_y}\right)^2\right]v = \hat{\Omega}_{D0}v$$

  $$\Omega_R(\vec{v}) = \frac{2a_{46}m^*}{\hbar^2}E_y v = \hat{\Omega}_{R0}v. \tag{8.49}$$

The spin precesses according to Equation (8.6):

$$\frac{d\vec{S}}{dt} = \hat{x}\frac{dS_x}{dt} + \hat{y}\frac{dS_y}{dt} + \hat{z}\frac{dS_z}{dt}$$

$$= \vec{\Omega}(\vec{v}) \times \vec{S}$$

$$= det \begin{bmatrix} \hat{x} & \hat{y} & \hat{z} \\ \Omega_D(\vec{v}) & 0 & \Omega_R(\vec{v}) \\ S_x & S_y & S_z \end{bmatrix}$$

$$= -\hat{x}\left(\Omega_R(\vec{v})S_y\right) - \hat{y}\left(\Omega_D(\vec{v})S_z - \Omega_R(\vec{v})S_x\right) + \hat{z}\left(\Omega_D(\vec{v})S_y\right) , \tag{8.50}$$

where S_n is the spin component along the n-axis.

Equating each component separately and using the chain rule of differentiation, we get

$$\frac{dS_x}{dx}\frac{dx}{dt} = \frac{dS_x}{dx}v = -\hat{\Omega}_{R0}vS_y$$

$$\frac{dS_y}{dx}\frac{dx}{dt} = \frac{dS_y}{dx}v = \hat{\Omega}_{R0}vS_x - \hat{\Omega}_{D0}vS_z$$

$$\frac{dS_z}{dx}\frac{dx}{dt} = \frac{dS_z}{dx}v = \hat{\Omega}_{D0}vS_y. \tag{8.51}$$

First, let us derive an expression for S_y as a function of position x. From the second equation above, we get by further differentiation

$$\frac{d^2S_y}{dx^2} = \hat{\Omega}_{R0}\frac{dS_x}{dx} - \hat{\Omega}_{D0}\frac{dS_z}{dx}$$

$$= -\hat{\Omega}_{R0}\left(\hat{\Omega}_{R0}S_y\right) - \hat{\Omega}_{D0}\left(\hat{\Omega}_{D0}S_y\right)$$

$$= -\left(\hat{\Omega}_{R0}^2 + \hat{\Omega}_{D0}^2\right)S_y$$

$$= \hat{\Omega}_0^2 S_y, \tag{8.52}$$

where $\hat{\Omega}_0^2 = \hat{\Omega}_{R0}^2 + \hat{\Omega}_{D0}^2$.

The solution of the above equation is

$$S_y(x) = A\cos\left(\hat{\Omega}_0 x\right) + B\sin\left(\hat{\Omega}_0 x\right). \tag{8.53}$$

Using the boundary condition in Equation (8.47), we reduce the above to

$$S_y(x) = B\sin\left(\hat{\Omega}_0 x\right). \tag{8.54}$$

Differentiating the above and using Equation (8.51), we get

$$\frac{dS_y}{dx} = B\hat{\Omega}_0\cos\left(\hat{\Omega}_0 x\right) = \hat{\Omega}_{R0}S_x - \hat{\Omega}_{D0}S_z. \tag{8.55}$$

Again, using the boundary condition at $x = 0$ from Equation (8.47) in the above, we get that $B = \hat{\Omega}_{R0}/\hat{\Omega}_0$, so that

$$S_y(x) = \frac{\hat{\Omega}_{R0}}{\hat{\Omega}_0} \sin\left(\hat{\Omega}_0 x\right). \tag{8.56}$$

Next, let us derive an expression for S_x as a function of x. From Equation (8.51) and the above solution for S_y, we get

$$\frac{dS_x}{dx} = -\hat{\Omega}_{R0} S_y = -\frac{\hat{\Omega}_{R0}^2}{\hat{\Omega}_0} \sin\left(\hat{\Omega}_0 x\right). \tag{8.57}$$

Integration of the above yields

$$S_x(x) = \frac{\hat{\Omega}_{R0}^2}{\hat{\Omega}_0^2} \cos\left(\hat{\Omega}_0 x\right) + C, \tag{8.58}$$

where C is a constant to be determined from the boundary condition in Equation (8.47). This determination yields $C = 1 - \hat{\Omega}_{R0}^2/\hat{\Omega}_0^2 = \hat{\Omega}_{D0}^2/\hat{\Omega}_0^2$. Therefore, the expression for S_x becomes

$$S_x(x) = \frac{\hat{\Omega}_{R0}^2}{\hat{\Omega}_0^2} \cos\left(\hat{\Omega}_0 x\right) + \frac{\hat{\Omega}_{D0}^2}{\hat{\Omega}_0^2} = 1 - 2\frac{\hat{\Omega}_{R0}^2}{\hat{\Omega}_0^2} \sin^2\left(\frac{\hat{\Omega}_0 x}{2}\right). \tag{8.59}$$

Finally, we derive an expression for S_z. Using Equation (8.51) and the expression for S_y, we get

$$\frac{dS_z}{dx} = \hat{\Omega}_{D0} S_y = \frac{\hat{\Omega}_{R0}\hat{\Omega}_{D0}}{\hat{\Omega}_0} \sin\left(\hat{\Omega}_0 x\right). \tag{8.60}$$

Integrating the above, we obtain

$$S_z(x) = -\frac{\hat{\Omega}_{R0}\hat{\Omega}_{D0}}{\hat{\Omega}_0^2} \cos\left(\hat{\Omega}_0 x\right) + D, \tag{8.61}$$

where D is a constant determined from boundary condition (Equation (8.47)) to be $\hat{\Omega}_{R0}\hat{\Omega}_{D0}/\hat{\Omega}_0^2$. Therefore,

$$S_z(x) = \frac{\hat{\Omega}_{R0}\hat{\Omega}_{D0}}{\hat{\Omega}_0^2} \left[1 - \cos\left(\hat{\Omega}_0 x\right)\right] = 2\frac{\hat{\Omega}_{R0}\hat{\Omega}_{D0}}{\hat{\Omega}_0^2} \sin^2\left(\frac{\hat{\Omega}_0 x}{2}\right). \tag{8.62}$$

Note from Equations (8.59), (8.56), and (8.62), that the spin components at any position in space are *independent* of the carrier velocity v since the velocity "canceled out" in Equation (8.51). Therefore, carrier scattering, which changes v, has no direct effect on the spin components.

It may have an indirect effect through $\hat{\Omega}_{R0}$ and $\hat{\Omega}_{D0}$ if these quantities are affected by scattering. $\hat{\Omega}_{R0}$ certainly is not affected and $\hat{\Omega}_{D0}$ is not either unless there is inter-subband scattering that changes the subband indices m and n in Equation (8.49) and causes multi-channeled transport where more than one subband will carry current. Therefore, we conclude that *intra-subband scattering does not affect the spin components in space*. This is a remarkable result, which leads to the absence of D-P spin relaxation in space as long as inter-subband scattering (leading to multi-channeled transport) is absent. We point out that this happens because the spin precession frequencies Ω_D and Ω_R, **which are spin-independent**, are each proportional to the spin-independent magnetic field and therefore proportional to the group velocity v and *not* wavevector k. If Ω_D and Ω_R were proportional to k instead of v, then v would not have canceled out in Equation (8.51) and the spin components at any coordinate position in space would not have been unaffected by scattering. As a result, there would have been D-P relaxation in space.

Examining the expressions for $S_x(x)$, $S_y(x)$, and $S_z(x)$ given by Equations (8.56), (8.59) and (8.62), we see immediately that if Rashba interaction is absent ($\hat{\Omega}_{R0} = 0$), then $S_x(x) = 1$, $S_y(x) = 0$ and $S_z(x) = 0$, even if there is inter-subband scattering. Therefore, the original spin orientation does not change in space at all and there is no D-P spin relaxation. Of course, it is not necessary that the original spin orientation remain intact for D-P relaxation to not occur. What is necessary is that even if the orientation changes in space, it must change in exactly the same way for every electron so that the ensemble averaged magnitude of the spin $|<S>|$ does not decay in space (here the angular bracket $<..>$ denotes ensemble averaging). This will happen if $|<S>|^2 = <S_x>(x)^2 + <S_y>(x)^2 + <S_z>(x)^2$ is independent of x.

Next we consider what happens if the Dresselhaus interaction is absent. In that case, $S_x(x) = cos\left(\hat{\Omega}_0 x\right)$, $S_y(x) = sin\left(\hat{\Omega}_0 x\right)$ and $S_z(x) = 0$. Here, the spin orientation does change in space, but the quantity $|S|^2 = S_x(x)^2 + S_y(x)^2 + S_z(x)^2 = cos^2\left(\hat{\Omega}_0 x\right) + sin^2\left(\hat{\Omega}_0 x\right) = 1$, independent of x. Therefore, there is no D-P spin relaxation, even if there were inter-subband scattering. Inter-subband scattering is relevant to Dresselhaus interaction only; therefore if the Dresselhaus interaction is absent, inter-subband scattering is immaterial.

Now let us consider the situation when both Dresselhaus and Rashba interaction are present. As long as transport is single channeled and there is no inter-subband scattering, $\hat{\Omega}_{D0}$ and $\hat{\Omega}_0$ are the same for every electron. In that case, it is easy to show that $<S_x>(x) = S_x(x)$, $<S_y>(x) = S_y(x)$ and $<S_z>(x) = S_z(x)$.

Therefore,

$$| < S > |^2 = < S_x > (x)^2 + < S_y > (x)^2 + < S_z > (x)^2$$

$$= S_x(x)^2 + S_y(x)^2 + S_z(x)^2$$

$$= 1 + 4 \left(\frac{\hat{\Omega}_{R0}}{\hat{\Omega}_0} \right)^4 sin^4 \left(\frac{\hat{\Omega}_0 x}{2} \right) - 4 \left(\frac{\hat{\Omega}_{R0}}{\hat{\Omega}_0} \right)^2 sin^2 \left(\frac{\hat{\Omega}_0 x}{2} \right)$$

$$+ 4 \left(\frac{\hat{\Omega}_{R0}}{\hat{\Omega}_0} \right)^2 sin^2 \left(\frac{\hat{\Omega}_0 x}{2} \right) cos^2 \left(\frac{\hat{\Omega}_0 x}{2} \right)$$

$$+ 4 \left(\frac{\sqrt{\hat{\Omega}_{R0} \hat{\Omega}_{D0}}}{\hat{\Omega}_0} \right)^4 sin^4 \left(\frac{\hat{\Omega}_0 x}{2} \right)$$

$$= 1 + 4 \left(\frac{\hat{\Omega}_{R0}}{\hat{\Omega}_0} \right)^4 sin^4 \left(\frac{\hat{\Omega}_0 x}{2} \right) - 4 \left(\frac{\hat{\Omega}_{R0}}{\hat{\Omega}_0} \right)^2 sin^4 \left(\frac{\hat{\Omega}_0 x}{2} \right)$$

$$+ 4 \frac{\hat{\Omega}_{R0}^2 (\hat{\Omega}_0^2 - \hat{\Omega}_{R0}^2)}{\hat{\Omega}_0^4} sin^4 \left(\frac{\hat{\Omega}_0 x}{2} \right)$$

$$= 1. \tag{8.63}$$

Thus, we have shown that as long as transport is single channeled, there is no D-P relaxation. Combining all of the above, we conclude that in a quantum wire, three ingredients must be simultaneously present to cause D-P spin relaxation in space: Rashba interaction, Dresselhaus interaction and multi-channeled transport. In our proof, we assume that the initial spin polarization at $x = 0$ is along the wire axis. The reader can easily repeat the proof if the initial polarization is along either of the other two directions, y or z. A slightly different (more involved) proof of this result can be found in [12].

We now revisit the issue of whether there is D-P relaxation in time. The difference between *time* and *space* is the following: in the case of "space," we ask what are the spin components $S_x(x)$, $S_y(x)$ and $S_z(x)$ at a position x, irrespective of how long it took the electron to get there. The electron may have visited the coordinate x at some earlier time, then gone forward under the action of an accelerating electric field, subsequently backscattered to a position behind x, and again gone forward under the electric field to reach the position x a second time. What we have shown here is that the spin components are identical during the first and second visits to the location x! In other words, the past history does not matter. An electron may visit the location x many times, and each time it will have exactly the same spin components at x as long as we have single channeled transport. No amount of intra-subband scattering can change this result! This is certainly counter-intuitive. Not only that, but every electron at positions x has exactly the same spin components, even though they may have had very different scattering histories! That too is counterintuitive.

In the case of "time," we ask what are the spin components of different electrons, irrespective of what their locations are in the quantum wire, if we take a snapshot at time t. We know that these spin components will be different, depending on the location, since we have derived the spin components as a function of x. Thus, if we ensemble average over all electrons, this average will change with time t. The change may not be a monotonic decay with time. There may be oscillations and non-monotonic behavior. But nontheless there is a temporal change and that can be interpreted as "spin relaxation."

The interesting feature is that there is a distinct difference between relaxation in *space* and relaxation in *time*. The relaxation length in space is infinity if transport is single channeled, but the relaxation time is finite. Drift-diffusion models of spin transport often designate the relaxation length as the spin diffusion length $L_s = \sqrt{D\tau_s}$ where D is the spin diffusion coefficient and τ_s is the spin relaxation time. We are now faced with the following dilemma: $L_s \to \infty$, but τ_s is finite. This can be resolved in only two ways: (i) $D_s \to \infty$, or (ii) the drift-diffusion model that stipulates $L_s = \sqrt{D\tau_s}$ is completely invalid. We will address the shortcomings of drift-diffusion models in Chapter 10, but for the time being let us ponder the possibility that $D_s \to \infty$. We can be certain that the "charge diffusion coefficient" D_c is finite, since there is plenty of scattering. Therefore, if the drift-diffusion model remains valid, we must conclude that $D_c \neq D_s$. In other words, charge and spin diffusion coefficients are very different. There seems to be scant appreciation of this fact within the spintronics community. Most researchers assume tacitly (and incorrectly) that the two diffusion coefficients are always the same. In the case of Elliott-Yafet relaxation, they are probably not too different since momentum relaxing scattering, which contributes to the charge diffusion coefficient, is also the cause of spin relaxation. Moreover, the charge and spin relaxation rates are proportional to each other. However, in the case of D'yakonov-Perel' relaxation, the two rates are *inversely* proportional to each other. There is no reason to assume a priori that if the D'yakonov-Perel' mechanism is the major cause of spin relaxation, the two diffusion coefficients will be equal.

- **Problem 8.4**

Electrons are injected into a quantum wire with a constant velocity. There is a constant electric field driving transport and there are no momentum randomizing collisions (transport is ballistic). A single subband is occupied. Show that there is no D'yakonov-Perel' relaxation in *time*.

Solution

Let the initial injection velocity $= u$. The velocity at any time t is then given by

$$v(t) = u + ft, \qquad (8.64)$$

where f is the acceleration $= \mathcal{E}/m^*$ with $\mathcal{E}$ being the electric field driving transport. The above equation is valid *for every electron* in the ensemble since there is no scattering to change the velocities of different electrons randomly. This shows that at any instant of time, *every* electron in the ensemble has exactly the same velocity $v(t)$.

From Equation (8.51),

$$\frac{dS_x}{dt} = -\hat{\Omega}_{R0}v(t)S_y$$

$$\frac{dS_y}{dt} = \hat{\Omega}_{R0}v(t)S_x - \hat{\Omega}_{D0}v(t)S_z$$

$$\frac{dS_z}{dt} = \hat{\Omega}_{D0}v(t)S_y. \qquad (8.65)$$

Since $v(t)$ is the same for every electron, we can replace the spin component S_n in the above equation with its ensemble averaged value $< S_n >$. Therefore,

$$\frac{d| < S > |^2}{dt} = \frac{d < S_x >^2}{dt} + \frac{d < S_y >^2}{dt} + \frac{d < S_z >^2}{dt}$$

$$= 2 < S_x > \frac{d < S_x >}{dt} + 2 < S_y > \frac{d < S_y >}{dt}$$

$$+ 2 < S_z > \frac{d < S_z >}{dt}$$

$$= -2\hat{\Omega}_{R0}v(t) < S_y >< S_x > +2\hat{\Omega}_{R0}v(t) < S_x >< S_y >$$

$$-2\hat{\Omega}_{D0}v(t) < S_z >< S_y > +2\hat{\Omega}_{D0}v(t) < S_y >< S_z >$$

$$= 0. \qquad (8.66)$$

Therefore, $| < S > |$ will not decay in time and there will be no D'yakonov-Perel' spin relaxation in time. This exercise will show the reader that there is a simpler solution to Problem 8.3, but we presented a more involved solution since we wanted to derive expressions for the individual spin components S_n.

- **Problem 8.5**

If the spin-orbit magnetic field B_{so} is proportional to wavevector k_x and not velocity v_x, then you can show easily that there will be D'yakonov-Perel' spin relaxation in space in a quantum wire with a single subband occupied. Use Equation (7.62) to relate v_x to k_x and derive this result. This result is of course unphysical since there should never be any

D'yakonov-Perel' relaxation in strictly single-channeled transport. Does this mean that the spin-independent field gives the right result and the spin-dependent field does not? Not necessarily. In the analysis, you would have assumed a spin-independent precession frequency Ω which is proportional to B_{so}. Therefore, B_{so} should also be spin-independent and you cannot use the spin-dependent B_{so} which is proportional to k.

- **Problem 8.6**

Consider a two-dimensional electron gas (2-DEG) formed in a non-centrosymmetric crystal in the $(x - y)$ plane. An electric field causing structural inversion asymmetry is applied in the z-direction to induce the Rashba interaction. Assume that the symmetry-breaking electric field's magnitude is such that the strengths of the Rashba and Dresselhaus interactions are equal, i.e., $\eta = \nu$.

First show that the spin-dependent pseudo-magnetic field $\vec{B}_{Dresselhaus}$ arising from the Dresselhaus spin-orbit interaction is given by

$$\vec{B}_{Dresselhaus} = \frac{2\nu}{g\mu_B} \left[k_x \hat{x} - k_y \hat{y} \right], \tag{8.67}$$

where k_x and k_y are the x- and y-components of the wavevector, and $\hat{x}$ and $\hat{y}$ are the unit vectors along x- and y-directions. (Hint: Use the relation $-(g/2)\mu_B \vec{B}_{Dresselhaus} \cdot \vec{\sigma} = a_{42}\vec{\kappa} \cdot \vec{\sigma}$. Remember that $\nu = a_{42} \langle p_z^2 \rangle /\hbar^2$.)

An expression for the pseudo-magnetic field $\vec{B}_{Rashba}$ arising from the Rashba spin-orbit interaction is derived in the next chapter. Using Equation (9.16) from the next chapter, which gives this expression, and the last equation, show that the total spin-dependent spin-orbit magnetic field $\vec{B}_{so}$, which is $\vec{B}_{Rashba} + \vec{B}_{Dresselhaus}$, is fixed, lies in the $(x-y)$ plane, and subtends an angle of $45°$ with either the x- or the y-axis.

Repeat the problem for the spin-independent spin-orbit magnetic field.

- **Problem 8.7**

Will there be any D'yakonov-Perel' relaxation in the two-dimensional electron gas (2-DEG) if the strengths of the Rashba and the Dresselhaus spin-orbit interactions are equal? Explain. There will be no Elliott-Yafet spin relaxation since the eigenspinors will be wavevector-independent as we have seen in Chapter 7.

8.5   References

[1] M. I. D'yakonov and V. I. Perel', "Orientation of electrons associated with the interband absorption of light in semiconductors", Sov. Phys. JETP, **33**, 1053 (1971).

[2] J. Fabian and S. Das Sarma, "Spin relaxation of conduction electrons", J. Vac. Sci. Technol. B, **17**, 1708 (1999).

[3] R. J. Elliott, "Theory of the effect of spin orbit coupling on magnetic resonances in some semiconductors", Phys. Rev., **96**, 266 (1954).

[4] G. E. Pikus and A. N. Titkov, in *Optical Orientation*, Eds. F. Meier and B. P. Zakharchenya (North Holland, Amsterdam, 1984).

[5] F. J. Jedema, M. S. Nijboer, A. T. Philip and B. J. Van Wees, "Spin injection and spin accumulation in all metal spin valves", Phys. Rev. B, **67**, 085319 (2003).

[6] Y. Yafet in *Solid State Physics*, Eds. F. Seitz and D. Turnbull, Vol. 14 (Academic, New York, 1963).

[7] F. Monod and P. Beunue, "Conduction electron spin flip by phonons in metals: Analysis of experimental data", Phys. Rev. B, **19**, 911 (1979).

[8] M. I. D'yakonov and V. I. Perel', "Orientation of electrons associated with the interband absorption of light in semiconductors", Sov. Phys. JETP, **33**, 1053 (1971).

[9] S. Pramanik, S. Bandyopadhyay and M. Cahay, "Spin relaxation of "upstream" electrons in quantum wires: Failure of the drift diffusion model", Phys. Rev. B, **73**, 125309 (2006).

[10] J. S. Sandhu, A. P. Heberle, J. J. Baumberg and J. R. A. Cleaver, "Gateable suppression of spin relaxation in semiconductors", Phys. Rev. Lett., **86**, 2150 (2001).

[11] J. Kainz, U. Rössler and R. Winkler, "Temperature dependence of D'yakonov-Perel' spin relaxation in zinc blende semiconductor quantum structures", Phys. Rev. B, **70**, 195322 (2004).

[12] S. Pramanik, S. Bandyopadhyay and M. Cahay, "Spin relaxation in the channel of a Spin Field Effect Transistor", IEEE Trans. Nanotech., **4**, 2 (2005).

[13] A. G. Mal'shukov and K. A. Chao, "Waveguide diffusion modes and slowdown of D'yakonov-Perel' spin relaxation in narrow two dimensional semiconductor channels", Phys. Rev. B, **61**, R2413 (2000).

[14] S. Pramanik, S. Bandyopadhyay and M. Cahay, "Decay of spin polarized hot carrier current in a quasi one-dimensional spin valve structure", Appl. Phys. Lett., **84**, 266 (2004).

[15] A. W. Holleitner, V. Sih, R. C. Myers, A. C. Gossard and D. D. Awschalom, "Suppression of spin relaxation in submicron InGaAs wires", Phys. Rev. Lett., **97**, 036805 (2006).

[16] S. Pramanik, S. Bandyopadhyay and M. Cahay, "Spin dephasing in quantum wires", Phys. Rev. B, **68**, 075313 (2003).

[17] T. P. Pareek and P. Bruno, "Spin coherence in two dimensional electron gas with Rashba spin orbit interaction", Phys. Rev. B, **65**, 241305(R) (2002).

[18] N. S. Averkiev and L. E. Golub, "Giant spin relaxation anisotropy in zinc blende heterostructures", Phys. Rev. B, **60**, 15582 (1999).

[19] Y. Ohno, R. Terauchi, T. Adachi, F. Matsukura and H. Ohno, "Spin relaxation in GaAs(110) quantum wells", Phys. Rev. Lett., **83**, 4196 (1999).

[20] A. V. Khaetski, D. Loss and L. Glazman, "Electron spin evolution induced by interaction with nuclei in a quantum dot", Phys. Rev. B, **67**, 195329 (2003).

[21] A. V. Khaetski and Y. V. Nazarov, "Spin flip transitions between Zeeman sublevels in semiconductor quantum dots", Phys. Rev. B, **61**, 12639 (2000).

[22] S. Amasha, K. MacLean, Iuliana Radu, D. M. Zümbuhl, M. A. Kastner, M. P. Hanson and A. C. Gossard, "Electrical control of spin relaxation in a quantum dot", arXiv:0707.1656.

[23] R. Hanson, et al., "Zeeman energy and spin relaxation in one electron quantum dot", Phys. Rev. Letters, **91**, 196802 (2003).

[24] M. Ikezawa, B. Pal, Y. Masumoto, I. V. Ignatiev, S. Yu Verbin and I. Ya. Gerlovin, "Sub-millisecond electron spin relaxation in InP quantum dots", Phys. Rev. B, **72**, 153302 (2005).

[25] R. Hanson, L. H. Willems vand Beveren, I. T. Vink, J. M. Elzerman, W. J. M. Naber, F. H. L. Koppens, L. P. Kouwenhoven and L. M. K. Vandersypen, "Single-shot readout of electron spin states in a quantum dot using spin dependent tunnel rates", Phys. Rev. Lett., **94**, 196802 (2005).

[26] M. Kroutvar, Y. Ducommun, D. Heiss, M. Bichler, D. Schuh, G. Abstreiter and J. J. Finley, "Optically programmable electron spin memory using semiconductor quantum dots", Nature (London), **432**, 81 (2004).

[27] S. Amasha, K. MacLean, Iuliana Radu, D. M. Zümbuhl, M. A. Kastner, M. P. Hanson and A. C. Gossard, "Measurements of the spin relaxation rate at Low magnetic fields in a quantum dot", www.arXiv.org/cond-mat/0607110.

[28] S. Pramanik, C-G Stefanita, S. Patibandla, S. Bandyopadhyay, K. Garre, N. Harth and M. Cahay, "Observation of extremely long spin relaxation time in an organic nanowire spin valve", Nature Nanotech., **2**, 216 (2007).

[29] S. Sanvito, "Organic electronics: Memoirs of a spin", Nature Nanotech., **2**, 204 (2007).

[30] S. Cortez, et al., "Optically driven spin memory in n-doped InAs-GaAs quantum dots", Phys. Rev. Letters, **89**, 207401 (2002).

[31] D. V. Bulaev, D. Trauzettel and D. Loss, "Spin orbit interaction and anomalous spin relaxation in carbon nanotube quantum dots", Phys. Rev. B., **77**, 235301 (2008).

[32] V. N. Golovach, A. Khaetski and D. Loss, "Phonon induced decay of the electron spin in quantum dots", Phys. Rev. Lett., **93**, 016601 (2004).

[33] R. de Sousa and S. Das Sarma, "Electron spin coherence in semiconductors: Considerations for a spin-based solid-state quantum computer architecture", Phys. Rev. B, **67**, 033301 (2003).

[34] S. Bandyopadhyay, et al., "Electrochemically assembled quasi perioidic quantum dot arrays", Nanotechnology, **7**, 360 (1996).

[35] A. Abragam, *The Principles of Nuclear Magnetism*, (Oxford University Press, London, 1961); C. P. Slichter, *Principles of Magnetic Resonance*, (Springer-Verlag, Berlin, 1996).

[36] T. A. Kennedy, J. S. Colton, J. E. Butler, R. C. Linares and P. J. Doering, "Long coherence times at 300 K for nitrogen-vacancy center spins in diamond grown by chemical vapor deposition", Appl. Phys. Lett., **83**, 4190 (2003).

[37] T. Gaebel, et al., "Room-temperature coherent coupling of single spins in diamond", Nature Physics, **2**, 408 (2006).

[38] M. V. Gurudev Dutt, L. Childress, L. Jiang, E. Togan, J. Maze, F. Jelezko, A. S. Zibrov, P. R. Hemmer and M. D. Lukin, "Quantum register based on individual electronic and nuclear spin qubits in diamond", Science, **316**, 1312 (2007).

[39] J. A. Gupta, D. D. Awschalom, X. Peng and A. P. Alivisatos, "Spin coherence in semiconductor quantum dots", Phys. Rev. B, **59**, R10421 (1999).

[40] R. I. Dzhioev, V. L. Korenev, I. A. Merkulov, B. P. Zakharchenya, D. Gammon, Al. L. Efros and D. S. Katzer, "Manipulation of the spin memory of electrons in n-GaAs", Phys. Rev. Lett., **88**, 256801 (2002).

[41] S. Pramanik, B. Kanchibotla and S. Bandyopadhyay, "Transverse spin relaxation time in an ensemble of electrochemically self assembled CdS quantum dots", Proc. of IEEE NANO 2006, Cincinnati, OH, July 2006 (IEEE Press).

[42] I. A. Merkulov, Al. L. Efros and M. Rosen, "Electron spin relaxation by nuclei in semiconductor quantum dots", Phys. Rev. B, **65**, 205309 (2002).

[43] A. V. Khaetski, D. Loss and L. Glazman, "Electron spin decoherence in quantum dots due to interaction with nuclei", Phys. Rev. Lett., **88**, 186802 (2002).

[44] A. Greilich, A. Shabaev, D. R. Yakovlev, Al. L. Efros, I. A. Yugova, D. Reuter, A. D. Wieck and M. Bayer, "Nuclei induced frequency focusing of electron spin coherence", Science, **317**, 1896 (2007).

[45] A. M. Tyryshkin, S. A. Lyon, W. Jantsch and F. Schaffler, "Spin manipulation of free two dimensional electrons in Si/SiGe quantum wells", Phys. Rev. Lett., **94**, 126802 (2005).

[46] S. Bandyopadhyay and M. Cahay, "Rashba effect in an asymmetric quantum dot in a magnetic field", Superlat. Microstruct., **32**, 171 (2002).

[47] A. S. Bracker, E. A. Stinaff, D. Gammon, M. E. Ware, J. G. Tischler, A. Shabaev, Al. L. Efros, D. Park, D. Gershoni, V. L. Korenev and I. A. Merkulov, "Optical pumping of the electronic and nuclear spin of single charge-tunable quantum dots", Phys. Rev. Lett., **94**, 047402 (2005).

[48] M. V. Gurudev Dutt, Jun Cheng, Bo Li, Xiadong Xu, P. R. Berman, D. G. Steel, A. S. Bracker, D. Gammon, Sophia E. Economou, Ren Bao Liu, and L. J. Sham, "Stimulated and spontaneous optical generation of electron spin coherence in charged GaAs quantum dots", Phys. Rev. Lett., **94**, 227403 (2005).

9

Some Spin Phenomena

The spin properties of electrons can give rise to certain observable phenomena that often have practical applications and frequently bring forth the special characteristics of spin transport. In this chapter, we will discuss a few of these phenomena that have evoked considerable interest.

9.1 The Spin Hall effect

In solid state physics, a student is introduced to the traditional Hall effect, which has nothing to do with an electron's spin. However, with the advent of spintronics, a plethora of "Hall effects" have come to the fore, which we discuss next. First, let us revisit the traditional Hall effect.

The normal Hall effect

Consider a solid structure subjected to a magnetic field of flux density B_z in the z-direction. An electric field is applied in the x-direction that causes an electron in the solid to accelerate in the $-x$-direction. However, scattering causes a frictional force, proportional to the electron velocity, which opposes the force due to the electric field. When the two forces balance, the electron reaches a constant velocity known as the "drift velocity". In the steady state, the electron drifts along the $-x$-direction with this constant velocity v_{drift}. The reader should be familiar with this model which is known as the *Drude model* of conduction.

The z-directed magnetic field exerts a Lorentz force $ev_{drift}B_z$ on the electron which pushes it in either the $+y$-direction or $-y$-direction depending on the direction of the magnetic field. If the sample has boundaries in the y-direction, then electrons will pile up at one edge (the edge that they are being pushed towards), causing a charge imbalance between this edge and the opposite edge. This charge imbalance will cause a y-directed electric field and an associated potential difference which can be measured by connecting a potentiometer between the two edges. This is the Hall voltage and this effect is the celebrated *Hall effect* predicted by Edwin Hall in 1879 [1]. It is routinely used today in semiconductor characterization to ascertain carrier concentration in a sample or its polarity (p-type or n-type). The sign of the Hall voltage depends on the polarity of charge carriers (whether they are electrons or holes). Its magnitude is inversely proportional to the carrier concentration and directly proportional to the magnetic field.

The anomalous Hall effect

In a ferromagnet, there can be an additional contribution to the Hall voltage that arises without any external magnetic field. Because of spin-dependent band structure [2] or spin-dependent scattering events due to spin-orbit coupling [3, 4, 5], electrons whose spins are polarized in the, say, $+z$-direction, are scattered to one edge of the sample and electrons whose spins are polarized in the $-z$-direction are scattered to the other edge [6]. Since the material is a *ferromagnet*, there are majority and minority spins, meaning that there can be more electrons with $+z$-polarized spins than $-z$-polarized spins. In that case, more electrons are scattered towards one edge than the other, leading once again to a charge imbalance between these two edges and a resulting Hall voltage. This Hall voltage does *not* require a magnetic field to be produced. This is the so-called anomalous Hall effect, whose existence is often used as proof of ferromagnetism in materials. Note that, not only does a charge imbalance exist between the two edges, but a spin imbalance exists as well. Therefore, both a charge current and a spin current can flow between the two edges.

The total Hall resistance R_H of a ferromagnet will have the form [7]

$$R_H = R_0 B + R_s M, \qquad (9.1)$$

where B is the magnitude of the magnetic flux density, M is the magnitude of the magnetization, R_0 is the ordinary Hall coefficient and R_s is the anomalous Hall coefficient.

The normal and anomalous Hall effects are explained in Fig. 9.1, which is adapted from [8].

The quantum anomalous Hall effect

The quantum Hall effect is a well-known phenomenon in a two-dimensional electron gas (2-DEG) subjected to a transverse magnetic field at low temperatures. The longitudinal resistance of the electron gas vanishes and the tranverse resistance (or Hall resistance) exhibits plateaus at certain values of the magnetic feld, the plateau resistance being quantized to h/e^2. This happens due to the formation of well-defined Landau levels and edge states that carry current. A similar phenomenon can occur *without a magnetic field* in a two-dimensional ferromagnetic insulator. In the limit of vanishing spin-orbit coupling (that couples the majority and minority spins) and large enough exchange splitting that separates the majority and minority spin bands in a ferromagnet, the majority spin band will be completeley full and the minority spin band completely empty at low temperatures. When the exchange splitting is gradually reduced, the two bands intersect each other in wavevector

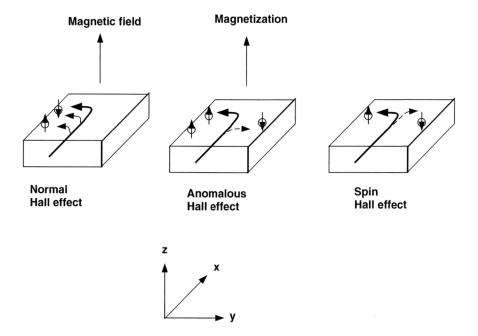

FIGURE 9.1

Pictorial depiction of the normal, the anomalous and the spin Hall effects.

space, leading to band inversion. The energy degeneracy at the intersection point can be removed by turning on the spin-orbit coupling, leading to the formation of an energy gap. When the Fermi level is in this gap, the 2-DEG is insulating since the density of states at the Fermi level is zero. Such an insulator, also known as a Chern insulator, should exhibit a quantized anomalous Hall resistance of h/e^2 and zero longitudinal resistance when the Fermi level is placed within the gap [9].

The quantized anomalous Hall effect was experimentally demonstrated in the magnetic topological insulator Cr-doped $(Bi,Sb)_2Te_3$ whose surface behaves like a two-dimensional ferromagnet [10]. The longitudinal and anomalous Hall resistances were measured in the absence of any magnetic field. The Fermi level was varied by applying a potential to the topological insulator with a gate terminal. At certain values of the gate voltage (which corresponded to placing the Fermi level in the gap), the longitudinal resistance nearly vanished and the anomalous Hall resistance exihibited a plateau quantized to h/e^2, indicating the formation of the quantum anomalous Hall state.

The Spin Hall effect

There are two types of Spin Hall effect: extrinsic and intrinsic. The extrinsic effect arises from spin-dependent scattering or band structure peculiarities, much like the anomalous Hall effect [11, 12]. A spin-unpolarized current flowing into a paramagnetic semiconductor slab in the absence of any magnetic field will inject electrons into the slab with spins pointing up and down with respect to the slab's plane, as shown in the far right figure of Fig. 9.1. Spin-dependent scattering will deflect spin-up electrons to one side of the slab and spin-down electrons to the other side, resulting in a spin imbalance between the right and left edges of the slab. Therefore, a spin current will flow between the two edges just as in the case of the anomalous Hall effect. However, since the material is *not* a ferromagnet and the current injected is *not* spin-polarized, there are *equal* numbers of spin-up and spin-down electrons in the slab. The result is that the number of spin-up electrons piling up at one edge is equal to the number of spin-down electrons piling up at the other edge. Therefore, there is no charge imbalance between the two edges and hence no charge current flows, unlike in the case of the anomalous Hall effect. The extrinsic Spin Hall effect has been experimentally demonstrated in paramagnets [13, 14].

Following [12], we will assume that the elastic scatterers in the paramagnet are spinless and described by a scattering potential of the form

$$V = V_c(\vec{r}) + V_s(\vec{r})\,\vec{\sigma} \cdot \vec{L}, \tag{9.2}$$

where $\vec{\sigma}$ and $\vec{L}$ are the spin and orbital angular momentum operators of an electron scattered by the scatterer. The scattered beam will be spin-polarized with a polarization vector given by

$$\vec{\nu} = \frac{fg^* + f^*g}{|f|^2 + |g|^2}\hat{n}, \tag{9.3}$$

where $\hat{n}$ is a unit vector perpendicular to the scattering plane defined by the initial and final wavevectors of the electron (before and after scattering), while f and g are the spin-independent and spin-dependent parts of the scattering amplitude.

The Spin Hall voltage generated due to spin-up electrons can be written in analogy to the anomalous Hall effect:

$$V_\uparrow = R_s W J_x n_\uparrow \mu_B, \qquad (9.4)$$

where W is the width of the slab, J_x is the magnitude of the (x-directed) current density flowing through the slab, $n_\uparrow$ is the density of excess spin-up electrons piling up at one edge, and μ_B is the Bohr magneton. A similar expression will hold for spin-down electrons with $n_\uparrow$ replaced by $n_\downarrow$. In Fig. 9.1, spin-up electrons will be flowing to the left and spin-down electrons to the right. These two spin current components *add* and hence the total Spin Hall voltage will be

$$V_{SH} = R_s W J_x \left(n_\uparrow + n_\downarrow\right) \mu_B = R_s W J_x n \mu_B. \qquad (9.5)$$

The spin current for each spin is given by

$$J_\uparrow = J_\downarrow = \frac{V_{SH}}{\rho W}, \qquad (9.6)$$

where ρ is the resistivity of the sample and we have assumed that the resistivity for the spin current is the same as that for the charge current [12]. The ratio of the spin current to the charge current (sometimes referred to as the "Spin Hall angle") is therefore

$$\theta_{SH} = \frac{J_{spin}}{J_{charge}} = \frac{J_\uparrow + J_\downarrow}{J_x} = \frac{R_s n \mu_B}{\rho} = R_s n^2 e \mu \mu_B, \qquad (9.7)$$

where e is the electronic charge and μ is the electron drift mobility in the sample.

Note that the spin current is *dissipationless*. The electric field $\vec{\mathcal{E}}$ causing the charge current injection must be collinear with the charge current and is hence perpendicular to the spin current $\vec{J}_{spin}$. As a result, $\vec{J}_{spin} \cdot \vec{\mathcal{E}} = 0$, so the spin current does not dissipate any energy.

The Spin Hall voltage of course cannot be measured with a voltmeter since there is no voltage difference between the two sides. The electrochemical potentials of the spin-up electrons at the two edges are different as are the electrochemical potentials for the spin-down electrons, but the total electrochemical potentials (due to both spin-up and spin-down electrons) at the two edges are the same.

However, if we connect the two edges with a conductor of width w to allow spin-up electrons to flow to the left and spin-down electrons to flow to the right, then something remarkable happens. The same spin-dependent

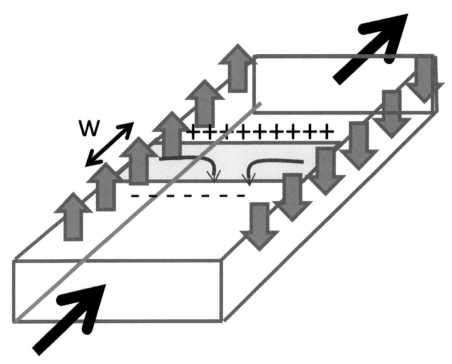

FIGURE 9.2
The Inverse Spin Hall effect

scatterers will deflect the left-flowing spin-up electrons toward one side and right-flowing spin-down electrons to the *same* side. In other words, electrons accumulate on one side and deplete on the other. Therefore, a charge imbalance will build up at the two edges of the conductor, as shown in Fig. 9.2, and the resulting potential drop can be measured with a voltmeter! These voltages have been experimentally measured [13, 15, 16]. This effect is sometimes called the *inverse* Spin Hall effect. The electric field associated with the potential drop is given by

$$\vec{\mathcal{E}}_{ISH} = \theta_{SH}\rho \left[\vec{J}_\sigma \times \vec{\sigma} \right], \tag{9.8}$$

where $\vec{\sigma}$ is the spin polarization of the spin current $\vec{J}_\sigma$.

The extrinsic Spin Hall effect has many applications, which we will discuss later in the context of the *giant* Spin Hall effect, but let us now discuss the physically intriguing *intrinsic* version of the Spin Hall effect.

9.1.1 The intrinsic Spin Hall effect

The intrinsic Spin Hall effect is an intriguing phenomenon that takes place in a two-dimensional electron gas (2-DEG) in *ballistic transport* (no scattering, whether spin-dependent or spin-independent), provided there is spin-orbit interaction, such as the Rashba effect, present in the 2-DEG. To an observer, there is no difference between the extrinsic and intrinsic Spin Hall effect; when a charge current is injected in the plane of the 2-DEG, a dissipationless spin current (consisting of spins polarized perpendicular to the plane of the 2-DEG) flows in the plane of the 2-DEG in a direction perpendicular to the charge current. However, it is not caused by spin-dependent scattering deflecting opposite spins to opposite sides. Instead, it is caused by the Rashba spin-orbit interaction [17, 18]. This phenomenon has been demonstrated experimentally [19].

We will derive the intrinsic Spin Hall effect along the lines of [17]. Consider a 2-DEG in the $(x - y)$ plane (see Fig. 9.1) and a symmetry-breaking electric field $\mathcal{E}_z$ is applied in the z-direction to induce the Rashba spin-orbit interaction in the electron gas. The Hamiltonian describing this system is

$$
\begin{aligned}
H &= \frac{|\vec{p}|^2}{2m^*} + V(\vec{r}) - \frac{a_{46}}{\hbar} \mathcal{E}_z \hat{z} \cdot [\vec{\sigma} \times \vec{p}] \\
&= \begin{bmatrix} \frac{p^2}{2m^*} + V(\vec{r}) & \frac{\eta}{\hbar}(p_y + ip_x) \\ \frac{\eta}{\hbar}(p_y - ip_x) & \frac{p^2}{2m^*} + V(\vec{r}) \end{bmatrix},
\end{aligned} \tag{9.9}
$$

where $V(\vec{r})$ is the potential energy, $\vec{p} = p_x \hat{x} + p_y \hat{y}$, and a_{46}, η are the Rashba interaction constants in the 2-DEG.

The Rashba spin-orbit interaction acts like an effective magnetic field $\vec{B}_{Rashba}$. We know that spin-orbit interaction gives rise to a magnetic field in the reference frame of a moving electron moving in an electric field. This magnetic field is (see Equation (6.2))

$$
\vec{B}_{Rashba} = \frac{\vec{E} \times \vec{v}}{2c^2}, \tag{9.10}
$$

where $\vec{v}$ is the electron velocity, $\vec{E}$ is the electric field, and c is the speed of light in vacuum. Here we have ignored the relativistic factor since we are dealing with a non-relativistic electron in a solid. Recently, the effective magnetic fields associated with Rashba and Dresselhaus spin-orbit interactions have been experimentally measured in a 2-DEG [20]. Note that, according to the above equation, the effective field depends on the velocity $\vec{v}$ and *not* the momentum $\vec{p}$ or $\hbar\vec{k}$. It *makes a difference*, since in the presence of spin-orbit interaction, $\vec{v}$ and $\vec{p}$ (or $\vec{k}$) are *not* proportional to each other, as we have already seen (recall Section 7.1.1 and Equation (7.38)). We can show this more directly by considering the fact that the Hamiltonian of an electron in a solid, in the presence of spin-orbit interaction, is given by

$$H = \frac{\hbar^2 k^2}{2m^*} + V(\vec{r}) + \frac{\hbar^2}{4(m^*)^2 c^2} \left[\frac{\partial V(\vec{r})}{\partial \vec{r}}\right] \cdot \left[\vec{\sigma} \times \vec{k}\right].$$ (9.11)

Hamiltonian mechanics for canonically conjugated variables [momentum $(\vec{p})$ and position $(\vec{r})$] dictate

$$\vec{v} = \frac{d\vec{r}}{dt} = \frac{\partial H}{\partial \vec{p}} = \frac{\partial H}{\partial(\hbar\vec{k})} = \frac{\hbar\vec{k}}{m^*} + \frac{\hbar}{4(m^*)^2 c^2} \left[\frac{\partial V(\vec{r})}{\partial \vec{r}} \times \vec{\sigma}\right].$$ (9.12)

Clearly, because of the second term on the right hand side, which is entirely due to spin-orbit interaction, $\vec{v}$ is not proportional to $\vec{k}$. They are related as $\vec{v} = A\vec{k} + B$, where A and B are constants. This is not a bandstructure issue; it is true even in a parabolic band. Therefore, B_{Rashba} cannot be simultaneously proportional to both v and k. It should be one or the other. Unfortunately, this issue is not always appreciated. One often finds in the literature that B_{Rashba} is assumed proportional to k [21], and sometimes it is assumed proportional to v. Obviously, both cannot be correct in every situation. Here, we will take the first steps toward resolving this issue.

One can appeal to the Ehrenfest theorem (which we saw in Chapter 4) to deduce whether B_{Rashba} is proportional to v or k. According to this theorem (see Chapter 18 for the exact statement of this theorem), we get

$$\frac{d\langle \vec{S}\rangle}{dt} = \frac{1}{i\hbar}\langle[H, \vec{S}]\rangle = \frac{1}{i\hbar}\langle[H, (\hbar/2)\vec{\sigma}]\rangle,$$ (9.13)

where the angular brakets denote the expectation values and the square bracket denotes the commutator. Here H is the Hamiltonian in Equation (9.9). The magnetic field B_{Rashba} will cause a spin to precess about it and this precession will be described by Equation (4.4). Therefore, Equation (4.4) tells us that the left hand side of the above equation is $(g\mu_B \vec{B}_{Rashba}/\hbar) \times \langle \vec{S}\rangle$. Consequently, we obtain that

$$\frac{g\mu_B \vec{B}_{Rashba}}{\hbar} \times \langle \vec{S}\rangle = \frac{1}{i\hbar}\langle[H, \vec{S}]\rangle = \frac{1}{i\hbar}\left\langle\left[-\frac{a_{46}}{\hbar}\mathcal{E}_z \hat{z} \cdot [\vec{\sigma} \times \vec{p}], (\hbar/2)\vec{\sigma}\right]\right\rangle.$$ (9.14)

The reader can easily show from the above equation that B_{Rashba} is proportional to p or k, and not v.* Thereafter, it is easy to show that the magnitude of B_{Rashba} can be obtained by equating the Zeeman splitting energy caused by B_{Rashba} to the spin-splitting energy $2\eta k$, which would yield that

$$|B_{Rashba}| = \frac{2\eta k}{g\mu_B} = \left|\frac{2a_{46}\mathcal{E}_z k}{g\mu_B}\right|.$$ (9.15)

---

*The authors are grateful to Prof. Yuli Lyanda-Geller for this discussion.

However, this result presents an immediate conundrum. First, note from Fig. 7.4(b)[†] that an electron with a given energy E does *not* have a unique wavevector k in the presence of spin-orbit interaction. The wavevector is *spin-dependent* since the electron will have two different wavevectors in the two spin-split bands for the same E. The wavevector will be $k_1(E)$ in one spin band and $k_2(E)$ in the other. Therefore, if $|B_{Rashba}|$ were proportional to k according to Equation (9.15), then it would be proportional to k_1 in one band and k_2 in the other. Thus, it will have two different magnitudes for two different spins. In other words, the spin-orbit magnetic field that an electron experiences *depends on what its spin orientation is*.

A *spin-dependent* $\vec{B}_{Rashba}$ is fine as long as we remember that it is different in the two spin-split bands and account for that fact explicitly in any analysis. However, there are situations when a spin-dependent field becomes inconvenient. This happens if spin polarizations are changing with time, which then necessitates a time-dependent $\vec{B}_{Rashba}$. If the spins are changing randomly because of spin flip scattering, then B_{Rashba} changes randomly with time. If the spins are precessing coherently in time, B_{Rashba} still changes continuously, albeit coherently. It is therefore much more convenient to deal with a *spin-independent* $\vec{B}_{Rashba}$ since that could be time-invariant even if the spins are time-varying. In the next chapter, we will show that such a $\vec{B}_{Rashba}$ or $\vec{B}_{Dresselhaus}$ can be formulated and will be proportional to $\vec{v}$ and not $\vec{k}$. The spin-independent fields are very convenient tools to explain the workings of the Spin Field Effect Transistor, or the absence of D'yakonov-Perel' spin relaxation in strictly single channeled transport in a [100]-oriented quantum wire.

In the discussion of the intrinsic Spin Hall effect, we can, however, work with a spin-dependent $\vec{B}_{Rashba}$ since we will account for effects arising from the two spin-split bands separately. Therefore, we will use Equation (9.15). The reader is warned that the magnetic field in Equation (9.15) is always *spin-dependent* and using this equation requires particular care; we must account for effects from the two spin-split bands separately.

Before we proceed further, we need to ascertain the direction of $\vec{B}_{Rashba}$, which is not given by Equation (9.15). Equation (7.37) and Equation (9.12) both tell us that v and k are collinear, i.e., they have the same direction. Next, we appeal to Equation (6.2), which tells us that the direction of $\vec{B}_{Rashba}$ (in either spin-split band) is perpendicular to both the direction of the effective electric field that causes the Rashba interaction and the velocity (or wavevector). Therefore, for our 2-DEG, we can write the *spin-dependent* B_{Rashba} as

$$\vec{B}_{Rashba} = -\frac{2a_{46}}{g\mu_B}\vec{\mathcal{E}} \times \vec{k} = \frac{2\eta}{g\mu_B}\hat{z} \times \vec{k}$$
$$= \Xi\left[-k_y\hat{x} + k_x\hat{y}\right] , \tag{9.16}$$

[†]The quantum wire can be thought of as a 2-DEG with edges.

where $\Xi = \frac{2\eta}{g\mu_B}$. Note that $\vec{k} \cdot \vec{B}_{Rashba} = 0$.

Next, using Equation (4.4), we get

$$\frac{d\vec{S}}{dt} = \vec{\Omega} \times \vec{S}$$

$$= \frac{g\mu_B \vec{B}_{Rashba}}{\hbar} \times \vec{S}$$

$$= det \begin{bmatrix} \hat{x} & \hat{y} & \hat{z} \\ \frac{g\mu_B}{\hbar} B_x & \frac{g\mu_B}{\hbar} B_y & 0 \\ S_x & S_y & S_z \end{bmatrix}$$

$$= \frac{2\eta}{\hbar} \left[k_x S_z \hat{x} + k_y S_z \hat{y} - (k_y S_y + k_x S_x)\hat{z} \right] , \qquad (9.17)$$

where B_x and B_y are the x- and y-components of $\vec{B}_{Rashba}$.

Therefore, equating each component separately, we obtain

$$\frac{dS_x}{dt} = \frac{g\mu_B}{\hbar} B_y S_z = \frac{2\eta}{\hbar} k_x S_z,$$

$$\frac{dS_y}{dt} = -\frac{g\mu_B}{\hbar} B_x S_z = \frac{2\eta}{\hbar} k_y S_z,$$

$$\frac{dS_z}{dt} = \frac{g\mu_B}{\hbar} (B_x S_y - B_y S_x) = -\frac{2\eta}{\hbar} (k_y S_y + k_x S_x). \qquad (9.18)$$

From Equation 9.16, we see that the only magnetic field in our 2-DEG, i.e., $\vec{B}_{Rashba}$, lies entirely in the plane of the 2-DEG and has no z-component. Since in the presence of dissipation the electron spins must align parallel or anti-parallel to the magnetic field, we conclude that the electron spins in the 2-DEG will have no z-component and $S_z = 0$. Consider now the situation when an electric field E_x is applied in the positive x-direction at time $t = 0$. This field will induce a drift of electrons in the $-x$-direction. We will show that three things result from this drift: (1) the spins develop a z-component; (2) $+z$-polarized and $-z$-polarized spins accumulate at opposite edges of the 2-DEG, causing a spin imbalance between the edges and hence a spin-current flow along the y-direction; and (3) the ratio of the spin current density to the electric field (i.e., the spin-Hall conductivity) is a universal constant. This is the intrinsic Spin Hall effect.

Let us focus on an electron with initial velocity $\vec{v}_0$ and wavevector $\vec{k}_0$ at time $t = 0$ and define a new set of coordinate axes (the "primed" coordinates) such that the y'-axis coincides with $\vec{v}_0$ or $\vec{k}_0$ (see Fig. 9.3). It should be obvious that Equations (9.16 – 9.18) will continue to hold if we replace the unprimed coordinates everywhere with the primed coordinates since the $x - y$ system in these equations is arbitrary. Let the y'-axis subtend an angle ϕ with the y-axis.

At time $t = 0$, $\vec{B}_{Rashba}(t = 0)$ is along the x' direction, since it is perpendicular to $\vec{k}_0$ and $\hat{z}$. Since $B_{x'} \neq 0$ and $B_{y'} = 0$, $S_{x'}(0) = -\hbar/2$ in the lower

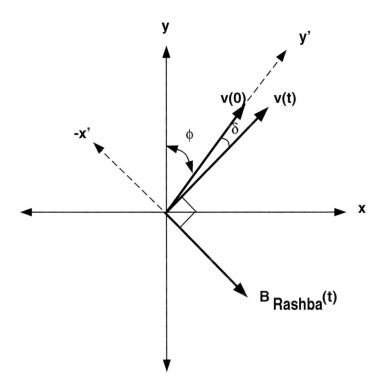

FIGURE 9.3
The coordinate axes in the two-dimensional electron gas (2-DEG) and the velocity vector, as well as the vector representing B_{Rashba}.

Rashba spin-split band, and $+\hbar/2$ in the upper spin-split band, while $S_{y'}(0)$ $= S_z(0) = 0$ in either band. The application of the electric field along the x-direction will change B_{Rashba} since it will accelerate the electron and change k_x (see Equation (9.16)). We will assume that the electric field E_x is so weak that B_{Rashba} changes slowly in time, allowing the spin to follow adiabatically and align itself parallel or anti-parallel to B_{Rashba} at every instant of time.

Immediately after the application of the electric field, i.e., at time $t = 0+ = \tau$, the change in the magnitude of B_{Rashba} can be ignored, but its direction will have changed slightly. Let us assume that it has rotated by a small angle $\delta(\tau)$ from its initial direction at time $t = 0$. Then the change in B_{Rashba} is $\Delta B_{Rashba} = \Delta B_{x'} = B_{x'}(cos\delta - 1) \approx -B_{x'}\delta^2$, while the change in $B_{y'}$ is $\Delta B_{y'} = B_{x'}sin\delta \approx B_{x'}\delta$. The ratio of the two changes is $\Delta B_{x'}/\Delta B_{y'} = -\delta$, which is a small quantity. Therefore, the former change is much smaller than the latter. Consequently, we can postulate that at $t = 0+ = \tau$,

$$\left.\frac{dB_{x'}(t)}{dt}\right|_{t=\tau} \approx 0,$$

$$\left.\frac{dB_{y'}(t)}{dt}\right|_{t=\tau} \neq 0,$$

$$\text{and } S_{x'}(\tau) \approx \pm\frac{\hbar}{2} , \qquad (9.19)$$

where the $+$ sign is for the upper Rashba spin-split band and the - sign is for the lower Rashba spin-split band.

At time $t = 0+ = \tau$, B_{Rashba} develops a small y'-component which is equal to $B_{y'}(0) + \Delta B_{y'} \approx B_{x'}\delta$. Since the spin follows B_{Rashba} adiabatically, we infer that

$$\left.\frac{dS_{y'}}{dt}\right|_{t=\tau} \neq 0. \qquad (9.20)$$

We will argue that the change is linear in time over the small duration τ so that

$$\left.\frac{d^2 S_{y'}}{dt^2}\right|_{t=\tau} = 0. \qquad (9.21)$$

From Equations (9.18) – (9.21), we get that at time $t = \tau$,

$$0 = \frac{d^2 S_{y'}}{dt^2} = -\frac{g\mu_B}{\hbar}\left[\frac{dB_{x'}(t)}{dt}S_z + \frac{dS_z}{dt}B_{x'}(t)\right] \approx -\frac{g\mu_B}{\hbar}\frac{dS_z}{dt}B_{x'}. \qquad (9.22)$$

Using Equation (9.18) in the above equation, we get

$$0 \approx B_{x'}(\tau)S_{y'}(\tau) - B_{y'}(\tau)S_{x'}(\tau)$$
$$\approx B_{x'}(\tau)S_{y'}(\tau) \mp B_{y'}(\tau)(\hbar/2) , \qquad (9.23)$$

which yields

$$S_{y'}(\tau) = \pm\frac{B_{y'}(\tau)}{B_{x'}(\tau)}\frac{\hbar}{2} , \qquad (9.24)$$

where the + sign is for the upper Rashba band and the - sign is for the lower Rashba band.

From Equation (9.18), we also get

$$\frac{dS_{y'}}{dt} = -\frac{g\mu_B}{\hbar}B_{x'}(t)S_z(t) .$$

(9.25)

Substitution of Equation (9.24) in Equation (9.25) yields that at $t = \tau$,

$$S_z(\tau) = \mp\frac{\hbar}{g\mu_B}\frac{1}{B_{x'}(\tau)}\frac{d}{dt}\left(\frac{B_{y'}(t)}{B_{x'}(t)}\right)\Bigg|_{t=\tau}\frac{\hbar}{2}$$

$$= \mp\frac{\hbar}{g\mu_B}\frac{1}{B_{x'}(\tau)}\left[\frac{B_{x'}(\tau)(dB_{y'}(t)/dt)_{t=\tau} - B_{y'}(\tau)(dB_{x'}(t)/dt)_{t=\tau}}{B_{x'}^2(\tau)}\right]\frac{\hbar}{2}$$

$$= \mp\frac{\hbar}{g\mu_B}\frac{1}{B_{x'}^2(\tau)}\frac{dB_{y'}(t)}{dt}\Bigg|_{t=\tau}\frac{\hbar}{2} ,$$

(9.26)

where we have used Equation (9.19) to arrive at the last equality above.

Recall that the change in $B_{y'}$ in time $t = \tau$ is

$$\Delta B_{y'}(\tau) \approx B_{x'}(\tau)\delta(\tau).$$

(9.27)

Substituting this in Equation (9.26), we obtain that, at $t = \tau$,

$$S_z(\tau) = \mp\frac{\hbar^2}{2g\mu_B}\frac{1}{B_{x'}^2(\tau)}\left[\frac{dB_{x'}(t)}{dt}\Bigg|_{t=\tau}\delta(\tau) + B_{x'}(\tau)\frac{d\delta(t)}{dt}\Bigg|_{t=\tau}\right]$$

$$\approx \mp\frac{\hbar^2}{2g\mu_B B_{x'}(\tau)}\frac{d\delta(t)}{dt}\Bigg|_{t=\tau} ,$$

(9.28)

where we have used the fact that $dB_{x'}/dt|_{t=\tau} \approx 0$.

Referring to Fig. 9.3,

$$sin(\phi + \delta(t)) = \frac{v_x(t)}{v(t)} = \frac{k_x(t)}{k(t)},$$

(9.29)

where the last equality above can be easily proved using Equation (7.37).

Therefore,

$$\frac{d\delta(t)}{dt} = \frac{d(\phi + \delta(t))}{dt} = \frac{d}{dt}\left[\sin^{-1}\left(\frac{k_x(t)}{k(t)}\right)\right]$$

$$= \frac{1}{\sqrt{1 - k_x^2(t)/k^2(t)}} \frac{d}{dt}\left(\frac{k_x(t)}{k(t)}\right)$$

$$= \frac{1}{k_y(0)/k(t)} \frac{k(t)(dk_x(t)/dt) - k_x(t)(dk(t)/dt)}{k^2(t)}$$

$$= \frac{1}{k_y(0)/k(t)} \frac{k(t)(dk_x(t)/dt) - (k_x^2(t)/k(t))(dk_x(t)/dt)}{k^2(t)}$$

$$= \frac{k(t)}{k_y(0)} \frac{k_y^2(0)}{k^3(t)} \frac{dk_x(t)}{dt}$$

$$= \frac{k_y(0)}{k^2(t)} \frac{dk_x(t)}{dt}, \tag{9.30}$$

where we have used the fact that $k^2(t) = k_x^2(t) + k_y^2(t)$ and $k_y(t) = k_y(0)$ because there is no electric field in the y-direction to change the y-component of the momentum.

Substituting Equation (9.30) in Equation (9.28), we obtain

$$S_z(\tau) = \mp \frac{\hbar^2}{2g\mu_B B_{x'}(\tau)} \frac{k_y(0)}{k^2(\tau)} \frac{dk_x}{dt}\bigg|_{t=\tau}. \tag{9.31}$$

Newton's law in a crystal dictates that

$$\frac{d(\hbar k_x)}{dt} = -eE_x, \tag{9.32}$$

where we have neglected forces due to scattering, assuming ballistic transport. The above equation is valid even in the presence of spin-orbit interaction as slong as the Rashba interaction strength is spatially invariant (see Problem 7.6).

Substituting the above result in Equation (9.31), we obtain

$$S_z(\tau) = \pm \frac{e\hbar}{2g\mu_B B_{x'}(\tau)} \frac{k_y(0)}{k^2(\tau)} E_x. \tag{9.33}$$

Now, from Fig. 9.3,

$$B_{x'}^2 = B_{Rashba}^2 = B_x^2 + B_y^2, \tag{9.34}$$

where B_x is the x-component and B_y is the y-component of $B_{x'}$ or B_{Rashba}. Using Equation (9.16) in the above equation, we get

$$B_{x'}^2 = \Xi^2(k_x^2 + k_y^2) = \Xi^2 k^2(t) = \left(\frac{2\eta}{g\mu_B}\right)^2 k^2(t). \tag{9.35}$$

Using this relation in Equation (9.33) yields

$$S_z(\tau) = \pm \frac{e\hbar}{4\eta} \frac{k_y(0)}{k^3(\tau)} E_x \;, \qquad (9.36)$$

where the - sign is for the lower spin band and the + sign is for the upper.

The last equation is instructive. Note that even though there would have been no z-component of the spin in the 2-DEG in the absence of the electric field (because B_{Rashba} would have been entirely in the x-y plane), a z-component develops when the electric field is present and it is directly proportional to the strength of the field. It is also inversely proportional to the strength of the spin-orbit interaction η. A stronger spin-orbit interaction would have resulted in a stronger B_{Rashba} in the plane of the 2-DEG, tending to keep the spin polarization constrained to the plane of the 2-DEG and decreasing the z-component of spin. That is the physics.

Let us now examine if there is any y-directed spin current density $J_{sy}(t)$ due to the z-component of the spins at temperature $T \to 0$. If it exists, then there must be spin accumulation at the edges of the 2-DEG since any spin current flowing in the y-direction will need a spin imbalance.

The spin current density operator associated with the flow of z-polarized spins is $j_{s_z} = (1/2)\{\sigma_z, \vec{v}\}$, where the curly brackets denote the anti-commutator. Its expected value is given by (see [17]) $< \psi^+|j_{s_z}|\psi^- >$, where $|\psi^\pm >$ are the eigenstates in the upper and lower spin bands. The reader can easily show from the 2-DEG Hamiltonian that the eigenstates are

$$|\psi^\pm >= \frac{1}{\sqrt{2}} \begin{pmatrix} \pm e^{i\theta} \\ 1 \end{pmatrix} e^{i\vec{k}\cdot\vec{r}}, \qquad (9.37)$$

where $\theta = arctan(k_x/k_y)$.

The operator for the y-component of velocity v_y is

$$v_y = \frac{\partial H}{\partial p_y} = \begin{bmatrix} \frac{p_y}{m^*} & \frac{\eta}{\hbar} \\ \frac{\eta}{\hbar} & \frac{p_y}{m^*} \end{bmatrix}. \qquad (9.38)$$

Therefore, the operator for the y-component of the z-polarized spin current density is

$$j_{s_z}^y = \frac{1}{2}\{\sigma_z, v_y\} = \begin{bmatrix} \frac{p_y}{m^*} & 0 \\ 0 & -\frac{p_y}{m^*} \end{bmatrix}. \qquad (9.39)$$

Consequently, the contribution to the y-component of the z-polarized spin current density from a wavevector state $\vec{k}$ in either spin-split band is $J_{sy}(k(\tau))$ $= S_z^\pm(k(\tau)) < \psi^+|j_{s_z}^y|\psi^- >= S_z^\pm(k(\tau)) < \psi^-|j_{s_z}^y|\psi^+ >= -S_z^\pm(k(\tau))(\hbar k_y(0)/m^*)$.

Therefore, the y-directed spin current density due to z-polarized spin at time $t = \tau$ is

$$J_{sy}(\tau) = -\frac{1}{4\pi^2} \int_0^{2\pi} \int_0^{k_F} d\phi k(\tau) dk(\tau) \frac{\hbar k_y(0)}{m^*} S_z^\pm(k(\tau)) \;, \qquad (9.40)$$

where k_F is the Fermi wavevector, which is the wavevector corresponding to the Fermi energy (see Fig. 9.4). The spin current density in each band is

$$J_{sy1}(\tau) = -\frac{e\hbar^2}{16\pi^2 m^* \eta} E_x \int_0^{2\pi} \int_0^{k_{F1}} d\phi k(\tau) dk(\tau) \frac{k_y^2(0)}{k^3(\tau)} \quad \text{(upper band)}$$

$$J_{sy2}(\tau) = \frac{e\hbar^2}{16\pi^2 m^* \eta} E_x \int_0^{2\pi} \int_0^{k_{F2}} d\phi k(\tau) dk(\tau) \frac{k_y^2(0)}{k^3(\tau)} \quad \text{(lower band)}.$$

$$(9.41)$$

Since $k_y(0) \approx k(\tau) cos\phi$ (see Fig. 9.3), we obtain

$$J_{sy1}(\tau) = -\frac{e\hbar^2}{16\pi^2 m^* \eta} E_x \int_0^{2\pi} \int_0^{k_{F1}} d\phi dk(\tau) cos^2\phi$$

$$= -\frac{e\hbar^2}{16\pi m^* \eta} E_x k_{F1} \quad \text{(upper band)}$$

$$J_{sy2}(\tau) = \frac{e\hbar^2}{16\pi^2 m^* \eta} E_x \int_0^{2\pi} \int_0^{k_{F2}} d\phi dk(\tau) cos^2\phi$$

$$= \frac{e\hbar^2}{16\pi m^* \eta} E_x k_{F2} \quad \text{(lower band)}. \quad (9.42)$$

Consequently, if both Rashba spin-split bands are occupied, then the total spin current is

$$J_{sy}(\tau) = J_{sy1}(\tau) + J_{sy2}(\tau) = \frac{e}{8\pi} E_x, \quad (9.43)$$

where we have used the fact that $k_{F2} - k_{F1} = 2m^* \eta / \hbar^2$. This result shows that a spin-current due to z-polarized spins flowing in the y-direction exists, which means that $+z$ and $-z$-polarized spins accumulate at opposite edges of the sample due to the combined actions of the spin-orbit interaction and the electric field causing drift (or the charge current resulting from this drift). This is the intrinsic spin-Hall effect, which does not require spin-dependent scattering at all, unlike the extrinsic spin-Hall effect.

The Spin Hall conductivity is

$$\sigma_{SH} = \frac{J_{sy}}{E_x} = \frac{e}{8\pi}. \quad (9.44)$$

Note that the Spin Hall conductivity depends only on the universal constant e, which is the charge of the electron. Hence the name "universal" intrinsic Spin Hall effect. Note also that we accounted for contributions from the two spin-split bands separately, which allowed us to work with the spin-dependent B_{Rashba}.

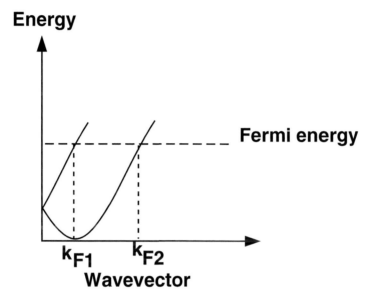

FIGURE 9.4
Energy dispersion relation in a two-dimensional electron gas (2-DEG) with
Rashba spin-orbit interaction. The two-dimensional wavevector in the x-y
plane is denoted by "Wavevector".

The giant Spin Hall effect

In most paramagnets, the Spin Hall angle θ_{SH} defined in Equation (9.7) is very small, so very little spin current is generated by reasonable charge currents, regardless of whether the Spin Hall effect is intrinsic or extrinsic. However, in some materials, like platinum-doped gold [22] and β-tantalum [23], the angle has been reported to be as large as 0.12 radians in very thin films. The large spin currents generated in tantalum have been used to switch magnets [23] using the spin-torque effect described later in this chapter. Such switching has been utilized to propagate logic information unidirectionally in arrays of dipole coupled nanomagnets, described in more detail in Chapter 17 [24]. The giant Spin Hall effect is a term used to describe Spin Hall effect with very large Spin Hall angles. Its discovery has renewed interest in the Spin Hall effect in view of its direct application in computing – logic and memory.

There are at least three possible reasons why the Spin Hall angle in certain thin film materials can be much larger than usual: (1) large spin-orbit interaction of impurities leading to strong spin-dependent deflection [25], (2) Coulomb correlation or spin fluctuation of impurities resulting in enhanced spin-orbit interaction at impurity sites accompanied by resonant skew scattering [26], and (3) surface assisted skew scattering from impurities which would be predominant in very thin films [22]. Recently, there has been a new twist to the Spin Hall effect. Generally, it is measured under very low charge injection current, which means that the electrostatic potential drop in the direction of the current is very small. As a result, only the electrons at or near the Fermi level in the paramagnet carry current. Since spin-orbit interaction is dependent on the electron's velocity (and hence kinetic energy), it would be illuminating to probe the Spin Hall effect while varying the electron energy sufficiently away from the Fermi level. Recently, this feat was accomplished by injecting electrons with specific energies through a tunneling contact and then varying the potential drop (or bias) across the sample to vary the electron's energy and carry out Spin Hall effect tunneling spectroscopy [27]. These studies allow one to find the energy dependence of the Spin Hall effect and open a route to engineering materials for optimum effect by placing the Fermi level close to the energies where the Spin Hall effect is largest. This can be accomplished by doping or backgating techniques.

Recently, there have been some exciting developments involving *topological insulators* whose bulk is insulating but the surface is not only conducting but is spin-polarized with the electron's spin orientation fixed with respect to its propagation direction. An example of such a material is Bi_2Se_3. In these materials, the Spin Hall angle can effectively exceed unity [28] and these materials can be used to switch the magnetization of a ferromagnet grown on top very efficiently.

The quantum Spin Hall effect

The integer quantum Hall effect [29] was discovered more than 30 years ago and resulted in the 1985 Nobel Prize in physics. This effect makes the Hall resistance of a sample (which is the Hall voltage divided by the current flowing through the sample) quantized. It occurs in a two-dimensional electron gas (2-DEG) at high magnetic fields and results from the formation of conducting one dimensional channels that develop at the edges of the sample (hence the name "edge states"). Each of these edge channels transmits with unit probability (regardless of scattering, which is the surprising result) and contributes a quantized conductance of $2e^2/h$.[‡] The Hall resistance is also quantized and given by

$$R_{quant-Hall} = \frac{h}{2e^2 N},\qquad(9.45)$$

where N is the number of edge channels.

Each edge channel conducts a dissipationless current in one direction. They form only at high magnetic fields, but recently there have been theoretical predictions that a different kind of edge channel can form even *without* a magnetic field [30, 31, 32]. In certain insulators with suitable electronic structures – called two-dimensional *topological insulators* – spins of opposite polarization will move along opposite directions along opposite edges of a sample because of spin-orbit interaction and without the presence of an actual magnetic field. These are "spin edge states" shown in Fig. 9.5 and will give rise to the "quantum Spin Hall effect".

Bernevig, et al. [33] predicted that a quantum well of (Hg,Cd)Te will exhibit the Quantum Spin Hall effect because of the peculiarities of its electronic structure. This has now been confirmed experimentally culminating in the demonstration of the Quantum Spin Hall effect [34]. Research in this field is just beginning to take root and exciting discoveries might be around the corner. Topological insulators are now a subject of intense research in their own right [37].

9.2 The Spin Galvanic effect

In a spin polarized two-dimensional electron gas (2-DEG), spin relaxing scattering can cause strange effects in the presence of spin-orbit interaction. One example is the *Spin Galvanic effect*, where an electric current (charge current)

[‡]We will ignore the fact that the magnetic field can lift the spin degeneracy of the edge states because of the Zeeman effect. The g-factors in most materials are small enough that they cause negligible Zeeman splitting at the magnetic fields where the quantum Hall effect is observed. Hence, we can assume that the edge states are spin-degenerate.

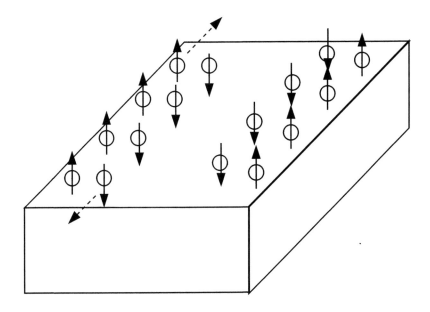

FIGURE 9.5
Depiction of the quantum Spin Hall effect where edge channels consist of spin
polarized carriers. In any given edge, channels with opposite spin polarization
travel in opposite directions, as shown by the broken arrows.

can flow without a battery [38, 39]! This phenomenon is explained below, following [39].

Electric current is usually generated by (i) electric and/or magnetic fields (drift current), or (ii) a spatial gradient of carrier concentration (diffusion current) or (iii) temperature (thermogalvanic current). In a spin-polarized electron gas, an electric current can be generated without any electric/magnetic fields, or without a concentration or thermal gradient. This is the Spin Galvanic effect.

Consider a two-dimensional electron gas in the $(x - z)$ plane. The energy dispersion relation (E versus k_x for a fixed k_z) is shown in Fig. 9.6. For the sake of simplicity, let us assume that there is only Rashba or Dresselhaus interaction (but not both) so that each band has a fixed spin quantization axis. The *E-k* plots are two displaced parabolas, as shown in Fig. 9.6. Since the electron gas is spin polarized, we show that one band is filled more than the other, meaning that it is occupied to higher energy than the other (assuming that both bands have the same density of states).

As long as the carrier distribution in either $E - k$ parabola is symmetric about the corresponding band minima at $k = k_{x1,x2}$, there can be no net current flow since for every carrier with a velocity v_x, there is a carrier with velocity $-v_x$. However, asymmetric k-dependent scattering can occur that will cause an imbalance in the carrier population, making the population of carriers with positive v_x slightly exceed that with negative v_x, or vice versa. In that case, a charge current flows without any electric or magnetic field, or without any concentration or thermal gradient. There is no violation of energy conservation since energy must be constantly supplied to maintain the non-equilibrium distribution of spin polarized carriers. Typically, the spin polarized carrier population is created by optical orientation using circularly polarized light [43].

Because of spin-orbit interaction, the energy dispersion parabolas are horizontally shifted from each other. As a result, the wavevector transfer Δk_x associated with a scattering event that flips a left traveling down-spin electron in one band to a right traveling up-spin electron in the other band (with a change in the direction of velocity) is different from that associated with the scattering event that flips a right traveling down-spin electron in the first band to a left traveling up-spin electron in the second band. In other words, the wavevector transfer Δk_x depends on the *initial direction of velocity*. Now, the spin flip scattering could be caused by a charged magnetic impurity that interacts with an electron via Coulomb interaction. The probability of that scattering is proportional to $1/(\Delta k_x)^2$, if we neglect screening. Therefore, the probability of spin flip scattering, with an associated change in the direction of velocity, depends on the *sign* of the initial velocity (i.e., whether the electron was initially left-traveling or right-traveling). In the example shown in Fig. 9.6, there is a higher probability of a right traveling electron flipping to a left traveling electron. Therefore, the population of left traveling electrons will increase at the expense of right traveling electrons, causing a charge imbal-

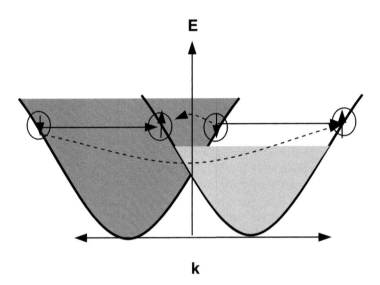

FIGURE 9.6

Understanding the Spin Galvanic effect. The energy dispersion relation in the presence of Rashba interaction in a spin polarized electron gas. We assume that the down-spin band has more carriers than the up-spin band, so that the down-spin band is occupied to higher energy. Pauli exclusion principle will allow an electron from a filled state in the down-spin band to scatter to an empty state in the up-spin band. The spin relaxing events are designated with arrows. We show an elastic scattering event, but the arguments presented here hold for inelastic events as well. Because of spin orbit interaction, the energy dispersion parabolas are horizontally shifted from each other. As a result, the wavevector transfer associated with scattering events that result in a change in the direction of velocity (shown with broken arrows) depends on the sign of the initial velocity. This gives rise to the Spin Galvanic effect. Adapted from ref. [39], Reprinted by permission from Macmillan Publishers Ltd: [Nature (London), **417**, 153, copyright 2002].

ance that results in a net charge current flow without a battery, concentration gradient, or thermal gradient! This is the basis of the Spin Galvanic effect.

Conversely, an electric current flowing through the two-dimensional electron gas can generate a spin polarization. This is sometimes called the *inverse spin galvanic effect* and has been described by a number of authors [40, 41, 42].

9.3   The Spin Capacitor

The "spin capacitor" [45, 46] is another example where a charge current can flow through a device without a bias voltage. There is no violation of energy conservation; the energy to drive the current comes not from a battery but from an external source (such as a magnetic field or electromagnetic radiation) that creates and maintains a *non-equilibrium* distribution of spins. The spin capacitor extracts energy from this source and delivers it to the load through which the zero-bias current flows.

Consider a Spin Field Effect Transistor structure[§] shown in Fig. 9.7(a). It has source and drain contacts made of Co and Fe, respectively. A strong magnetic field applied perpendicular to the channel (in the "down" direction) magnetizes the contacts, making down-spins majority spins and up-spins minority spins in both contacts. The field is then removed, but the contacts are magnetized permanently.

When the source (Co contact) is connected to the negative terminal of a battery, it preferentially injects its *minority* spins (instead of majority spins) into the transistor's channel through the Schottky barrier that forms at the source/channel interface since Co has a *negative* tunneling spin polarization at the Fermi energy [44]. These spins cannot tunnel into the right contact (Fe) through the Schottky barrier formed at the drain/channel interface since Fe has a *positive* tunneling spin polarization at the Fermi energy [6]. This is why the Co contact is shown magnetized "up" and the Fe contact shown magnetized "down" in the figure. Therefore, under ordinary circumstances, the resistance of the channel will be infinite if we assume that the tunneling spin polarizations of Co and Fe are −100% and +100%, respectively, and spins do not flip while traversing the channel. The same will be true if we change the polarity of the bias and connect the negative terminal of the battery to the drain.

The situation changes dramatically if we allow the spins in the channel to flip by interacting with localized electrons trapped in the gate insulator. Consider a scenario where the trapped electrons were spin polarized in the "up" direction by a polarizing magnetic field that did not demagnetize the source

---

[§]The Spin Field Effect Transistor is discussed in Chapter 13.

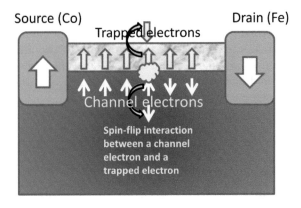

(a)

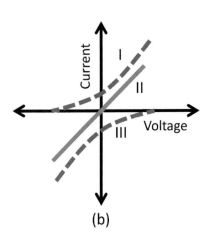

(b)

FIGURE 9.7

(a) A spin capacitor that can convert energy stored in an array of polarized spin (in a non-equilibrium distribution) into electrical energy. Since the spins injected by the source and transmitted by the drain are mutually anti-parallel, we depict the two contacts as having effectively anti-parallel magnetizations. Forward bias is defined as the bias polarity where the source contact is attached to the negative terminal of an external battery. Reverse bias corresponds to the source contact being connected to the positive terminal of the external battery. (b) The source-to-drain current versus the source-to-drain voltage characteristic. Characteristic I refers to the situation when all the trapped spins in the gate insulator are polarized "up" and maintained in that polarization state by some external energy source, characteristic II refers to the situation when the trapped spins are in thermal equilibrium (equal population of up- and down-spins), and characteristic III refers to the situation when all the trapped spins are polarized "down" and maintained in that state by an energy source. Adapted from [46].

and drain contacts because it was much weaker than the coercive fields of the contacts. After the polarizing field is removed, the trapped electrons' spins will spontaneously randomize since there is insufficient exchange interaction between them (unlike in a ferromagnet) that will keep them permanently polarized in one direction. We will assume that the time for depolarization is much longer than the time it takes for the trapped electrons' spins to flip by interacting with the spins of the channel electrons.

When the source (Co) is connected to the negative terminal of the battery, it injects up-spin electrons into the channel because of its negative tunneling spin polarization. Let us consider the scenario when the injected channel electrons interact with the electrons trapped in the insulator through exchange interaction of the form $\sum_j J\delta\left(\vec{r} - \vec{R}_j\right) \vec{S}_e \cdot \vec{S}_j$, where J is the strength of the exchange interaction between a channel electron located at $\vec{r}$ in the channel and the j-th trapped electron located at $\vec{R}_j$ in the insulator, $\vec{S}_e$ is the spinor for the channel electron, and $\vec{S}_j$ is the spinor for the j-th trapped electron in the gate insulator. This interaction exists only between parallel spins; hence, only up-spin electrons in the channel will experience it since the trapped spins have been polarized in the "up" direction. This interaction will flip the spin of an up-spin electron in the channel (while simultaneously flipping the spin of trapped electron in the insulator), as shown in Fig. 9.7(a). The flipped spin in the channel can now transmit through the drain, which only allows down-spin electrons to go through. Consequently, current flows and the resistance is no longer infinity.

The current flow cannot continue forever. After a while, all the trapped electrons would have flipped down and thereafter the injected up-spin electrons will no longer interact with any trapped electron since $\vec{S}_e \cdot \vec{S}_j = 0$. Hence the injected electrons will no longer flip and be transmitted by the drain. As a result, the current will gradually die down with time and finally the resistance will go back up to infinity (zero current flow).

The analogy with the "capacitor" is now obvious. The process of flowing current through the channel caused the trapped spins to flip down and when all of them flipped, the current went to zero. It is as if the current was a charging current that was charging up a capacitor. It died down to zero when the capacitor got fully charged, i.e., when all the trapped spins flipped down. Just as a charged capacitor stores energy, the fully spin-polarized traps also store energy.

Next, if we remove the battery and electrically short the Co and Fe contacts, a current will begin to flow through the channel but gradually die down. The net current through the channel is the *difference* between the current injected into the channel by the Co contact and the current injected in the opposite direction by the Fe contact. The former component cannot flow since the electrons injected by Co cannot transmit through the channel, but the latter component can flow because the down-spin electrons injected by Fe will interact with the trapped spins, flip up, and transmit through the Co contact. Therefore, a current will flow through the channel until 50% of the trapped spins have been flipped back up by interacting with channel electrons. At that point one-half of the trapped spins are polarized up and the rest are polarized down, so that the net spin polarization of the trapped ensemble is zero. At that point, currents injected by Co and Fe are equal and opposite, so they cancel and the net current becomes zero. The reader will understand that this was the process of discharging the capacitor. The spin capacitor gets fully discharged when the net spin polarization of the trapped electrons goes to zero.

The charging/discharging phenomena have been observed experimentally (although the capacitor did not fully discharge) in a spin capacitor [47].

Let us now examine another feature of the spin capacitor. Consider the scenario when the spins of the trapped electrons have been left pointing "up" by the polarizing field, which was later removed. If we had connected the negative terminal of the battery to Co and the positive terminal to Fe, then a current would have flowed (albeit temporarily). But if the polarity of the battery were reversed (negative terminal connected to Fe and positive to Co), spins would have been injected from the Fe contact into the channel with down-polarization, which would not have interacted with the trapped electrons' spins, not flipped, and hence not been transmitted by the Co contact. Clearly, the device is *non-reciprocal* and rectifies like a diode. Current can flow in forward bias (negative terminal connected to Co), but not in reverse bias (positive terminal connected to Co).

Consider now the scenario when a *permanent* polarizing magnetic field is kept on in the "up"-direction which will continuously re-polarize the trapped spins in the "up" direction if they flip down by interacting with channel electrons. We will assume that the polarizing magnetic field has no effect on the channel electrons because, say, the g-factor of the channel material is zero. In this case, when the negative polarity of the battery is connected to Co and positive to Fe, a *steady-state* current can flow. A channel electron enters from Co with up-spin, interacts with a trapped spin while traversing the channel, flips down, and transmits through the Fe contact, resulting in a channel current. The flipped trapped spin is immediately repolarized "up" by the continuously-on polarizing magnetic field so that it can interact with the next channel electron. Since the entire population of trapped spins always remains polarized "up" because of the always-present polarizing magnetic field, the channel current never dies down. It is as if the capacitor can never be fully charged and hence the charging current never dies down. Of course, if we reverse the bias polarity, then no current will flow. So, the rectifying behavior remains intact.

What is even more surprising is that a steady-state current can flow at zero bias voltage! To understand how that can happen, one needs to understand that the net current through the channel is the current injected by the source and transmitted by the drain *minus* the current injected by the drain and transmitted by the source. If the channel is one-dimensional, then we can write the net current as

$$I_{net} = \frac{2e}{h} \int_0^\infty dE \left[T_{1\to 2}(E) f(E) - T_{2\to 1}(E) f(E + eV) \right], \quad (9.46)$$

where $T_{i\to j}$ is the transmission probability of an electron from contact i to contact j at an energy E, $f(\epsilon)$ is the Fermi-Dirac occupation probability at energy ϵ, and V is the voltage drop or bias across the device. Normally, reciprocity holds, i.e., $T_{1\to 2} = T_{2\to 1}$, which will make $I_{net} = 0$ when $V = 0$. However, in this case it obviously does not hold ($T_{1\to 2} \neq T_{2\to 1}$) as long as the trapped spins have a non-zero spin polarization. As a result, $I_{net} \neq 0$ when $V = 0$. This is not so surprising in non-steady-state situations since a displacement current can flow temporarily through a capacitor even when the voltage across it is zero, since the displacement current is CdV/dt ($C =$ capacitance) which need not be zero when the voltage V across the capacitor is zero. However, it is surprising in steady-state situations. If there is a current flow, there is energy dissipated in the load. This energy cannot come from the bias source since the bias voltage is zero. It can only come from the agent that maintains the non-zero spin polarization (which is a non-equilibrium state). That agent in this case happens to be the polarizing magnetic field.

A few points to note about the spin capacitor:

1. In the non-ideal case, the trapped spins are not going to be 100% spin polarized but as long as the polarization is non-zero, i.e., there are more up-spins than down-spins, we will see the non-reciprocal diode-like behavior.

2. Just like any capacitor, this device can act as "memory." One can "write" binary bit information into the trapped spins by flipping them either all up (bit 1), or all down (bit 0), by an external source or even by interacting with spin-polarized channel electrons. In the latter case, writing is complete when the current stops flowing. We can read the stored information by measuring the conductance under forward and reverse bias. Let us say that we encode the logic bit 0 in down-spin polarization (of the trapped electrons) and the logic bit 1 in up-spin polarization. If the device conducts under forward bias (source connected to negative polarity of the battery), then the stored spins are "up" and the stored bit is 1; otherwise, the stored spins are "down" and the stored bit is 0. Such a device can also be viewed as a "spin-charge converter" which converts spin information into charge information and acts as a spin reader. Spin readers are extremely valuable as we will see in Chapter 15 when we discuss Single Spin Logic.

The above features discussed are captured in the current-voltage characteristic cartoon in Fig. 9.7(b).

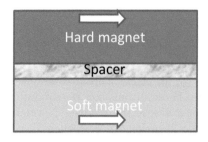
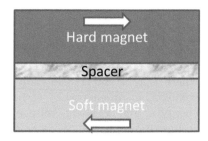

Low resistance state   High resistance state

The block arrows denote the direction of magnetization

FIGURE 9.8

A typical magneto-tunneling junction consisting of a hard and soft magnet separated by a non-magnetic spacer layer. The resistance between the two magnets is low when their magnetizations are mutually parallel and high when they are mutually anti-parallel.

9.4   The Spin Transfer Torque

The spin transfer torque effect has become immensely important in view of its application in Spin Transfer Torque Random Access Memory (STT-RAM) which could emerge as the mainstay of non-volatile magnetic memory in near future. In an STT-RAM, bit information (0 or 1) is stored in the resistance state of a magneto-tunneling junction (MTJ) consisting of two ferromagnets sandwiching a thin spacer layer. One of the ferromagnets has a very high coercivity and its magnetization is stiff, meaning that if it is magnetized in one direction, it stays magnetized in that direction. This is called the fixed or hard magnet. The other feromagnetic layer is a free or soft magnet whose magnetization can be rotated or flipped easily. A typical MTJ structure is shown in Fig. 9.8.

When the hard and soft magnets have anti-parallel magnetizations, the MTJ resistance measured between the two ferromagnets is typically large because the spins injected by one layer into the spacer is blocked by the other layer, resulting in negligible tunneling probability through the spacer. However, when the magnetizations of the two ferromagnets are mutually parallel, the tunneling probability increases and the resistance drops. The ratio of the two resistances is called the tunneling magnetoresistance ratio (TMR) and could be considerably larger than 1. Therefore, a bit can be written and stored in the MTJ by simply making the soft layer's magnetization parallel or anti-parallel to that of the hard layer, and then reading the bit by measuring the MTJ resistance.

The obvious way to switch the magnetization of the soft layer would be to use a magnetic field generated by a current. This has too many disadvantages. First, a lot of current is needed to generate the magnetic field, which results in intolerable power dissipation during the write cycle. Second, magnetic fields cannot be easily localized over a very small area to switch only the target MTJ to the exclusion of all other neighboring MTJs. This results in large spacing between MTJs and a resulting drop in bit density in the memory.

A better solution is to pass a current through the MTJ (current can be completely localized within the MTJ). Let us consider the situation when the soft layer's magnetization is originally anti-parallel to that of the hard layer and it has to be switched to make it *parallel* to that of the hard layer. In this case, the negative terminal of the battery is connected to the hard layer and the positive terminal to the soft layer. Electrons with majority spins from the hard layer are injected into the soft layer after tunneling through the spacer. Choice of the right spacer/magnet combination ensures that a significant amount of spin-polarized injection occurs, resulting in a large TMR [48]. As the injected electrons pass through the soft layer, they transfer their spin angular momenta to the electrons in the soft layer and this flips the latters' spins. Gradually the population of flipped spins builds up and this ultimately flips the magnetization of the soft layer to make it parallel to that of the hard layer. It is as if the transfer of angular momentum from the injected spins generated a torque on the magnetization vector of the soft magnet to make it rotate and flip.

Next, consider the situation when the two layers are initially in the parallel configuration and are to be switched to the anti-parallel configuration. In this case, the battery polarity is reversed. Now the soft layer becomes the electron injector and the hard layer the receiver. The soft layer preferentially injects its majority spins since the hard layer will block transmission of the minority spins (remember that the two layers were intially magnetized parallel to each other so that the majority spins in the soft layer are also majority spins in the hard layer). This injection depletes the supply of majority spins in the soft layer and ultimately there is a role reversal of spins in the soft layer whereby the majority spins become minority and vice versa. At that point, the magnetization of the soft layer flips to assume a configuration anti-parallel to that of the hard layer.

The spin transfer torque effect was independently explained by Berger [49] and Slonczewski [50] and a review of this phenomenon can be found in [51]. The precession of the magnetization vector of a magnet during switching with a spin polarized current generating a spin transfer torque could become the basis of novel microwave oscillators [52] and has received some attention.

The spin transfer torque in an MTJ's soft layer has some intriguing properties. It actually has two components. One (called the Slonczewski torque) is non-conservative while the other (called the field-like torque) is conservative [53]. The expression for the torque can be found in Chapter 17 and is given by

$$\vec{T}_{STT}(t) = \frac{\hbar}{2e}\zeta I\left[b\left(\hat{\mathbf{m}}\times\hat{\mathbf{s}}\right) + c\left(\hat{\mathbf{m}}\times\left(\hat{\mathbf{m}}\times\hat{\mathbf{s}}\right)\right)\right],\qquad(9.47)$$

where I is the current with spin polarization ζ being passed through the magnets, c is the coefficient of the in-plane (or Slonczewski) component of the torque, and b is the out-of-plane (or field-like) component of the torque. The quantities $\hat{\mathbf{m}}$ and $\hat{\mathbf{s}}$ are the unit vectors in the directions of the magnetization vector and spin polarization, respectively. More about this in Chapter 17.

9.5   The Spin Hanle effect

The spin Hanle effect is based on precession of a spin about a magnetic field. Consider a one-dimensional spin valve structure shown in Fig. 9.9 consisting of two ferromagnetic contacts magnetized in the direction of current flow. The left contact is a spin-polarizer that injects electrons with spins polarized in the direction of current flow into the spacer layer and the right contact is a spin analyzer that selectively transmits electrons whose spins are aligned along the direction of its own magnetization. We will assume that there is no spin relaxation in the spacer layer. In that case, the conductance of the structure will vary as $cos^2(\theta/2)$ where θ is the angle between the spin polarization of the electrons arriving at the right contact and the magnetization of the right contact. We will see a derivation of this relation when we discuss the Spin Field Effect Transistor.

In the absence of any magnetic field, spin-orbit interaction, and spin relaxing events, the injected spins will arrive intact at the right contact. Since both contacts are magnetized in the same direction, $\theta = 0°$ in this case and the conductance will be maximum.

Next, imagine that a magnetic field is applied perpendicular to the direction of current flow (and hence perpendicular to the injected spin polarization). The spins will then execute Larmor precession about this magnetic field as they traverse the spacer layer. The angle by which they will precess will determine the angle θ:

$$\theta = \Omega\tau_t = \Omega(L/v) = \frac{g\mu_B B}{\hbar}\frac{L}{v},\qquad(9.48)$$

where Ω is the spin precession frequency, B is the magnetic flux density, τ_t is the transit time through the spacer, L is the spacer region's length, and v is the electron velocity.

Clearly, the conductance, which is proportional to $cos^2(\theta/2)$, will oscillate as B is varied, and these are called *Hanle oscillations*. However, because θ depends on v, different electrons having different velocities precess by different

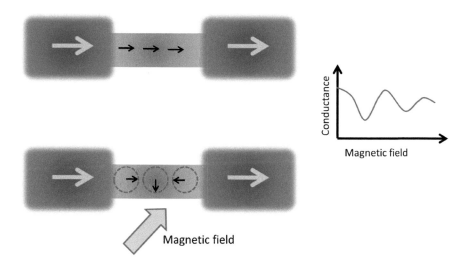

FIGURE 9.9

The Spin Hanle effect. Spins are injected into the spacer layer of a spin valve structure from a ferromagnetic contact (spin polarizer) with their polarizations in the direction of current flow. In the absence of any magnetic field and spin relaxation events, the spins arrive intact at the right ferromagnetic contact, which acts as a spin analyzer and transmits electrons with a probability of $cos^2(\theta/2)$, where θ is the angle between the magnetization of the contact and the polarization of arriving spins. A magnetic field applied transverse to the direction of current flow causes the spins to precess about the field as they travel between the contacts and this changes θ. As a result, the current through the spin valve (and hence its conductance) oscillates as a function of the magnetic field. The oscillation damps with increasing magnetic field since θ depends also on the velocity of the electrons and is hence different for different electrons. Ensemble averaging over all electrons causes the damping of the conductance oscillation.

angles. Ensemble averging over the electron velocities will make the conductance oscillation amplitude decay with increasing magnetic field strength. The Hanle oscillation is shown in Fig. 9.9.

Hanle oscillations have been observed at room temperature in one-dimensional spin-valve structures [55, 56] and also observed in two-dimensional structures [57, 58, 59, 60]. In two-dimensional structures, the Hanle conductance/resistance oscillation measurement is usually made with non-local probes [58] and the non-local resistance $R_{\text{non-local}}$ is proportional to:

$$R_{\text{non-local}} \propto \int_0^\infty \sqrt{\frac{1}{4\pi Dt}} \cos(\Omega t) e^{-t/\tau_s} dt$$

$$= Re \left[\frac{1}{2\sqrt{D}} \frac{exp\left(-L\sqrt{\frac{1}{D\tau_s} - i\frac{\Omega}{D}}\right)}{\sqrt{\frac{1}{\tau_s} - i\Omega}} \right], \qquad (9.49)$$

where L is the distance traversed by electrons between injecting and detecting contacts, D is the diffusion coefficient of the electrons, τ_s is the spin relaxation time, and Re stands for the real part. Clearly, the oscillation amplitude decays with increasing spin relaxation rate $1/\tau_s$.

9.6   The Spin Seebeck effect

The generation of an electrical voltage by placing a conductor in a thermal gradient is the well-known Seebeck effect. The generation of a spin voltage by placing a sample in a thermal gradient is the Spin Seebeck effect. In a metallic magnet, spin-up and spin-down electrons have different densities and different scattering rates. Hence when the magnet is placed in a thermal gradient, the rate at which spin-up electrons are driven by the gradient is different from that at which spin-down electrons are driven. Hence a spin imbalance will build up downstream, resulting in a spin voltage and an accompanying spin current. The spin voltage is measured via the inverse Spin Hall effect.

The Spin Seebeck effect has been observed experimentally in a $Ni_{81}Fe_{19}$ layer [61] and opens the door to thermo-spintronics - the technology of generating spin currents and voltages by thermal gradients. It has also been observed in insulators like $LaY_2Fe_5O_{12}$ where there are no conduction electrons [62]. In the latter case, the spin imbalance is thought to be created not by unequal flow of spin-up and spin-down electrons, but rather by a complicated spin-wave phenomenon. A temperature difference between the magnons in the magnetic insulator and the electrons in the metallic contacts pumps a spin current [63, 64].

The *Nernst effect* is a variation of the Seebeck effect. When a conductor is placed in a thermal gradient with a magnetic field perpendicular to the gradi-

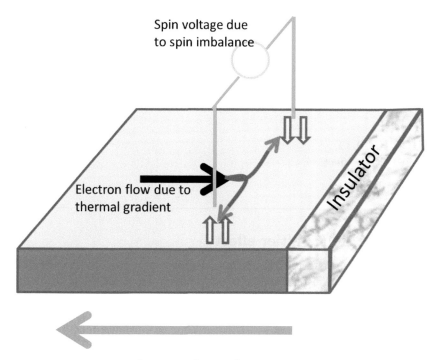

FIGURE 9.10
The Spin Nernst effect in a two-dimensional electron gas (2-DEG) with an insulating barrier.

ent, an electric field develops that is perpendicular to both the magnetic field and the thermal gradient. The Spin Nernst effect occurs in a non-magnetic semiconductor without any magnetic field as long as there is spin-dependent skew scattering. Consider the situation in Fig. 9.10, where a thermal gradient is imposed in the x-direction, but no conduction current can flow in this direction because of the insulating barriers. The thermal gradient still causes of flow of electrons in the x-direction, except the electrons are deflected in the y-direction by the barriers. If there are spin-dependent scatterers in the medium, as in the case of the extrinsic Spin Hall effect, then up-spin electrons will be deflected to one side $(+y)$ and down-spin electrons to the other side $(-y)$, leading to a spin imbalance and a spin voltage drop in the y-direction, perpendicular to the thermal gradient. This is the Spin Nernst effect.

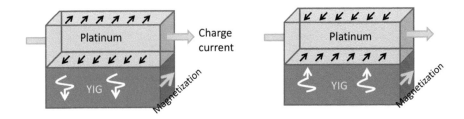

FIGURE 9.11
Spin Peltier Effect.

9.7   The Spin Peltier effect

The Spin Peltier effect is the Onsager reciprocal of the Spin Seebeck effect. It is the generation of a magnon heat current in a magnetic insulator by the flow of a spin current through the interface with the metallic contact. This phenomenon has been demonstrated experimentally in a Pt-YIG (platinum yttrium-iron-garnet) heterostructure [65]. Consider Fig. 9.11. A charge current through the platinum induces a transverse spin current through a magnetized YIG layer due to the Spin Hall effect. When the spins at the interface between Pt and YIG are anti-parallel to the magnetization of the YIG, the spin torque transfers energy and momentum from the electrons in the Pt to the magnons in the YIG, thereby cooling the electrons and heating the magnons. This raises the magnon temperature over the electron temperature. When the spins are parallel to the magnetization (right panel), the spin torque transfers energy and momentum from the magnons in the YIG to the electrons in the Pt, thereby cooling the magnons and heating the electrons.

9.8   Problems

- **Problem 9.1**

  Show that Equation (9.32) is correct in the presence of spin-orbit interaction as long as the Rashba interaction strength is spatially invariant. Show also that if the Rashba interaction strength is space and time invariant, then

  $$\frac{dv_x}{dt} = -\frac{eE_x}{m^*}, \tag{9.50}$$

  even though $\hbar k_x \neq m^* v_x$.

Hint: Use the Ehrenfest theorem.

Show that if you used Equations (7.37) and (9.32) to derive dv_x/dt, you would not have obtained the same result as above. In fact, you would have obtained a wrong result. This is an example of **pitfalls** that lurk when spin-orbit interaction is present.

If the Rashba interaction strength is space and time dependent, then the above equations are not correct and a new spin-dependent force term should be added to modify Newton's law to read

$$\frac{dv_x}{dt} = -\frac{eE_x}{m^*} + F_{spin}. \tag{9.51}$$

For a derivation of F_{spin}, see [35].

- **Problem 9.2**

Show that if the *spin-dependent* B_{Rashba} were proportional to velocity v instead of wavevector k, then the intrinsic Spin Hall effect would *not* be universal and the Spin Hall conductivity would be

$$\sigma_{SH} = \frac{e}{8\pi}\left[1 + ln\left(\frac{\hbar v_F}{\eta}\right)\right], \tag{9.52}$$

where v_F is the Fermi velocity. You will end up with an improper integral $\int_{-\eta/\hbar}^{\eta/\hbar} \frac{1}{v(\tau)}dv(\tau)$. This has a singularity at $v(\tau) = 0$, but since the integrand is an odd function of the integration variable, this integral is exactly zero.

- **Problem 9.3**

Since the B_{Rashba} that we used in the derivation of the universal intrinsic Spin Hall effect is spin-dependent, we should have accounted for two different B_{Rashba}-s (i.e., B_{Rashba}^+ and B_{Rashba}^-). Repeat the derivation using two different B_{Rashba}-s for the two spin-split bands and show that the same result emerges.

- **Problem 9.4**

Derive Equation (9.15) from Equation (9.14).

- **Problem 9.5**

Suppose you want to flow an alternating current through a spin capacitor. Is there is a cut-off frequency below which the capacitor will not allow a current to flow through it? What will this cut-off frequency be?

- **Problem 9.6**

Consider the Hamiltonian of a single electron placed in a static magnetic field.

(a) Show that the stable and metastable spin polarizations are parallel and anti-parallel to the magnetic field, respectively.

(b) The spin should ultimately align along the stable orientation, i.e., parallel to the magnetic field. This is what we assumed tacitly while deriving the Spin Hall effect. However, the equation describing Larmor precession,

$$\frac{d\vec{S}}{dt} = \frac{g\mu_B \vec{B}}{\hbar} \times \vec{S}, \tag{9.53}$$

does not predict that this will happen and instead predicts that the spin will not align along the magnetic field, but instead precess around it in a cone (see Chapter 3). How do you reconcile these two different pictures?

Solution

The Hamiltonian of a single electron in a $\hat{z}$-directed magnetic field is

$$H = H_0 - (g/2)\mu_B \sigma_z B_z = H_0[\mathbf{I}] - (g/2)\mu_B \begin{pmatrix} B_z & 0 \\ 0 & -B_z \end{pmatrix}, \tag{9.54}$$

where H_0 is the spin-independent part of the Hamiltonian and we have assumed that the g-factor is positive.

Diagonalization of this Hamiltonian immediately shows that the eigenstate energies are

$$E = \langle H_0 \rangle \pm (g/2)\mu_B B_z, \tag{9.55}$$

and the corresponding eigenspinors are

$$\Psi^+ = \begin{pmatrix} 0 \\ 1 \end{pmatrix}$$

$$\Psi^- = \begin{pmatrix} 1 \\ 0 \end{pmatrix}. \tag{9.56}$$

This implies that the eigenspin states are parallel and anti-parallel to the z-directed magnetic field. Note that the "parallel" state has lower energy than the "anti-parallel" state (the two differ by the Zeeman splitting energy $g\mu_B B_z$). Hence, the former is the stable state and the latter is the metastable state.

The Larmor precession equation was derived from the Ehrenfest theorem, which does not incorporate any dissipation since it is derived from the Schrödinger equation, which cannot handle dissipation in a straight-forward way (see also Problem 8.1). Without dissipation, the spin will continue to precess about the magnetic field forever. However, if dissipation is present, then the spin will gradually decay to the lowest energy state and ultimately align parallel to the magnetic field. The latter physics can be included phenomenologically within the Larmor precession equation by modifying it as

$$\frac{d\vec{S}}{dt} = \frac{g\mu_B \vec{B}}{\hbar} \times \vec{S} - \alpha \vec{S} \times \left(\vec{B} \times \vec{S}\right),$$ (9.57)

where α is a phenomenological damping constant.

9.9 References

[1] E. H. Hall, "On a new action of the magnet on electric currents", Am. J. Math., **2**, 287 (1879).

[2] M. Onoda and N. Nagaosa, "Topological nature of anomalous Hall effect in ferromagnets", J. Phys. Soc. Jpn., **71**, 19 (2002).

[3] N. F. Mott and H. S. Massey, *The Theory of Atomic Collisions*, (Clarendon Press, Oxford, 1965).

[4] J. Smit, "The spontaneous Hall effect in ferromagnetics", Physica, **21**, 877 (1955).

[5] L. Berger, "Side jump mechanism for the Hall effect of ferromagnets", Phys. Rev. B, **2**, 4559 (1970).

[6] A. Fert, A. Friederich and A. Hamzic, "Hall effect in dilute magnetic alloys", J. Magn. Magn. Mater., **24**, 231 (1981).

[7] E. M. Pugh and N. Rostoker, "Hall effect in ferromagnetic mateials", Rev. Mod. Phys., **25**, 151 (1953).

[8] J. Inoue and H. Ohno, "Taking the Hall effect for a spin", Science, **309**, 2004 (2005).

[9] R. Yu, W. Zhang, H-J Zhang, S-C Zhang, X. Die and Z. Fang, "Quantized anomalous Hall effect in magnetic topological insulators", Science, **329**, 61 (2010).

[10] C-Z Chang, et al., "Experimental observation of the quantum anomalous Hall effect in a magnetic topological insulator", Science, **340**, 168 (2013).

[11] M. I. D'yakonov and V. I. Perel'. "Possibility of orienting electron spins with current", JETP Lett., **13**, 467 (1971).

[12] J. E. Hirsch, "Spin Hall Effect", Phys. Rev. Lett., **83**, 1834 (1999).

[13] S. O. Valenzuela and M. Tinkham, "Direct electronic measurement of the spin Hall effect", Nature, **442**, 176 (2006).

[14] Y. K. Kato, R. C. Myers, A. C. Gossard and D. D. Awschalom, "Observation of the Spin Hall Effect in semiconductors", Science, **306**, 1910 (2004).

[15] K. Ando and E. Saitoh, "Observation of the inverse Spin Hall Effect in silicon", Nature Commun., DOI: 10.1038/ncomms1640.

[16] E. Saitoh, M. Ueda, H. Miyajima and G. Tatara, "Conversion of spin current into charge current at room temperature: Inverse spin Hall effect", Appl. Phys. Lett., **88**, 182509 (2006).

[17] J. Sinova, D. Culcer, Q. Niu, N. A. Sinitsyn, T. Jungwirth and A. H. MacDonald, "Universal intrinsic Spin Hall Effect", Phys. Rev. Lett., **92**, 126603 (2004).

[18] S. Murakami, N. Nagaosa and S-C Zhang, "Dissipationless quantum spin current at room temperature", Science, **301**, 1348 (2003).

[19] J. Wunderlich, B. Kaestner, J. Sinova and T. Jungwirth, "Experimental observation of the spin Hall effect in a two-dimensional spin-orbit coupled semiconductor system", Phys. Rev. Lett., **94**, 047204 (2005).

[20] L. Meier, G. Salis, I. Shorubalko, E. Gini, S. Schön and K. Ensslin, "Measurement of Rashba and Dreselhaus spin-orbit magnetic fields", Nature Phys., **3**, 650 (2007).

[21] H. A. Engel, E. I. Rashba and B. I. Halperin, "Out-of-plane spin polarization from in-plane electric and magnetic fields", Phys. Rev. Lett., **98**, 036602 (2007).

[22] B. Gu, I. Sugai, T. Ziman, G-Y Guo, N. Nagaosa. T. Seki, K. Takanashi and S. Maekawa, "Surface assisted Spin Hall Effect in Au films with Pt impurities", Phys. Rev. Lett., **105**, 216401 (2010).

[23] L. Liu, C-F Pai, Y. Li, H. W. Tseng, D. C. Ralph and R. A. Buhrman, "Spin torque switching with the giant Spin Hall Effect of Tantalum", Science, **336**, 555 (2012).

[24] D. Bhowmik, L. You and S. Salahuddin, "Spin Hall effect clocking of nanomagnetic logic without a magnetic field", Nature Nanotechnol., **9**, 59 (2014).

[25] N. Nagaosa, J. Sinova, S. Onada, A. H. MacDonald and n. P. Ong, "Anomalous Hall effect", Rev. Mod. Phys., **82**, 1539 (2010).

[26] G-Y Guo, S. Maekawa and N. Nagaosa, "Enhanced Spin Hall Effect by resonant skew scattering in the orbital-dependent Kondo effect", Phys. Rev. Lett., **102**, 036401 (2009).

[27] L. Liu, C-T Chen and J. Z. Sun, "Spin Hall effect tunneling spectroscopy", Nature Phys., **10**, 561 (2014).

[28] A. R. Mellnik, et al., "Spin-transfer torque generated by a topological insulator", Nature, **511**, 449 (2014).

[29] K. v. Klitzing, G. Dorda and M. Pepper, "New method for high-accuracy determination of the fine structure constant based on quantized Hall resistance", Phys. Rev. Lett., **45**, 494 (1980).

[30] S. Murakami, N. Nagaosa and S-C Zhang, "Spin Hall insulator", Phys. Rev. Lett., **93**, 156804 (2004).

[31] C. L. Kane and E. J. Mele, "Z_2 topological order and the quantum spin Hall effect", Phys. Rev. Lett., **95**, 146802 (2005).

[32] B. A. Bernevig and S-C Zhang, "Quantum spin Hall effect", Phys. Rev. Lett., **96**, 106802 (2006).

[33] B. A. Bernevig, T. L. Hughes and S-C Zhang, "Quantum spin Hall effect and topological phase transition in HgTe quantum wells", Science, **314**, 1757 (2006).

[34] M. König, S. Wiedmann, C. Brüne, A. Roth, H. Buhmann, L. W. Molenkamp, X-L Qi and S-C Zhang, "Quantum spin Hall insulator state in HgTe quantum wells", Science, **318**, 766 (2007).

[35] E. M. Chudnovsky, "Theory of Spin Hall Effect: Extension of the Drude Model", Phys. Rev. Lett., **99**, 206601 (2007).

[36] T. L. Gilbert, "A phenomenological theory of damping in ferromagnetic materials", IEEE Trans. Magn., **40**, 3443 (2004).

[37] Joel E. Moore, "The birth of topological insulators", Nature, **464**, 194 (2010).

[38] E. L. Ivchenko, Yu. B. Lyanda-Geller and G. E. Pikus, "Photocurrent in structures with quantum wells with an optical orientation of free carriers", JETP Lett., **50**, 175 (1989).

[39] S. D. Ganichev, E. L. Ivchenko, V. V. Bel'kov, S. A. Tarasenko, M. Sollinger, D. Weiss, W. Wegscheider and W. Prettl, "Spin-galvanic effect", Nature (London), **417**, 153 (2002).

[40] A. G. Aronov and Yu. B. Lyanda-Geller, "Nuclear electric resonance and orientation of carrier spins by an electric field", JETP Lett., **50**, 431 (1989).

[41] V. M. Edelstein, Solid State Commun., "Spin polarization of conduction electrons induced by electric current in two-dimensional asymmetric electron systems", **73**, 233 (1990).

[42] A. G. Mal'shukov and K. A Chao, "Optoelectronic spin injection in a semiconductor heterostructure without a ferromagnet", Phys. Rev. B, **65**, 241308(R), (2002).

[43] F. Meier and B. P. Zakharchenya, Eds. *Optical Orientation*, (Elsevier Science, Amsterdam, 1984).

[44] E. Y. Tsymbal, O. N. Mryasov and P. R. LeClair, "Spin-dependent tunneling in magnetic tunnel junctions", J. Phys.: Condens. Matter, **15**, R109 (2003).

[45] S. Datta, "Proposal for a 'spin capacitor'", Appl. Phys. Lett., **87**, 013115 (2005).

[46] S. Salahuddin and S. Datta, "Electrical detection of spin excitations", Phys. Rev. B, **73**, 081301 (2006).

[47] D. Saha, L. Siddiqui, P. Bhattacharya, S. Datta, D. Basu and M. Holub, "Electrically driven spin dynamics of paramagnetic impurities", Phys. Rev. Lett., **100**, 196603 (2008).

[48] X. G. Zhang and W. H. Butler, "Large magnetoresistance in bcc Co/MgO/Co and FeCo/MgO/FeCo tunnel junctions", Phys. Rev. B., **70**, 172407 (2004).

[49] L. Berger, "Emission of spin waves by a magnetic multilayer traversed by a current", Phys. Rev. B., **54**, 9353 (1996).

[50] J. C. Slonczewski, "Current driven excitation of magnetic multilayers", J. Magn. Magn. Mater., **159**, L1 (1996).

[51] D. C Ralph and M. D. Stiles, "Spin transfer torques", J. Magn. Magn. Mater., **320**, 1190 (2008).

[52] S. I. Kiselev, J. C. Sankey, I. N. Krivorotov, N. C. Emley, R. J. Schoelkopf, R. A. Buhrman and D. C. Ralph, "Microwave oscillations of a nanomagnet driven by a spin-polarized current", Nature, **425**, 380 (2003).

[53] H. Kubota, et al., "Quantitative measurement of voltage dependence of spin-transfer torque in MgO-based magnetic tunnel junctions", Nature Phys., **4**, 37 (2008).

[54] S. Salahuddin, D. Datta and S. Datta, "Spin-transfer torque as a non-conservative pseudo-field", arXiv:0811:3472.

[55] S. Bandyopadhyay, Md. I Hossain, H. Ahmad, J, Atulasimha and S. Bandyopadhyay, "Coherent spin transport and suppression of spin relaxation in InSb nanowires with single subband occupancy at room temperature", Small, **10**, 4379 (2014).

[56] H. Kum, J. Heo, S. Jahangir, A. Banerjee, W. Guo and P. Bhattacharya, "Room temperature single GaN nanowire spin valves with FeCo/MgO tunnel contacts", Appl. Phys. Lett., **100**, 182407 (2012).

[57] X. Lou, C. Adelmann, S. A. Crooker, E. S. Garlid, J. Zhang, S. M. Reddy, S. D. Flexner, C. J. Palmstrom and P. A. Crowell, "Electrical detection of spin transport in lateral ferromagnet-semiconductor devices", Nature Phys., **3**, 197 (2007).

[58] F. J. Jedema, M. V. Costache, H. B. Heershce, J. J. A. Baselmans and B. J. van Wees, "Electrical detection of spin accumulation and spin precession at room temperature in metallic spin valves", Appl. Phys. Lett., **81**, 5162 (2002).

[59] I. Appelbaum, B. Huang and D. J. Monsma, "Electronic measurement and control of spin transport in silicon", Nature, **447**, 295 (2007).

[60] Y. Fukuma, L. Wang, H. Idzuchi, S. Takahasi, S. Maekawa and Y. C. Otani, "Giant enhancement of spin accumulation and long distance spin precession in metallic lateral spin valves", Mature Mater., **10**, 527 (2011).

[61] K. Uchida, S. Takahashi, K. Harii, W. Koshibae, K. Ando, S. Maekawa and E. Saitoh, "Observation of the spin Seebeck effect", Nature, **455**, 778 (2008).

[62] K. Uchida, et al., "Spin Seebeck insulator", Nature Mater., **9**, 894 (2010).

[63] J. Xiao, G. E. W. Bauer, K. C. Uchida, E. Saitoh and S. Maekawa, "Theory of magnon-driven Spin Seebeck effect", Phys. Rev. B., **81**, 214418 (2010).

[64] S. Hoffman, K. Sato and Y. Tserkovnyak, "Landau-Lifshitz theory of the longitudinal Spin Seebeck effect", Phys. Rev. B., **88**, 064408 (2013).

[65] J. Flipse, F. K. Dejene, D. Wagenaar, G. E. W. Bauer, J. Ben Youssef and B. J. van Wees, "Observation of the Spin Peltier Effect for magnetic insulators", Phys. Rev. B., **113**, 027601 (2014).

10

Exchange Interaction

In this chapter, we discuss the extremely important notion of exchange interaction between electrons, which plays a major role in the operation of many spin-based devices (both classical – see Chapter 15 – and quantum mechanical – see Chapter 16). We start by describing the formulation of the *Exclusion Principle* enunciated by Pauli in 1924. We then use this principle to calculate the energy eigenvalues of the Helium atom based on perturbation theory. This leads to the concept of exchange energy which was extended by Heisenberg to a system of spins to propose a model for the origin of ferromagnetism. The concept of exchange coupling and Heisenberg Hamiltonian are crucial in understanding the operation of many solid state spin-based devices described in this book and elsewhere.

10.1   Identical particles and the Pauli exclusion principle

In the hydrogen atom (recall Chapter 1), if we neglect spin-orbit coupling, then the behavior of the single electron is completely described by specifying four quantum numbers (n, l, m, s) that are associated with total energy, orbital angular momentum, the z-component of the angular momentum in a magnetic field, and the spin, respectively. When these four quantum numbers are given, the state of the one-electron system is precisely specified. This is one of the few examples in quantum mechanics where the problem is exactly solvable and one can describe the dynamics of the lone electron rigorously.

The quantum mechanics of multi-electron systems do not typically lend themselves to an exact solution. The description of all such systems is imprecise in the sense that the problem cannot be solved exactly. There are, however, important principles and theorems that govern many electron systems. Probably, the most fundamental of them is the Pauli Exclusion Principle that applies to any two fermions.

As early as 1924, while analyzing spectroscopic data of atoms with more than one electron, Pauli conjectured that no two electrons can occupy the same quantum state if their wavefunctions overlap. This led him to formulate the famous *Exclusion Principle* that bears his name. This Principle states that *no two fermions, whose wavefunctions have non-zero overlap, can have exactly the same set of quantum numbers.* One of the early successes of the Pauli Exclusion Principle was to provide an explanation for the periodic table of the elements.

The concept of exchange energy follows from the Pauli Exclusion Principle. Any determination of the energy eigenvalues and corresponding eigenstates of a system composed of more than one electron must take into account the Coulomb interaction between electrons since electrons are charged particles. Furthermore, for such a system of identical and indistinguishable Fermi particles, the *Symmetry Principle* dictates that the overall wavefunction of the system must be antisymmetric under the operation of swapping the indices of any two electrons. This principle, along with the Pauli Exclusion Principle, gives rise to the exchange interaction.

10.1.1   The helium atom

The Pauli Exclusion Principle was first applied to study the simplest many-electron system, namely, the helium atom, which has two electrons orbiting a nucleus. Using first-order perturbation theory, one can estimate the energy levels of the helium atom. Although the use of perturbation theory is crude, it is the simplest way to introduce the concept of exchange energy and illustrate its fundamental connection to "spin" and the Pauli Exclusion Principle. This approach clearly brings out the truly quantum-mechanical origin of the exchange energy concept.

Consider a helium atom composed of a heavy nucleus with charge $+2e$ and two surrounding electrons, as shown in Figure 10.1. Neglecting the motion of the nucleus and using the latter as the origin of a fixed frame of reference, the Hamiltonian of the helium atom, taking into account the Coulomb attraction between each electron and the nucleus and the Coulomb repulsion between the electrons, is given by

$$H_{He} = \frac{|\vec{p}_1|^2}{2m_0} + \frac{|\vec{p}_2|^2}{2m_0} - \frac{2e^2}{4\pi\epsilon_0|\vec{r}_1|} - \frac{2e^2}{4\pi\epsilon_0|\vec{r}_2|} + \frac{e^2}{4\pi\epsilon_0|\vec{r}_{12}|}, \qquad (10.1)$$

where $\vec{p}_1$ and $\vec{p}_2$ are the momentum operators associated with electrons 1 and 2, respectively. Their spatial coordinates are denoted by $\vec{r}_1$ and $\vec{r}_2$, and $\vec{r}_{12} = \vec{r}_1 - \vec{r}_2$.

Let P be the operator which permutes the two identical electrons. It must commute with the Hamiltonian H_{He} since the latter is invariant upon interchange of the indices 1 and 2; therefore,

$$[H_{He}, P] = 0. \qquad (10.2)$$

This implies that the wavefunctions of the helium atom must be both eigenstates of H_{He} and P. If we call $\phi(e_1, e_2)$ the wavefunction describing the

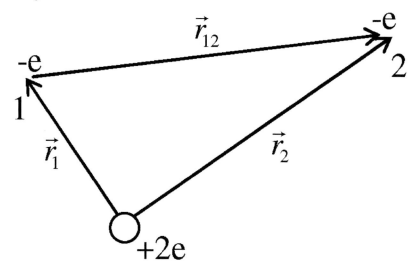

FIGURE 10.1
The helium atom: The positions of the nucleus and the two electrons are represented.

system of two electrons so that $\phi(e_1, e_2)$ satisfies the Schrödinger equation

$$H_{He}\phi(e_1, e_2) = E\phi(e_1, e_2), \tag{10.3}$$

then we must have

$$P\phi(e_1, e_2) = \phi(e_2, e_1) = \lambda\phi(e_1, e_2), \tag{10.4}$$

where the labels (e_1, e_2) are the set of variables $(\vec{r_1}, s_1)$ and $(\vec{r_2}, s_2)$ representing the position and spin coordinates of electrons 1 and 2, respectively.

Applying the permutation operator to both sides of the last equation, we get

$$P^2\phi(e_1, e_2) = \lambda P\phi(e_1, e_2) = \lambda^2\phi(e_1, e_2) = \phi(e_1, e_2), \tag{10.5}$$

which leads to

$$\lambda^2 = 1, \tag{10.6}$$

or

$$\lambda = \pm 1. \tag{10.7}$$

If $\lambda = +1$, the overall wavefunction $\phi(e_1, e_2)$ is symmetric. On the other hand, if $\lambda = -1$, the wavefunction is antisymmetric. Since we are dealing with a system of fermions, the Symmetry Principle constrains us to accept the antisymmetric solution only. Furthermore, since we have neglected any spin-orbit interaction, which is the only term which would couple the position and spin components (i.e., the orbital and spin degrees of freedom), the overall wavefunction can be written as a tensor product of a purely spatial part and a purely spinor part (since the Hamiltonian does not contain any spin-dependent operator). This is the same "variable separation" approach that we visited in Chapter 7. There, it was inexact since we were dealing with spin-orbit interaction. Here, it is permissible since spin-orbit interaction is ignored.

The overall wavefunction $\phi_A(\vec{r_1}, s_1; \vec{r_2}, s_2)$ must be antisymmetric upon interchanging the indices 1 and 2, i.e., either

$$\phi_A(\vec{r_1}, s_1; \vec{r_2}, s_2) = \psi_S(\vec{r_1}, \vec{r_2}) \Xi_A(s_1, s_2), \tag{10.8}$$

or

$$\phi_A(\vec{r_1}, s_1; \vec{r_2}, s_2) = \psi_A(\vec{r_1}, \vec{r_2}) \Xi_S(s_1, s_2), \tag{10.9}$$

where the subscripts S and A stand for "symmetric" and "antisymmetric," respectively.

Since each component of the total spin operator

$$\vec{S} = \vec{S_1} + \vec{S_2} = \frac{\hbar}{2}(\vec{\sigma_1} + \vec{\sigma_1}) \tag{10.10}$$

and its square magnitude $S^2 = \vec{S}.\vec{S}$ are conserved quantities (recall Chapter 2), the spinorial part must be eigenstates of these operators. However, since the three individual components of $\vec{S}$ do not commute, but each of them commutes with S^2 (recall Chapter 2), we can only select one component of $\vec{S}$ to form a complete set of spinorial operators with S^2. Hereafter, we select the z-component, but the ensuing arguments would hold true if we had selected either S_x or S_y instead of S_z.

As shown in Chapter 2, the eigenvectors of S_z or the Pauli σ_z matrix are given by

$$|0>= \begin{bmatrix} 1 \\ 0 \end{bmatrix}, \tag{10.11}$$

with eigenvalue $+1$, and

$$|1>= \begin{bmatrix} 0 \\ 1 \end{bmatrix}, \tag{10.12}$$

with eigenvalue -1. We need to add an extra index to these eigenvectors to label which one is associated with which electron in the subsequent steps.

The eigenstates of the spinorial part of the wavefunction for the two electron system are given explicitly in Table 10.1. In this table, the single wavefunction $\Xi_A(s_1, s_2)$ is referred to as the *singlet* state whereas the three wavefunctions $\Xi_S(s_1, s_2)$ are referred to as the *triplet* states of the helium atom.

Exercise: Show explicitly that the application of the operators S_z and S^2 to the spinorial states of the two-electron system described here are the values listed in columns 2 and 3 in Table 10.1.

TABLE 10.1

Eigenstates of the spinorial part

Spinorial part	S_z	S^2
$\Xi_S(s_1, s_2) = \|0>_1 \|0>_2$	$+\hbar$	$2\hbar^2$
$\Xi_S(s_1, s_2) = \frac{1}{\sqrt{2}}\left[\|0>_1 \|1>_2 + \|0>_2 \|1>_1\right]$	0	$2\hbar^2$
$\Xi_S(s_1, s_2) = \|1>_1 \|1>_2$	$-\hbar$	$2\hbar^2$
$\Xi_A(s_1, s_2) = \frac{1}{\sqrt{2}}\left[\|0>_1 \|1>_2 - \|0>_2 \|1>_1\right]$	0	0

The total spin operator associated with the two-electron system is given by

$$\vec{S} = \vec{S_1} + \vec{S_2} = \frac{\hbar}{2}\vec{\sigma_1} + \frac{\hbar}{2}\vec{\sigma_2}, \tag{10.13}$$

with

$$\vec{\sigma} = (\sigma_x, \sigma_y, \sigma_z), \tag{10.14}$$

and one extra label must be added to distinguish between electron 1 and 2. Therefore,

$$S^2 = \vec{S} \cdot \vec{S} = \frac{\hbar^2}{4}\left(\sigma_1^2 + \sigma_2^2 + 2\sigma_1 \cdot \sigma_2\right) = \frac{\hbar^2}{2}\left(3 \times I + \vec{\sigma_1} \cdot \vec{\sigma_2}\right), \tag{10.15}$$

where

$$\vec{\sigma_1} \cdot \vec{\sigma_2} = \sigma_{1x}\sigma_{2x} + \sigma_{1y}\sigma_{2y} + \sigma_{1z}\sigma_{2z}. \tag{10.16}$$

Since, according to the properties of the Pauli spin matrices derived in Chapter 2, we have

$$\sigma_x|0\rangle = |1\rangle; \sigma_y|0\rangle = i|1\rangle; \sigma_z|0\rangle = |0\rangle;$$
$$\sigma_x|1\rangle = |0\rangle; \sigma_y|1\rangle = -i|0\rangle; \sigma_z|1\rangle = -|1\rangle, \tag{10.17}$$

we get

$$\vec{\sigma_1} \cdot \vec{\sigma_2}\{|0\rangle_1|0\rangle_2\} = |0\rangle_1|0\rangle_2. \tag{10.18}$$

Hence

$$S^2|0\rangle_1|0\rangle_2 = \frac{\hbar^2}{2}\left(3 \times I + \vec{\sigma_1} \cdot \vec{\sigma_2}\right)|0\rangle_1|0\rangle_2$$
$$= 2\hbar^2|0\rangle_1|0\rangle_2. \tag{10.19}$$

Furthermore,

$$S_z|0\rangle_1|0\rangle_2 = \frac{\hbar}{2}[\sigma_{1z}|0\rangle_1|0\rangle_2 + \sigma_{2z}|0\rangle_1|0\rangle_2] = \hbar|+\rangle_1|+\rangle_2. \tag{10.20}$$

Similar derivations lead to the other eigenvalues listed in Table 10.1.

In reality, the spatial part of the wavefunction must be found through a rigorous solution of the Schrödinger equation

$$H_{He}\left(\vec{r_1},\vec{r_2}\right)\psi_{S,A}\left(\vec{r_1},\vec{r_2}\right) = E\psi_{S,A}\left(\vec{r_1},\vec{r_2},\right). \tag{10.21}$$

Despite the rather simple form of the Hamiltonian, this problem has no known exact solution. Hereafter, we seek solutions to this equation using perturbation theory under the assumption that the Coulomb repulsion between the two electrons is a weak perturbation on the unperturbed Hamiltonian H_0 given by

$$H_0 = H_1 + H_2, \tag{10.22}$$

where

$$H_1 = \frac{|\vec{p_1}|^2}{2m_0} - \frac{2e^2}{4\pi\epsilon_0|\vec{r_1}|}, \tag{10.23}$$

and

$$H_2 = \frac{|\vec{p_2}|^2}{2m_0} - \frac{2e^2}{4\pi\epsilon_0|\vec{r_2}|}. \tag{10.24}$$

Here, H_1 and H_2 are the Hamiltonians of the individual electrons with a kinetic term and a potential energy term due to the attraction of the nucleus with charge $Z = +2e$.

The eigenfunctions and corresponding eigenvalues of H_1 and H_2 can be easily found. They are the well-known wavefunctions for the hydrogen atom (the Laguerre polynomials for the radial part and the spherical harmonics for the angular parts, if we adopt spherical coordinates) [1] with the only difference being that the charge of the nucleus is $+2e$ rather than $+e$. Once that difference is accounted for, if we label the eigenfunctions and corresponding eigenvalues as $\phi_i(\vec{r})$ and ϵ_i, respectively, then any product of the form

$$\phi_i(\vec{r_1})\phi_j(\vec{r_2}) \tag{10.25}$$

is an eigenfunction of H_0 with eigenvalue $\epsilon_i + \epsilon_j$. A solution of this type is often called the *Hartree* solution after Hartree.

There is one serious problem with the Hartee solution. It does not account for the Symmetry Principle which mandates that the total wavefunction (spatial and spinorial parts together) must be antisymmetric under exchange of indices. In order to reconcile with the Symmetry Principle, we must select a proper linear combination of products of the form (10.25) such that the overall spatial part of the wavefunction of H_0 is either symmetric or antisymmetric. Of course, when the spatial part is symmetric, the spin part must be anti-symmetric, and vice versa, so that the overall wavefunction (spatial + spin) is antisymmetric, as mandated by the Symmetry Principle.

If $i = j$, the only possibility is a symmetric spatial wavefunction of the form

$$\psi_S(\vec{r_1},\vec{r_2}) = \phi_i(\vec{r_1})\phi_i(\vec{r_2}). \tag{10.26}$$

If $i \neq j$, the spatial part of the wavefunction can be either symmetric

$$\psi_S(\vec{r_1}, \vec{r_2}) = \frac{1}{\sqrt{2}} [\phi_i(\vec{r_1})\phi_j(\vec{r_2}) + \phi_i(\vec{r_2})\phi_j(\vec{r_1})], \qquad (10.27)$$

or antisymmetric

$$\psi_A(\vec{r_1}, \vec{r_2}) = \frac{1}{\sqrt{2}} [\phi_i(\vec{r_1})\phi_j(\vec{r_2}) - \phi_i(\vec{r_2})\phi_j(\vec{r_1})]. \qquad (10.28)$$

Exercise: (a) Prove that the factors $1/\sqrt{2}$ appearing in Equations (10.27) and (10.28) are needed to normalize $\psi_S(\vec{r_1}, \vec{r_2})$ and $\psi_A(\vec{r_1}, \vec{r_2})$. In other words,

$$\int d\vec{r_1} \int d\vec{r_2} \, \psi_s^\star (\vec{r_1}, \vec{r_2}) \, \psi_s (\vec{r_1}, \vec{r_2}) = 1 \qquad (10.29)$$

and a similar relation holds for $\psi_A (\vec{r_1}, \vec{r_2})$, if the individual $\phi_i's$ are normalized, i.e.,

$$\int d\vec{r_1} \phi_i^\star (\vec{r_1}) \, \phi_i (\vec{r_1}) = \int d\vec{r_2} \phi_i^\star (\vec{r_2}) \, \phi_i (\vec{r_2}) = 1. \qquad (10.30)$$

(b) Show that $\psi_s (\vec{r_1}, \vec{r_2})$ and $\psi_A (\vec{r_1}, \vec{r_2})$ are eigenstates of H_0 with energy $\varepsilon_i + \varepsilon_j$. The solution of this exercise is left to the reader.

If the two electrons have the same spatial part, the only possible choice for the spinorial part of the wavefunction which will make the overall wavefunction antisymmetric is the singlet state in Table 10.1, i.e.,

$$\phi_A(\vec{r_1}, s_1; \vec{r_2}, s_2) = \frac{1}{\sqrt{2}} \phi_i(\vec{r_1})\phi_i(\vec{r_2}) [|0>_1 |0>_2 - |0>_2 |0>_1]. \qquad (10.31)$$

Using first order perturbation theory to find the approximate values of the energies of the helium atom for which the Coulomb repulsion between electrons is assumed to be small, the energy associated with $\phi_A(\vec{r_1}, s_1; \vec{r_2}, s_2)$ is given by

$$E_i = 2\epsilon_i + E_C, \qquad (10.32)$$

where the correcting term

$$E_C = \left\langle \phi_A \left| \frac{e^2}{4\pi\epsilon_0 |\vec{r}_{12}|} \right| \phi_A \right\rangle = \left\langle \phi_A \left| \frac{e^2}{4\pi\epsilon_0 |\vec{r_1} - \vec{r_2}|} \right| \phi_A \right\rangle \qquad (10.33)$$

is the Coulomb energy

$$E_C = \frac{e^2}{2} \int d\vec{r_1} \int d\vec{r_2} \, [_1\langle 0|_2\langle 1| - _2\langle 0|_1\langle 1|] \frac{|\phi_i(\vec{r_1})|^2 |\phi_i(\vec{r_2})|^2}{4\pi\epsilon_0 |\vec{r}_{12}|} [|0\rangle_1 |1\rangle_2 - |0\rangle_2 |1\rangle_1]. \qquad (10.34)$$

As easily checked, the inner product involving the spinorial part of the wavefunction is

$$({}_1\langle 0|_2\langle 1| -_2 \langle 0|_1\langle 1|) \, (|0\rangle_1|1\rangle_2 - |0\rangle_2|1\rangle_1) = 2. \tag{10.35}$$

Therefore, E_C becomes

$$E_C = \int d\vec{r_1} \int d\vec{r_2} \frac{e|\phi_i(\vec{r_1})|^2 \cdot e\phi_i(\vec{r_2})|^2}{4\pi\epsilon|\vec{r}_{12}|} = \int d\vec{r_1} \int d\vec{r_2} \frac{e|\phi_i(\vec{r_1})|^2 \cdot e\phi_i(\vec{r_2})|^2}{4\pi\epsilon|\vec{r}_1 - \vec{r}_2|}, \tag{10.36}$$

which is clearly an energy resulting from the Coulomb repulsion between the charge distributions associated with the probability densities $|\phi_i(\vec{r_1})|^2$ and $|\phi_i(\vec{r_2})|^2$.

Exercise: Calculate the value of the Coulomb term E_C if $\phi_i(\vec{r_1})$, $\phi_i(\vec{r_2})$ are given by the ground state wavefunctions of the hydrogen atom. Do not forget to replace the atomic number of the nucleus by $Z = +2e$.

Solution

The integral appearing in Equation (10.36) is more readily performed using the Fourier transform

$$\frac{1}{r_{12}} = \frac{1}{2\pi^2} \int d\vec{k} \, \frac{e^{i\vec{k} \cdot (\vec{r_1} - \vec{r_2})}}{k^2}. \tag{10.37}$$

The wavefunction for the ground state of the Hamiltonians H_1 and H_2 in Equations (10.23) and (10.24) is

$$\phi_{1s}(r) = \frac{1}{\sqrt{\pi a_{He}^3}} e^{-\frac{r}{a_{He}}}, \tag{10.38}$$

where $a_{He} = a_0/2$ and a_0 is the Bohr radius.

Hence

$$E_C = \frac{1}{4\pi\varepsilon_0} \frac{e^2}{2\pi^2} \int \frac{d\vec{k}}{k^2} \left(\int d\vec{r_1} e^{i\vec{k} \cdot \vec{r_1}} |\phi_{1s}(r_1)|^2 \right) \left(\int d\vec{r_2} e^{-i\vec{k} \cdot \vec{r_2}} |\phi_{1s}(r_2)|^2 \right), \tag{10.39}$$

or

$$E_C = \frac{1}{4\pi\varepsilon_0} \frac{e^2}{2\pi^2} \int \frac{d\vec{k}}{k^2} g\left(\vec{k}\right) g^\star\left(\vec{k}\right), \tag{10.40}$$

where the asterisk represents complex conjugate and $g(\vec{k})$ is given by

$$g\left(\vec{k}\right) = \int d\vec{r} \, e^{i\vec{k} \cdot \vec{r}} |\phi_{1s}(r)|^2. \tag{10.41}$$

Since $|\phi_{1s}(r)|^2$ depends only on the magnitude of r, the vector can be selected to be along the z-axis and we get

$$g\left(\overrightarrow{k}\right) = \int_0^{+\infty} dr r^2 2\pi \int_0^\pi d\theta \sin\theta e^{ikr\cos\theta} |\phi_{1s}(r)|^2. \tag{10.42}$$

The latter can be easily calculated using Equation (10.38) and is found to be

$$g\left(\overrightarrow{k}\right) = \frac{16}{\left(4 + a_{He}^2 k^2\right)^2}. \tag{10.43}$$

Therefore, the Coulomb energy is given by

$$E_C = \frac{1}{4\pi\varepsilon_0} \frac{e^2}{2\pi^2} \int \frac{d\overrightarrow{k}}{k^2} \left(\frac{16}{\left(4 + a_{He}^2 k^2\right)^2}\right)^2$$

$$= \frac{5e^2}{32\pi\varepsilon_0 a_0}. \tag{10.44}$$

Since the ground state of the Hamiltonian H_0 in Equation (10.22) has an energy equal to $+8E_{1s}$, where E_{1s} is the energy of the ground state of the hydrogen atom

$$E_{1s} = -\frac{1}{4\pi\varepsilon_0} \frac{e^2}{2a_0}, \tag{10.45}$$

we have

$$\frac{E_C}{8E_{1s}} = \frac{-5}{16} \approx -0.3. \tag{10.46}$$

Despite the fact that E_C is not quite a small correction to the unperturbed value $8E_{1s}$, the derivation given above is well suited to describing the concept of exchange energy when considering the case where the spatial parts of the electron wavefunctions are different, as described next.

If the electron wavefunctions $\phi_i(\overrightarrow{r})$, $\phi_j(\overrightarrow{r})$ are different, i.e., $i \neq j$, then the overall wavefunction of the total Hamiltonian H_{He} can be of two forms. If we choose the spinorial part of the overall wavefunction to be one of the triplet state, the spatial part must be antisymmetric to keep the overall wavefunction antisymmetric. The simplest way to write the spatial part is

$$\psi_{2-electron} = \frac{1}{\sqrt{2}} \left[\phi_i(\overrightarrow{r_1})\phi_j(\overrightarrow{r_2}) - \phi_i(\overrightarrow{r_2})\phi_j(\overrightarrow{r_1})\right]. \tag{10.47}$$

This type of wavefunction is often referred to as the *Slater determinant* after Slater:

$$\psi_{2-electron} = \frac{1}{\sqrt{2}} det \begin{vmatrix} \phi_i(\overrightarrow{r_1}) & \phi_i(\overrightarrow{r_2}) \\ \phi_j(\overrightarrow{r_1}) & \phi_j(\overrightarrow{r_2}) \end{vmatrix}. \tag{10.48}$$

In the theory of the helium atom, the triplet states composed of the product of this spatial part and the three spin states described as triplet states in Table 10.1 are referred to as the *orthohelium* states.

If, on the other hand, we select the spinorial part of the overall wavefunction to be the singlet state in Table 10.1, the spatial part must be symmetric and the simplest way to achieve this is to write the spatial part as follows

$$\frac{1}{\sqrt{2}}\left[\phi_i(\vec{r_1})\phi_j(\vec{r_2}) + \phi_i(\vec{r_2})\phi_j(\vec{r_1})\right]. \tag{10.49}$$

In the theory of the helium atom, the states written as the product of the symmetrical spatial part and the antisymmetric spinorial part are referred to as *parahelium* states.

When $i \neq j$, the energy associated with the Hamiltonian H_0 is given by $\epsilon_i + \epsilon_j$. For the overall Hamiltonian containing the effects of the Coulomb repulsion, we have, to the first order in perturbation theory, a correction given by

$$E_S = \frac{e^2}{2}\int d\vec{r_1}\int d\vec{r_2}\frac{\left(\phi_i^*(\vec{r_1})\phi_j^*(\vec{r_2}) + \phi_j^*(\vec{r_1})\phi_i^*(\vec{r_2})\right)\left(\phi_i(\vec{r_1})\phi_j(\vec{r_2}) + \phi_j(\vec{r_1})\phi_i(\vec{r_2})\right)}{4\pi\epsilon_0|\vec{r}_{12}|}$$
$$\times \langle\Xi_A(\vec{r}_1,\vec{r}_2)|\Xi_A(r_1,r_2)\rangle \tag{10.50}$$

when the two electrons are in the singlet state.

Using $\Xi_A(\lambda_1, \lambda_2)$ listed in Table 10.1, the vector product is easily shown to be equal to unity, leading to

$$E_S = K_{ij} + J_{ij}, \tag{10.51}$$

where

$$K_{ij} = \int d\vec{r_1}\int d\vec{r_2}\frac{e|\phi_i(\vec{r_1})|^2 \cdot e\phi_j(\vec{r_2})|^2}{4\pi\epsilon_0|\vec{r}_{12}|} \tag{10.52}$$

is the Coulomb energy term found earlier for the case $i = j$, and

$$J_{ij} = e^2\int d\vec{r_1}\int d\vec{r_2}\frac{\phi_i^*(\vec{r_1})\phi_j^*(\vec{r_2})\phi_j(\vec{r_1})\phi_i(\vec{r_2})}{4\pi\epsilon_0|\vec{r}_{12}|}. \tag{10.53}$$

This last term is referred to as the *exchange energy* term and is truly quantum-mechanical in origin. Its appearance can be traced back to the Pauli exclusion principle and to the fact that the two electrons in the helium atom are indistinguishable and therefore must obey the Symmetry Principle. For two such particles in close proximity, their wavefunctions will overlap and if we introduce an overlap density for the system of these two electrons as

$$\rho_{ij}(\vec{r_1}) = \phi_i^*(\vec{r_1})\phi_j(\vec{r_1}), \tag{10.54}$$

and

$$\rho_{ij}(\vec{r_2}) = \phi_j^*(\vec{r_2})\phi_i(\vec{r_2}), \tag{10.55}$$

then J_{ij} can be written as

$$J_{ij} = e^2\int d\vec{r_1}\int d\vec{r_2}\frac{\rho_{ij}(\vec{r_1})\rho_{ij}(\vec{r_2})}{4\pi\epsilon|\vec{r}_{12}|}, \tag{10.56}$$

which appears in a form similar to the Coulomb term K_{ij} where the true charge density of individual charges is replaced by the overlap charge densities ρ_{ij}.

A similar calculation starting with the overall wavefunction associated with the triplet states leads to

$$E_T = K_{ij} - J_{ij}. \tag{10.57}$$

For the case of the helium atom, the exchange energy term J_{ij} is positive and the triplet states (orthohelium) possess *lower* energy than the singlet states (parahelium).

Exercise: For the case where the wavefunction of the two electrons is written as a product of Equation (10.47) and $\Xi_S(s_1, s_2)$ in Table 10.1, show that the average value of the Coulomb repulsion term is indeed given by Equation (10.52).

10.1.2   The Heitler–London model of the hydrogen molecule

In the early days of quantum mechanics, Heitler and London [2] used an approach similar to the one described for the He atom to calculate the energy eigenvalues of the H_2 molecule. Note that the hydrogen molecule is different from the helium atom in that here there are two nuclei instead of one. There is one other very significant difference. Contrary to the case of the helium atom, the lowest energy state is the singlet state, referred to as *parahydrogen*, rather than the triplet state. This is an intriguing difference between the He atom and the H_2 molecule, both of which have two electrons. It is believed that if two electrons are bound to the *same* potential (the lone nucleus in the case of the helium atom), then the triplet state has lower energy than the singlet state. If, on the other hand, the two electrons are bound to two *different* potentials (two different nuclei in the case of the hydrogen molecule), then the singlet state has the lower energy.

A solid state analog of the H_2 molecule is a system of two electrons confined to two coupled quantum dots. The confining potential of a quantum dot acts like the confining potential of a proton. Loss and DiVincenzo [3] have used the Heitler-London approach to determine the energy eigenstates and eigenfunctions of two electrons bound to two coupled quantum dots (separated by a tunnel barrier). These dots are delineated electrostatically in a two-dimensional electron gas. Since the exchange energy depends on the overlap between the wavefunctions of the two electrons, it can be tuned by adjusting the width and height of the tunnel barrier between the quantum dots using metallic Schottky gates. Modulation of the exchange energy in this fashion forms the basis of quantum logic gates (see Chapter 16), particularly the two qubit square-root-of-swap operation. The validity and limitation of the Heitler-London approach to estimate the exchange coupling in semi-

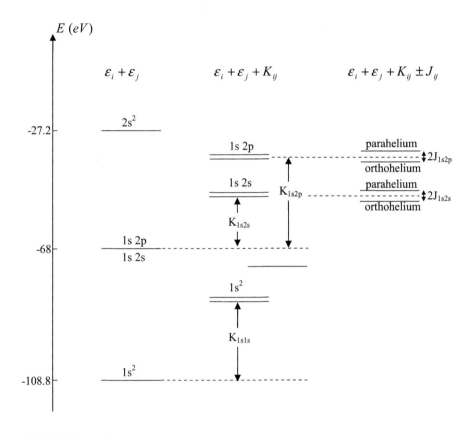

FIGURE 10.2

Schematic representation of the energy level of the helium atom. From left to right, the energy levels neglecting the effects of the Coulomb interaction between electrons, with the effects of the Coulomb repulsion, and including both the effects of the Coulomb repulsion and the exchange interaction.

conductor nanostructures have been assessed recently by several authors [4]. More rigorous numerical calculations of the exchange energy in coupled quantum dots have been presented by Melnikov and Leburton [5]. They solve the two-electron Schrödinger equation self-consistently with the Poisson equation in the presence of a magnetic field:

$$H\psi\left(\vec{r}_1, \vec{r}_2\right) = \left[\sum_{i=1}^{2} h_i + \frac{e^2}{4\pi\epsilon\left|\vec{r}_1 - \vec{r}_2\right|}\right]\psi\left(\vec{r}_1, \vec{r}_2\right)$$

$$h_i = \left(\vec{p}_i - e\vec{A}_i\right)^2 / 2m^* + V\left(\vec{r}_i\right)$$

$$\vec{\nabla}\cdot\left(\epsilon\vec{\nabla}V(\vec{r})\right) = \rho(|\psi|^2), \tag{10.58}$$

where ϵ is the dielectric constant of the medium, $\vec{A}$ is the vector potential due to any external magnetic field, and ρ is the charge density, which depends on the squared magnitude of the wavefunction. The above equations are solved numerically in three dimensions. Melkinov and Leburton [5] found that the exchange energy is very sensitive to the confining potentials of the quantum dots and barely exceeds 1 meV in realistic quantum dots defined by gates on a two-dimensional electron gas. This does not bode well for spin devices that depend on the modulation of a large exchange energy for their operation.

Since (in units of $\hbar$)

$$\vec{S}_1 \cdot \vec{S}_2 = \frac{1}{2}\left[\left(\vec{S}_1 + \vec{S}_2\right)^2 - S_2^2 - S_1^2\right]$$

$$= \frac{1}{2}\left[S(S+1) - s_1(s_1+1) - s_2(s_2+1)\right], \tag{10.59}$$

where $s_1 = s_2 = \frac{1}{2}$.

For triplet states,

$$\left(\vec{S}_1 \cdot \vec{S}_2\right)_T = \frac{1}{4}, \tag{10.60}$$

and for singlet states

$$\left(\vec{S}_1 \cdot \vec{S}_2\right)_S = -\frac{3}{4}. \tag{10.61}$$

Hence

$$E_T - E_S = -2J_{12}\left[\frac{1}{4} - \left(\frac{-3}{4}\right)\right] = -2J_{12}. \tag{10.62}$$

Neglecting the energy corresponding to the unperturbed portion of the Hamiltonian H_0 (column 1 in Table 10.1), the correction to first order perturbation theory can be written as

$$\Delta_{ij} = K_{ij} - 2\left(\vec{S}_1 \cdot \vec{S}_2 + \frac{1}{4}\right)J_{ij}, \tag{10.63}$$

or

$$\Delta_{ij} = \left(K_{ij} - \frac{1}{2} J_{ij} \right) - 2 \vec{S}_1 \cdot \vec{S}_2 J_{ij}. \qquad (10.64)$$

10.2 Hartree and Hartree–Fock approximations

So far, we have mostly dealt with two electrons. But now, let us take a look at true many-body systems comprising numerous electrons. The many-body wavefunction of N interacting electrons is written as

$$\Psi = \Psi \left(\vec{r}_1 s_1, \vec{r}_2 s_2, ... \vec{r}_N s_N \right). \qquad (10.65)$$

The *Variational Principle* states that if Ψ is a solution of the many-body Schrödinger equation, then the quantity

$$\langle H \rangle = \frac{(\Psi, H\Psi)}{(\Psi, \Psi)} \qquad (10.66)$$

must be stationary. Here (A, B) stands for the integral $\int d\vec{r} A^* B$. In fact, the ground state wavefunction is the Ψ which minimizes $\langle H \rangle$.

The Hartree wavefunction will be a product wavefunction:

$$\Psi = \Psi \left(\vec{r}_1 s_1, \vec{r}_2 s_2, ... \vec{r}_N s_N \right) = \psi_1 \left(\vec{r}_1 s_1 \right) \psi_2 \left(\vec{r}_2 s_2 \right) ... \psi_N \left(\vec{r}_N s_N \right). \qquad (10.67)$$

This wavefunction, however, does *not* satisfy the Symmetry Principle which mandates that the sign of Ψ should change if we interchange any two of its arguments, i.e.,

$$\Psi \left(\vec{r}_1 s_1, ... \vec{r}_i s_i, ... \vec{r}_j s_j, ... \vec{r}_N s_N \right) = -\Psi \left(\vec{r}_1 s_1, ... \vec{r}_j s_j, ... \vec{r}_i s_i, ... \vec{r}_N s_N \right). \qquad (10.68)$$

However, a wavefunction that does satisfy the Symmetry Principle is the *Slater determinant* that we encountered earlier:

$$\Psi \left(\vec{r}_1 s_1, \vec{r}_2 s_2, ... \vec{r}_N s_N \right) = det \begin{pmatrix} \psi_1 \left(\vec{r}_1 s_1 \right) & \psi_1 \left(\vec{r}_2 s_2 \right) & \cdots & \psi_1 \left(\vec{r}_N s_N \right) \\ \psi_2 \left(\vec{r}_1 s_1 \right) & \psi_2 \left(\vec{r}_2 s_2 \right) & \cdots & \psi_2 \left(\vec{r}_N s_N \right) \\ . & . & \cdots & . \\ . & . & \cdots & . \\ \psi_N \left(\vec{r}_1 s_1 \right) & \psi_N \left(\vec{r}_2 s_2 \right) & \cdots & \psi_N \left(\vec{r}_N s_N \right) \end{pmatrix}. \qquad (10.69)$$

This wavefunction automatically satisfies the Pauli principle since it vanishes if $\psi_i = \psi_j$, showing that no two electrons can be put in the same state.

It can be shown that if the energy in Equation (10.66) is evaluated with Ψ given by Equation (10.69) where the ψ-s are orthonormal, then the result is

$$
\langle H \rangle = \sum_i \int d\vec{r}\, \psi_i^* \vec{r} \left(-\frac{\hbar^2}{2m^*}\right) \psi_i \vec{r}
$$

$$
\frac{1}{2} \sum_{i,j} \int d\vec{r} d\vec{r}'\, \frac{e^2}{4\pi\epsilon |\vec{r} - \vec{r}'|} |\psi_i(\vec{r})|^2 \, |\psi_j(\vec{r}')|^2
$$

$$
-\frac{1}{2} \sum_{i,j} \int d\vec{r} d\vec{r}'\, \frac{e^2}{4\pi\epsilon |\vec{r} - \vec{r}'|} \delta_{s_i s_j} \psi_i^*(\vec{r}) \psi_i(\vec{r}') \psi_j^*(\vec{r}') \psi_j(\vec{r}), \quad (10.70)
$$

where the δ is a Krönicker delta indicating that the last term, which is the exchange term, is non-zero only between parallel spins. The second term is the Coulomb term. The exchange term is *negative* and hence tends to lower the total energy. Evaluating $< H >$ with the Slater determinant is known as the *Hartee Fock* approximation. Unlike the bare *Hartree* approximation, it accounts for the Symmetry Principle.

Minimizing the above equation with respect to ψ_i^* leads to the *Hartree-Fock equation* for ψ_i:

$$
-\frac{\hbar^2}{2m^*}\psi_i(\vec{r}) + \left[\int d\vec{r}'\, |\psi_i \vec{r}'|^2 \, \frac{e^2}{4\pi\epsilon |\vec{r} - \vec{r}'|}\right] \psi_i(\vec{r})
$$

$$
-\sum_j \int d\vec{r}'\, \frac{e^2}{4\pi\epsilon |\vec{r} - \vec{r}'|} \psi_j^*(\vec{r}') \psi_i(\vec{r}') \psi_j(\vec{r}) \delta_{s_i s_j} = E_i \psi_i(\vec{r}). \quad (10.71)
$$

If the ψ-s are plane waves, then the right hand side of the above equation assumes the form $E(\vec{k}_i)\psi_i$, where

$$
E(\vec{k}) = \frac{\hbar^2 k^2}{2m^*} - \frac{1}{V} \sum_{k' \leq k_F} \frac{e^2}{\epsilon \left|\vec{k} - \vec{k}'\right|^2}
$$

$$
= \frac{\hbar^2 k^2}{2m^*} - \int_0^k F \frac{d\vec{k}'}{(2\pi)^3} \frac{e^2}{\epsilon \left|\vec{k} - \vec{k}'\right|^2}
$$

$$
= \frac{\hbar^2 k^2}{2m^*} - \frac{e^2}{2\pi^2\epsilon} k_F f\left(\frac{k}{k_F}\right), \quad (10.72)
$$

where V is the normalizing volume, k_F is the Fermi wavevector and

$$
f(x) = \frac{1}{2} + \frac{1 - x^2}{4x} ln \left|\frac{1 + x}{1 - x}\right|. \quad (10.73)
$$

The total energy of N interacting electrons is obtained by summing the energy given by Equation (10.72) over all wavevector states $\vec{k}$ up to the Fermi wavevector k_F, multiplying by 2 to account for the spin degeneracy, and then dividing the last term by 2 to correct for double counting. This yields

$$E_N = 2\left\{\sum_{k<k_F} \frac{\hbar^2 k^2}{2m^*} - \frac{e^2 k_F}{4\pi^2 \epsilon} \sum_{k<k_F}\left[1 + \frac{k_F^2 - k^2}{2kk_F}ln\left|\frac{k_F + k}{k_F - k}\right|\right]\right\}. \quad (10.74)$$

The first term in the equation above can be found by converting the summation to an integral:

$$2\sum_{k<k_F} \frac{\hbar^2 k^2}{2m^*} = 2\frac{V}{8\pi^3}\int_0^{k_F} \frac{\hbar^2}{2m^*}4\pi k^2 dk = \frac{V}{5\pi^2}\frac{\hbar^2}{2m^*}k_F^5. \quad (10.75)$$

Further, noting that the electron density $N/V = k_F^3/3\pi^2$ (this is left as an exercise for the reader), we obtain that

$$2\sum_{k<k_F} \frac{\hbar^2 k^2}{2m^*} = \frac{3}{5}\frac{\hbar^2 k_F^2}{2m^*}N = \frac{3}{5}NE_F, \quad (10.76)$$

where E_F is the Fermi energy.

The second term in Equation (10.74) can be evaluated by converting the summation to an integral. This exercise is also left to the reader. The final result is

$$E_N = N\left[\frac{3}{5}E_F - \frac{3e^2 k_F}{16\pi^2\epsilon}\right]. \quad (10.77)$$

10.3   The role of exchange in ferromagnetism

10.3.1   The Bloch model of ferromagnetism

Felix Bloch [6] pointed out that the last equation, the result of Hartree-Fock approximation, can explain ferromagnetism in a gas of electrons interacting through mutual Coulomb repulsion. In deriving Equation (10.77), we assumed that every state (labeled by the wavevector k) is filled by two electrons of opposite spins, in keeping with the Pauli principle. A more general approach (also consistent with the Pauli principle) is to fill each one-electron level $k < k_\uparrow$ with up-spin electrons and $k < k_\downarrow$ with down-spin electrons. Then,

$$E_\uparrow = N_\uparrow\left[\frac{3}{5}\frac{\hbar^2 k_\uparrow^2}{2m^*} - \frac{3e^2 k_\uparrow}{16\pi^2\epsilon}\right]$$

$$E_\downarrow = N_\downarrow\left[\frac{3}{5}\frac{\hbar^2 k_\downarrow^2}{2m^*} - \frac{3e^2 k_\downarrow}{16\pi^2\epsilon}\right], \quad (10.78)$$

where

$$E_N = E_\uparrow + E_\downarrow$$

$$\frac{N}{V} = \frac{N_\uparrow}{V} + \frac{N_\downarrow}{V} = \frac{k_\uparrow^3}{6\pi^2} + \frac{k_\uparrow^3}{6\pi^2} = \frac{k_F^3}{3\pi^2}. \tag{10.79}$$

Equation (10.77) is the form E_N takes if $N_\uparrow = N_\downarrow = N/2$. But now let us assume that the electron gas is completely spin polarized so that $N_\uparrow = N$ and $N_\downarrow = 0$. Then $E_N = E_\uparrow$ and $k_\uparrow = 2^{1/3} k_F$. Therefore, in a completely spin polarized gas,

$$E_N^{polarized} = N \left[\frac{3}{5} 2^{2/3} \frac{\hbar^2 k_F^2}{2m^*} - 2^{1/3} \frac{3e^2 k_F}{16\pi^2 \epsilon} \right]. \tag{10.80}$$

Comparing the above equation with Equation (10.77), we obtain

$$E_N^{polarized} - E_N^{unpolarized} = N \left[\frac{3}{5} \left(2^{2/3} - 1 \right) \frac{\hbar^2 k_F^2}{2m^*} - \left(2^{1/3} - 1 \right) \frac{3e^2 k_F}{16\pi^2 \epsilon} \right]. \tag{10.81}$$

If this quantity is negative, i.e., if the exchange term (second term) is larger than the first term, then the polarized state, which is *ferromagnetic*, is preferred as the ground state of the electron gas. Generally speaking, low densities (smaller k_F) will favor the ferromagnetic state. Thus, Bloch was able to explain ferromagnetism in an interacting electron gas where electrons interact only via Coulomb repulsion.

10.3.2 The Heisenberg model of ferromagnetism

Independent of Bloch, Heisenberg had postulated another scenario where ferromagnetism could be traced to exchange interaction between electrons. He was the first to propose that the internal magnetic field in ferromagnets, arising from all spins within a domain lining up parallel to each other, could be linked to the presence of exchange coupling between neighboring ionic spins. The basic argument is as follows: From the discussion of the energy eigenstates of the helium atoms, the difference between energy states (triplet and singlet) associated with orbital (i, j) characterizing the spatial parts of the orbitals is given by

$$E_T - E_S = -2J_{ij}. \tag{10.82}$$

The latter can be formally obtained as eigenvalues of a Hamiltonian

$$H_{ij} = -2J_{ij} \vec{S}_1 \cdot \vec{S}_2, \tag{10.83}$$

where $\vec{S}_1$ and $\vec{S}_2$ are the spin operators (in units of $\hbar$) associated with electrons 1 and 2, i.e., H_{ij} can also be written as

$$H_{ij} = -\frac{J_{ij}}{2} \vec{\sigma}_1 \cdot \vec{\sigma}_2. \tag{10.84}$$

Since, when J_{ij} is positive, a parallel spin configuration (triplet state) has lower energy than the anti-parallel configuration (singlet state), Heisenberg postulated that a generalization of the exchange concept is well suited to describe the onset of ferromagnetism in a system of ionic spins on a lattice. In Heisenberg's model, an ensemble of ionic spins is characterized by a Hamiltonian of the form

$$\widehat{H} = -\frac{1}{2} \sum_{i<j} J_{ij} \vec{\sigma}_i \cdot \vec{\sigma}_j, \tag{10.85}$$

where the sum is carried out over both indices with $i < j$ to avoid double counting and self-interacting terms.

For given $J'_{ij}s$, if the lowest state of the Hamiltonian corresponds to all spins aligned, then this will account for the onset of ferromagnetism, since the lowest state corresponds to the ground state.

The Heisenberg Hamiltonian is a fundamental concept in the theory of magnetism. Early solutions for the case of interacting spins with only one spin component, i.e., $\vec{\sigma}_1 \cdot \vec{\sigma}_2$ replaced by $\sigma_{1z}\sigma_{2z}$ and nearest neighbor interaction ($J_{ij} = J$ for $|i - j| = 1$ and 0 otherwise), were worked out by Ising [7] and by Onsager [8] for one and two-dimensional lattices of interacting spins, respectively. A complete solution to the full three-dimensional Heisenberg Hamiltonian with the spin operators is still lacking.

10.4   The Heisenberg Hamiltonian

The Heisenberg Hamiltonian in Equation (10.85) is an important theoretical tool in many areas of spintronics. We will have occasion to use it in the description of Single Spin Logic gates in Chapter 15. It is also widely used in the theoretical description of many other spin-based devices that rely on the manipulation of exchange interaction for their operation. One example of this is 2-qubit quantum logic gates discussed in Chapter 16.

This chapter has introduced the notion of exchange interaction and showed how this notion is at the heart of fundamental physical phenomena such as ferromagnetism. In later chapters, we will show how this is also at the heart of single spin devices that perform useful operations such as classical or quantum logic.

10.5 Problems

- ### Problem 10.1

  The following problem should be solved using material from Section 2.7.1 in Chapter 2.

  Show that when there are two electrons (or any two spin-1/2 particles), the 2-body wavefunction for which the sum of the two spins equals 1 does not change its value when the spin variables of the electrons are exchanged (that is, the function is symmetric). The wavefunction for which the sum becomes 0 changes sign when the spin variables are interchanged (i.e., the function is antisymmetric).

  The proof of this theorem is taken from S. I. Tomonoga, *The Story of Spin* (The University of Chicago Press, Chicago, 1997).

  Let the spin operators of the two electrons be $\mathbf{s}_1 = (s_{1x}, s_{1y}, s_{1z})$ and $\mathbf{s}_2 = (s_{2x}, s_{2y}, s_{2z})$. The squared magnitude of the sum is

  $$|\mathbf{s}_1 + \mathbf{s}_2|^2 = |\mathbf{s}_1|^2 + |\mathbf{s}_2|^2 + 2(\mathbf{s}_1 \cdot \mathbf{s}_2), = \frac{1}{2}\left[3 + 4(\mathbf{s}_1 \cdot \mathbf{s}_2)\right], \quad (10.86)$$

  where we have used Equation (2.17) to derive the last equality.

  The two body wavefunction $\psi(s_{z1}, s_{z2})$ can be written as a column vector spanning all possible spin configurations:

  $$\psi_{two-body} = \begin{bmatrix} \psi(+1/2, +1/2) \\ \psi(+1/2, -1/2) \\ \psi(-1/2, +1/2) \\ \psi(-1/2, -1/2) \end{bmatrix}. \quad (10.87)$$

  In the above equation, the second coordinate of ψ refers to the spin of the second electron and the first coordinate refers to the spin of the first.* We have dropped the coordinate $\mathbf{x}$ since it is not important here.

  We need to calculate $[\mathbf{s}_1 \cdot \mathbf{s}_2]\psi$. Note that $\mathbf{s}_1 \cdot \mathbf{s}_2 = s_{1x}s_{2x} + s_{1y}s_{2y} + s_{1z}s_{2z}$, so that $\mathbf{s}_1 \cdot \mathbf{s}_2\psi = [s_{1x}s_{2x} + s_{1y}s_{2y} + s_{1z}s_{2z}]\psi$.

  We now apply the "rules" of Equation (2.76):

---

*It may appear to the student that when we label one electron as the "first" and the other as the "second" electron, we are somehow distinguishing between indistinguishable particles, which is not allowed. This is actually not the case. We are using these labels only as a matter of convenience. The two electrons are still treated as equals and are indistinguishable.

$$s_{2z}\psi_{two-body} = \begin{bmatrix} (1/2)\psi(+1/2,+1/2) \\ -(1/2)\psi(+1/2,-1/2) \\ (1/2)\psi(-1/2,+1/2) \\ -(1/2)\psi(-1/2,-1/2) \end{bmatrix}, \tag{10.88}$$

$$s_{1z}s_{2z}\psi_{two-body} = \begin{bmatrix} (1/4)\psi(+1/2,+1/2) \\ -(1/4)\psi(+1/2,-1/2) \\ -(1/4)\psi(-1/2,+1/2) \\ (1/4)\psi(-1/2,-1/2) \end{bmatrix}. \tag{10.89}$$

Similarly,

$$s_{1y}s_{2y}\psi_{two-body} = \begin{bmatrix} -(1/4)\psi(-1/2,-1/2) \\ (1/4)\psi(-1/2,+1/2) \\ (1/4)\psi(+1/2,-1/2) \\ -(1/4)\psi(1/2,1/2) \end{bmatrix} \tag{10.90}$$

and

$$s_{1x}s_{x}\psi_{two-body} = \begin{bmatrix} (1/4)\psi(-1/2,-1/2) \\ (1/4)\psi(-1/2,+1/2) \\ (1/4)\psi(+1/2,-1/2) \\ (1/4)\psi(1/2,1/2) \end{bmatrix}. \tag{10.91}$$

Therefore

$$[\mathbf{s_1} \cdot \mathbf{s_2}]\psi_{two-body} = \begin{bmatrix} (1/4)\psi(1/2,1/2) \\ (1/2)\psi(-1/2,+1/2) - (1/4)\psi(+1/2,-1/2) \\ (1/2)\psi(+1/2,-1/2) - (1/4)\psi(-1/2,+1/2) \\ (1/4)\psi(-1/2,-1/2) \end{bmatrix}. \tag{10.92}$$

If we use Equation (10.86) here, then we get

$$|\mathbf{s_1} + \mathbf{s_2}|^2\psi_{two-body} = \begin{bmatrix} 2\psi(1/2,1/2) \\ \psi(-1/2,+1/2) + \psi(+1/2,-1/2) \\ \psi(+1/2,-1/2) + \psi(-1/2,+1/2) \\ 2\psi(-1/2,-1/2) \end{bmatrix}. \tag{10.93}$$

Now, the total spin $= 1$ implies that $|\mathbf{s_1} + \mathbf{s_2}|^2\psi_{two-body} = 1(1+1)$ $\psi_{two-body} = 2\psi_{two-body}$, and the total spin $= 0$ implies $|\mathbf{s_1}+\mathbf{s_2}|^2\psi_{two-body}$

$= 0$. In the former case, if we compare Equations (10.87) and (10.93), then we get

$$\psi(+1/2, -1/2) + \psi(-1/2, +1/2) = 2\psi(+1/2, -1/2)$$
$$or, \quad \psi(-1/2, +1/2) = \psi(+1/2, -1/2). \quad (10.94)$$

Therefore,

$$\psi(s_{1z}, s_{2z}) = \psi(s_{2z}, s_{1z}). \quad (10.95)$$

In the latter case, we need

$$\psi(1/2, 1/2) = \psi(-1/2, -1/2) = 0$$
$$\psi(+1/2, -1/2) + \psi(-1/2, +1/2) = 0. \quad (10.96)$$

.

The last equality gives

$$\psi(s_{1z}, s_{2z}) = -\psi(s_{2z}, s_{1z}). \quad (10.97)$$

Equations (10.95) and (10.97) prove the theorem.

- **Problem 10.2**

Using the theory of the helium atom in Section 10.1.1, calculate the Coulomb (K_{1s2s}) and exchange (J_{1s2s}) integrals. The ground state wavefunction is given by Equation (10.38) and the first excited state wavefunction is

$$\phi_{2s}(r) = \frac{1}{\sqrt{8\pi a_{He}^3}} \left(1 - \frac{r}{2a_{He}}\right) e^{-\frac{r}{2a_{He}}}, \quad (10.98)$$

where $a_{He} = a_0/2$ and a_0 is the Bohr radius ($= 0.529$ Å).

Solution

$$K_{1s2s} = \frac{1}{4\pi\varepsilon_0} \frac{e^2}{2\pi^2} \int \frac{d\vec{k}}{k^2} \left(\int d\vec{r}_1 e^{i\vec{k}\cdot\vec{r}_1} |\phi_{1s}(r_1)|^2 \right)$$
$$\times \left(\int d\vec{r}_2 e^{-i\vec{k}\cdot\vec{r}_2} |\phi_{2s}(r_2)|^2 \right), \quad (10.99)$$

which yields

$$K_{1s2s} = \frac{1}{4\pi\varepsilon_0} \frac{e^2}{2\pi^2} \int \frac{d\vec{k}}{k^2} g_1\left(\vec{k}\right) g_2\left(\vec{k}\right), \quad (10.100)$$

where

$$g_1\left(\vec{k}\right) = \int d\vec{r} e^{i\vec{k}\cdot\vec{r}} |\phi_{1s}(r)|^2$$

$$= \int_0^{+\infty} dr r^2 2\pi \int_0^\pi d\theta \sin\theta e^{ikr\cos\theta} |\phi_{1s}(r)|^2$$

$$= \frac{16}{\left(4 + a_0^2 k^2\right)^2} \tag{10.101}$$

and

$$g_2\left(\vec{k}\right) = \int d\vec{r} e^{-i\vec{k}\cdot\vec{r}} |\phi_{2s}(r)|^2$$

$$= \int_0^{+\infty} dr r^2 2\pi \int_0^\pi d\theta \sin\theta e^{ikr\cos\theta} |\phi_{2s}(r)|^2$$

$$= \frac{1 - 3a_{He}^2 k^2 + 2a_{He}^4 k^4}{\left(1 + 2a_{He}^4 k^4\right)^2}. \tag{10.102}$$

Therefore

$$K_{1s2s} = \frac{1}{4\pi\varepsilon_0} \frac{e^2}{2\pi^2} \int \frac{d\vec{k}}{k^2} \left(\frac{16}{\left(4 + a_0^2 k^2\right)^2}\right) \left(\frac{1 - 3a_{He}^2 k^2 + 2a_{He}^4 k^4}{\left(1 + 2a_{He}^4 k^4\right)^2}\right)$$

$$= \frac{17e^2}{162a_0} \pi\varepsilon_0. \tag{10.103}$$

The exchange energy is given by

$$J_{1s2s} = \frac{1}{4\pi\varepsilon_0} \frac{e^2}{2\pi^2} \int \frac{d\vec{k}}{k^2} \left(\int d\vec{r}_1 e^{i\vec{k}\cdot\vec{r}_1} \phi_{1s}(r_1)\phi_{2s}^\star(r_1)\right)$$

$$\times \left(\int d\vec{r}_2 e^{-i\vec{k}\cdot\vec{r}_2} \phi_{2s}(r_2)\phi_{1s}^\star(r_2)\right). \tag{10.104}$$

Using the intermediate result,

$$g_{12}\left(\vec{k}\right) = \int d\vec{r} e^{i\vec{k}\cdot\vec{r}} \phi_{1s}(r)\phi_{2s}^\star(r)$$

$$= \frac{256\sqrt{2}a_{He}^2 k^2}{\left(9 + 4a_{He}^2 k^2\right)^3}, \tag{10.105}$$

we finally get

$$J_{1s2s} = \frac{8e^2}{729a_0\pi\varepsilon_0}. \tag{10.106}$$

10.6 References

[1] M. Karplus and R. N. Porter, *Atoms and Molecules*, (The Benjamin Cummings Publishing Company, Menlo Park, CA, 1970).

[2] W. Heitler and F. London, Z. Phys. **44**, 455 (1927).

[3] D. Loss and D.P. DiVincenzo, "Quantum computation with quantum dots", Phys. Rev. A, **57**, 120 (1998).

[4] M. J. Calderón, B. Keiller and S. Das Sarma, "Exchange coupling in semiconductor nanostructures: validity and limitations of the Heitler-London approach", Phys. Rev. B, **74**, 045310 (2006).

[5] D. Melnikov, J-P Leburton, A. Taha and N. Sobh, "Coulomb localization and exchange modulation in two electron coupled quantum dots", Phys. Rev. B, **74**, 041309(R) (2006); D. V. Melnikov and J-P Leburton, "Dimensionality effects in the two-electron system in circular and elliptic quantum dots", Phys. Rev. B, **73**, 085320 (2006).

[6] F. Bloch, "Remarks on the electron theory of ferromagnetism and of electrical conductivity", Z. Physik, **57**, 545 (1929).

[7] E. Ising, "Beitrag zur theorie des ferromagnetismus", Z. Phys., **31**, 253 (1925).

[8] L. Onsager, "Crystal statistics. I: A two-dimensional model with an order-disorder transition", Phys. Rev., Ser. II, **65**, 117 (1944).

11

Spin Transport in Solids

One of the most important topics in spintronics is spin transport, i.e., how do spin polarized carriers travel through a piece of metal and semiconductor and what happens to their spin polarizations in the presence of various spin relaxing mechanisms such as D'yakonov-Perel', Elliott-Yafet, Bir-Aronov-Pikus and hyperfine interactions with nuclear spins. This topic has been dealt with extensively in the literature leading to a hierarchy of *spin transport models*. The two primary models are the so-called *drift-diffusion model* and the more sophisticated *semi-classical model*. In this chapter, we briefly discuss these two models. There are of course other models that have been used in the literature, but these two are the most popular and have emerged as the mainstays of spintronic device simulation.

11.1 The drift-diffusion model

A rigorous derivation of the drift-diffusion equations that describe spin transport in a semiconductor or metal can be found in [1] which derives this equation starting from the density matrix and Wigner distribution function. That rigorous derivation is quite complex and the interested reader is referred to [1] for the details. Here, we first present the final results from that work, and follow that with a much simpler (easier to understand) derivation.

Let n_Σ be the density of carriers with spin Σ in a two-dimensional electron gas (2-DEG). The equation that describes the spatio-temporal evolution of n_Σ in a driving electric field $-E$ was shown in [1] to be

$$\frac{\partial n_\Sigma}{\partial t} - \mathbf{D_s}\frac{\partial^2 n_\Sigma}{\partial x^2} - \mathbf{A}\frac{\partial n_\Sigma}{\partial x} + \mathbf{B}n_\Sigma = 0, \tag{11.1}$$

where

$$\mathbf{D_s} = \begin{pmatrix} D_s & 0 & 0 \\ 0 & D_s & 0 \\ 0 & 0 & D_s \end{pmatrix}, \tag{11.2}$$

$$\mathbf{A} = \begin{pmatrix} \mu_s E & 2\beta_{xz}D_s & 0 \\ -2\beta_{xz}D_s & \mu_s E & 0 \\ 0 & 0 & \mu_s E \end{pmatrix}, \tag{11.3}$$

301

$$\mathbf{B} = \begin{pmatrix} D_s \left(\beta_{xz}^2 + \beta_{yz}^2 \right) & -\mu_s E \beta_{xz} & -\beta_{yx}\beta_{yz}D_s \\ \mu_s E \beta_{xz} & D_s \left(\beta_{xz}^2 + \beta_{yz}^2 + \beta_{yx}^2 \right) & 0 \\ -\beta_{yx}\beta_{yz}D_s & 0 & D_s\beta_{yx}^2 \end{pmatrix}, \qquad (11.4)$$

$$D_s = \frac{kT\tau_s}{m^*} \qquad \mu_s = -\frac{e\tau_s}{m^*}. \qquad (11.5)$$

The quantities β_{lk} are determined by spin-orbit interaction strengths, and τ_s is the spin relaxation time (determined by, among other things, the spin-orbit interaction strengths). Note that the spin diffusion coefficient D_s need not be equal to the charge diffusion coefficient D and the spin mobility μ_s need not be equal to the charge mobility μ since the spin relaxation time τ_s is generally not equal to the momentum relaxation time of charge carriers τ_m. Unfortunately, many published works in the spintronics literature have made the tacit assumption that $D = D_s$ and $\mu = \mu_s$. This is, of course, not correct, and any conclusions based on this assumption need to be carefully reexamined.

Saikin [1] has solved Equation (11.1) for the special case of spin injection in a two-dimensional electron gas residing in the $(x - y)$ plane. Electrons are injected with their spins polarized in the x-direction at $x = 0$. We reproduce those solutions here since we wish to highlight the difference between upstream transport (when carriers travel against the force exerted on them by the driving electric field) and downstream transport (carriers traveling along the force). The solutions of Equation (11.1) for this special case were found in [1] to be

$$n_{\Sigma_x}(x) = n_{\Sigma_x}^0 \exp\left[-\left(\frac{\mu E}{2D_s} + \sqrt{\left(\frac{\mu E}{2D_s} \right)^2 + B_{yz}^2} \right) x \right] \cos\left(B_{xz}x \right),$$

$$n_{\Sigma_y}(x) = n_{\Sigma_y}^0 \exp\left[-\left(\frac{\mu E}{2D_s} + \sqrt{\left(\frac{\mu E}{2D_s} \right)^2 + B_{yz}^2} \right) x \right] \sin\left(B_{xz}x \right),$$

$$n_{\Sigma_z}(x) = n_{\Sigma_z}^0 \exp\left[-\left(\frac{\mu E}{2D_s} + \left| \frac{\mu E}{2D_s} \right| \right) x \right], \qquad (11.6)$$

where $n_{\Sigma_x}^0$, $n_{\Sigma_y}^0$ and $n_{\Sigma_z}^0$ are three constants.

The above equation shows that the ensemble averaged x-component of the spin oscillates with distance x with an exponentially decaying envelope $exp(-x/L_x)$. Similarly, the ensemble averaged y-component of the spin oscillates with distance x with an exponentially decaying envelope $exp(-x/L_y)$. The ensemble averaged z-component has no oscillatory component and decays exponentially with decay constant L_z. The decay constants L_x, L_y, and L_z all depend on the driving electric field E. What is most interesting is that the decay constants depend on the *sign* of the electric field. If carriers are traveling "upstream" against the force exerted on them by the electric field, then the sign of E is positive and the decay lengths are much smaller than in the case when carriers are traveling "downstream" and the sign of E is negative. This means that spin polarization is randomized (or lost) much more rapidly in upstream transport compared to downstream transport. In fact, the z-component of the spin does not decay at all (only in this special case) if transport is downstream (since $L_z \to \infty$), but decays rapidly if transport is upstream.

Equation (11.1), albeit rigorously derived, is not particularly user-friendly since it involves dyadics (9-component tensors that are 3×3 matrices). A simpler (and less rigorous) form of the one-dimensional drift diffusion equation that involves only scalars is [2]

$$\frac{\partial n_\uparrow}{\partial t} = -\frac{n_\uparrow}{\tau_{\uparrow\downarrow}} + \frac{n_\downarrow}{\tau_{\downarrow\uparrow}} - \frac{\partial(n_\uparrow v_\uparrow)}{\partial x} + D_\uparrow \frac{\partial^2 n_\uparrow}{\partial x^2},$$

$$\frac{\partial n_\downarrow}{\partial t} = -\frac{n_\downarrow}{\tau_{\downarrow\uparrow}} + \frac{n_\uparrow}{\tau_{\uparrow\downarrow}} + \frac{n_\downarrow}{\tau_{\downarrow\uparrow}} - \frac{\partial(n_\downarrow v_\downarrow)}{\partial x} + D_\downarrow \frac{\partial^2 n_\downarrow}{\partial x^2}, \tag{11.7}$$

where the n-s are the concentrations, D-s are the diffusion coefficients, and v-s are the drift velocities of up-spin and down-spin carriers. The quantity $n_{uparrow}/\tau_{\uparrow\downarrow}$ is the rate at which the up-spin population is converting to the down-spin population via spin flips, and $n_\downarrow/\tau_{\downarrow\uparrow}$ is the rate at which the down-spin population is converting to up-spin. In general, these two rates need not be equal.

The above equation pictures spin transport as transport of two coupled channels, as shown in Figure 11.1. The up-spin and down-spin channels are coupled by spin flip processes described by the rates $1/\tau_{\uparrow\downarrow}$ and $1/\tau_{\downarrow\uparrow}$.

The disturbing feature of Equation (11.7) is that it deals with up-spin and down-spin electrons only. It really does not account correctly for superposition states which are superpositions of up-spin and down-spin. An electron can obviously exist in a superposition of up-spin and down-spin polarizations (e.g., if it is precessing about a magnetic field). Furthermore, the concept of spin flipping from up to down, or vice versa, cannot account for spin relaxation via mechanisms such as the D'yakonov-Perel' mode, since that mechanism is not a discrete spin flip event in time. It comes about because of asynchronous spin precession about the pseudo magnetic fields caused by spin-orbit interactions. Such a process, which is continuous in time, cannot be described satisfactorily by a relaxation time involving discrete transitions between up-spin and down-spin states. The reader should bear in mind these issues when dealing with equations such as Equation (11.7).

Equation (11.7) is reminiscent of the standard drift-diffusion equations of charge transport [3] used to model bipolar transport (transport of both electrons and holes) in a semiconductor. The up-spin and down-spin electrons play the roles of electrons and holes in bipolar transport. The spin flip rates $1/\tau_{\uparrow\downarrow}$ and $1/\tau_{\downarrow\uparrow}$ act like electron-hole recombination rates. In the following

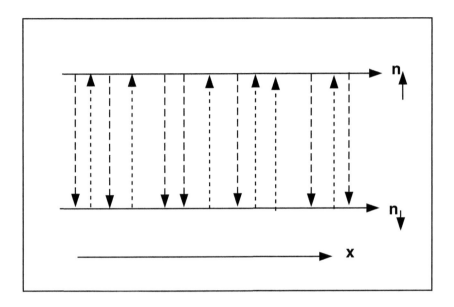

FIGURE 11.1

Pictorial depiction of the simplest spin drift-diffusion model. The up-spin and down-spin channels are coupled by spin flip processes. The long dashed vertical arrows depict spin flip processes that convert an up-spin to a down-spin and the short dashed vertical arrows depict the opposite process of flipping from down-spin to up-spin.

subsection, we derive the steady-state version of Equation (11.7) by analogy with bipolar charge transport.

11.1.1 Derivation of the simplified steady-state spin drift-diffusion equation

Analogous to the general current density expressions for electrons and holes, which include drift and diffusion components, the spin-polarized current densities of up-spin and down-spin populations are written as

$$J_\uparrow = n_\uparrow e \mu_\uparrow E + eD_\uparrow \frac{dn_\uparrow}{dx}, \tag{11.8}$$

and

$$J_\downarrow = n_\downarrow e \mu_\downarrow E + eD_\downarrow \frac{dn_\downarrow}{dx}, \tag{11.9}$$

where the first and second terms in each equation above are drift and diffusion components, respectively. The driving electric field is E, the electron charge is e, and the mobilities of up-spin and down-spin populations are $\mu_\uparrow$ and $\mu_\downarrow$, respectively.

The "continuity equations" for the up-spin and down-spin electron densities are given by

$$\frac{\partial n_\uparrow}{\partial t} = -\frac{n_\uparrow}{\tau_{\uparrow\downarrow}} + \frac{n_\downarrow}{\tau_{\downarrow\uparrow}} + \frac{1}{e}\frac{d}{dx}J_\uparrow, \tag{11.10}$$

$$\frac{\partial n_\downarrow}{\partial t} = -\frac{n_\downarrow}{\tau_{\downarrow\uparrow}} + \frac{n_\uparrow}{\tau_{\uparrow\downarrow}} + \frac{1}{e}\frac{d}{dx}J_\downarrow, \tag{11.11}$$

which includes in- and out-scattering rates for both populations, by analogy with electron-hole generation and recombination rates in the description of bipolar charge transport in semiconductors. Note that the above two equations are the same as Equation (11.7) if, and only if, $\tau_{\uparrow\downarrow} = \tau_{\downarrow\uparrow}$.

In steady state, the last two equations become

$$\frac{1}{e}\frac{dJ_\uparrow}{dx} - \frac{n_\uparrow}{\tau_{\uparrow\downarrow}} + \frac{n_\downarrow}{\tau_{\downarrow\uparrow}} = 0, \tag{11.12}$$

$$\frac{1}{e}\frac{dJ_\downarrow}{dx} - \frac{n_\downarrow}{\tau_{\downarrow\uparrow}} + \frac{n_\uparrow}{\tau_{\uparrow\downarrow}} = 0. \tag{11.13}$$

In the continuity equations, charge recombination processes are neglected since we consider unipolar spin transport.

Inserting the current density expressions (11.8) and (11.9) in these two last equations, we get

$$(\frac{d}{dx}\sigma_\uparrow)E + \sigma_\uparrow(\frac{dE}{dx}) + eD_\uparrow\frac{d^2 n_\uparrow}{dx^2} = e(\frac{n_\uparrow}{\tau_{\uparrow\downarrow}} - \frac{n_\downarrow}{\tau_{\downarrow\uparrow}}), \tag{11.14}$$

and

$$\left(\frac{d}{dx}\sigma_\downarrow\right)E + \sigma_\downarrow\left(\frac{dE}{dx}\right) + eD_\downarrow\frac{d^2 n_\downarrow}{dx^2} = e\left(\frac{n_\downarrow}{\tau_{\downarrow\uparrow}} - \frac{n_\uparrow}{\tau_{\uparrow\downarrow}}\right), \tag{11.15}$$

where $\sigma_{\uparrow(\downarrow)} = n_{\uparrow(\downarrow)}e\mu_{\uparrow(\downarrow)}$ are the up-spin (down-spin) conductivities.

Multiplying both sides of Equation (11.14) by $\sigma_\downarrow$ and both sides of (11.15) by $\sigma_\uparrow$ and subtracting, we get

$$X + Y = Z, \tag{11.16}$$

where

$$X = \left(\sigma_\downarrow\frac{d}{dx}\sigma_\uparrow - \sigma_\uparrow\frac{d}{dx}\sigma_\downarrow\right)E, \tag{11.17}$$

$$Y = eD_\uparrow\sigma_\downarrow\frac{d^2 n_\uparrow}{dx^2} - eD_\downarrow\sigma_\uparrow\frac{d^2 n_\downarrow}{dx^2}, \tag{11.18}$$

and

$$Z = en_\uparrow\left(\frac{\sigma_\downarrow}{\tau_{\uparrow\downarrow}} + \frac{\sigma_\uparrow}{\tau_{\uparrow\downarrow}}\right) - en_\downarrow\left(\frac{\sigma_\uparrow}{\tau_{\downarrow\uparrow}} + \frac{\sigma_\downarrow}{\tau_{\downarrow\uparrow}}\right). \tag{11.19}$$

With the use of the conductivity expressions above, the expression for X becomes

$$X = \left(e\sigma_\downarrow\mu_\uparrow\frac{dn_\uparrow}{dx} - e\sigma_\uparrow\mu_\downarrow\frac{dn_\downarrow}{dx}\right)E. \tag{11.20}$$

For a homogeneous system, in which space-charge build up is neglected, any local electron charge due to $n_\uparrow(x)+n_\downarrow(x)$ must be balanced by a corresponding local hole charge. This does not preclude a net *spin polarization* of carriers, i.e., $n_\uparrow - n_\downarrow \neq 0$. In other words, spin polarization can be created without changing electron or hole densities. Charge conservation in unipolar materials dictates

$$\frac{d}{dt}(n_\uparrow + n_\downarrow) = 0, \tag{11.21}$$

but

$$\frac{d}{dt}(n_\uparrow - n_\downarrow) \neq 0. \tag{11.22}$$

If we rewrite the up-spin and down-spin electron densities as

$$n_\uparrow = \left(\frac{n_\uparrow + n_\downarrow}{2}\right) + \left(\frac{n_\uparrow - n_\downarrow}{2}\right), \tag{11.23}$$

and

$$n_\downarrow = \left(\frac{n_\uparrow + n_\downarrow}{2}\right) - \left(\frac{n_\uparrow - n_\downarrow}{2}\right), \tag{11.24}$$

then the expression for X becomes

$$X = \left[e\sigma_\downarrow\mu_\uparrow\frac{d}{dx}\left(\frac{n_\uparrow - n_\downarrow}{2}\right) + e\sigma_\uparrow\mu_\downarrow\frac{d}{dx}\left(\frac{n_\uparrow - n_\downarrow}{2}\right)\right]E \tag{11.25}$$

or

$$X = \left[\frac{e}{2}(\sigma_\uparrow \mu_\uparrow + \sigma_\uparrow \mu_\downarrow)\frac{d}{dx}(n_\uparrow - n_\downarrow)\right]E. \tag{11.26}$$

The expression for Y is

$$Y = e(D_\downarrow \sigma_\uparrow + D_\uparrow \sigma_\downarrow)\frac{d^2}{dx^2}(\frac{n_\uparrow - n_\downarrow}{2}), \tag{11.27}$$

while the expression for Z is

$$Z = e(\frac{n_\uparrow - n_\downarrow}{2})(\sigma_\downarrow + \sigma_\uparrow)\frac{1}{\tau_{\uparrow\downarrow}} - e(\frac{n_\downarrow - n_\uparrow}{2})(\sigma_\downarrow + \sigma_\uparrow)\frac{1}{\tau_{\downarrow\uparrow}} \tag{11.28}$$

or

$$Z = e(\frac{n_\uparrow - n_\downarrow}{2})(\sigma_\uparrow + \sigma_\downarrow)(\frac{1}{\tau_{\uparrow\downarrow}} + \frac{1}{\tau_{\downarrow\uparrow}}). \tag{11.29}$$

Substituting the expressions for X, Y, and Z in Equation (11.16), we get

$$\frac{eE}{2}(\sigma_\uparrow \mu_\downarrow + \sigma_\downarrow \mu_\uparrow)\frac{d}{dx}(n_\uparrow - n_\downarrow) + \frac{e}{2}(D_\uparrow \sigma_\downarrow + D_\downarrow \sigma_\uparrow)\frac{d^2}{dx^2}(n_\uparrow - n_\downarrow)$$
$$= \frac{e}{2}(n_\uparrow - n_\downarrow)(\sigma_\uparrow + \sigma_\downarrow)\frac{1}{\tau_s}, \tag{11.30}$$

where we have introduced the average spin relaxation time

$$\frac{1}{\tau_s} = \frac{1}{\tau_{\uparrow\downarrow}} + \frac{1}{\tau_{\downarrow\uparrow}}. \tag{11.31}$$

By introducing the effective mobility

$$\mu_{eff} = \frac{\sigma_\uparrow \mu_\downarrow + \sigma_\downarrow \mu_\uparrow}{\sigma_\uparrow + \sigma_\downarrow} \tag{11.32}$$

and the effective diffusion constant

$$D_{eff} = \frac{\sigma_\uparrow D_\downarrow + \sigma_\downarrow D_\uparrow}{\sigma_\uparrow + \sigma_\downarrow} \tag{11.33}$$

and the intrinsic spin diffusion length

$$L_s = \sqrt{D_{eff}\tau_s}, \tag{11.34}$$

Equation (11.30) can be written in the compact form

$$\frac{d^2}{dx^2}(n_\uparrow - n_\downarrow) + \frac{eE\mu_{eff}}{eD_{eff}}\frac{d}{dx}(n_\uparrow - n_\downarrow) - \frac{(n_\uparrow - n_\downarrow)}{L_s^2} = 0, \tag{11.35}$$

which should be compared with the *steady state* version of Equation (11.7).

We search for general solutions to Equation (11.35) of the form

$$n_\uparrow - n_\downarrow = Ae^{-\lambda x} \tag{11.36}$$

corresponding to spin polarization decaying exponentially in space.

Substitution of this solution in Equation (11.35) yields

$$\lambda^2 - \lambda \left(\frac{\mu_{eff} E}{D_{eff}} \right) - \frac{1}{L_s^2} = 0 \tag{11.37}$$

or

$$\lambda = \frac{\mu_{eff} E}{2D_{eff}} \pm \sqrt{\left(\frac{\mu_{eff} e}{2D_{eff}} \right)^2 + \frac{1}{L_s^2}}. \tag{11.38}$$

We want to avoid solutions which blow up at $x = \pm\infty$ since these are unstable and inadmissible. Suppose a continuous spin imbalance $(n_\uparrow - n_\downarrow)|_o$ is injected at $x = 0$, and the electric field is along the $-x$ direction. Since the spin polarization must gradually decay with distance from the point of injection, the distribution of the spin polarization is given by

$$(n_\uparrow - n_\downarrow)(x) = (n_\uparrow - n_\downarrow)|_o e^{-\lambda x}, \tag{11.39}$$

where $(n_\uparrow - n_\downarrow)|_o$ is the spin polarization at the point of injection at $x = 0$.

For "downstream transport," when electrons are traveling in the direction opposite to the electric field and are being accelerated by the field, we have for $x > 0$

$$(n_\uparrow - n_\downarrow)(x) = (n_\uparrow - n_\downarrow)|_o e^{-\frac{x}{L_d}}, \tag{11.40}$$

where

$$L_d = \left[-\frac{|E|}{2} \frac{\mu_{eff}}{D_{eff}} + \sqrt{\left(\frac{|E|}{2} \frac{\mu_{eff}}{D_{eff}} \right)^2 + \frac{1}{L_s^2}} \right]^{-1}. \tag{11.41}$$

For "upstream transport," when electrons are traveling in the direction of the electric field and are hence decelerated by the field, we have for $x < 0$

$$(n_\uparrow - n_\downarrow)(x) = (n_\uparrow - n_\downarrow)|_o e^{\frac{x}{L_u}}, \tag{11.42}$$

where

$$L_u = \left[\frac{|E|}{2} \frac{\mu_{eff}}{D_{eff}} + \sqrt{\left(\frac{|E|}{2} \frac{\mu_{eff}}{eD_{eff}} \right)^2 + \frac{1}{L^2}} \right]^{-1}. \tag{11.43}$$

These expressions should be compared with the expressions in Equation (11.6).

Furthermore, we have

$$L_u L_d = L_s^2. \tag{11.44}$$

11.2 The semiclassical model

The drift-diffusion model just described has several intrinsic shortcomings. The first is that it is obviously incapable of describing any non-linear effects since the drift-diffusion equations are linear differential equations. Non-linearities in spin transport have been observed experimentally [4, 5] and therefore are a reality. The second is that the drift-diffusion model cannot handle superposition of up-spin and down-spin states satisfactorily [6]. As a result, it cannot treat coherence effects arising from superposition of up-spin and down-spin states. Recently, these superpositions were treated in a Bloch equation approach [7] and a generalized drift-diffusion type approach derived from the Boltzmann transport equation [8]. These are beyond the normal drift-diffusion formalism and are not discussed in this book. The interested reader can peruse the pertinent references. Third, the drift-diffusion formalism cannot handle hot electron effects in spin transport. These are important in many devices, such as the Spin Valve Transistor discussed in Chapter 14. Finally, the normal drift-diffusion formalism is not capable of incorporating spin relaxation via the D'yakonov-Perel' mechanism (as stated before) since that mechanism does not involve discrete spin flip events in time and therefore cannot be captured by a simple rate such as $1/\tau_{\uparrow\downarrow}$ depicting flipping from an up-spin state to a down-spin state. The D'yakonov-Perel' (D-P) relaxation is a continuous process in time. The D-P relaxation also has complex connections with momentum relaxation since the pseudo-magnetic fields associated with spin-orbit interactions – which cause the D-P relaxation – are velocity dependent (recall Chapter 8). Therefore, such a process cannot be described by a constant relaxation rate. Any model capable of handling the D-P process must be able to treat the momentum dependence of the spin-orbit coupling (the primary cause of D-P relaxation) self-consistently.

In reality, the temporal evolution of spin and the temporal evolution of the momentum of an electron cannot be separated when there is transport. A moment's reflection will convince the reader that the D-P spin relaxation (or depolarization) rates must be functionals of the electron distribution function in momentum space, which continuously evolves with time when an electric field is applied to drive transport and there is random carrier scattering. Thus, the D-P relaxation rate is not a constant but is a dynamic variable. To account for all this, one must self-consistently solve the Boltzmann Transport equation describing the spatio temporal evolution of the electron distribution function and the Liouville equation describing the spatio temporal evolution of the spin density matrix. The Liouville equation will contain a momentum dependent damping term; hence, the need to simultaneously solve the Boltzmann equation and the Liouville equation. Such situations are best treated by Monte Carlo simulation, which has been recently adopted by a number of groups to study spin transport in quasi-two-dimensional structures (see, for example, [6, 9] among many).

11.2.1   Spin transport in a quantum wire: Monte Carlo simulation

In this subsection, we will provide the basics of Monte Carlo methods. Consider the quasi one-dimensional semiconductor structure of Fig. 8.4. An electric field $\mathcal{E}_\S$ is applied along the axis of the quantum wire to induce charge flow. In addition, there is another transverse electric field E_y that causes Rashba spin-orbit interaction.

Following Saikin et al. [6], we treat the spin using the standard spin density matrix

$$\rho_\sigma(t) = \begin{bmatrix} \rho_{\uparrow\uparrow}(t) & \rho_{\uparrow\downarrow}(t) \\ \rho_{\downarrow\uparrow}(t) & \rho_{\downarrow\downarrow}(t) \end{bmatrix} \quad , \tag{11.45}$$

which is related to the spin polarization component as $S_n(t) = Tr(\sigma_n\rho_\sigma(t))$ $(n = x, y, z)$. Over a small time interval δt (much smaller than the mean time between collisions), we will assume that no scattering takes place and that the electron's momentum changes slowly enough (in other words $\mathcal{E}_\S \delta t$ is small enough) that transport can be described by a constant (time-independent) momentum or velocity within this time interval.* We take this velocity to be the average of the velocity at the beginning and end of the time interval:

$$v = (v_{initial} + v_{final})/2 = v_{initial} + (q\mathcal{E}_\S \delta t)/(2m^*). \tag{11.46}$$

During this interval, the spin density matrix undergoes a unitary evolution according to

$$\rho_\sigma(t + \delta t) = e^{-iH_{so}(k)\delta t/\hbar}\rho_\sigma(t)e^{iH_{so}(k)\delta t/\hbar} \quad , \tag{11.47}$$

where $H_{so}(k)$ is the wavevector dependent spin-orbit interaction Hamiltonian.

Equation (11.47) describes the coherent rotation or precession of the spin vector about an effective magnetic field (or pseudo magnetic field) $\vec{B}_{so}$ caused by spin-orbit interaction. This spin precession is described by the equation

$$\frac{d\vec{S}}{dt} = \vec{\Omega} \times \vec{S},$$

$$\vec{\Omega} = \frac{g\mu_B\vec{B}_{so}}{\hbar}. \tag{11.48}$$

The reader should be familiar with this notion because it was discussed in Chapter 8, and is the basis of D-P relaxation. The strength of the pseudo magnetic field is proportional to the electron's velocity (recall Equation (8.23)),

---

*Within such an interval, the electron's momentum changes only because of the electric field accelerating the electron.

which changes with time. Within a time interval δt, the strength of the pseudo-magnetic field is determined by the magnitude of the average velocity during that time interval. Note that, during this time interval, the spin dynamics is coherent and there is no "relaxation" since the evolution is *unitary*. However, there are two agents that ultimately cause relaxation via the D-P mechanism. The first is the electric field $\mathcal{E}_§$, which changes the "average velocity" from one time interval δt to the next, thereby changing the effective pseudo magnetic field that an electron sees. The second is stochastic scattering, which changes the "average velocity" between two successive intervals (separated by a scattering event) *randomly*. These two causative agents produce a distribution of spin states that results in effective relaxation. The evolution of the spin polarization vector $\mathbf{S}$ ($= S_x \mathbf{u}_x + S_y \mathbf{u}_y + S_z \mathbf{u}_z$; where $\mathbf{u}_n$ is the unit vector along the n-direction) can be viewed as coherent motion (rotation) and relaxation/depolarization (reduction in magnitude). The coherent dynamics causes the spin to oscillate and the relaxation causes the amplitude of the oscillation to decay in time. This type of relaxation is the D'yakonov-Perel' relaxation.

In Monte Carlo models, one assumes that, over the short time interval δt, the electron's velocity is time-invariant and given by the average velocity in Equation (11.46). Consequently, the spin precession frequencies Ω are constant and independent of time in the interval δt. Accordingly, the solution of Equation (11.48) for the spin components yields

$$S_x(t + \delta t) = -\frac{\Omega_R}{\Omega_T} \left[S_y(t)sin(\Omega_T \delta t) + \left(\frac{\Omega_D}{\Omega_T} S_z(t) - \frac{\Omega_R}{\Omega_T} S_x(t) \right) cos(\Omega_T \delta t) \right]$$
$$+ \left(\frac{\Omega_D}{\Omega_T} \right)^2 S_x(t) + \frac{\Omega_D \Omega_R}{\Omega_T^2} S_z(t),$$

$$S_y(t + \delta t) = S_y(t)cos(\Omega_T \delta t) + \left(\frac{\Omega_R}{\Omega_T} S_x(t) - \frac{\Omega_D}{\Omega_T} S_z(t) \right) sin(\Omega_T \delta t),$$

$$S_z(t + \delta t) = \frac{\Omega_D}{\Omega_T} \left[S_y(t)sin(\Omega_T \delta t) + \left(\frac{\Omega_D}{\Omega_T} S_z(t) - \frac{\Omega_R}{\Omega_T} S_x(t) \right) cos(\Omega_T \delta t) \right]$$
$$+ \left(\frac{\Omega_R}{\Omega_T} \right)^2 S_z(t) + \frac{\Omega_D \Omega_R}{\Omega_T^2} S_x(t), \tag{11.49}$$

where Ω_R and Ω_D are given in Equation (8.49) for a quantum wire and $\Omega_T = \sqrt{\Omega_R^2 + \Omega_D^2}$. All Ω-s are calculated at the average value of the velocities in the interval δt given by Equation (11.46). It is easy to verify from the above equations that spin is conserved for every individual electron, i.e.,

$$S_x^2(t + \delta t) + S_y^2(t + \delta t) + S_z^2(t + \delta t) = S_x^2(t) + S_y^2(t) + S_z^2(t) = |\mathbf{S}|. \tag{11.50}$$

11.2.2 Monte Carlo simulation

The average value of the velocity in an interval δt depends on the initial value of the wavevector $v_{initial}$ at the beginning of the interval. Intervals are chosen such that a scattering event can occur only at the beginning or end of an interval. The choice of scattering events and the velocity state after the event (i.e., the time evolution of the velocity) are found from a Monte Carlo solution of the Boltzmann Transport equation in a quantum wire [10, 11, 12].

Equations (11.49) are solved directly in the Monte Carlo simulator. One typically uses a time step δt of few femtoseconds and a $1{,}000 - 10{,}000$ electron ensemble to collect spin statistics. The Monte Carlo solution yields the spin components $S_x(t_n)$, $S_y(t_n)$, and $S_z(t_n)$ of every one of the electrons at every sampled instant of time t_n. From this distribution, one calculates how the magnitude of the ensemble averaged spin vector $| < \mathbf{S} > |$ decays with time. The characteristic time constant of this decay is sometimes interpreted as the spin relaxation time associated with D'yakonov-Perel' relaxation.

11.2.3 Specific examples: Temporal decay of spin polarization

In Fig. 11.2, we show the results of a specific Monte Carlo simulation. Here we study the time evolution of an ensemble of spins in a quantum wire of width 30 nm and thickness 4 nm (refer to Fig. 8.4). The electrons travel along the axis of the wire under the influence of a driving electric field, suffer random scattering events due to phonons, while all the time precessing about the pseudo magnetic fields due to spin-orbit interaction. The symmetry-breaking electric field E_y inducing the Rashba effect was assumed to be 100 kV/cm and the material was GaAs. Spins were injected from the left contact $(x = 0)$ at time $t = 0$, polarized in the x-direction (along the wire axis). The lattice temperature was 30 K. Electron scattering due to confined optical phonons and bulk acoustic phonons was considered.

Fig. 11.2 shows how the magnitude of the average (ensemble averaged over all electrons) spin vector $\mathbf{S}$ decays with time for four different values of the driving electric field $\mathcal{E}_\S$ applied along the axis of the wire. This field drives transport. Only D-P relaxation is considered here since it is known to be dominant over Elliott-Yafet or any other mechanism in semiconductor structures at the temperatures considered.

As expected, the decay time decreases with increasing electric field. There are two contributing factors for this trend. First, the average velocity v given by Equation (11.46) increases with increasing electric field $\mathcal{E}_\S$. Hence the spins precess more rapidly since Ω_T increases with increasing v. Now consider the random effect of scattering which results in a different initial velocity $v_{initial}$ for different electrons in a given time interval. This results in a larger standard deviation in v at higher electric fields $\mathcal{E}_\S$ (and hence different Ω-s, or different precession rates) for different electrons in the same time interval. Ensemble averaging over the electrons therefore causes the magnitude of the spin vector $\mathbf{S}$ to decay more rapidly, resulting in faster spin relaxation (or depolarization). The second factor that contributes to this trend is that the frequency of scattering itself increases with increasing electric field. Scattering randomizes the Ω-s since it randomizes v. This also results in faster relaxation at stronger electric fields.

At first glance, it may be troubling to assimilate the fact that the magnitude of the ensemble averaged $\mathbf{S}$ $(= | < \mathbf{S} > |)$ can decay with time. One should remember that the magnitude of $\mathbf{S}$ $(= |\mathbf{S}|)$ is conserved only for an *individual* electron, as shown by Equation (11.50), but the magnitude is not invariant when we ensemble average over *many* electrons. It is this *ensemble averaging* that results in effective spin relaxation, as mentioned before.

Decay of spin components in time

In Fig. 11.3 we show how the average (ensemble averaged over all electrons) x-, y- and z-components of the spin decay with time. The driving electric field $\mathcal{E}_\S = 2\,\text{kV/cm}$ and the lattice temperature is 30 K. Since, initially, the spin was polarized along the x-direction, the ensemble averaged y- and z-components remain near zero and the ensemble averaged x-component decays with time. The decay of the ensemble averaged x-component $< S_x >$ is indistinguishable from that of $| < \mathbf{S} > |$ in this case.

11.2.4 Specific examples: Spatial decay of spin polarization

To study *spatial* decay of spin polarization (as opposed to *temporal* decay), Monte Carlo simulations are carried out where carriers are injected into a structure with a fixed spin polarization and then their spins are tracked in space as they transport under the influence of a driving electric field. We present here the example of a GaAs quantum wire of rectangular cross section 30 nm × 4 nm. We assume that the left contact is an ideal half metal magnetized along the x-direction that injects spins into the quantum wire with 100% efficiency. Therefore, electrons are injected into the quantum wire with injection velocities picked from a Fermi-Dirac distribution, but all spins are aligned along the channel axis (x-axis) at $x = 0$ in order to simulate 100% efficient spin injection from the half metallic ferromagnetic contact.

In Fig. 11.4, we show the spatial decay of the ensemble averaged spin $| < \mathbf{S} > |$ for two different temperatures of 30 K and 77 K. The spin decays faster at the higher temperature because of increased phonon scattering. If we (heuristically) define the spin relaxation length L_s as the distance it takes for $| < \mathbf{S} > |$ to decay to $1/e$ times its initial value, then the relaxation length is quite large – greater than 10 μm at 77 K and 20 μm at 30 K.

11.2.5 Upstream transport

In Figs. 11.2, 11.3, and 11.4, we present results for *downstream* spin transport. Note that the decays of $| < \mathbf{S} > |$ in both time and space are approximately exponential, so that the drift-diffusion model is at least qualitatively admissible, even if it is not quantitatively correct. We now present results for *upstream* transport and show that the drift-diffusion model completely breaks down

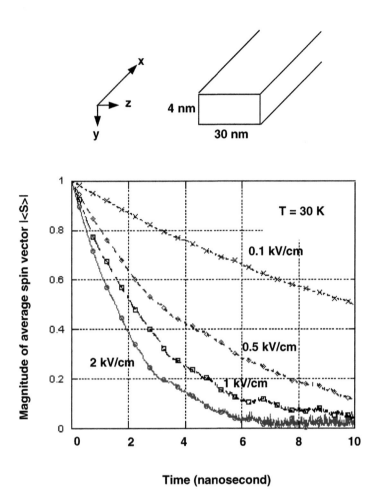

FIGURE 11.2

Temporal decay of the ensemble average spin vector with time in a GaAs quantum wire of dimension 4 nm × 30 nm at a lattice temperature of 30 K. The results are shown for various driving electric fields $\mathcal{E}_\S$. The spins are injected with their polarization initially aligned along the wire axis. At the top, we show the geometry of the quantum wire. Reprinted with permission from S. Pramanik, S. Bandyopadhyay and M. Cahay, Phys. Rev. B, **68**, 075313 (2003). Copyright (2003) American Physical Society. http://link.aps.org/abstract/PRB/v68/p075313.

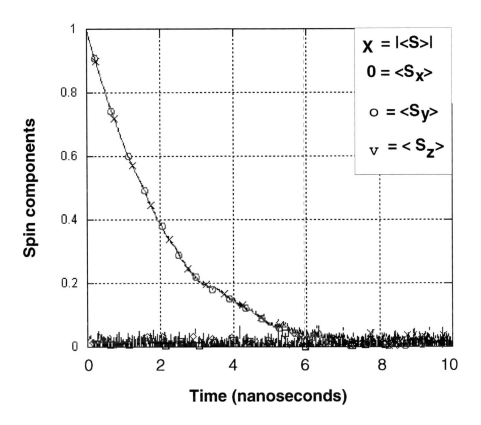

FIGURE 11.3

Temporal decay of the x-, y- and z-components of spin in the same GaAs wire at 30 K. The driving electric field is 2 kV/cm and the spins are injected with their polarization initially aligned along the wire axis (x-axis). The y- and z-components of the spin remain zero throughout. Reprinted with permission from S. Pramanik, S. Bandyopadhyay and M. Cahay, Phys. Rev. B, **68**, 075313 (2003). Copyright (2003) American Physical Society. http://link.aps.org/abstract/PRB/v68/p075313.

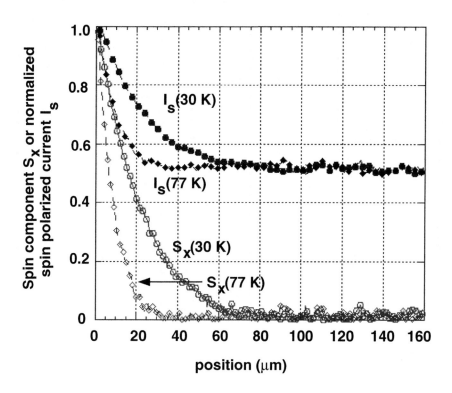

FIGURE 11.4

Spatial decay of the ensemble average spin vector in a GaAs quantum wire of cross-section 4 nm × 30 nm at lattice temperatures of 30 K and 77 K. The spins are injected into the quantum wire at $x = 0$ from a perfect spin injector with their polarization initially aligned along the wire axis. The driving electric field is 2 kV/cm. Also shown is the current I_s transmitted by a spin detector that is magnetized in the same direction (parallel) as the spin injector. When the spins are completely randomized, I_s will drop to 50% of its initial value. Reprinted with permission from S. Pramanik, S. Bandyopadhyay and M. Cahay, Applied Physics Letters, **84**, 266 (2004). Copyright (2004) American Institute of Physics.

in this case. For upstream motion, the decay of spin in space is not even monotonic, let alone exponential. Such dynamics cannot be described even qualitatively by the drift-diffusion model.

In Fig. 11.5, we show the spatial evolution of the ensemble averaged spin $| <S> |$ of upstream electrons that were injected into a quantum wire of cross section 30 nm × 4 nm against the force exerted on them by a driving electric field of 2 kV/cm. The injection energy is such that an injected electron can access the three lowest transverse subbands in the wire. Initially, each subband is populated equally, i.e., an injected electron has equal probability of going into any of the three subbands. A constant stream of injected electrons is maintained by continued injection to supplant those that exhaust their initial injection energies traveling upstream and finally turn around to become downstream travelers. The wire has Rashba and Dresselhaus interactions (exactly similar to the wire of Figs. 11.2, 11.3, or 11.4). We consider only D-P relaxation.

Note that, at first, $| <S> |$ decays as electrons travel upstream. But then $| <S> |$ recovers, and finally climbs up to 1 and settles down there with no further change! Thus, we do not see a monotonic decay unlike in the case of downstream transport. What has happened is that two out of the three subbands have been depopulated of upstream electrons after they travel some distance. Surprisingly, the lowest two subbands depopulate and all the upstream electrons end up in the highest subband. This strange "population inversion" that happens in upstream transport has been explained in [13]. When a single subband is occupied in a quantum wire (in this case the highest subband), there can be no further D-P relaxation as was shown in Problem 8.3. That is the reason $| <S> |$ saturates at 1. Of course, if we included the Elliott-Yafet mechanism, then there would have been some additional spin relaxation and $| <S> |$ would be less than unity. Nonetheless, since the D-P relaxation is dominant, we would still most likely have seen the non-monotonic spin evolution in space. This behavior is completely beyond the drift-diffusion model.

11.3 Concluding remarks

In this chapter, we have introduced two models that are used to describe spin transport in solids. The drift-diffusion model is simple, user friendly, but often inadequate and inaccurate, particularly in upstream transport, where it breaks down completely. A more advanced model is the semi-classical formalism based on Monte Carlo simulation which captures many of the subtle features that are beyond the drift-diffusion model. However, it is computationally much more demanding. The Monte Carlo technique has currently (at the time

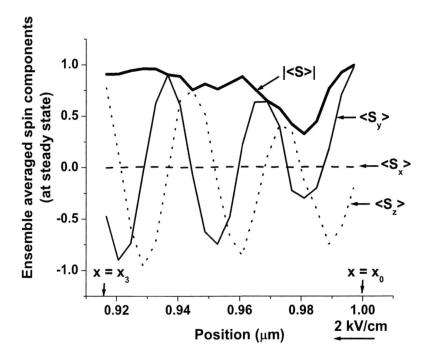

FIGURE 11.5

Spatial variation of ensemble averaged spin components for driving electric field $E_x = 2\text{kV/cm}$ at steady state. Transport is upstream. Lattice temperature is 30 K and injection energy $E_0 = 426$ meV. Electrons are injected with velocities against the force exerted on them by the driving electric field. They are injected equally into the three lowest subbands, but ultimately end up in the highest subband (subband 3). This peculiar population inversion, which is very counter-intuitive and well beyond the drift-diffusion model, has been explained in Phys. Rev. B, **73**, 125309 (2006). The classical turning point of electrons traveling upstream in the highest subband (subband 3) is denoted by x_3 and x_{scat} indicates the point along the channel axis where subbands 1 and 2 get virtually depopulated. The classical turning point is the point where injected electrons have exhausted all the injection energy by traveling against the electric field, and therefore must turn around to travel in the direction of the force exerted by the electric field. A few electrons are found beyond the classical turning point since backscattering can push them beyond that point. This feature, too, is beyond the drift-diffusion picture. Injected electrons are $\hat{y}$ polarized and $x = x_0 = 1$ μm is the point of injection. Reprinted with permission from S. Pramanik, S. Bandyopadhyay and M. Cahay, Phys. Rev. B, **73**, 125309 (2006). Copyright (2006) American Physical Society. http://link.aps.org/abstract/PRB/v73/p125309.

of writing this book) become quite popular and is widely used for spintronic device simulation (see, for example, [6]). There are, of course, other models for simulating spin transport. This is a rapidly developing field and new models, or old models with new twists, appear in the literature frequently. The reader may scour the literature to find the other relevant models that have been proposed and developed to describe propagation of spins in solids. They all have their strengths and weaknesses.

11.4   Problems

- **Problem 11.1**

  Consider a unipolar material and neglect space charge build up. Furthermore assume that $D_\uparrow = D_\downarrow$. Show using the simple drift-diffusion equation (11.7) that the total drift current $J_{drift} = e\left(n_\uparrow v_\uparrow + n_\downarrow v_\downarrow\right)$ is spatially invariant, i.e., $\partial J_{drift}/\partial x = 0$ in the steady state.

  If $D_\uparrow \neq D_\downarrow$, then under what circumstance will J_{drift} be spatially invariant?

- **Problem 11.2**

  Consider a spatially homogeneous system in which $n_\uparrow/\tau_{\uparrow\downarrow} = n_\downarrow/\tau_{\downarrow\uparrow}$. Using the drift-diffusion equation (11.7), show that the spin polarization density $n_\uparrow - n_\downarrow$ cannot change with time.

11.5   References

[1] S. Saikin, "A drift diffusion model for spin polarized transport in a two-dimensional non-degenerate electron gas controlled by spin-orbit interaction", J. Phys: Condens. Matt., **16**, 5-71 (2004).

[2] Y. Qi, Z. G. Yu and M. E. Flatte, "Spin Gunn Effect", Phys. Rev. Lett., **96**, 026602 (2006); Z. G. Yu and M. E. Flatté, "Electric field dependent spin diffusion and spin injection into semiconductors", Phys. Rev. B, **66**, 201202 (2002).

[3] R. F. Pierret, *Semiconductor Fundamentals*, Vol. 1, Purdue Modular Series on Solid State Devices, 2nd edition, (Addison-Wesley, Reading, MA, 1989).

[4] G. Schmidt, C. Gould, P. Grabs, A. M. Lunde, G. Richter, A. Slobodsky and L. W. Molenkamp, "Spin injection in the non-linear regime: band bending effects", www.arXiv.org/cond-mat/0206347.

[5] H. Sanada, Y. Arata, Y. Ohno, Z. Chen, K. Kayanuma, Y. Oka, F. Matsukura and H. Ohno, "Relaxation of photo-injected spins during drift transport in GaAs", Appl. Phys. Lett., **81**, 2788 (2002).

[6] S. Saikin, M. Shen, M. C. Cheng and V. Privman, "Semiclassical model for in-plane transport of spin polarized electrons in III-V heterostructures", www.arXiv.org/cond-mat/0212610; M. Chen, S. Saikin, M. C. Cheng and V. Privman, "Monte Carlo modeling of spin polarized transport", www.arXiv.org/cond-mat/0302395; "Monte Carlo modeling of Spin FETs controlled by spin-orbit interaction", www.arXiv.org/cond-mat/0309118; M. Shen, S. Saikin and M. C. Cheng, "Monte Carlo modeling of spin injection through a Schottky barrier and spin transport in a semiconductor quantum well", www.arXiv.org/cond-mat/0405270.

[7] M. Q. Weng and M. W. Wu, "Kinetic theory of spin transport in n-type semiconductor quantum wells", J. Appl. Phys., **93**, 410 (2003).

[8] Y. Qi and S. Zhang, "Spin diffusion at finite electric and magnetic fields", Phys. Rev. B., **67**, 052407 (2003).

[9] A. Bournel, P. Dollfus, S. Galdin, F-X Musalem and P. Hesto, "Modelling of gate induced spin precession in a striped channel high electron mobility transistor", Solid State Commun., **104**, 85 (1997); A. Bournel, V. Delmouly, P. Dollfus, G. Tremblay and P. Hesto, "Theoretical and experimental considerations on the spin field effect transistor", Physica E, **10**, 86 (2001).

[10] D. Jovanovich and J-P Leburton, in *Monte Carlo Device Simulation: Full Band and Beyond*, Ed. K. Hess, (Kluwer Academic, Boston, 1991) pp. 191-218.

[11] N. Telang and S. Bandyopadhyay, "Effects of a magnetic field on hot electron transport in quantum wires", Appl. Phys. Lett., **66**, 1623 (1995).

[12] N. Telang and S. Bandyopadhyay, "Hot electron magnetotransport in quantum wires", Phys. Rev. B, **51**, 9728 (1995).

[13] S. Pramanik, S. Bandyopadhyay and M. Cahay, "Spin relaxation of upstream electrons in a quantum wire: Failure of the drift-diffusion models", Phys. Rev. B, **73**, 125309 (2006).

12

Passive Spintronic Devices and Related Concepts

In this chapter, we introduce the reader to the physics of important *passive* spin-based devices such as sensors and read heads for reading tiny magnetic fields associated with magnetically stored data. *Active* spin based devices (such as spin transistors) are discussed in the next chapter. The reader is introduced to the concepts of "spin valve," "giant magnetoresistance," and the notion of "spin injection efficiency," which affects the performance of almost every spintronic device extant. This chapter will also familiarize the reader with frequently found terminologies and show that several ideas that may appear unrelated at first are actually interlinked and could be understood in terms of a few simple principles.

12.1 Spin valve

The "spin valve" is a device that is widely used to study and measure properties associated with spin transport in paramagnetic metals or semiconductors. One important parameter that can be experimentally measured with a spin valve device is the *spin diffusion length* L_s that we discussed in Chapter 11. Knowledge of this parameter and the spin diffusion coefficient D_s yields the spin relaxation time τ_s based on the relation $L_s = \sqrt{D_s \tau_s}$. The spin valve is therefore an important measurement device. It is also at the heart of addressable data storage devices, such as magnetic random access memory (MRAM).

The basic spin valve device is simply a trilayered structure consisting of a paramagnetic layer sandwiched between two ferromagnetic layers. We describe several models to understand spin transport in this three-layer system.

First, a simple two resistor model, based on the two-current model of charge transport in ferromagnets, is introduced for the ferromagnetic layers. This model was originally proposed by Mott [1]. It has now been superseded by a four-current model – one current for charge and three currents for the x-, y- and z-components of the spin [2, 3, 4], but we will not be discussing that model in this book, although the reader should be aware of it. We will continue with a discussion of the Stoner-Wohlfarth model for the band structure of ferromagnets. This will allow us to determine the relative magnitudes of the two currents in the Mott model and is a simple extension of the Sommerfeld model of metals [5]. Finally, we extend the two-current model to account for the presence of a paramagnetic layer between the two ferromagnetic layers. Putting all these together, we can come up with a tractable model to describe the behavior of a "spin valve." A slightly more advanced treatment is then presented based on the transfer matrix formalism described in Chapter 18. The transfer matrix method allows us to include the effects of simple spin-independent and spin-dependent scattering mechanisms at the interfaces between the outer ferromagnetic layers (or contacts) and the inner paramagnetic layer, as well as scattering within the layers themselves. Such an analysis will highlight the highly sensitive nature of spin valve characteristics to the details of the contact's electron density of states (for a quick primer on the "density of states", see Chapter 18), scattering at the interface between the contacts and the paramagnetic layer, and the potential barrier heights at the ferromagnet/paramagnet interfaces. Later, using the transfer matrix method, we will derive the *Jullière formula* used to relate the magnetoresistance of a spin valve to the spin polarizations of carriers at the Fermi energy in the contacts [6]. An extension of the Jullière formula can be used to relate the magnetoresistance (change of resistance in a magnetic field) of a spin valve structure to the spin diffusion length L_s in the paramagnetic layer, provided we assume that carriers enter the paramagnet via tunneling from the injecting ferromagnet and then drift/diffuse through the paramagnet with exponentially decaying spin polarization, before tunneling into the second (detecting) ferromagnetic layer.

The spin valve concept can be also applied to understand spin transport in multilayered structures composed of alternating layers of ferromagnetic and paramagnetic materials. This led to the discovery of the phenomenon of *giant magnetoresistance* (GMR) in the late 1980s [7, 8]. A simple resistor network theory of GMR is discussed in this chapter. The GMR effect is one of the early successes of spintronics, and devices based on the GMR effect are used as "read head sensors" for reading data stored in magnetic storage media (computer hard disks, Apple iPods, etc.). The read heads are passive devices, but an active device, namely, a "spin valve junction transistor," has also been proposed based on GMR in the base region. This is discussed in Chapter 14. The discovery of GMR has been recognized as an important advance in basic science by the 2007 Nobel Prize in Physics. Another application of spin valves is in magnetic random access memory (MRAM) which is a widely available commercial product based on spintronics. Suffice it to say then that the spin valve is an extremely important device, from the perspective of both fundamental spin physics and device applications.

12.2 Spin injection efficiency

Paramagnets do not intrinsically have any net spin polarization. In other words, the populations of up-spin and down-spin electrons are equal under equilibrium. Only ferromagnets have a non-zero spin polarization of carriers under equilibrium conditions. Therefore, the only way that a net spin polarization Ξ given by

$$\Xi = \frac{n_\uparrow - n_\downarrow}{n_\uparrow + n_\downarrow} \tag{12.1}$$

(where $n_\uparrow$ is the concentration of up-spin electrons and $n_\downarrow$ is the concentration of down-spin electrons) can be generated in a paramagnet is either via the use of *electrical spin injection* of charge carriers from a ferromagnet or *optical spin injection*. The former (more relevant to electrical devices as opposed to optical and electro-optic devices) requires successfully injecting an imbalance of spin from a ferromagnet into the paramagnet in the form of a current.

This brings into focus the issue of *spin injection efficiency* at the ferromagnet/paramagnet interface. This is a measure of how efficiently a ferromagnet can inject spin into a paramagnet that it is in electrical contact with. A related quantity is the *spin detection efficiency*, which is a measure of how efficiently a ferromagnet can filter spin, i.e., how selectively it can transmit electrons of a particular spin orientation impinging on it from the paramagnet. If a ferromagnet injector injects only one kind of spin – either the majority or the minority spin exclusively– into a paramagnet, then it is an ideal spin injector and the spin injection efficiency is 100%. Similarly, if a ferromagnet detector transmits only one kind of spin, then the spin detection efficiency is 100%. Real ferromagnets will typically inject/transmit both majority and minority spins, albeit not equally. Therefore, spin injection and detection efficiencies tend to be less than 100%. Spin injectors are also sometimes referred to as "spin polarizers" in analogy with optics, and spin detectors are similarly referred to as "spin analyzers."

The spin-valve, GMR devices and the Spin Field Effect Transistors (discussed in the next chapter) all rely on high (ideally 100%) efficiency of electrical spin injection/detection across a ferromagnet/paramagnet interface for their operation. The spin injection efficiency is defined as

$$\zeta = \frac{J_\uparrow - J_\downarrow}{J_\uparrow + J_\downarrow}, \tag{12.2}$$

where $J_\uparrow$ is the current density due to up-spin electrons and $J_\downarrow$ is that due to down-spin electrons, at the interface. In the ferromagnet, $J_\uparrow \neq J_\downarrow$ since $n_\uparrow \neq n_\downarrow$. But can this inequality be sustained in the paramagnet as well? The current continuity equation, familiar to all students of device physics and engineering, mandates that in steady state, $\vec{\nabla} \cdot \vec{J} = 0$. Therefore,

$$J^\uparrow_{ferromagnet} + J^\downarrow_{ferromagnet} = J^\uparrow_{paramagnet} + J^\downarrow_{paramagnet}. \tag{12.3}$$

The above equation does not require that $J^{\uparrow}_{paramagnet} \neq J^{\downarrow}_{paramagnet}$. In other words, there is no guarantee that there will be a spin polarized current in the paramagnet. What would favor the inequality $J^{\uparrow}_{paramagnet} \neq J^{\downarrow}_{paramagnet}$ in the paramagnet (and therefore result in a spin polarized current) is "resistivity matching," i.e., the resistance of the ferromagnet and paramagnet should be about equal [9]. This would, of course, be impossible to attain when the ferromagnet happens to be a metal (e.g., cobalt or iron) and the paramagnet happens to be a semiconductor (e.g., GaAs or Si) or an insulator (e.g., Al_2O_3). Fortunately, there is a way out, which requires interposing a tunnel barrier between the ferromagnet and paramagnet [10]. This can circumvent the resistivity mismatch problem.

Much effort has been expended to improve spin injection efficiency at ferromagnet/paramagnet interfaces, culminating in the demonstration of $\zeta = 90\%$ at low temperatures [11] and 70% at room temperature [12]. Feiderling, et al. [11] actually used a semiconducting paramagnet that allowed good resistance matching, and Salis et al. [12] used a tunnel barrier with a metallic ferromagnet. These two strategies can result in relatively high spin injection efficiencies. However, for many applications, such as Spin Field Effect Transistors, even 90% is not enough, and more than 99.999% is required for respectable performance, as we will show in Chapter 15. This matter remains one of the major challenges in the entire field of spintronics.

12.2.1 Stoner–Wohlfarth model of a ferromagnet

It is imperative to have a basic understanding of the band structure of ferromagnets when attempting to understand such phenomena as spin injection and detection, as well as the Mott two-current model. The simplest band structure model of a ferromagnet is the so-called Stoner-Wohlfarth model. It was motivated by the two-current model of Mott [1] to describe spin transport in ferromagnets. Mott suggested that at temperatures well below the Curie temperature, most scattering mechanisms that an electron will encounter in a ferromagnet will preserve the direction of the electron spin. In other words, scattering will generally not cause a spin flip or spin relaxation. As a result, the majority and minority spin currents $J^{\uparrow}_{ferromagnet}$ and $J^{\downarrow}_{ferromagnet}$ can be thought of as two independent components of the total current since no mixing occurs between them in the absence of spin flip scattering. The simplest way to implement this two-fluid or two-current model is to use an extension of the Sommerfeld model of metals [5] in which the energy-momentum relationship of both spin subbands is assumed to be parabolic and the mass of the carriers is the free electron mass.* The Stoner-Wohlfarth bandstructure (or dispersion relation) for a ferromagnet is depicted in Figure 12.1, in which the minimum of the conduction band for the minority spins is displaced upward

---

*It could be replaced by "effective mass," if needed.

(along the energy-axis) from the majority spin band by an amount equal to the exchange splitting energy Δ. The latter can be several electron-volts in many ferromagnets. The majority spins are those whose spin polarization is aligned parallel to the internal magnetic field in the ferromagnet[†], whereas the minority spins are the ones aligned opposite to this field. In this simple model, the $E - k$ relationship for majority spins is given by

$$E_\uparrow = \frac{\hbar^2}{2m_0} k^2,$$ (12.4)

whereas, for minority spins, it is equal to

$$E_\downarrow = \frac{\hbar^2}{2m_0} k^2 + \Delta.$$ (12.5)

At the Fermi energy, we define the Fermi wavevectors for the majority (up-spin) and minority (down-spin) electrons as

$$\hbar k_\uparrow^F = \sqrt{2m_0 E_F},$$ (12.6)

and

$$\hbar k_\downarrow^F = \sqrt{2m_0(E_F - \Delta)}.$$ (12.7)

The band structure (or energy dispersion relations) in real ferromagnets is, of course, far more complicated than the one described by the Stoner-Wohlfarth model. In fact, for some of the most widely used ferromagnetic metals, the density of states at the Fermi level can be larger for the minority spin band than for the majority spin band. Two well known examples of this are cobalt and nickel. In a metal like cobalt, d-electrons are more numerous at the Fermi level, but there are also s-electrons. The d- and s-electrons tend to have anti-parallel spins. Using tunnel barriers of varying thickness, one can electrically inject either d- or s-electrons from cobalt into a paramagnet [13].

There is also a great deal of effort to create ferromagnets which are 100% spin polarized, i.e., for which the exchange splitting energy is larger than the Fermi energy measured from the bottom of the majority spin band. For such ferromagnets, the Fermi level will be placed below the bottom of the minority spin band, but above the bottom of the majority spin band. Assuming low enough temperatures such that the energy difference between the bottom of the minority spin band and the Fermi level is much larger than the thermal energy kT, we can assume that the minority band is completely unoccupied. In that case, we only have majority spins, and the spin polarization Ξ given by Equation (12.1) is exactly 100%. Such materials are called ferromagnetic

---

[†]Within each 'domain' of the ferromagnet, there is an internal magnetic field which aligns spins either parallel or anti-parallel to it. In the unmagnetized state, the internal fields in different domains point in different directions. In the magnetized state, the field in every domain points in the direction of the magnetization.

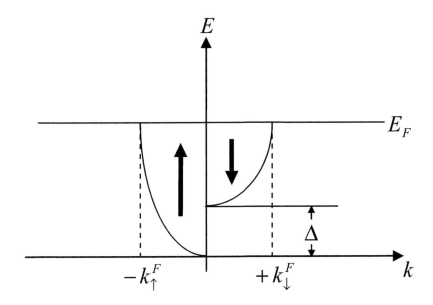

FIGURE 12.1

The Stoner-Wohlfarth model of a ferromagnet in which the energy dispersion relationships are assumed to be parabolic. The bottom of the minority spin subband is above that of the majority spin subband by an amount Δ (the exchange splitting energy). The effective mass of the carriers is assumed to be the free electron mass in both subbands. If the Fermi level is above the bottom of both spin subbands, then both subbands are occupied and carriers at the Fermi energy have a mixture of majority and minority spins. This is a normal ferromagnet. If the Fermi level is below the bottom of the minority spin subband but above that of the majority spin subband, then only the majority spin subband is occupied and the minority spin subband is unoccupied. Carriers with the Fermi energy have only majority spins, and, at absolute zero temperature, the spin polarization in the ferromagnet is ideally 100 % since all carriers have majority spin. This is an ideal half metallic ferromagnet.

half metals [14]. One well-known example of a half metal is lanthanum stron-
tium manganate (LSMO), which has nearly 100% spin polarization at low
temperatures. Some so-called Heusler alloys are also reportedly half metals.

Of course, perfectly ideal half metals may not actually exist. It has been
argued that at any temperature above absolute zero, phonons and magnons
will excite carriers to the minority spin band and destroy the 100% spin polarization [15]. Even at absolute zero, boundaries and inhomogeneities in a
real sample will preclude 100% spin polarization. Nonetheless, the concept of
half metals is important. In ideal half metals, $J^{\downarrow}_{ferromagnet} = 0$. Therefore,
theoretically speaking, these materials should be capable of 100% efficient
spin injection into a paramagnet. Insofar as 100% spin injection efficiency
is an important target in spintronics, the study of half metals has received
considerable attention from spintronics enthusiasts.

Exercises:

- In the Sommerfeld theory of a metal, the energy dependence of three-
  dimensional density of states is given by (see Chapter 18)

$$g_{3D}(E) = \frac{m_0}{\pi^2 \hbar^3} \sqrt{2m_0 E}. \tag{12.8}$$

Prove that, at zero temperature, $g_{3D}(E)$ can be rewritten as

$$g_{3D}(E) = \frac{3}{2} \frac{n}{E_F} \sqrt{\frac{E}{E_F}}, \tag{12.9}$$

where E_F and n are the Fermi energy and the electron density of the
metal.

Solution: At zero temperature, the electron density of the metal is
given by

$$n = \int_0^{E_F} dE g_{3D}(E). \tag{12.10}$$

Performing the integration using Equation (12.8),

$$n = \frac{\sqrt{2}m_0^{3/2}}{\pi^2 \hbar^3} \frac{2}{3} E_F^{3/2}. \tag{12.11}$$

According to Equation (12.8),

$$g_{3D}(E) = \frac{\sqrt{2}m_0^{3/2}}{\pi^2 \hbar^3} \frac{E_F^{3/2}}{E_F} \sqrt{\frac{E}{E_F}}, \tag{12.12}$$

which, with the use of Equation (12.11), can be rewritten as

$$g_{3D}(E) = \frac{3}{2} \frac{n}{E_F} \sqrt{\frac{E}{E_F}}. \tag{12.13}$$

- Using the results of the previous exercise, show that the density of majority spins $n_\uparrow$ is lower than the density of minority spins $n_\downarrow$. Calculate the ratio of $n_\uparrow/n_\downarrow$ in terms of E_F and Δ. Assume absolute zero temperature for simplicity.

- Show that if the Fermi level is well below the bottoms of both majority and minority spin subbands (the carrier populations are non-degenerate), so that Boltzmann statistics can replace Fermi-Dirac statistics, then the spin polarization is given by

$$\Xi = \frac{n_\uparrow - n_\downarrow}{n_\uparrow + n_\downarrow} = \frac{e^{\Delta/kT} - 1}{e^{\Delta/kT} + 1} = tanh\left(\frac{\Delta}{2kT}\right). \tag{12.14}$$

12.2.2   A simple two-resistor model to understand the spin valve

Consider two ferromagnetic contacts (F) in close proximity and separated by a narrow barrier of a paramagnetic material (N), as shown in Fig. 12.2. In the two-fluid (or -current) model of Mott, we can expect a difference between the resistance of the F/N/F junction when the magnetizations (indicated by arrows in the ferromagnetic layers) are either parallel (Fig.11.2(a)) or anti-parallel (Fig.11.2(b)). An electron whose spin is parallel to the magnetization of a ferromagnet will be a "majority spin" in the ferromagnet and will travel more easily through that region. In other words, it encounters minimal resistance. On the other hand, an electron whose spin is anti-parallel to the magnetization will be a minority spin and will encounter greater resistance in traversing that region. All this can be understood from the Stoner-Wohlfarth band structure in Fig. 12.1. For a fixed energy, an electron with majority spin has higher velocity and therefore encounters less scattering from impurities and defects. Hence it encounters less resistance.

Based on this simple picture, we can estimate the resistance of the F/N/F junction using the resistor network model depicted in Fig. 12.2(c) and 12.2(d). For the case of parallel magnetization, the resistance for the up-spin (majority) band is given by $2R_\uparrow$ and the resistance of the down-spin (minority) band is given by $2R_\downarrow$ ($R_\downarrow \gg R_\uparrow$). Now, in the Mott two-current model, the up-spin and down-spin channels are independent, so that the total resistance of the device is the parallel combination of the resistances $2R_\uparrow$ and $2R_\downarrow$. This is indicated in Fig. 12.2(c). As a result, the resistance R_P of the junction (here the subscript "P" denotes the fact that the two ferromagnetic layers have parallel magnetizations) is given by

$$R_P = \frac{(2R_\uparrow)(2R_\downarrow)}{2R_\uparrow + 2R_\downarrow} = \frac{2R_\uparrow R_\downarrow}{R_\uparrow + R_\downarrow}. \tag{12.15}$$

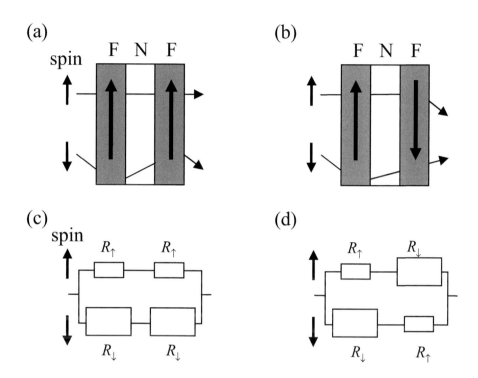

FIGURE 12.2

An F/N/F junction in which two ferromagnetic contacts (F) sandwich a paramagnetic (N) region a few nanometers thick. A spin valve effect comes from
the difference of the overall resistance of the junction as the relative magnetization orientation of the two contacts is changed from parallel to anti-parallel.
Panels (c) and (d) illustrate the simple two resistor network model of the junction and spin valve effect based on the two-current model of the ferromagnetic
contacts originally proposed by Mott [1]. Reproduced with permission from
M. A. M. Gijs and G. E. W. Bauer, Adv. Phys., **46**, 285 (1997). Copyright
(1997) Taylor & Francis.

When the two ferromagnetic layers have anti-parallel magnetizations, the two-resistor model depicted in Fig. 12.2(d) leads to a resistance R_{AP} for the F/N/F junction given by

$$R_{AP} = \frac{R_\uparrow + R_\downarrow}{2}. \tag{12.16}$$

Therefore,

$$R_{AP} - R_P = \frac{(R_\downarrow - R_\uparrow)^2}{2(R_\downarrow + R_\uparrow)} > 0. \tag{12.17}$$

Consequently, the anti-parallel configuration has a higher resistance. Here we have assumed that the majority spins in one ferromagnet are also majority spins in the other. However, if that were not true, and the majority spin in one ferromagnetic layer is a minority spin in the other (e.g., in the case of iron and cobalt as the two contacts), then the anti-parallel configuration may have the lower resistance.

The spin valve effect has been investigated for a wide variety of ferromagnetic contacts and intermediate paramagnetic layers. Three magnetoresistance ratios are commonly used to characterize the difference between the resistances R_P and R_{AP}:

(a) The tunneling magnetoresistance ratio

$$MR_1 = \frac{R_{AP} - R_P}{R_P} = \frac{G_P - G_{AP}}{G_{AP}}, \tag{12.18}$$

where $G_P = 1/R_P$ and $G_{AP} = 1/R_{AP}$ are the conductance of the junction in the parallel and anti-parallel configurations, respectively. This ratio is most often labeled using the acronym **TMR** in the literature.

(b) The junction magnetoresistance ratio

$$MR_2 = \frac{R_{AP} - R_P}{R_{AP}} = \frac{G_P - G_{AP}}{G_P}, \tag{12.19}$$

most often labeled with the acronym **JMR** in the literature, and finally,

(c) The spin conductance ratio

$$MR_3 = \frac{G_P - G_{AP}}{G_P + G_{AP}}. \tag{12.20}$$

These three ratios are related to each other.

Exercises:

- Establish the relationships between the different ratios MR_1, MR_2, MR_3 defined above, i.e., express MR_1 in terms of MR_2 or MR_3, and so on.

  Solution

  This is based on simple algebra and is left to the reader.

- Using the expressions for R_P and R_{AP} given above, express the ratios MR_1, MR_2, MR_3 in terms of the ratio $R_\uparrow/R_\downarrow$. Show that

$$MR_1 = \frac{\left(\frac{R_\uparrow}{R_\downarrow} - 1\right)^2}{4\frac{R_\uparrow}{R_\downarrow}},$$

$$MR_2 = \frac{\left(\frac{R_\uparrow}{R_\downarrow} - 1\right)^2}{\left(\frac{R_\uparrow}{R_\downarrow} + 1\right)^2},$$

$$MR_3 = \frac{\left(\frac{R_\uparrow}{R_\downarrow} - 1\right)^2}{\left(\frac{R_\uparrow}{R_\downarrow}\right)^2 + 6\frac{R_\uparrow}{R_\downarrow} + 1}. \tag{12.21}$$

12.2.3 More advanced treatment of the spin valve

The treatment in the previous section was based on the two-independent-current model of the ferromagnetic contacts due to Mott. No consideration was given to the paramagnetic layer in between. Obviously the paramagnetic layer must play a critical role since spin can relax in this layer, causing mixing between the two channels (up-spin channel and down-spin channel), which will no longer remain independent. The quality of the interfaces between the paramagnet and the ferromagnets must also play a role in the magnetoresistance of the spin valve. Some (but not all) of these additional features are better described by the resistor network model shown in Fig. 12.3.

The resistance of a piece of metal with resistivity ρ, length l and cross-sectional area A is given by

$$R = \rho l / A. \tag{12.22}$$

In the Drude model for simple metals and semiconductors, ρ is given by

$$\rho = \frac{m^*}{ne^2\tau_m}, \tag{12.23}$$

where n is the electron density, m^* the electron (effective) mass, and τ_m is the mean free time or average time between collisions for the carriers (actually, this is the "momentum relaxation time" in the parlance of transport theory [16]).

In the two-current model of Mott, the spin-dependent resistance of a ferromagnet $R_{fm,\uparrow}$ and $R_{fm,\downarrow}$ can be assumed to be inversely proportional to the conductivity for the up-spin channel ($\sigma_\uparrow$) and the down-spin channel ($\sigma_\downarrow$). We therefore have

$$R_{fm,\uparrow}/R_{fm,\downarrow} = \sigma_\downarrow/\sigma_\uparrow. \tag{12.24}$$

(a)

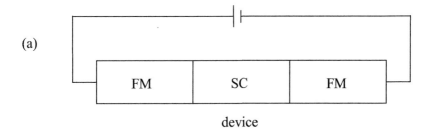

device

(b)

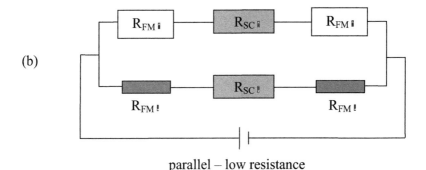

parallel – low resistance

(c)

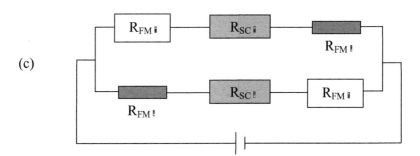

antiparallel – high resistance

FIGURE 12.3
Advanced model of the spin valve effect taking into account the non-magnetic semiconductor sandwiched between two ferromagnetic contacts. Panels (b) and (c) show the resistor network model used to calculate the resistance of the spin valve for the case of parallel and anti-parallel magnetizations of the contacts, respectively. Reproduced with permission from G. Schmidt and L. W. Molenkamp, Semicond. Sci. Technol., **17**, 310 (2002).Copyright(2002) Institute of Physics.

Introducing a parameter β_p, which is the "conductivity spin polarization in the ferromagnet"[‡], given by

$$\beta_p = \frac{\sigma_\uparrow - \sigma_\downarrow}{\sigma_\uparrow + \sigma_\downarrow}, \tag{12.25}$$

we can then rewrite

$$R_{fm,\uparrow} = \frac{2R_{fm}}{1 + \beta_p}, \tag{12.26}$$

$$R_{fm,\downarrow} = \frac{2R_{fm}}{1 - \beta_p}, \tag{12.27}$$

where $R_{fm} = \frac{l}{(\sigma_\uparrow + \sigma_\downarrow)A}$.

If $\beta_p = 100\%$, i.e., only majority spins (up-spins) are present in the ferromagnet (half metallic ferromagnets), then $R_{fm,\downarrow} \to \infty$, indicating that the resistance encountered by (the absent) minority spins (down-spins) is infinity. This is consistent with the Mott picture.

A more realistic treatment of the spin valve effect must take into account the presence of the non-magnetic layer between the two contacts. The latter is assumed to be a paramagnetic semiconducting layer (SC) in Figure 12.3 and the same two-current fluid model of Mott is used for the spin channels in the paramagnet as well. In the paramagnetic semiconductor, there are no "majority" or "minority" spins (both spins are equal in population). Therefore,

$$R_{sc,\uparrow} = R_{sc,\downarrow} = 2R_{sc}, \tag{12.28}$$

where R_{sc} is the resistance of the semiconducting channel which can be simply approximated by $R_{sc} = \rho_{sc}d/A$, ρ is the resistivity of the semiconductor, d is the separation between the contacts, and A is the cross-sectional area of the junction.

The resistance of the spin valve in the parallel and anti-parallel configurations can then be calculated using the resistor network model depicted in Figure 12.3(b) and 12.3(c).

Referring to Fig. 12.3(c), we see that, in the anti-parallel configuration, the resistances of both spin channels are equal and given by

$$R_\uparrow = R_\downarrow = R_{fm,\downarrow} + 2R_{sc} + R_{fm,\uparrow}. \tag{12.29}$$

As a result, the polarization of the spin current defined by the quantity

$$\alpha_p = \frac{R_\downarrow - R_\uparrow}{R_\downarrow + R_\uparrow} \tag{12.30}$$

[‡]Note that $\beta_p = \Xi$ if the mobilities of up-spin and down-spin electrons are equal, but not otherwise.

is equal to zero.

Referring to Fig. 12.3(b), the situation is different for the parallel case. Here, the polarization of the spin current is given by

$$\alpha_p = \frac{R_\downarrow - R_\uparrow}{R_\uparrow + R_\downarrow} = \frac{2(R_{fm,\uparrow} - R_{fm,\downarrow})}{2(R_{fm,\uparrow} + R_{fm,\downarrow}) + 4R_{sc}}. \tag{12.31}$$

This quantity is also sometimes referred to as the *spin (current) injection efficiency.*

Using the expressions for $R_{fm,\uparrow}$ and $R_{fm,\downarrow}$ given above, this leads to

$$\alpha_p = \beta_p \left(\frac{R_{fm}}{R_{sc}} \right) \frac{2}{\left[\frac{2R_{fm}}{R_{sc}} + \left(1 - \beta_p^2 \right) \right]}. \tag{12.32}$$

This last equation shows that the spin polarization of the current is proportional to the spin polarization β_p of the ferromagnetic contacts (for small β_p).

A plot of α_p versus β_p for different values of the ratio R_{sc}/R_{fm} is shown in Fig. 12.4. Obviously, the highest α_p is obtained when the spin polarization in the ferromagnets, β_p, is 100%. This shows that *half metals* are the best spin injectors and detectors, as we had surmised earlier.

In actual spin valves, the ratio R_{sc}/R_{fm} can be of the order of 10^4, leading to a very low spin polarization of the current in the semiconductor. It is therefore important that this ratio be close to unity if significant spin polarization in the semiconductor is desired. This is the basis of the infamous *resistance mismatch* problem pointed out in [9] which insisted that in order to observe significant spin polarization in the semiconductor (i.e., a large value of α_p), the resistance of the semiconductor and ferromagnet should be "matched," i.e., $R_{sc} \approx R_{fm}$. Actually, from Equation (12.32), it appears that, if $R_{fm} \gg R_{sc}$, then α_p approaches β_p. However, the ferromagnet is typically a metal and therefore unlikely to be more resistive than the semiconductor. Hence, this condition is impossible to meet with a metallic ferromagnet. This stimulated much research in semiconducting ferromagnets (or dilute magnetic semiconductor materials) until Rashba pointed out that the interposition of a tunnel barrier between the ferromagnet and the semiconductor can improve spin injection into the semiconductor [10] in spite of a resistance mismatch. However, tunnel barriers are inconvenient and reduce the overall conductance of the spin valve, diminishing its current-carrying capability. Consequently, research in dilute magnetic semiconductors has remained a big enterprise.

For a spin valve, the overall resistance in the parallel configuration is

$$R_P = \frac{(2R_{fm,\uparrow} + 2R_{sc})(2R_{fm,\downarrow} + 2R_{sc})}{2R_{fm,\uparrow} + 2R_{fm,\downarrow} + 4R_{sc}}, \tag{12.33}$$

while for the anti-parallel configuration, we get

$$R_{AP} = \frac{R_{fm,\uparrow} + R_{fm,\downarrow} + 2R_{sc}}{2}. \tag{12.34}$$

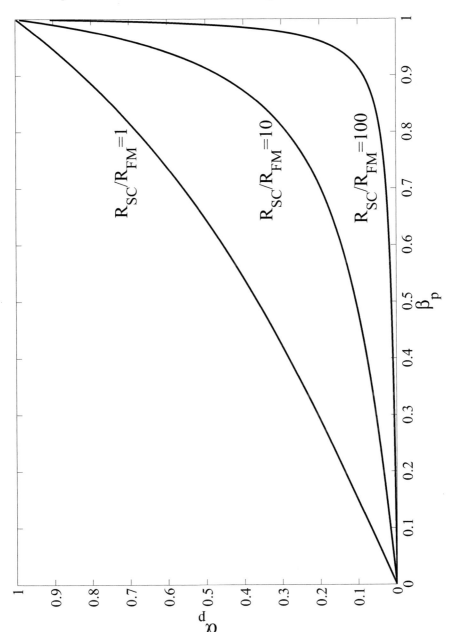

FIGURE 12.4

Plot of the spin current injection efficiency α_P versus the spin polarization β_P in the ferromagnetic contacts for different ratios of the resistances of the ferromagnetic contacts and semiconductor layer.

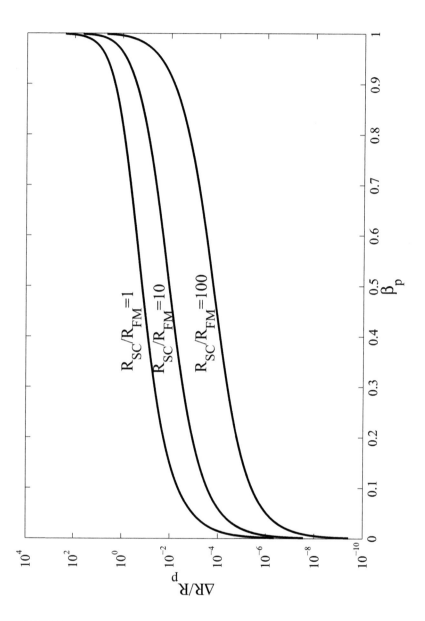

FIGURE 12.5
Plot of the tunnel magnetoresistance $\Delta R/R_p$ versus the spin polarization β_P of the ferromagnetic contacts for different ratios of the resistances of the ferromagnetic contacts and semiconductor channel.

Hence, the tunneling magnetoresistance ratio MR_1 defined earlier is given by

$$TMR = MR_1 = \frac{\Delta R}{R_P} = \frac{R_{AP} - R_P}{R_P} = \frac{(R_{fm,\uparrow} - R_{fm,\downarrow})^2}{4\,(R_{fm,\uparrow} + R_{sc})\,(R_{fm,\downarrow} + R_{sc})}.$$
(12.35)

Using the values of $R_{fm,\uparrow}$ and $R_{fm,\downarrow}$ defined above, we get

$$MR_1 = \frac{\Delta R}{R_P} = \frac{\beta_p^2}{1 - \beta_p^2} \left(\frac{R_{fm}}{R_{sc}}\right)^2 \left[\frac{4}{4\left(\frac{R_{fm}}{R_{sc}}\right)^2 + 4\left(\frac{R_{fm}}{R_{sc}}\right) + (1 - \beta_p^2)}\right].$$
(12.36)

Since MR_1 is proportional to R_{fm}^2/R_{sc}^2, it should be almost impossible to detect in an experiment if R_{fm}/R_{sc} is of the order of 10^{-4}. A plot of $MR_1 = \Delta R/R_P$ versus β_p for different values of the ratio R_{sc}/R_{fm} is shown in Fig. 12.5. Since β_p is on the order of 0.3–0.5 for normal metallic ferromagnets (*not* half metals) and R_{sc}/R_{fm} is typically on the order of 10^4, because of the large difference in the conductivities between non-magnetic semiconductor layers and metallic ferromagnetic contacts, $\Delta R/R_P$ is expected to be very small and therefore very difficult to detect in an experiment. Surely this could have had a chilling effect on research in spin valves. Fortunately, this did not happen.

The simple resistor network model seems to tell us that spin valves are doomed from the very start by the resistance mismatch problem. Fortunately, this is not the case, and spin valves are now widely used as basic components of hard-drive read heads and magnetic random access memories. Therefore, clearly, the simple resistor network model does not always represent reality. There is a need for more advanced theoretical treatments of the spin valve effect, beyond the resistor network model, which we discuss in the next subsections.

Exercise:

- Using the resistor network model shown in Fig. 12.3, derive an expression for the other magnetoresistance ratios MR_2 and MR_3 in terms of β_p and the ratios R_{fm}/R_{sc}.

- Derive analytical expressions for α_p and MR_1 if the two ferromagnetic contacts are characterized by two different spin polarization factors β_{p1} and β_{p2}.

- If we define the parameter $\gamma = R_{fm}/R_{sc}$, show that

$$\alpha_p = \frac{2\beta_p \gamma}{2\gamma + 1 - \beta_p^2}$$
(12.37)

and

$$MR_1 = \frac{\Delta R}{R_P} = \frac{\beta_p^2}{1 - \beta_p^2}\gamma^2 \left[\frac{4}{4\gamma^2 + 4\gamma + (1 - \beta_p^2)}\right]. \qquad (12.38)$$

Compute $d\alpha_p/d\gamma$ and $d(\Delta R/R_P)/d\gamma$ for $\beta_p = 0.5$ and plot those quantities as a function of γ.

12.2.4 A transfer matrix model

So far, the spin valve effect was modeled using a resistor network model, which did not include the effects of the interface between the contacts and the non-magnetic layer. In this section, we describe a simple one-dimensional model of the spin valve effect, in which the intermediate non-magnetic semiconductor layer is assumed to be so thin that it can be approximated as a delta scatterer, as shown in Figure 12.6. The scatterer has a spin-independent part of strength Γ and a spin-dependent part of strength Γ'. Furthermore, we assume that the scattering potential has a spin-independent and spin-dependent part. The magnetoresistance of the spin valve is calculated using the transfer matrix formalism described in Chapter 18 based on a purely one-dimensional picture of spin transport. We assume that there is only one occupied subband in the one-dimensional (quantum wire) semiconductor layer. Furthermore, we assume absolute zero temperature and calculate the conductance of the majority and minority spin subbands using the Landauer conductance formula [17].

Case 1: Both Γ and Γ' are set equal to zero

In the parallel configuration, since there is no spin scattering at the interface, the conductance for both the majority and minority spins is given by

$$G_\uparrow = G_\downarrow = G_P, \qquad (12.39)$$

and

$$G_P = 2e^2/h. \qquad (12.40)$$

In the anti-parallel configuration, the majority spins incident from the left contact see a step of size Δ, the exchange splitting energy, as they transmit across the interface. For the minority spin band, incident from the left, this appears as a step down of size Δ, as shown in the middle panel of Fig. 12.6. The transmission coefficients for the majority and minority spin subbands are

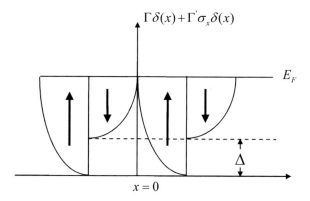

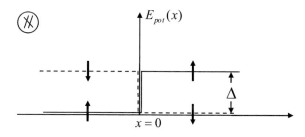

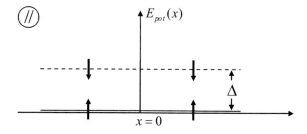

FIGURE 12.6

(Top panel) Simple one-dimensional model of a barrier between two identical ferromagnetic contacts. The scattering potential contains a spin-independent part of strength Γ and a spin-dependent part of strength Γ'. (Middle panel) In the absence of any scattering potential at the interface, when the magnetizations of the contacts are anti-parallel, the majority spins impinging from the left see a step barrier of height Δ (full line). The dashed line is the step down barrier for the minority spins incident from the left contact. (Bottom panel) When the magnetizations of both contacts are parallel, there is no potential step in the path of either spin flowing from one contact to the other.

identical and each is equal to

$$T_\uparrow = T_\downarrow = \frac{4k_\uparrow^F k_\downarrow^F}{\left[k_\uparrow^F + k_\downarrow^F\right]^2}. \tag{12.41}$$

As a result, the conductance of the spin valve in the anti-parallel configuration is given by

$$G_{AP} = \frac{8k_\uparrow^F k_\downarrow^F}{\left[k_\uparrow^F + k_\downarrow^F\right]^2} \frac{e^2}{h}. \tag{12.42}$$

A plot of the three different magnetoresistance ratios MR_1, MR_2 and MR_3 is shown in Figure 12.7 as a function of the ratio $\lambda = k_\uparrow^F / k_\downarrow^F$. This simple model reveals that the values of MR_1, MR_2, MR_3 can be quite different. Much more sophisticated treatments than the one described here also reveal this difference. Therefore, one must be very careful when comparing experimental data with theoretical models since all three definitions (MR_1, MR_2, MR_3) proliferate in the literature.

Case 2: Γ and $\Gamma' \neq 0$

First, we calculate the conductance for majority spin incident from the left contact for which the incoming and outgoing components are depicted in Figure 12.8.

To calculate the reflection ($R_\uparrow$, $R_\downarrow$) and transmission ($T_\uparrow$, $T_\downarrow$) amplitudes, we must solve the Pauli equation subject to the following boundary equations for the spinor at the interface:

$$\psi(0_-) = \psi(0_+)$$
$$\frac{d\psi}{dx}(0_+) = \frac{d\psi}{dx}(0_-) + \frac{2m_0\Gamma}{\hbar^2}\left[1 + \left(\frac{\Gamma'}{\Gamma}\right)\sigma_x\right]\psi(0). \tag{12.43}$$

Exercise: Starting with the Pauli equation and the scattering potential shown in Fig. 12.6, prove that the boundary condition (12.43) must be satisfied by integrating the Pauli equation across the interface.

Upon substituting the expressions for the spinors shown in Fig. 12.8, we get a set of four equations for the four unknowns ($R_\uparrow$, $R_\downarrow$, $T_\uparrow$, $T_\downarrow$). This system

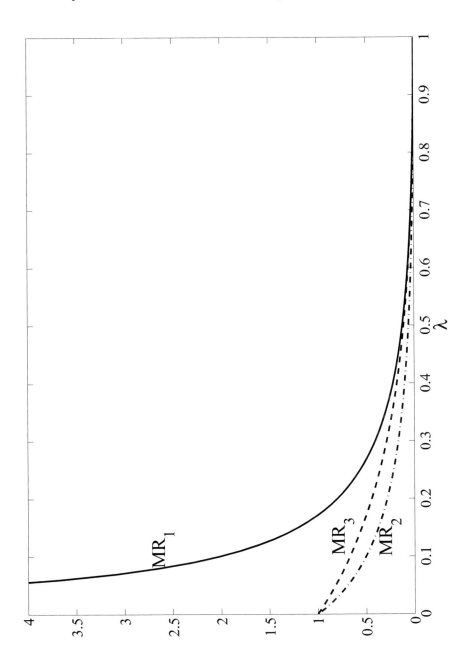

FIGURE 12.7

Plot of the three magnetoresistance ratios MR_1, MR_2 and MR_3 as functions of $\lambda = k_\downarrow^F / k_\uparrow^F$, where $k_\downarrow^F$ and $k_\uparrow^F$ are defined in section 11.2.1.

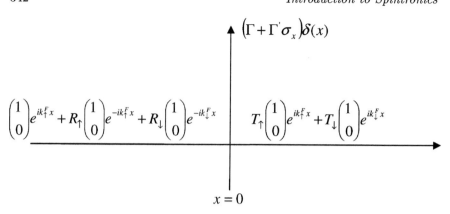

FIGURE 12.8

The scattering problem for a majority spin incident from the left contact. The magnetizations in the two contacts are assumed to be parallel.

of equations can be solved easily, leading to

$$R_\uparrow = \frac{-i\hbar^2 k_\uparrow^F m_0 \Gamma + m_0^2 \left(\Gamma^2 - \Gamma'^2\right)}{A},$$

$$R_\downarrow = \frac{-i\hbar^2 k_\uparrow^F m_0 \Gamma'}{A},$$

$$T_\uparrow = \frac{\hbar^2 k_\uparrow^F \left(\hbar^2 k_\downarrow^F + i m_0 \Gamma\right)}{A},$$

$$T_\downarrow = \frac{-i\hbar^2 k_\uparrow^F m_0 \Gamma'}{A}, \tag{12.44}$$

where

$$A = \hbar^4 k_\uparrow^F k_\downarrow^F + i\hbar^2 m_0 \Gamma \left(k_\uparrow^F + k_\downarrow^F\right) + m_0^2 \left(\Gamma'^2 - \Gamma^2\right). \tag{12.45}$$

In units of e^2/h, the conductance of the majority spin band is given by

$$G_\uparrow = |T_\uparrow|^2 + \frac{k_\downarrow^F}{k_\uparrow^F} |T_\downarrow|^2. \tag{12.46}$$

It can also be shown that

$$|T_\uparrow|^2 + \frac{k_\downarrow^F}{k_\uparrow^F} |T_\downarrow|^2 + |R_\uparrow|^2 + \frac{k_\downarrow^F}{k_\uparrow^F} |R_\downarrow|^2 = 1, \tag{12.47}$$

which merely expresses the obvious fact that the sum of reflection and transmission probabilities, with and without spin flip, must add up to unity.

The scattering problem described above must be repeated for electrons incident from the minority spin subband. In this case, the reflection and transmission amplitudes are

$$R'_\uparrow = \frac{-i\hbar^2 k_\downarrow^F m_0 \Gamma'}{A},$$

$$R'_\downarrow = -1 + \frac{\hbar^4 k_\uparrow^F k_\downarrow^F + i\hbar^2 m_0 \Gamma k_\downarrow^F}{A},$$

$$T'_\uparrow = \frac{-i\hbar^2 k_\downarrow^F m_0 \Gamma'}{A},$$

$$T'_\downarrow = \frac{\hbar^4 k_\uparrow^F k_\downarrow^F + i\hbar^2 m_0 \Gamma k_\downarrow^F}{A}, \tag{12.48}$$

from which the conductance (in units of e^2/h) of the minority spin subband is found to be

$$G_\downarrow = \frac{k_\uparrow^F}{k_\downarrow^F}|T_\uparrow|^2 + |T_\downarrow|^2. \tag{12.49}$$

Since the majority and minority spins are *orthogonal* in the ferromagnetic contacts, the total conductance (in units of e^2/h) is

$$G_{total} = G_\uparrow + G_\downarrow = \left(\frac{k_\uparrow^F}{k_\downarrow^F} + 1\right)|T_\uparrow|^2 + \left(\frac{k_\downarrow^F}{k_\uparrow^F} + 1\right)|T_\downarrow|^2. \tag{12.50}$$

Exercise: Using the boundary conditions (12.43), show that the reflection and transmission amplitudes of a majority spin incident from the left contact, as shown in Figure 12.8, are indeed given by equation (12.44). Repeat the exercise for electrons incident from the minority spin subband. This problem can be solved using software tools such as MATLAB or MATHEMATICA.

The total conductance in the parallel configuration G_P can then be calculated by adding the conductances $G_\uparrow$ and $G_\downarrow$ found above. The scattering problem must then be repeated for the two contacts in the anti-parallel configuration to derive G_{AP} to finally calculate the magnetoresistance ratios MR_1, MR_2, and MR_3.

Numerical examples:

(a) $\Gamma \neq 0$ and $\Gamma' = 0$

When the magnetizations of the two ferromagnetic contacts are parallel and spin-dependent scattering at the interface is neglected, the Landauer conductance of majority and minority spin is equal to 1 (in units of e^2/h) since there is no spin mixing and no interface potential step.

If the magnetization of the right contact is parallel to that of the left contact, show that the transmission coefficient for a majority spin incident from the left contact is given by

$$|T_\uparrow|^2 = k_\uparrow^{F2} / \left(k_\uparrow^{F2} + k_B^2 \right), \tag{12.51}$$

where $k_B = m\Gamma/\hbar^2$.

For the minority spin incident from the left contact, we have

$$|T_\downarrow|^2 = k_\downarrow^{F2} / \left(k_\downarrow^{F2} + k_B^2 \right). \tag{12.52}$$

The Landauer conductance (in units of e^2/h) for the anti-parallel configuration is therefore given by

$$G_P = |T_\uparrow|^2 + |T_\downarrow|^2 = \frac{k_\uparrow^{F2}}{k_\uparrow^{F2} + k_B^2} + \frac{k_\downarrow^{F2}}{k_\downarrow^{F2} + k_B^2}. \tag{12.53}$$

In the anti-parallel configuration, both $|T_\uparrow|^2$ and $|T_\downarrow|^2$ are given by

$$|T_\uparrow|^2 = \frac{4k_\uparrow^F k_\downarrow^F}{\left[k_\uparrow^F + k_\downarrow^F \right]^2 + 4k_B^2}, \tag{12.54}$$

and therefore the Landauer conductance for the anti-parallel case is given by

$$G_{AP} = \frac{8k_\uparrow^F k_\downarrow^F}{\left[k_\uparrow^F + k_\downarrow^F \right]^2 + 4k_B^2}. \tag{12.55}$$

Using the results above, we plot the three magnetoresistance ratios MR_1, MR_2, and MR_3 in terms of the quantity P defined as

$$P = \frac{k_\uparrow^F - k_\downarrow^F}{k_\uparrow^F + k_\downarrow^F}, \tag{12.56}$$

which characterizes the polarization of the contacts. In Figures 12.9 – 12.11, the different curves in each plot correspond to different values of the ratio $k_B/k_\uparrow^F$, where $k_B = m_0\Gamma/\hbar^2$ characterizes the strength of the spin-independent delta scatterer.

The magnetoresistance ratio MR_1 is very small unless P approaches unity. This means MR_1 is tiny unless $k_\downarrow^F$ is close to zero, or the exchange splitting energy Δ is close to the Fermi energy E_F (recall the Stoner-Wohlfarth model of the ferromagnets in Section 11.2.1). For $P = 1$, neither up-spin nor down-spin can be transmitted across the interface and the conductance for the anti-parallel magnetizations is equal to zero, which leads to a divergence of MR_1. This is why MR_1 is sometimes referred to as "inflationary definition of the magnetoresistance" since it is unbounded from above.

In contrast, the magnetoresistance ratios MR_2 and MR_3 have a maximum value of 1 when $P = 1$. As Figures 12.9 – 12.11 indicate, the ratios MR_1, MR_2, and MR_3 are fairly sensitive functions of the strength of the scattering potential representing the barrier between the two ferromagnetic contacts. This trend has been confirmed with much more sophisticated forms of the scattering potential than the one considered here. It has also been found to be practically universal according to detailed models developed to treat metallic, semiconducting, insulating, and organic materials as the intermediate layer of a spin valve.

(b) $\Gamma \neq 0$ and $\Gamma' \neq 0$

In Figure 12.12, we plot the magnetoresistance ratio MR_3 in Equation (12.20) as a function of the square of the polarization factor P for different values of the ratios $k_B/k_\uparrow^F$ and $k_B'/k_\uparrow^F$, where $k_B = m_0 \Gamma/\hbar^2$ and $k_B' = m_0 \Gamma'/\hbar^2$ characterize the strength of the spin-independent and spin-dependent scattering potentials at the interface between the two ferromagnetic contacts. Figure 12.12 has a total of nine curves corresponding to values of $k_B/k_\uparrow^F$ and $k_B'/k_\uparrow^F$ equal to $0.1, 0.5$, and 0.9.

Figure 12.12 shows that MR_3 is highly sensitive to the presence of both spin-independent and spin-dependent scattering mechanisms at the interface. Actually, the magnetoresistance MR_3 can become negative, which means that $G_P < G_{AP}$ in this case.

There have been reports of both positive and negative magnetoresistances in the literature. These reports can be confusing since different workers adopt different definitions of the magnetoresistance ratios. For instance, if the interaction between two ferromagnetic contacts is such that their magnetization is anti-parallel for a specific barrier, the spin valve will exhibit a lower resistance when parallel magnetization is induced via an external magnetic field. This switching from high to low resistance in the presence of an external magnetic field is referred to as negative magnetoresistance. The results of the simple model studied in this section show that positive magnetoresistance is also possible. Thus, extended measurements over a wide variety of junctions can be very informative about the details of the physical mechanisms involved in spin conduction through the junction.

The reader may have noticed that everything we have discussed so far invloves the conductances G_P and G_{AP} corresponding to the two ferromagnetic contacts of a spin valve having parallel and anti-parallel orientations. What about the case when the orientations are at an angle Θ where $0° < \Theta < 180°$? The two-current model cannot tell us what the conductance will be in that situation since it deals with only up- and down-spins and not spins of any arbitrary orientation, but the more recent four-current model [2, 3, 4] can because it deals with any arbitrary orientation. We have not discussed that model here, but the reader can follow the evolving literature in that field.

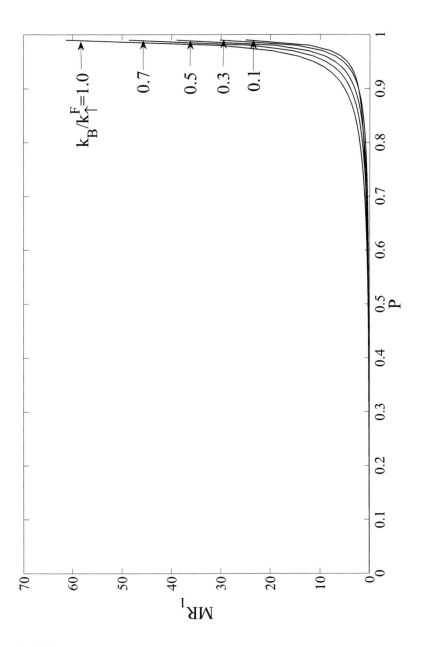

FIGURE 12.9

Plot of the magnetoresistance ratio $MR_1 = (R_{AP} - R_P)/R_P$ as a function of $P = \left(k_\uparrow^F - k_\downarrow^F\right) / \left(k_\uparrow^F + k_\downarrow^F\right)$ for different values of the ratio $k_B/k_\downarrow^F$.

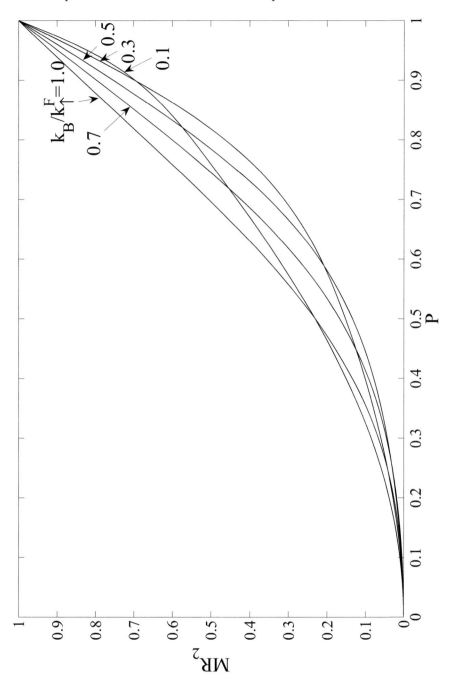

FIGURE 12.10
Plot of the magnetoresistance ratio MR_2 as a function of P.

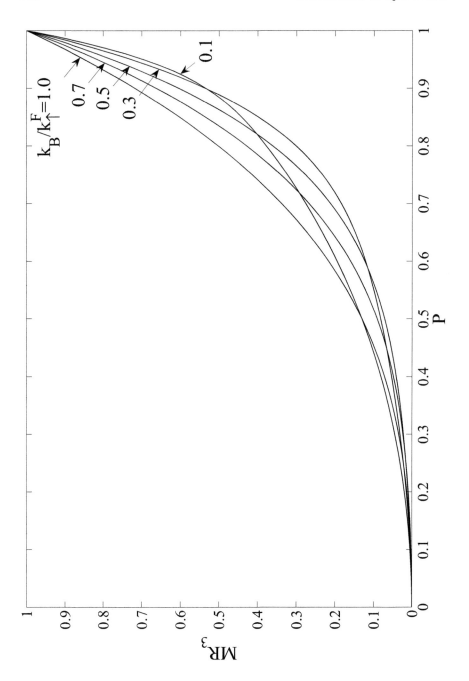

FIGURE 12.11
Plot of the magnetoresistance ratio MR_3 as a function of P.

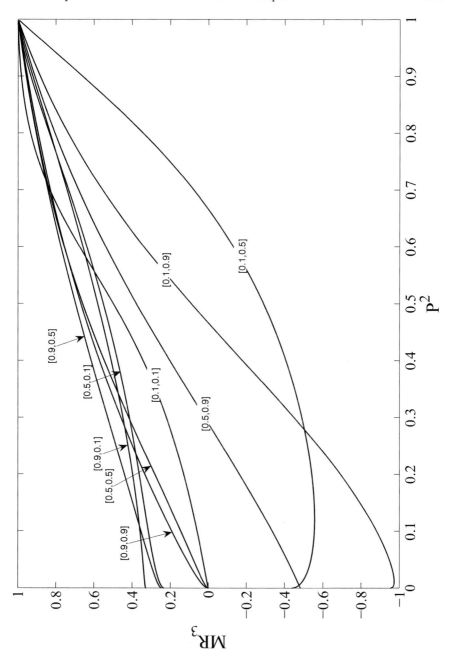

FIGURE 12.12

Plot of the ratio $MR_3 = (G_P - G_{AP})/(G_P + G_{AP})$ versus P^2 for different values of the parameters $\left[k_B/k_\uparrow^F, k_B'/k_\uparrow^F \right]$ characterizing the strength of the spin-independent and spin-dependent scattering potentials between the two ferromagnetic contacts.

Jullière formula

In the early years, the magnetoresistance of spin valves was studied systematically by Jullière [6]. In his pioneering work, Jullière prescribed a formula for the junction magnetoresistance ratio (definition MR_2) based on a generalization of the transfer matrix model discussed above. The Jullière formula relates MR_2 to the spin polarizations of carriers at the Fermi level in the ferromagnetic contacts. This formula is often used in the analysis of spin valve experiments, and it has been extended to account for spin relaxation in the intermediate paramagnetic layer. Because of its historical importance, we outline the derivation of the Jullière formula here. However, the reader should be warned that the Jullière formula is often used without a careful assessment of its applicability for the specific structure under investigation. In fact, there are some recent reports addressing the limits of validity of the Jullière formula, as increasing numbers of spin valve experiments are being reported.

In 1975, Jullière experimented with the magnetoresistance of Fe-Ge-Co junctions at $T \leq 4.2K$ as a function of the applied bias between the two ferromagnetic contacts made of two different materials.

To interpret his experimental results, he modified Bardeen's derivation of the tunneling current between two normal paramagnetic metals separated by a thin insulating barrier [18]. Bardeen's formula is very similar to the Tsu-Esaki formula routinely used by device engineers and physicists in the study of resonant tunneling diodes and devices based on resonant tunneling. The basic ingredients are the following: In the absence of an applied voltage, the Fermi levels of the two electrodes flanking a tunnel barrier must be aligned (no current flows). An applied voltage V shifts all the energy levels in one of the electrodes relative to the other by eV. This amounts to an energy shift eV between the two Fermi levels that then causes the tunneling current. In the Bardeen approach, one evaluates the current $I(V, E)$ flowing at a given energy E between the left (L) and right (R) electrodes from Fermi's golden rule, i.e.,

$$I(V, E) \propto |T(E)|^2 D_L(E) D_R(E + eV_{bias})[f(E) - f(E + eV_{bias})], \quad (12.57)$$

where $D_L(E)$, $D_R(E)$ are the densities of states of the left and right electrodes, $|T(E)|^2$ is the square of the tunneling matrix element, and $f(E)$ is the Fermi function.

If the tunneling matrix element is independent of energy over the relevant energy range $\approx eV_{bias}$, then the total tunneling current is obtained by integrating Equation (12.57) with respect to the energy

$$I(V) \propto |T|^2 \int_{-\infty}^{\infty} D_L(E) D_R(E + eV_{bias})[f(E) - f(E + eV_{bias})]dE. \quad (12.58)$$

In the low-bias regime $(V \to 0)$ and at low temperatures $(T \to 0)$,

$$\lim_{V_{bias},T \to 0} \frac{[f(E) - f(E + eV_{bias})]}{eV_{bias}} = \delta(E - E_F), \qquad (12.59)$$

where E_F is the Fermi energy, so that we find

$$G = \frac{dI}{dV} \propto |T|^2 D_L(E_F) D_R(E_F), \qquad (12.60)$$

where $dI/dV = G$ is the zero-bias conductance of the tunneling junction. It is clear from the above equation that in the low-bias regime, $I \propto V$, i.e., the junction displays an ohmic behavior.

Equation (12.60) was applied by Jullière [6] to discuss the magnitude of the observed tunneling magnetoresistance of an FM-I-FM junction with two ferromagnetic (FM) electrodes separated by an insulating (I) barrier. In the case of magnetic electrodes, the densities of states for electrons with spin parallel $D^{\uparrow}(E)$ and anti-parallel $D^{\downarrow}(E)$ to the magnetization are different because the energy bands of up-spin and down-spin electrons are split by the exchange interaction (recall the Stoner-Wohlfarth bandstructure in Fig. 12.1).

To take into account the dependence of the tunneling current on the relative orientation of the magnetic moments of the two electrodes, Jullière made an additional assumption that electron spin is conserved in tunneling. It then follows that tunneling of up-spin and down-spin electrons are two independent processes, i.e., the tunneling current flows in the up-spin and down-spin channels as if in two wires connected in parallel. Such a two-current model is also used to interpret the closely related giant magnetoresistance effect in magnetic multilayers. Assuming that the magnetic moments of the two electrodes are anti-parallel at zero magnetic field and parallel in a saturating applied field H_s, we find using Equation (12.60) that the conductance of the junction in zero field is

$$G(0) \propto D_L^{\uparrow}(E_F) D_R^{\downarrow}(E_F) + D_L^{\downarrow}(E_F) D_R^{\uparrow}(E_F), \qquad (12.61)$$

and its conductance at the saturating field is

$$G(H_s) \propto D_L^{\uparrow}(E_F) D_R^{\uparrow}(E_F) + D_L^{\downarrow}(E_F) D_R^{\downarrow}(E_F). \qquad (12.62)$$

As a result, the magnetoresistance ratio is given by

$$MR_2 = JMR = \frac{G(0)^{-1} - G(H)^{-1}}{G(0)^{-1}}. \qquad (12.63)$$

Introducing two new parameters P_1 and P_2 to characterize the spin polarization of left and right electrodes,

$$P_1 = [D_L^{\uparrow}(E_F) - D_L^{\downarrow}(E_F)]/[D_L^{\uparrow}(E_F) + D_L^{\downarrow}(E_F)] \qquad (12.64)$$

and

$$P_2 = [D_R^\uparrow(E_F) - D_R^\downarrow(E_F)]/[D_R^\uparrow(E_F) + D_R^\downarrow(E_F)], \qquad (12.65)$$

we find that the magnetoresistance ratio is given by the Jullière formula

$$JMR = \frac{2P_1P_2}{1 + P_1P_2}, \qquad (12.66)$$

which we show next.

The parameters P_1 and P_2 can be determined independently from measurements of the tunneling current in FM-I-S junctions, in which one of the electrodes is a ferromagnet (FM) and the other is a superconductor (S).

Substituting the values $G(0)$ and $G(H)$ in Equation (12.63), the junction magnetoresistance ratio is found to be

$$
\begin{aligned}
JMR &= \frac{\left(\frac{1}{D_L^\uparrow(E_F)D_R^\downarrow(E_F)+D_L^\downarrow(E_F)D_R^\uparrow(E_F)}\right) - \left(\frac{1}{D_L^\uparrow(E_F)D_R^\uparrow(E_F)+D_L^\downarrow(E_F)D_R^\downarrow(E_F)}\right)}{\frac{1}{\left(D_L^\uparrow(E_F)D_R^\downarrow(E_F)+D_L^\downarrow(E_F)D_R^\uparrow(E_F)\right)}} \\
&= \frac{D_L^\uparrow(E_F)[D_R^\uparrow(E_F) - D_R^\downarrow(E_F)] - D_L^\downarrow(E_F)[D_R^\uparrow(E_F) - D_R^\downarrow(E_F)]}{D_L^\uparrow(E_F)D_R^\uparrow(E_F) + D_L^\downarrow(E_F)D_R^\downarrow(E_F)}.
\end{aligned}
$$

$$(12.67)$$

Dividing numerator and denominator by $[D_L^\uparrow(E_F) + D_L^\downarrow(E_F)] \times [D_R^\uparrow(E_F) + D_R^\downarrow(E_F)]$ and rearranging, we find

$$JMR = \frac{P_1P_2}{\frac{D_L^\uparrow(E_F)D_R^\uparrow(E_F)+D_L^\downarrow(E_F)D_R^\downarrow(E_F)}{[D_L^\uparrow(E_F)+D_L^\downarrow(E_F)]\cdot[D_R^\uparrow(E_F)+D_R^\downarrow(E_F)]}}. \qquad (12.68)$$

Finally, multiplying and dividing the last expression by 2 and adding and subtracting 1 in the denominator and simplifying, we get

$$
\begin{aligned}
JMR &= \frac{2P_1P_2}{\frac{2D_L^\uparrow(E_F)D_R^\uparrow(E_F)+2D_L^\downarrow(E_F)D_R^\downarrow(E_F)}{[D_L^\uparrow(E_F)+D_L^\downarrow(E_F)][D_R^\uparrow(E_F)+D_R^\downarrow(E_F)]} - 1 + 1} \\
&= \frac{2P_1P_2}{\frac{D_L^\uparrow(E_F)D_R^\uparrow(E_F)+D_L^\downarrow(E_F)D_R^\downarrow(E_F)-D_L^\uparrow(E_F)D_R^\downarrow(E_F)-D_L^\downarrow(E_F)D_R^\uparrow(E_F)}{[D_L^\uparrow(E_F)+D_L^\downarrow(E_F)][D_R^\uparrow(E_F)+D_R^\downarrow(E_F)]} + 1} \\
&= \frac{2P_1P_2}{\frac{D_L^\uparrow(E_F)[D_R^\uparrow(E_F)-D_R^\downarrow(E_F)]-D_L^\downarrow(E_F)[D_R^\uparrow(E_F)-D_R^\downarrow(E_F)]}{[D_L^\uparrow(E_F)+D_L^\downarrow(E_F)][D_R^\uparrow(E_F)+D_R^\downarrow(E_F)]} + 1}.
\end{aligned}
$$

$$(12.69)$$

Using the definitions of the spin polarization (P_1, P_2) of the contacts, we finally get the compact result

$$JMR = \frac{2P_1P_2}{1 + P_1P_2}, \qquad (12.70)$$

which is the famous Jullière formula.

Exercise:

- It is customary to estimate the difference in resistance between the parallel and anti-parallel magnetizations of a tunnel junction (spin valve in which transport takes place through the intermediate layer via tunneling) using the tunneling magnetoresistance ratio (TMR) defined as

$$TMR = \frac{R_A - R_P}{R_P}. \tag{12.71}$$

Using the Jullière formula, show that

$$TMR = \frac{2P_1P_2}{1 - P_1P_2}. \tag{12.72}$$

- Starting with the Jullière formula, find the expression of the magnetoresistance ratio $(G_P - G_{AP}) / (G_P + G_{AP})$ in terms of P_1 and P_2, the spin polarizations of the ferromagnetic contacts.

12.2.5 Application of the Jullière formula to extract the spin diffusion length in a paramagnet from spin valve experiments

Consider a spin valve consisting of a ferromagnet-paramagnet-ferromagnet structure. Electrons are injected from the left ferromagnetic contact into the paramagnet with a polarization P_1, which is the spin polarization of carriers at the Fermi energy in the left ferromagnet. We assume that there is no loss of spin polarization at the interface. This is actually a pretty good approximation in some cases, e.g., when the paramagnet is a conjugated (organic) polymer that exhibits a "proximity effect," whereby a few atomic layers in contact with the ferromagnet can get spin polarized [19]. The polarization of the injected spins in the paramagnet decays with distance x as $exp[-x/L_s]$ as the spins drift and diffuse (see Chapter 11) across the paramagnetic layer. When the spins arrive at the right ferromagnetic contact, the polarization has decayed to P_1e^{-d/L_s} where d is the width of the paramagnetic layer. We then assume that there is a tunnel barrier at the right contact through which the spins tunnel. The TMR associated with tunneling through this barrier is given by the Jullière formula (Equation (12.72)) with P_1 replaced by P_1e^{-d/L_s}. Therefore

$$TMR = \frac{2P_1e^{-d/L_s}P_2}{1 - P_1e^{-d/L_s}P_2}. \tag{12.73}$$

Thus, from the measured TMR and knowledge of P_1, P_2, and d, we can find the spin diffusion length L_s. This method has been widely used to find the spin diffusion length in many materials.

12.2.6   Spin valve experiments

The typical magnetoresistance trace of a spin valve is shown in Fig. 12.13. One first applies a very high magnetic field on the structure which magnetizes both ferromagnetic contacts in the direction of the field. In this state, the contacts are magnetized parallel to each other, so that the device resistance will be low, assuming that the spin polarizations in the two contacts have the same sign (e.g., cobalt and nickel). The magnetic field is then decreased, swept past zero and made negative (i.e., its direction reversed). At some point, when the coercive field H_{c1} of one of the ferromagnets is reached, this ferromagnet flips magnetization, while the other ferromagnet (which has the higher coercivity H_{c2}) still retains its original direction of magnetization. At this point, the two magnetizations are anti-parallel and the resistance increases. Finally, when the magnetic field in the reverse direction increases further, the coercive field of the second ferromagnet is reached and this one flips too. Now, both magnetizations are again parallel and the resistance drops. Therefore, we should see a resistance peak between the coercive fields H_{c1} and H_{c2}. The height of this peak is $\Delta R = R_{AP} - R_P$. If the magnetic field is scanned in the reverse direction, we will see the same peak between the coercive fields. The background resistance is R_P.

From the measured data, we can find $TMR = \Delta R / R_P$, which then yields the spin diffusion length in the paramagnet from Equation (12.73).

The reader can easily infer that if the two ferromagnetic contacts have opposite signs of spin polarization, then instead of a spin valve "peak," one will observe a spin valve "trough."

12.3   Hysteresis in spin valve magnetoresistance

In real experiments, spin valve effects are sometimes masked by spurious phenomena such as anisotropic magnetoresistance (AMR), which was discovered by Lord Kelvin in 1851 and arises whenever the magnetizations of the ferromagnets are not collinear with the direction of current flow. Hysteresis of the ferromagnets can also complicate matters. These spurious effects are easily eliminated if the combined resistance of the ferromagnets is much smaller than the resistance of the paramagnet. Note that the spin valve device has three resistors in series: the first ferromagnet, the paramagnet, and the second ferromagnet. The paramagnet is by far the most resistive if it is a semiconductor or insulator, and the ferromagnets are metals. Therefore, the paramagnet's resistance will dominate. If the resistance peak ΔR is much larger than the total resistance of the ferromagnets, it is most likely not associated with AMR or hysteresis and can be unambiguously ascribed to the "spin valve effect."

In this section, we discuss the hysteresis effects in a ferromagnet since they

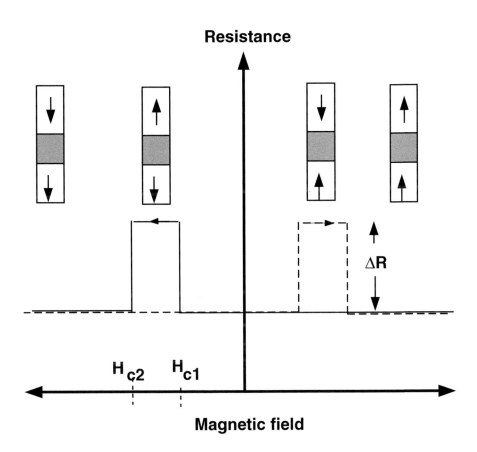

FIGURE 12.13

The ideal magnetoresistance trace measured in a spin valve experiment. A resistance peak is observed between the coercive fields of the two ferromagnets. The spin orientations in the ferromagnetic layers are shown in each region of the magnetic field. We assume that the lower layer has the lower coercivity. The solid trace is for the reverse scan of the magnetic field and the broken trace is for the forward scan.

are unavoidable features of any ferromagnet. A piece of bulk ferromagnet is made up of "domains" that have sizes varying between fractions of a micron to a few microns. Within each domain, the magnetic moments of all the atoms (those that have any residual magnetic moment) point in the same direction, so that each domain has complete spontaneous magnetic ordering and therefore an internal magnetic field. In the "unmagnetized" state, the magnetic moments of different domains point in different directions, so that they vectorially add up to zero and there is no net magnetization of the ferromagnet as a whole. When a sufficiently strong external magnetic field is applied to magnetize the ferromagnet, the magnetic moment of every domain turns in the direction of the field, so that ultimately all domains are magnetized parallel to the field. This is the saturated "magnetized" state.

There are two types of interactions between spins: "exchange", which we have discussed in Chapter 10, and "dipolar" interaction, which we have not discussed. A discussion of dipolar interaction can be found in [5]. It is much weaker, typically a thousand times weaker between neighboring spins in a metal, than exchange. However, the exchange interaction is short-range since it depends on the overlap between the wavefunctions of the interacting electrons. It falls off exponentially with increasing separation between the interacting electrons. In contrast, the dipolar interaction is long range and falls off as the inverse cube of the separation. As a result, the dipolar interaction becomes quite important if many spins are involved and may overwhelm the exchange interaction. This fact is responsible for the formation of "domains" in a ferromagnet.

Neighboring domains typically have oppositely directed magnetizations so that electrons bordering a domain wall on two sides have anti-parallel spins. Recall from the Hartree-Fock model of Chapter 10 that the exchange energy is non-zero and negative only between parallel spins. Therefore, the spins near the domain boundaries have their total energy raised by the vanishing of the exchange interaction. Spins far from the domain walls are not subjected to this effect since the exchange energy is short range. Therefore, only a small price in exchange energy is paid to form domains. On the other hand, because of the long range nature, the dipolar energy of every spin drops when domains are formed. Therefore, provided the domains are not too small, domain formation is energetically favored. Every spin can lower its (tiny) dipolar energy by forming domains, even though few spins have their (large) exchange energy raised in the process [5].

When a magnetizing field is applied, domains turn in the direction of the field, or domains that are already aligned along the field grow at the expense of unaligned domains. When the magnetizing field is removed, the domains do not necessarily return to their original alignments because crystalline defects may prevent domain wall motion. It then becomes necessary to apply a strong field in the opposite direction to overcome this pinning and restore the original unmagnetized state. This phenomenon is known as *hysteresis*, and the magnetic field required to reduce the magnetic moment to zero (corresponding to the original unmagnetized state) is known as the *coercive field*. In the magnetic recording industry, there is usually a great deal of desire to increase the coercive field since then small magnetic perturbations due to noise could not flip stored data and corrupt the memory. However, too large a coercive field is also not a good thing since that requires a strong magnetic field (and hence significant power consumption) to erase old data or write new data.

Hysteresis is usually manifested in magnetic moment versus magnetic field (M-H) measurements. A typical curve for cylindrical iron nanodots of 10 nm diameter and length varying from 200-600 nm is shown in Fig. 12.14. The value of the coercive field, the saturation magnetization and the shape of the hysteresis loop ("squareness") are determined by the precise domain mechanics and magnetic reversal mechanisms.

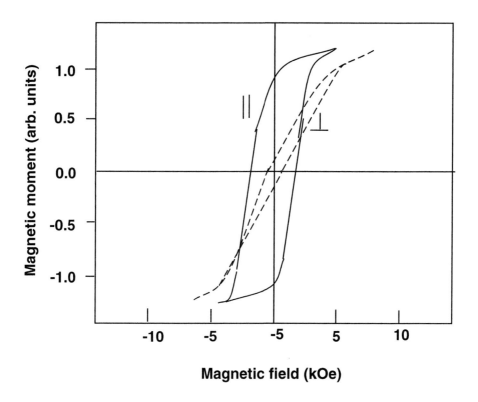

FIGURE 12.14

The approximate M-H loop for iron nano-cylinders of diameter 10 nm and length varying between 100 and 250 nm, showing hysteresis. The solid lines denote the loop obtained when the applied magnetic field is along the axis of the cylinder and the broken lines denote the loop obtained when the magnetic field is transverse to the axis.

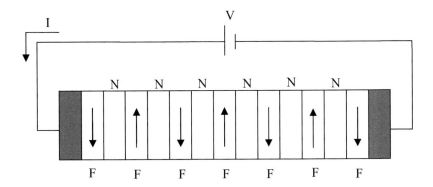

FIGURE 12.15

Spin-valve-superlattice shown in the anti-parallel configuration.

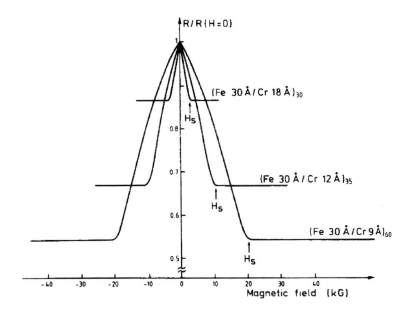

FIGURE 12.16

Magnetoresistance of three Fe/Cr superlattice units at a temperature of 4.2 K. The current and applied magnetic field are along the same [110] axis in the plane of the layers (CIP geometry). Reprinted with permission from M. N. Baibich, et al., Phys. Rev. Lett., **61**, 2472 (1988). Copyright (1988) American Physical Society. http://link.aps.org/abstract/PRL/v61/p2472.

12.4   Giant magnetoresistance

Giant magnetoresistance devices are an extension of spin valves. One might wonder if any benefit would accrue from increasing the number of layers in the basic unit cell of a spin valve, while keeping the width of each layer a few nanometers thick. This would result in forming a "spin-valve-superlattice," as depicted in Fig. 12.15.

Let us use the simple two-resistor model discussed earlier to calculate the resistance of the finite superlattice composed of N different layers.

If N is odd, i.e., the end layers are ferromagnetic, it is easy to show that, in the parallel configuration,

$$R_P = \frac{NR_\uparrow NR_\downarrow}{NR_\uparrow + NR_\downarrow} = N\frac{R_\uparrow R_\downarrow}{R_\uparrow + R_\downarrow}. \tag{12.74}$$

In the anti-parallel configuration, i.e., when the magnetization directions of the magnetic layers in adjacent ferromagnetic layers are anti-parallel

$$R_{AP} = \frac{\left[\frac{N-1}{2}R_\uparrow + \frac{N+1}{2}R_\downarrow\right]\left[\frac{N-1}{2}R_\downarrow + \frac{N+1}{2}R_\uparrow\right]}{N(R_\uparrow + R_\downarrow)}, \tag{12.75}$$

and if N is large, then

$$R_{AP} = \frac{N}{4}(R_\uparrow + R_\downarrow). \tag{12.76}$$

It is not difficult to prove that similar expressions hold if N is even and large.

Using the above results for R_P and R_{AP}, it is easy to show that the values of the magnetoresistance ratios MR_1, MR_2, and MR_3 are identical to those derived for the basic spin valve composed of just three layers.

Based on the simple two-resistor model, it would appear that there is no additional advantage in building a spin-valve-superlattice and studying its magnetoresistance properties. However, we know that the two-resistor model is oversimplified. Therefore, its predictions are dubious.

In the late 1980s, two groups performed magnetoresistance measurements on multilayered spin-valve-superlattices and came across a rather surprising result [7, 8]. Some of the experimental data from [7] are shown in Fig. 12.16.

The resistance of magnetic metallic multilayers dropped to less than one-half of their original value in a weak magnetic field. This huge negative magnetoresistance (MR) of multilayers is called *giant magnetoresistance* (GMR) because it is much larger than the intrinsic MR of the constituent bulk materials. The discovery of the GMR effect in Fe/Cr magnetic multilayers in 1988 [7, 8] spawned a great deal of research on layered magnetic materials systems. Its importance has now been recognized by the Nobel Prize in Physics awarded to authors from the two groups in [7, 8].

The basic explanation for the GMR phenomenon is qualitatively similar to the one given earlier for the operation of a spin valve composed of just three layers. It is a consequence of the non-equivalence of the two *spin-channels* and the resulting change in the magnetic state of the sample in the presence of a small external magnetic field. The presence of a small external magnetic field forces the magnetizations of the FM layers to be oriented in parallel and consequently the resistance of the superlattice can drop to less than one-half its value at zero external magnetic field.

Most of the GMR experiments were originally carried out with the current flowing in the plane of the multi-layers, a geometry referred to as current-in-plane (CIP) geometry, as illustrated in Fig.11.17(b). This configuration results in fairly large values of the sample resistance since the sample length (between the current injecting and detecting contacts) is typically orders of magnitude larger than the layer thicknesses. On the other hand, in the so-called current-perpendicular-to-plane (CPP) geometry illustrated in Fig.11.17(a), the length of the sample is just the sum total of the layer thicknesses, which is typically much smaller than the film's lateral dimensions. Hence, this configuration results in ultra-low sample resistances, which can only be measured with extremely sensitive equipment. Therefore, the CIP configuration is preferred over the CPP configuration, even though the latter may result in larger relative changes in the resistance in a magnetic field.

Resistor Network Theory of GMR

For a sophisticated treatment of GMR, which is capable of explaining the experimentally observed strong dependence of magnetoresistance on the number of layers in a GMR sample, we need to incorporate the effect of spin-dependent scattering to determine the resistance of a magnetic multilayer in its "ferromagnetic" (neighboring layers parallel) and "anti-ferromagnetic" (neighboring layers anti-parallel and alternating layers parallel) configurations. The reader is referred to Chapter 4 of [21], where Mathon has developed an extension of the resistor network theory of GMR which includes the effect of bulk and interface scattering for both the CPP and CIP configurations based on the two-fluid model described earlier. Increasingly more sophisticated models of GMR have been developed over the years which are fairly good at explaining most experimental data [20, 22, 23].

12.4.1 Applications of the spin valve and GMR effects

The demonstration of the GMR effect has been one of the early successes of spintronics. It has made a rapid transition from discovery to commercialization [24, 25]. The first commercial product based on GMR was a magnetic field sensor released in 1994 [26], followed shortly in 1997 by the production by IBM of the first GMR read head for reading data stored in magnetic hard disks [27]. In the 1990s, GMR-based read heads replaced earlier read heads

(a)

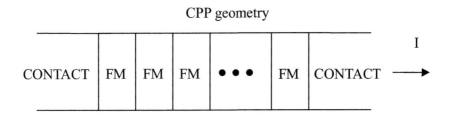

CPP geometry

(b)

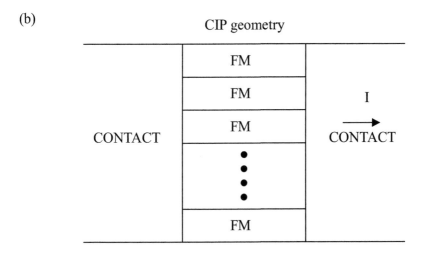

CIP geometry

FIGURE 12.17

Magnetoresistance measurements are performed in either the (a) current perpendicular to the plane (CPP) or in the (b) current parallel to the plane (CIP) geometry. Reproduced with permission from J. Mathon, "Phenomenological Theory of Giant Magnetoresistance" in ***Spin Electronics***, Eds. M. Ziese and M. J. Thornton, (Springer Verlag, New York, 2001). Copyright(2001) Springer Verlag.

based on the anisotropic magnetoresistance effect.[§] Today, GMR-based read heads are frequently found in laptop or desktop computers and music/video storage devices such as Apple iPods. In the 21st century, TMR-based read heads began to displace GMR based read heads. The typical TMR device is essentially a thin insulating layer (of thickness about 1 nm) sandwiched between two ferromagnetic contacts. At the time of writing this textbook, the record magnetoresistance exhibited by a TMR device is an astounding 472% at room temperature [28].

The first GMR-based random access memory (RAM) chips were produced by Honeywell in early 1997. MRAM chips based on TMR devices are now marketed by several companies, such as Freescale Semiconductor in the US. The impact of GMR/TMR technology on the sensor market is currently estimated worldwide at several $100 million a year [29].

Magnetic read heads and magnetic random access memories

Practical spin valves are typically constructed such that the magnetic moment in one of the ferromagnetic layers is more easily reversed than in the other by an external magnetic field. These materials are referred to as *soft* and *hard* magnetic layers, respectively. The soft layer can be made very sensitive to the presence of a small external magnetic field and acts as the control terminal to allow or block the current flow through the spin valve.

A schematic representation of one of the first magnetic read heads based on the GMR phenomenon is shown in Figure 12.18. The goal of the read head is to sense the magnetic bits that are stored in the storage media (tape or disk). The bits consist of a sequence of magnetized regions or domains whose magnetization directions correspond to either bit 0 or bit 1. The two magnetization directions corresponding to the two bits are anti-parallel. The magnetic field lines emanating from the magnetic domains are only substantial in the vicinity of the domain walls separating two adjacent domains. When the heads (north poles) of two adjacent domains meet (domain wall 2 in Fig. 12.18), uncompensated positive poles in the domains generate magnetic field lines pointing upwards on the surface passing under the GMR read head. In contrast, the magnetic field lines point downward (near domain wall 1 in Fig. 12.18) where the tails (south poles) of adjacent domains meet.

The GMR sensing element is fabricated in such a way that the magnetic moment in the soft ferromagnetic layers is parallel to the plane of the storage media in the absence of any external magnetic field. The magnetic moments in the hard ferromagnetic layers are oriented perpendicular to the plane of the media. When the GMR read head passes over domain wall 1, the field lines in its vicinity push the magnetization of the soft layers down, whereas it is pushed up when passing over domain wall 2. As a result, there is a change

---

[§]The anisotropic magnetoresistance effect typically produces a magnetoresistance of about 2%.

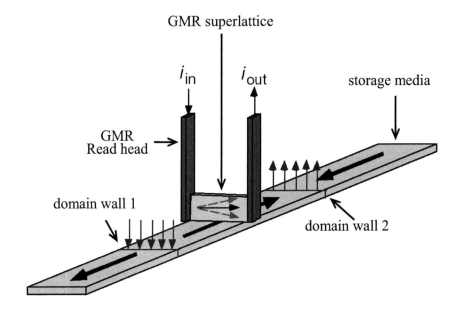

FIGURE 12.18
Schematic of a GMR read head composed of a ferromagnetic/normal-metal superlattice in which the adjacent ferromagnetic layers consist of hard and soft magnetic materials. The magnetization direction of the soft layers is slightly shifted either up or down in response to the magnetic field lines emanating from the domain walls existing at the interface between the magnetic domains in the storage media. The resulting small change in the resistance of the GMR read head is sensed as a small increase or decrease in the current flowing through the GMR element. Reproduced from ref. [27] with permission from the American Association for the Advancement of Science (copyright American Association for the Advancement of Science).

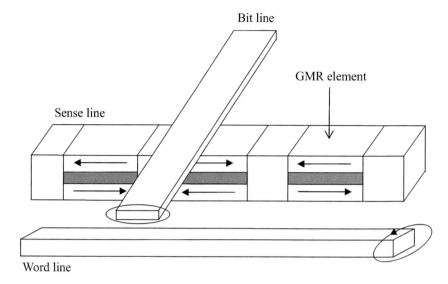

Bit line

GMR element

Sense line

Word line

FIGURE 12.19

Schematic illustration of a magnetic random access memory (MRAM) consisting of GMR elements connected in series. The writing and reading of each element is performed via the combination of applied magnetic fields generated by currents driven through the **Word** and **Bit** lines below and above the GMR elements. Reproduced from [27] with permission from the American Association for the Advancement of Science (copyright American Association for the Advancement of Science).

in the resistance of the GMR element. An increase is associated with more anti-aligned ferromagnetic layers, while a decrease is associated with more aligned layers. The best design of such GMR read heads is one for which a maximum rate of change in resistance can be observed for a small change in the sense field. In state-of-the-art GMR read heads, changes in resistance of about 1% per Oersted are readily achieved.

In 1997, Honeywell demonstrated the first *non-volatile* MRAM based on GMR elements fabricated with standard lithography. These devices exhibited access (read/write) speed and density comparable to semiconductor memory at that time. The basic structure is shown in Fig. 12.19. It consists of GMR elements placed in series and separated by lithographically delineated regions to form a *sense line*. The sense line is the information storing media (tape or disk) with a resistance that is the sum of the resistance of its GMR elements. When active, a current is passed through the sense line, and amplifiers at the ends of the line detect the change in resistance in the GMR elements. The resistance of each individual element can be changed through an array of overlay lines (*write* lines) and underlay lines (*bit* lines) to address each individual GMR component. The networks of lines are electrically insulated, but when current pulses are run through them, they generate magnetic fields that can act on the GMR elements. A typical addressing scheme uses "half-select" pulses in the overlay and underlay lines, i.e., the magnetic field produced by a word-line pulse is one-half of that needed to reverse the magnetization in the soft layers of each GMR element. Hence, wherever two lines (word and bit) in the grid overlap, the two half-select pulses they produce generate a combined magnetic field sufficient to selectively reverse a soft layer (or, at higher current levels, sufficient to reverse a hard layer also). Typically, a half-select pulse rotates the magnetization in the soft magnetic layer by 90° and the half-select pulse in the other line completes the magnetization rotation to 180° compared to its original state. Through this grid of word and bit lines, any GMR element in the array can be addressed either to store information (write data) or to read its magnetic state (read data) [27].

12.5   Spin accumulation

In this chapter, we introduced the concept of spin injection from a ferromagnet into a paramagnet. Spin injection will accumulate spin in the paramagnet near the interface, and the accumulated spin will gradually decay with distance away from the interface. We will now examine this decay and extract a "spin accumulation length," which is the characteristic distance over which the accumulated spin decays to $1/e$ times its magnitude at the interface.

Let us consider a bulk metal or semiconductor in which the energy dispersion relation is assumed to be parabolic for the sake of simplicity. If the material is non-magnetic, the two spin channels have the same mobility and, if we focus on the (k_x, k_y) plane, the Fermi energy surface for both spin channels will be shifted, as shown in the top part of Fig. 12.20 in the presence of a small external electric field. The shift Δk in momentum space of the up-spin and down-spin Fermi surfaces must be such that the following equation is satisfied:

$$F = -eE = \hbar \frac{dk}{dt} = \hbar \frac{\Delta k}{\tau_m}, \qquad (12.77)$$

where F is the force on the carrier, E is the external electric field, e is the magnitude of the electronic charge, τ_m is the electron scattering time, which is related to the mobility of the carrier by $\mu = e\tau_m/m^*$, and m^* is the electron effective mass.

In a ferromagnetic metal, the two spin channels will have different mobilities (recall Mott's two-fluid model of spin transport discussed earlier) and the shift of the Fermi surface associated with each spin channel will be different, as illustrated in the bottom part of Fig. 12.20.

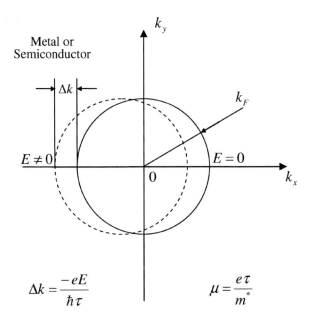

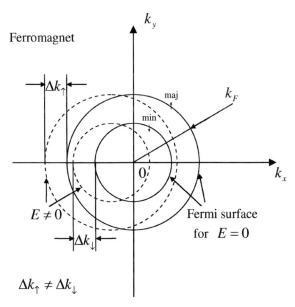

FIGURE 12.20

(Top) Shift of the Fermi surface in a paramagnetic metal or semiconductor in the presence of a constant external electric field leading to current conduction. (Bottom) Generalization to the two-fluid (or Mott model) of the ferromagnet in which the up-spin and down-spin electrons have different mobilities.

Let us consider a ferromagnet/paramagnet heterostructure. We will model the ferromagnet based on the Stoner-Wohlfarth model. We will assume that the up-spin electrons are the majority in the ferromagnet so that the density of up-spin electrons in the ferromagnet is much larger than that of the down-spin electrons. Consequently, the up-spin electrons will be the major contributors to any current injected by the ferromagnet. As a result, there will be a surplus of up-spin electrons in the paramagnet near the interface with the ferromagnet. This will cause *spin accumulation* in the paramagnet near the interface and an associated magnetic moment per unit volume. Under steady-state condition, the spin accumulation cannot extend into the paramagnet indefinitely since some of the up-spin electrons will eventually be converted into down-spin electrons as a result of spin-flip scattering events. In fact, far into the bulk of the paramagnet, the population of up-spin and down-spin electrons should be the same; therefore, we expect the spin accumulation to decay with distance as we move away from the interface. Hereafter, we give a simple argument to derive an expression for the length scale over which the spin accumulation decays away from the interface.

Based on the drift-diffusion model discussed in Chapter 11, we expect the spin accumulation to decay exponentially away from the interface as $\sim e^{-x/\lambda_{sd}}$ where λ_{sd} is the *spin accumulation length*. The latter can be estimated using the following simple method.

As illustrated in Fig. 12.21, an up-spin electron crossing from the ferromagnet into the non-magnetic material will undergo N momentum-changing collisions before being flipped. We label this average spin flip time $\tau_{\uparrow\downarrow}$. By definition, the average distance between momentum scattering collisions is λ_f, the mean free path. If the injected electron is allowed to move at random equally in all three dimensions after each collision, then the average distance which the spin penetrates into the non-magnetic material (perpendicular to the interface) or spin accumulation length λ_{sd} is given by

$$\lambda_{sd} = \lambda_f \sqrt{N/3}. \tag{12.78}$$

On the other hand, the total distance traveled by the injected up-spin electron is $N\lambda_f$, which in turn equals the injected electron's velocity at the Fermi level (Fermi velocity, v_F) times the spin-flip time τ_s $(=\tau_{\uparrow\downarrow} = \tau_{\downarrow\uparrow})$[¶], i.e.,

$$N\lambda_f = v_F \tau_s. \tag{12.79}$$

Eliminating the number N of collisions between the last two equations yields the expression

$$\lambda_{sd} = \sqrt{\frac{\lambda_f v_F \tau_s}{3}}. \tag{12.80}$$

[¶] We are assuming elastic collisions here, so that the magnitude of the electron's velocity is invariant. We are also assuming that the velocity with which the carriers are injected is the Fermi velocity v_F.

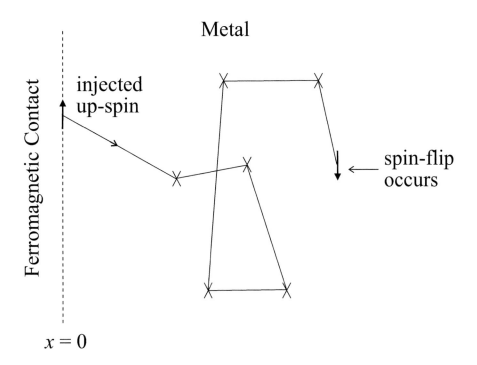

FIGURE 12.21
Illustration of the spin injection mechanism for an up-spin electron injected from a ferromagnetic contact into a paramagnet. After a random walk in the paramagnet, where it suffers N momentum-relaxing scattering events (indicated by crosses), the spin of the injected electron is eventually flipped. The average (over an ensemble of many injected carriers) distance from the interface for spin flip to occur is roughly the *spin diffusion length*.

This last relation clearly shows that λ_{sd} and λ_f are related but not equal. For example, if we dope silver with increasing levels of a non-magnetic impurity (such as gold), λ_{sd} will drop not only because of a decrease in the electron mean free path λ_f with increasing impurity concentration, but also because a larger concentration of impurities will lead to a reduced spin-flip time τ_s as a result of increased spin-orbit scattering due to the heavy gold atoms [30].

Warning: The reader who has already perused and understood the material in Chapter 8 will be immediately bothered by the preceding discussion. It appears that

$$\frac{\tau_s}{\tau_m} = N, \tag{12.81}$$

where τ_m is the momentum relaxation time. The last equation tells us that the spin relaxation time is directly proportional to the momentum relaxation time. This may be true for the Elliott-Yafet mechanism, but *is certainly not true* for the D'yakonov-Perel' mechanism. Spin relaxation due to the Elliott-Yafet mechanism is, in fact, associated with momentum relaxation and it is known that the spin relaxation time and momentum relaxation time are proportional to each other (recall Chapter 8)). On the other hand, the spin relaxation time associated with D'yakonov-Perel' relaxation is inversely proportional to the momentum relaxation time, which immediately contradicts Equation (12.81). Therefore, the picture presented here has no applicability to D'yakonov-Perel' relaxation. In fact, as we have seen in Chapter 11, the entire drift-diffusion model of spin transport is practically invalid if the primary spin relaxation mechanism is D'yakonov-Perel'. The reader should bear this in mind whenever the drift-diffusion model is encountered.

From the previous analysis, we can estimate the importance of spin accumulation in a metal for typical current densities across a ferromagnet/paramagnet interface. We define the spin polarization α of the injected current as

$$\alpha(x) = \left(J_\uparrow(x) - J_\downarrow(x)\right)/J, \tag{12.82}$$

where J is the total current density $J_\uparrow(x) + J_\downarrow(x)$ across the interface, with x being the direction of the normal to the interface. Since J does not depend on x, the net spin accumulation in the metal for a current J flowing through the junction can be found by equating the net spin injection across the interface

$$\left[\frac{d}{dt}\left(n_\uparrow - n_\downarrow\right)\right]_{x=0} = \frac{A\alpha(0)J}{e} \tag{12.83}$$

(where A is the cross-sectional area of the junction) to the decay rate of the overall spin concentration in the entire volume in the paramagnet where spin accumulation occurs. Assuming an exponential decay of the spin accumulation with characteristic decay constant λ_{sd}, we get

$$\left[\frac{d}{dt}(n_\uparrow - n_\downarrow)\right]_{x=0} = \frac{A}{\tau_s}\int_0^\infty (n_\downarrow - n_\uparrow)\,dx. \tag{12.84}$$

Since

$$n_\uparrow(x) - n_\downarrow(x) = n_0 e^{-x/\lambda_{sd}}, \tag{12.85}$$

where $n_0 = n_\uparrow(0) - n_\downarrow(0)$, we get

$$n_0 = \frac{\alpha(0)J\tau_s}{e\lambda_{sd}} = \frac{3\alpha(0)J\lambda_{sd}}{ev_F\lambda_f}. \tag{12.86}$$

.

For $\alpha(0) = 1$ (half-metallic contact), $v_F = 10^6$ m/s, $\lambda_f = 5$ nm, $\lambda_{sd} = 100$ nm, and a typical current density $J = 10^3$ A/cm^2, we get $n_0 = 4 \times 10^{22}$ m^{-3}. Since in normal metals, the electron concentration is on the order of several 10^{28} m^{-3}, the net spin accumulation in a normal metal (paramagnet) is typically small, with only one part in 10^6 of the electrons being spin polarized. The magnetic field B associated with this spin accumulation is given by

$$B = \mu_0 M = \mu_0 n_0 \mu_B \simeq 10^{-9} \text{ Tesla}, \tag{12.87}$$

which is very small compared to the magnetic field due to the current flowing through the interface and generating the spin accumulation.

12.6 Spin injection across a ferromagnet/metal interface

In 1987, van Son et al. [31] first pointed out that, because of spin accumulation, whenever a current flows from a ferromagnet into a paramagnet, the distribution of up-spin and down-spin carriers has to change. A similar phenomenon occurs across a superconductor-normal metal interface, where a conversion between normal current and supercurrent is accompanied by an electrochemical potential difference between quasiparticles and Cooper pairs near the interface.

For simplicity, we consider a ferromagnet/normal metal interface in the plane $x = 0$ and assume that the system is homogeneous in the y- and z-directions. The ferromagnet fills the half-space $(x < 0)$ and the normal metal is located in the half-space $(x > 0)$, as shown in Fig. 12.22. The thickness of each material is assumed to be greater than their respective spin diffusion length, and a current of electron is assumed to flow from the left to right in the x-direction only, i.e., the positive terminal of the voltage source (battery) is connected to the normal metal and the negative terminal to the ferromagnetic contact.

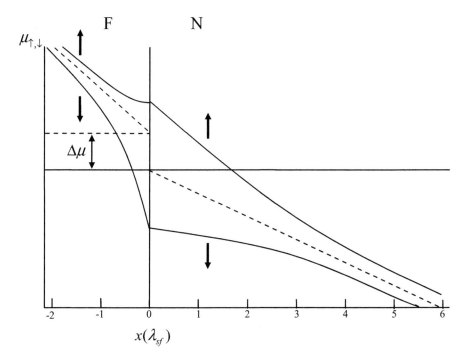

FIGURE 12.22

Illustration of the electrochemical potential difference for up-spin and down-spin electrons near a ferromagnetic metal/non-magnetic metal interface when a current flows through the junction. The tilted dashed line represents the weighted electrochemical potential, $\mu_0 = \alpha(x)\mu_\uparrow(x) + (1 - \alpha)\mu_\downarrow(x)$. Reproduced with permission from H. X. Tang et al., "Spin Injection and Transport in Micro- and Nano-scale Devices" in ***Semiconductor spintronics and Quantum Computation***, Eds. D. D. Awschalom, N. Samarth and D. Loss (Springer-Verlag, Berlin, 2002). Copyright(2002) Springer-Verlag.

If, in both contacts, the rate of scattering events that do *not* flip spin far exceeds the spin-flip rates at any coordinate point x, we can define individual electrochemical potential for the up-spin ($\mu_\uparrow$) and down-spin ($\mu_\downarrow$) channels. These electrochemical potentials are expected to be spatially varying and quite different from each other close to the interface. To quantify the difference between them, the conductivity and current density are separated into two components by introducing the variables $\alpha(x)$ and $\beta(x)$, such that

$$\sigma_\uparrow(x) = \alpha(x)\sigma, \tag{12.88}$$

$$\sigma_\downarrow(x) = (1 - \alpha(x))\sigma, \tag{12.89}$$

and

$$J_\uparrow(x) = \beta(x)J, \tag{12.90}$$

$$J_\downarrow(x) = (1 - \beta(x))J, \tag{12.91}$$

where J is the total current density flowing across the interface. In steady state, J is independent of x as a result of current continuity.

At a distance far from the interface (exceeding the spin accumulation length in either material), the two electrochemical potentials $\mu_\uparrow(x)$ and $\mu_\downarrow(x)$ will converge toward each other since the up-spin and down-spin populations should be near their equilibrium values, i.e.,

$$\beta(-\infty) = \alpha_F, \tag{12.92}$$

and

$$\beta(+\infty) = \alpha_N = 0.5, \tag{12.93}$$

where the subscripts "F" and "N" stand for ferromagnet and normal metal (or non-magnet), respectively.

Furthermore, $\alpha(x)$ is expected to change abruptly at the interface due to the difference in conductivities of the two materials, but $\beta(x)$ must be continuous in the absence of spin-flip scattering mechanisms at $x = 0$. Consequently, in regions that are a few diffusion lengths away from the interface, $d\beta/dx \neq 0$ and $\mu_\uparrow(x) \neq \mu_\downarrow(x)$. This electrochemical potential difference is the driving force for the spin current conversion across the interface.

At steady state, the electrochemical potential difference obeys the diffusion equation

$$\frac{\mu_\uparrow - \mu_\downarrow}{\tau_s} = D\frac{d^2(\mu_\uparrow - \mu_\downarrow)}{dx^2}, \tag{12.94}$$

where τ_s is the spin-flip scattering rate and D is the *charge* diffusion coefficient which, in the non-magnetic metal, is given by

$$D = \frac{1}{3}v_F\lambda_f. \tag{12.95}$$

In the ferromagnet, the charge diffusion coefficient D_F is estimated using a weighted average

$$D_F = (1 - \alpha_F)D_{F,\uparrow} + \alpha_F D_{F,\downarrow}, \tag{12.96}$$

where $D_{F,\uparrow}$ and $D_{F,\downarrow}$ are the charge diffusion coefficients for up-spin and down-spin electrons that are not necessarily equal to each other. A derivation of Equation (12.94) is given in a series of problems at the end of the chapter.

When a small bias is applied between the contacts to the ferromagnet and normal metal, a general solution to the diffusion equation (12.94) can be found by assuming the following spatial dependence of the electrochemical potentials.

In the ferromagnetic contact, the electrochemical potentials for the up-spin and down-spin channels are of the form

$$\mu_{F\uparrow}(x) = \mu_1 + a_1 x + c_{\uparrow}^F e^{x/\lambda_{sd,F}},$$
$$\mu_{F\downarrow}(x) = \mu_1 + a_1 x + c_{\downarrow}^F e^{x/\lambda_{sd,F}}, \tag{12.97}$$

while in the non-magnetic metal, they have the form

$$\mu_{N,\uparrow}(x) = \mu_2 + a_2 x + c_{\uparrow}^N e^{-x/\lambda_{sd,N}},$$
$$\mu_{N,\downarrow}(x) = \mu_2 + a_2 x + c_{\downarrow}^N e^{-x/\lambda_{sd,N}}, \tag{12.98}$$

where, again, the subscripts "F" and "N" refer to ferromagnet and non-magnet, respectively.

The spin current densities in both spin channels are related to the electrochemical potentials according to the relations [32]

$$J_{F,\uparrow}(x) = -\frac{\beta_F}{e} \sigma_F \frac{d}{dx} \mu_{F,\uparrow}(x), \tag{12.99}$$

and

$$J_{F,\downarrow}(x) = -\frac{1 - \beta_F}{e} \sigma_F \frac{d}{dx} \mu_{F,\downarrow}(x), \tag{12.100}$$

in the ferromagnetic contact, while in the normal metal, these relations are [32]

$$J_{N,\uparrow}(x) = -\frac{\sigma_N}{2e} \frac{d}{dx} \mu_{N,\uparrow}(x), \tag{12.101}$$

and

$$J_{N,\downarrow}(x) = -\frac{\sigma_N}{2e} \frac{d}{dx} \mu_{N,\downarrow}(x). \tag{12.102}$$

If there is no spin-flip at the interface, then the following boundary conditions must be satisfied:

$$J_{F,\uparrow}(x = 0) = J_{N,\uparrow}(x = 0),$$
$$J_{F,\downarrow}(x = 0) = J_{N,\downarrow}(x = 0),$$
$$\mu_{F,\uparrow}(x = 0) = \mu_{N,\uparrow}(x = 0),$$
$$\mu_{F,\downarrow}(x = 0) = \mu_{N,\downarrow}(x = 0). \tag{12.103}$$

Furthermore, the following condition must be satisfied for all x under steady state:

$$J = J_{F,\uparrow}(x) + J_{F,\downarrow}(x) = J_{N,\uparrow}(x) + J_{N,\downarrow}(x). \tag{12.104}$$

In the set of equations (12.97 – 12.104), we have the eight unknowns $(\mu_1, \mu_2, a_1, a_2, c_\uparrow^F, c_\downarrow^F, c_\uparrow^N, c_\downarrow^N)$. We can impose the condition that the total current densities in the two materials must be independent of x by setting their derivatives with respect to x equal to zero. One of the two current density expressions must be set equal to the current density flowing through the device J. This leaves us with seven equations and eight unknowns. However, a little reflection will reveal that the difference between the two unknowns μ_1 and μ_2 is an independent parameter proportional to the current density flowing through the junction. In other words, the quantities μ_1 and μ_2 can not be determined individually, but their difference is related to the bias applied between the contacts to the ferromagnet and metal.

With the use of software tools such as MATHEMATICA or MATLAB, the system of Equations (12.97 – 12.104) can be solved by taking the difference $\mu_1 - \mu_2$ as a single variable. Using these results, the value of β at the interface, $\beta_I = \beta(x = 0)$, is found to satisfy the equation

$$2\beta_I - 1 = \frac{2\alpha_F - 1}{1 + 4\alpha_F(1 - \alpha_F)(\sigma_N^{-1} - \lambda_{sd,N})/(\sigma_F^{-1}\lambda_{sd,F})}. \tag{12.105}$$

This last equation predicts that $\beta_I = \alpha_F = 1$ for half-metallic contact because any spin flip process can only occur on the metallic side of the junction. Furthermore, it can be easily shown that the following inequality holds:

$$\alpha_N = 0.5 \le \beta_I \le \alpha_F. \tag{12.106}$$

The proof of this inequality is left to the reader as an exercise.

Introducing the weighted electrochemical potential

$$\mu_0 = \alpha(x)\mu_\uparrow(x) + (1 - \alpha)\mu_\downarrow(x), \tag{12.107}$$

it can be shown that

$$\frac{d}{dx}\mu_0 = -\frac{e}{\sigma(x)}J, \tag{12.108}$$

where $\sigma = \sigma_\uparrow(x) + \sigma_\downarrow(x)$ and J is the total current density flowing through the junction. In other words, $\mu_0(x)$ is the value the electrochemical potential would have had without a non-equilibrium redistribution between the up-spin and down-spin components of the current. The spatial dependence of μ_0 is shown as a tilted dashed line on either side of the junction in Fig. 12.22.

The discontinuity of μ_0 at the interface leads to a boundary resistance defined as

$$R_b = \frac{\Delta\mu}{eJ} = \frac{\mu_{F,0}(x = 0_-) - \mu_{N,0}(x = 0_+)}{eJ}, \tag{12.109}$$

where

$$\mu_{F,0}(x = 0_-) = \alpha_F \mu_{F,\uparrow}(x = 0_-) + (1 - \alpha_F)\mu_{F,\downarrow}(x = 0_-),$$
$$\mu_{N,0}(x = 0_+) = \alpha_N \mu_{N,\uparrow}(x = 0_+) + (1 - \alpha_N)\mu_{N,\downarrow}(x = 0_+). \quad (12.110)$$

The results obtained using MATHEMATICA yield

$$R_b = \frac{(2\alpha_F - 1)^2 (\sigma_N^{-1}\lambda_N)(\sigma_F^{-1}\lambda_F)}{(\sigma_F^{-1}\lambda_F) + 4\alpha_F(1 - \alpha_F)(\sigma_N^{-1}\lambda_N)}. \quad (12.111)$$

If $\sigma_F^{-1}\lambda_F$ is much smaller than $4\alpha_F(1 - \alpha_F)(\sigma_N^{-1}\lambda_N)$, then

$$R_b \sim \frac{(2\alpha_F - 1)^2 \sigma_F^{-1}\lambda_F}{4\alpha_F(1 - \alpha_F)}. \quad (12.112)$$

In that case, the spin current conversion mostly takes place in the ferromagnetic contact. For a half-metallic ferromagnet, $\alpha_F = 1$ and spin current conversion must occur on the metallic side, no matter the value of $\sigma_N^{-1}\lambda_N$.

Next, we use the concept of electrochemical potentials for the up-spin and down-spin channels to calculate the amount of spin polarization of the current in a semiconductor of width x_0 sandwiched between two identical ferromagnetic contacts. The efficiency of this structure as a spin valve is also analyzed.

12.7   Spin injection in a spin valve

The theory outlined in the previous section can be extended to study the efficiency of a spin valve formed of a semiconductor sandwiched between two ferromagnetic contacts separated by a distance x_0, as shown in Fig. 12.23.

The equations characterizing spin transport in all three regions assume the diffusive regime [32]

$$\frac{d}{dx}\mu_{\uparrow,\downarrow} = -\frac{e}{\sigma_{\uparrow,\downarrow}(x)}J_{\uparrow,\downarrow}, \quad (12.113)$$

and the diffusion equation (12.94).

As in the previous section, spin flip scattering mechanisms are neglected at both interfaces with the contacts, and the electrochemical potentials and current densities in the two spin channels are assumed to be continuous throughout the structure.

Far into the bulk of both contacts, the electrochemical potentials of both spin channels must be equal. By analogy with the case of the ferromagnet/metal interface, the difference $\mu_\uparrow - \mu_\downarrow$ is also expected to reach a maximum at both interfaces and to be proportional to the total current density flowing through the structure.

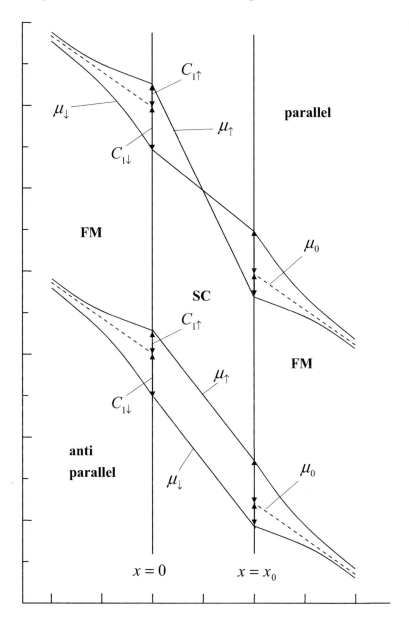

FIGURE 12.23

Illustration of the spatial dependence of the electrochemical potential in the two spin channels when the magnetizations in the identical ferromagnetic contacts are either parallel or anti-parallel. Reproduced with permission from G. Schmidt, et al., Phys. Rev. B, **62**, R4790 (2000). Copyright (2000) American Physical Society. http://link.aps.org/abstract/PRB/v62/p4790.

According to the diffusion equation (12.94), the splitting $\mu_\uparrow - \mu_\downarrow$ decays over the spin accumulation length $\lambda_{sd} = \sqrt{D\tau_s}$ of the material (where D is the charge diffusion constant).[||]

The spin accumulation length in the semiconductor is typically several orders of magnitude larger than the spin diffusion length in the ferromagnetic contacts [33]. Assuming an infinite spin accumulation length in the semiconductor, the diffusion equation (12.94) predicts a constant value (equal to the value at the interface) for $\mu_\uparrow - \mu_\downarrow$ inside the semiconductor. If the latter extends to infinity and the conductivities in both spin channels are equal in the semiconductor, the current densities in both channels are identical and according to Equation (12.94) no spin injection is possible, i.e., the spin-polarization of the current density $\alpha(x)$ is identically zero. However, the addition of a second ferromagnetic contact at a distance x_0 from the first one may lead to a different spatial dependence of the electrochemical potentials across the semiconductor, as illustrated qualitatively in Fig. 12.23. This is the problem that we formulate next and solve using the program MATHEMATICA. We will consider the cases where the magnetizations in the two contacts are either parallel or anti-parallel. Hereafter, any physical quantity in the left ferromagnetic contact, semiconductor channel and right ferromagnetic contact are labeled with the subscripts 1, 2 and 3, respectively.

By analogy with Equations (12.92–12.93), the bulk spin polarization in the two ferromagnetic contacts must be such that

$$\alpha_{1,3}(\pm\infty) = \beta_{1,3} = \beta, \qquad (12.114)$$

where β characterizes the fraction of the total spin current in the up-spin channel. Using the definitions (12.88) and (12.89), we then have

$$\sigma_{1,3}^\uparrow = \sigma_{1,3}(1 + \beta_{1,3})/2, \qquad (12.115)$$

and

$$\sigma_{1,3}^\downarrow = \sigma_{1,3}(1 - \beta_{1,3})/2. \qquad (12.116)$$

If the magnetizations of the two contacts are parallel, $\beta_1 = \beta_3$, and, if they are anti-parallel, $\beta_1 = -\beta_3$. Furthermore, if the applied bias between the two contacts is small (linear response regime), the difference in conductivity for the up-spin and the down-spin channel in the ferromagnets can easily be deduced from the Einstein relation with $D_{i\uparrow} \neq D_{i\downarrow} (i = 1, 3)$ and $\rho_{i\uparrow}(E_F) \neq \rho_{i\downarrow}(E_F)(i = 1, 3)$, where $\rho(E_F)$ is the density of states of the Fermi energy.

To separate the spin polarization effects from the normal current flow, we write the electrochemical potential in the ferromagnets for both spin directions as

---

[||]Note that the spin accumulation length λ_{sd} and the spin diffusion length L_s will be equal only if the charge diffusion coefficient D and the spin diffusion coefficient D_s are equal. They are generally *not* equal and therefore $\lambda_{sd} \neq L_s$.

$$\mu_{\uparrow,\downarrow} = \mu_0 + \mu_{\uparrow,\downarrow}^*, \tag{12.117}$$

μ_0 being the electrochemical potential without spin effects.

In the ferromagnetic contacts, we seek solutions of the diffusion equations as follows:

$$\mu_{i;\uparrow,\downarrow} = \mu_i^0 + \mu_{i\uparrow,\downarrow}^* = \mu_i^0(x) + C_{i;\uparrow,\downarrow}e^{\pm(x-x_i)/\lambda_{sd,F}}, \tag{12.118}$$

for $i = 1, 3$ with $x_1 = 0$, $x_3 = x_0$ and the $+(-)$ sign refers to the left and right contact, respectively.

From the boundary conditions

$$\mu_{1\uparrow}(-\infty) = \mu_{1\downarrow}(-\infty), \tag{12.119}$$

and

$$\mu_{3\uparrow}(+\infty) = \mu_{3\downarrow}(+\infty), \tag{12.120}$$

the slope of μ^0 is identical for both spin directions and also equal in regions 1 and 3 if the conductivity σ is identical in both regions.

In the semiconductor, we assume that the spin diffusion length L_s is several orders of magnitude longer than in the ferromagnet and much larger than the spacing between the two contacts and set $\tau_s = \infty$ in Equation (12.94). In this limit, the electrochemical potentials for up-spin and down-spin channels will vary linearly in the semiconductor channel

$$\mu_{2;\uparrow,\downarrow}(x) = \mu_{1;\uparrow,\downarrow}(0) + \gamma_{\uparrow,\downarrow}x, \tag{12.121}$$

where $\gamma_{\uparrow,\downarrow}$ are constants to be determined.

The last equation implies that $J_{2\uparrow}$ and $J_{2\downarrow}$ remain constant through the semiconductor. Furthermore, since we assume identical material for the two ferromagnetic contacts, the following equations must hold by symmetry:

$$\mu_{1\uparrow}(0) - \mu_{1\downarrow}(0) = \pm\left[\mu_{3\downarrow}(x_0) - \mu_{3\uparrow}(x_0)\right], \tag{12.122}$$

where the $+(-)$ sign corresponds to the parallel (anti-parallel) magnetizations in the contacts, respectively.

Parallel configuration

In the left ferromagnetic contact, the spatial dependence of the electro-chemical potentials in both spin channels must be of the form

$$\mu_{1\uparrow}(x) = \mu_1^0 + \delta_{1\uparrow}x + C_{1\uparrow}e^{\frac{x}{\lambda_{sd,F}}},$$
$$\mu_{1\downarrow}(x) = \mu_1^0 + \delta_{1\downarrow}x + C_{1\downarrow}e^{\frac{x}{\lambda_{sd,F}}}. \tag{12.123}$$

The corresponding spin current densities are given by [32]

$$J_{1\uparrow}(x) = -\frac{\sigma_{F,\uparrow}}{e}\frac{\partial \mu_{1\uparrow}}{\partial x} = -\frac{1+\beta}{2}\sigma_F\left(\delta_{1\uparrow} + \frac{C_{1\uparrow}}{\lambda_F}e^{\frac{x}{\lambda_{sd,F}}}\right),$$

$$J_{1\downarrow}(x) = -\frac{\sigma_{F,\downarrow}}{e}\frac{\partial \mu_{1\downarrow}}{\partial x} = -\frac{1-\beta}{2}\sigma_F\left(\delta_{1\downarrow} + \frac{C_{1\downarrow}}{\lambda_F}e^{\frac{x}{\lambda_{sd,F}}}\right). \quad (12.124)$$

In the semiconductor channel, we have

$$\mu_{2\uparrow}(x) = \mu_{1\uparrow}(0) + \gamma_\uparrow x = \mu_1^0 + C_{1\uparrow} + \gamma_\uparrow x,$$
$$\mu_{2\downarrow}(x) = \mu_{1\downarrow}(0) + \gamma_\downarrow x = \mu_1^0 + C_{1\downarrow} + \gamma_\downarrow x, \quad (12.125)$$

and the associated current densities are

$$J_{2\uparrow}(x) = -\frac{\sigma_{sc,\uparrow}}{e}\frac{\partial \mu_{2\uparrow}}{\partial x} = -\frac{\sigma_{sc}}{2e}\gamma_\uparrow,$$

$$J_{2\downarrow}(x) = -\frac{\sigma_{sc,\downarrow}}{e}\frac{\partial \mu_{2\downarrow}}{\partial x} = -\frac{\sigma_{sc}}{2e}\gamma_\downarrow.$$

$$(12.126)$$

Finally, in the right contact, we have

$$\mu_{3\uparrow}(x) = \mu_3^0 + \delta_{3\uparrow}x + C_{3\uparrow}e^{\frac{x_0-x}{\lambda_{sd,F}}},$$

$$\mu_{3\downarrow}(x) = \mu_3^0 + \delta_{3\downarrow}x + C_{3\downarrow}e^{\frac{x_0-x}{\lambda_{sd,F}}}, \quad (12.127)$$

and

$$J_{3\uparrow}(x) = -\frac{\sigma_{F,\uparrow}}{e}\frac{\partial \mu_{3\uparrow}}{\partial x} = -\frac{1+\beta}{2}\sigma_F\left(\delta_{3\uparrow} - \frac{C_{3\uparrow}}{\lambda_{sd,F}}e^{\frac{x_0-x}{\lambda_{sd,F}}}\right),$$

$$J_{3\downarrow}(x) = -\frac{\sigma_{F,\downarrow}}{e}\frac{\partial \mu_{3\downarrow}}{\partial x} = -\frac{1-\beta}{2}\sigma_{F,\downarrow}\left(\delta_{3\downarrow} - \frac{C_{3\downarrow}}{\lambda_{sd,F}}e^{\frac{x_0-x}{\lambda_{sd,F}}}\right). \quad (12.128)$$

Equation (12.122) leads to $C_{1\uparrow} = -C_{3\uparrow}$ and $C_{1\downarrow} = -C_{3\downarrow}$. For this case, the qualitative dependence of the electrochemical potentials is shown in the top portion of Fig. 12.23. As a result of the anti-symmetric splitting of the electrochemical potentials at the interfaces, there is a different slope for $\mu_\uparrow(x)$ and $\mu_\downarrow(x)$ in the semiconductor leading to a crossing of the two electrochemical potentials half-way through the semiconductor.

Anti-parallel configuration

In this case, the down-spin (up-spin) electrons in the left contact couple to the up-spin (down-spin) electrons in the right contact, as shown schematically in the bottom part of Fig. 12.23. The electrochemical potential curves are parallel in the semiconductor channel, and the total current is unpolarized. The general expressions for the electrochemical potentials and the spin current densities are still valid, but the following additional condition must be enforced: $C_{1\uparrow} = -C_{3\downarrow}$ and $C_{1\downarrow} = -C_{3\uparrow}$, with $J_\uparrow = J_\downarrow$ in the semiconductor channel. This leads to a value of α_2 exactly equal to zero.

For the parallel configuration ($\beta_1 = \beta_3$), the set of Equations (12.123–12.128) must be solved together with the following set of boundary conditions.

Boundary conditions

$$\frac{J_{1\uparrow}(-\infty) - J_{1\downarrow}(-\infty)}{J_{1\uparrow}(-\infty) + J_{1\downarrow}(-\infty)} = \beta,$$

$$\frac{J_{3\uparrow}(\infty) - J_{3\downarrow}(\infty)}{J_{3\uparrow}(\infty) + J_{3\downarrow}(\infty)} = \beta,$$

$$\mu_{2\uparrow,\downarrow}(x_0) - \mu_{3\uparrow,\downarrow}(x_0) = 0,$$

$$J_{1\uparrow,\downarrow}(0) - J_{2\uparrow,\downarrow}(0) = 0 \tag{12.129}$$

$$J_{2\uparrow,\downarrow}(x_0) - J_{3\uparrow,\downarrow}(x_0) = 0,$$

$$\frac{\partial \left(J_{1\uparrow}(x) + J_{1\downarrow}(x) \right)}{\partial x} = 0,$$

$$\frac{\partial \left(J_{3\uparrow}(x) + J_{3\downarrow}(x) \right)}{\partial x} = 0,$$

$$J_{1\uparrow}(0) + J_{1\downarrow}(0) = J_{tot}.$$

Setting $\beta_1 = \beta_3$ and using MATHEMATICA or MATLAB to solve the system of equations derived above, the finite spin polarization α_2, which is independent of position in the semiconductor, can easily be calculated at $x = 0$ and is found to be equal to

$$\alpha = \beta \left(\frac{\lambda_{sd,F}}{\sigma_F} \right) \left(\frac{\sigma_F}{x_0} \right) \frac{2}{\left[2 \left(\frac{\lambda_{sd,F}}{\sigma_F} \right) \left(\frac{\sigma_F}{x_0} \right) + 1 - \beta^2 \right]}. \tag{12.130}$$

This expression is formally equivalent to the expression found in the resistor model equivalent circuit of the spin valve in Section 11.2.3 if we make the substitution

$$R_{fm}/R_{sc} \rightarrow \left(\frac{\lambda_{sd,F}}{\sigma_F} \right) \left(\frac{\sigma_F}{x_0} \right). \tag{12.131}$$

In the current approach, the quantities x_0/σ_F and $\lambda_{sd,F}/\sigma_F$ are proportional to the resistance of the semiconductor channel and the relevant part of the resistance of the ferromagnetic contact (which is mostly located within one spin relaxation length from each interface), respectively.

As shown in the series of exercises at the end of section 11.2.3, α_2 reaches a maximum value of β. However, this only occurs for device parameters which are virtually impossible to reach in practice, i.e., when $x_0 \rightarrow \infty$, $\sigma_{sc}/\sigma_F \rightarrow \infty$, or $\lambda_{sd,F} \rightarrow \infty$. If we use realistic values such as $\beta = 0.6$, $x_0 = 1~\mu m$, $\lambda_{sd,F} = 10$ nm, and $\sigma_F = 10^4 \sigma_{sc}$, the value of α_2 turns out to be extremely small, $\sim 0.002\%$!

As an exercise, the reader is asked to reproduce Figures 2 and 3 of [9]. These figures show that, even for $\beta > 0.8$, $\lambda_{sd,F}$ would need to be larger than 100 nm or x_0 well below 10 nm to obtain a spin current polarization in excess of 1 %. Furthermore, when plotted as a function of β for different values of σ_F/σ_{sc}, α_2 is still below 1%, even when the $\sigma_F/\sigma_{sc} = 10$ for $\beta < 0.98$.

To assess the possibility of using a ferromagnet/semiconductor/ferromagnet junction as a spin valve, we estimate the magnetoresistance ratio $MR_1 = (R_{AP} - R_P)/R_P$ using Equation (12.133) and making the substitution (12.36). This leads to

$$MR_1 = \frac{\beta^2}{1-\beta^2} \left(\frac{\lambda_{sd,F}}{\sigma_F}\right)^2 \left(\frac{\sigma_{sc}}{x_0}\right)^2 \left[\frac{4}{2\left(\frac{\lambda_{sd,F}\sigma_{sc}}{\sigma_F x_0} + 1\right)^2 - \beta^2}\right]. \qquad (12.132)$$

Exercise:

For contacts made of metallic ferromagnets, MR_1 is dominated by $\left(\frac{\lambda_{sd,F}}{\sigma_F}\right)^2 \left(\frac{\sigma_{sc}}{x_0}\right)^2$. Show that, in this case, $MR_1 \sim \alpha_2{}^2$.

Since α_2 is very small in practical structures, this exercise shows that MR_1 will be very difficult to detect experimentally. The prediction of small values for α_2 and MR_1 by Schmidt et al. [9] is referred to in the literature as the *conductivity mismatch* or *resistivity mismatch* problem since it is mainly due to the large difference between σ_{sc} and σ_F.

12.8 Spin extraction at the interface between a ferromagnet and a semiconductor

The study of the physics of spin extraction from a ferromagnet into a paramagnet is still in its infancy, both experimentally and theoretically [34, 35]. In this section, we present a recent analysis by Pershin and Di Ventra of electrical spin extraction from a ferromagnet into a semiconducting paramagnet in contact with it. The analysis is based on the drift-diffusion model of spin transport introduced in Chapter 11. This simple model provides a brief glimpse of the plethora of new physical phenomena to be expected in this relatively new unexplored field.

Following Pershin and DiVentra, we consider the problem of spin extraction for the contact geometry shown in Figure 12.24, in which a half-metallic ferromagnet (i.e., one with 100% spin polarization) is in contact with a para-

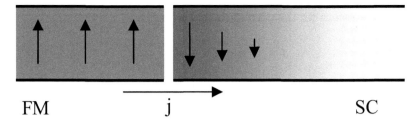

FIGURE 12.24
Illustration of the Spin Blockade phenomenon as a result of spin current extraction at a ferromagnetic contact/semiconductor interface. The flow of up-spin electrons from the semiconductor (SC) into the ferromagnet (FM) results in a higher concentration of down-spin electrons in the semiconductor channel near the interface. Reproduced with permission from Y. V. Pershin and M. Di Ventra, Phys. Rev. B, **75**, 193301 (2007). Copyright (2007) American Physical Society. http://link.aps.org/abstract/PRB/v75/p193301.

magnetic semiconductor with a non-degenerate electron population, i.e., a population well described by Maxwell-Boltzmann statistics.

If a bias is applied across the interface such that electrons flow from the semiconductor into the ferromagnetic contact, electrons incoming from the bulk of the semiconductor are spin-unpolarized. Since the half-metallic contact accepts only up-spin electrons (the majority spins in the half-metal), a cloud of down-spin electrons must accumulate in the semiconductor close to the interface since down-spin cannot enter the ferromagnetic contact. As a result, a local spin-dipole configuration forms close to the interface. Pershin and Di Ventra refer to this phenomenon as a *Spin Blockade* which occurs at a critical current magnitude when the semiconductor region near the interface becomes completely depleted of electrons with spins of the same polarization as the majority spin in the half-metallic contact. In the analysis below, spin flip scattering events at the interface are neglected and the interface is modeled as a planar junction. The one-dimensional drift-diffusion equations are used to study spin extraction from the paramagnetic semiconductor.

The continuity equations are given by

$$e\frac{\partial n_{\uparrow(\downarrow)}}{\partial t} = \frac{d}{dx}J_{\uparrow(\downarrow)} + \frac{e}{2\tau_s}\left(n_{\downarrow(\uparrow)} - n_{\uparrow(\downarrow)}\right), \qquad (12.133)$$

where we have introduced the spin relaxation time

$$\frac{1}{\tau_s} = \frac{1}{\tau_{\downarrow\uparrow}} + \frac{1}{\tau_{\uparrow\downarrow}}, \qquad (12.134)$$

with the spin flip time $\tau_{\uparrow\downarrow}$ indicating the average time for an up-spin to flip, and $\tau_{\downarrow\uparrow}$ for the reverse process. In the paramagnetic semiconductor, we further assume that $\tau_{\uparrow\downarrow} = \tau_{\downarrow\uparrow}$.

The expressions for the current densities are

$$J_{\uparrow(\downarrow)} = \sigma_{\uparrow(\downarrow)} E_0 + eD\frac{d}{dx}n_{\uparrow(\downarrow)}, \tag{12.135}$$

where $\sigma_{\uparrow(\downarrow)}$ $(= en_{\uparrow(\downarrow)}\mu)$ are the spin conductivities and the mobility is defined such that $v_{drift} = \mu E_0$, E_0 being the assumed homogeneous electric field in the semiconductor as a result of the applied bias. Additionally, D is the diffusion coefficient in the semiconductor assumed equal for charge and spin, and also assumed equal for up-spin and down-spin electrons.

If the total electron density in the semiconductor is constant, i.e., $n_{\uparrow(\downarrow)} + n_{\downarrow(\uparrow)} = N_0$, the total current density and the homogeneous electric field are related as

$$J = e\mu N_0 E_0. \tag{12.136}$$

Substituting the current density expressions into the continuity equation, we obtain a set of coupled equations.

$$\frac{\partial n_{\uparrow(\downarrow)}}{\partial t} = D\frac{\partial^2 n_{\uparrow(\downarrow)}}{\partial x^2} + \mu E_0\frac{\partial n_{\uparrow(\downarrow)}}{\partial x} + \frac{n_{\downarrow(\uparrow)} - n_{\uparrow(\downarrow)}}{2\tau_s}. \tag{12.137}$$

Under steady state, we seek solutions to the previous equation of the form

$$n_{\uparrow}(x) = \frac{N_0}{2} - Ae^{-\alpha_p x}, \tag{12.138}$$

and

$$n_{\downarrow}(x) = \frac{N_0}{2} + Ae^{-\alpha_p x}. \tag{12.139}$$

Substituting the last two expressions into the steady-state version of Equation (12.133), we get a quadratic equation for α_p, whose only physically acceptable root is given by

$$\alpha_p = \frac{\mu E_0 + \sqrt{\mu^2 E_0^2 + 4\frac{D}{\tau_s}}}{2D}. \tag{12.140}$$

To find the constant A, we impose the following boundary condition far into the semiconductor

$$J_{\uparrow}(x \to \infty) = J_{\downarrow}(x \to \infty) = J/2, \tag{12.141}$$

which states that the current must be unpolarized far into the bulk. Furthermore, if we assume a half-metallic contact and full polarization at the interface, we must have

$$J_{\uparrow}(x = 0) = J, \tag{12.142}$$

and

$$J_\downarrow(x = 0) = 0, \tag{12.143}$$

with $x = 0$ being the location of the ferromagnet/semiconductor interface.

Since $J = e\mu \left[n_\uparrow(x) + n_\downarrow(x) \right] E_0$, it can be easily shown that Equation (12.136) is satisfied if we use the solutions (12.138) and (12.139) in the expression of the majority-spin and minority-spin current densities. Furthermore, using the expressions for the current densities to enforce the boundary conditions leads to

$$A = \frac{N_0}{\sqrt{1 + 4\frac{D}{\tau_s \mu^2 E_0^2}} - 1}. \tag{12.144}$$

This last expression increases monotonically with the total current J. The quantity A gives the deviation of $n_\uparrow(x)$ and $n_\downarrow(x)$ from their equilibrium value $N_0/2$ at the interface. It can only reach a maximum value of $N_0/2$, which occurs for a critical current density J_c given by

$$J_c = eN_0 \sqrt{\frac{D}{2\tau_s}}, \tag{12.145}$$

corresponding to the maximum spin polarization (100%) corresponding to $n_\downarrow(0) = N_0$ and $n_\uparrow(0) = 0$.

Exercises:

- For a GaAs structure, assuming that $D = D_s = 200 \text{ cm}^2/\text{s}$, $N_0 = 10^{15}$ cm^{-3}, and $\tau_s = 10$ ns, plot α_p^{-1}, which is the inverse of the up-stream spin diffusion length, as a function of E_0 for E_0 ranging from 0 to 40 V/cm. What is the value of α_p^{-1} for $E_0 = 0$? Calculate the critical current density J_c in A/cm^2.

Answer: $1/\alpha_p^{-1} = 14 \ \mu m$ and $J_c = 17A/cm^2$.

The critical current density J_c found in the exercise above is well within reach of actual experiments. This spin-blockade critical current is the maximum current allowed through the interface between a perfect (half-metallic) ferromagnet and a non-magnetic semiconductor. Your plot should show that the critical current density increases slowly as η decreases from 1 to ~ 0.3. The spin blockade phenomenon discussed above is therefore also expected to be important in junctions with ordinary ferromagnets.

Pershin and Di Ventra recently extended the above analysis to take into account the presence of a contact resistance and calculated the current-voltage characteristics of the junction. They showed that, in the spin blockade regime, the latter saturates with increasing applied bias across the junction. A potential application of such junctions is therefore as a spin-based current stabilizer [35].

• The derivation above can be extended to the case of non-ideal ferromagnets. Defining the quantity

$$\eta = \frac{J_\uparrow(x = 0) - J_\downarrow(x = 0)}{J}, \qquad (12.146)$$

to characterize the level of spin polarization of the current and assuming that η does not depend on J and the ratio $J_\uparrow(x = 0)/J_\downarrow(x = 0)$ is a constant, prove that in this case

$$A = \frac{\eta N_0}{\sqrt{1 + 4\frac{D}{\tau_s \mu^2 E_0^2}} - 1}, \qquad (12.147)$$

and

$$J_c = eN_0\sqrt{\frac{D}{(\eta^2 + \eta)\tau_s}}. \qquad (12.148)$$

Plot $J_c(\eta)/J_c(0)$ as a function of η as η varies from 1 (fully polarized spin current) to 0 (fully unpolarized spin current).

12.9 Problems

• **Problem 12.1**

Reproduce the plots shown in Figures 12.4 and 12.5 based on the theory of Section 12.2.3.

• **Problem 12.2**

Reproduce the plots shown in Figure 12.7 and Figures 12.9 through 11.12 based on the theory of Section 12.2.4.

• **Problem 12.3**

Equation (12.43) is the boundary condition expressing the discontinuity of the derivative of the wavefunction across the delta-scatterer. Show that the final expressions for the transmission and reflection coefficients across the delta scatterer are the same if the boundary condition is selected to be

$$\left[\frac{d\psi}{dx}\right]_{0_+} - \left[\frac{d\psi}{dx}\right]_{0_-} = \frac{2m}{\hbar^2}\Gamma\psi(0_+). \qquad (12.149)$$

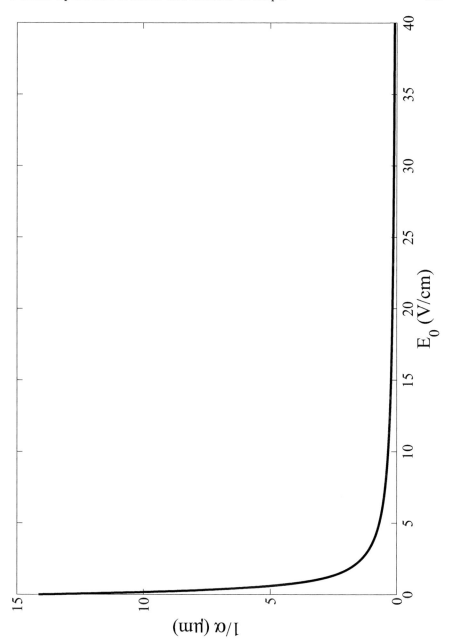

FIGURE 12.25
Plot of the up-stream spin diffusion length α_p^{-1} for a GaAs bulk sample for which $D = 200$ cm^2/s, $N_0 = 10^{15}$ cm^{-3} and the spin relaxation time $\tau_s = 10$ ns. The relation between the mobility and diffusion coefficients was assumed to be given by Einstein's relation and the temperature was set at 4.2 K.

The only difference with Equation (12.43) is that the last term is selected to be $\frac{2m}{\hbar^2}\Gamma\psi(0_+)$ rather than $\frac{2m}{\hbar^2}\Gamma\psi(0_-)$. Because of the continuity of $\phi(0_+)$, the goal of this exercise is to show that this does not influence the calculation of the transmission and reflection coefficients.

- **Problem 12.4**

Repeat the scattering problem in Section 12.2.4 and derive the explicit expression of the transfer matrix across the interface at $x = 0$ if the scattering potential at this point is given by

$$\Gamma + \Gamma' H\delta(x), \qquad (12.150)$$

where H is the Hadamard matrix.

- **Problem 12.5**

In this problem, we consider spin transport in degenerate conductors and introduce the concept of spin-dependent electrochemical potential following Hershfield and Zhao [36]. This concept is used in Section 11.6 to study the problem of spin injection across a ferromagnet/non-magnetic metal interface. We only consider the case of spatial variation in one direction (x) for simplicity. In such a system, if a small bias is applied across the junction, the equations describing the steady-state are the continuity equations (12.133), the current density expressions (12.135), and Poisson's equation

$$\frac{dE}{dx} = -\frac{q}{\varepsilon_0}\left(n_\uparrow + n_\downarrow\right). \qquad (12.151)$$

Use those equations to prove that the up-spin and down-spin electron concentrations obey the coupled equations (referred to as the diffusion equation)

$$\frac{d^2}{dx^2}\begin{pmatrix} n_\uparrow \\ n_\downarrow \end{pmatrix} = \begin{bmatrix} \kappa_\uparrow^2 + \frac{1}{D_\uparrow\tau_{\uparrow\downarrow}} & \kappa_\uparrow^2 - \frac{1}{D_\uparrow\tau_{\downarrow\uparrow}} \\ \kappa_\downarrow^2 - \frac{1}{D_\downarrow\tau_{\uparrow\downarrow}} & \kappa_\downarrow^2 - \frac{1}{D_\downarrow\tau_{\downarrow\uparrow}} \end{bmatrix}\begin{pmatrix} n_\uparrow \\ n_\downarrow \end{pmatrix}, \qquad (12.152)$$

where

$$\kappa_\uparrow^2 = \sigma_\uparrow/\left(D_\uparrow\varepsilon_0\right), \qquad (12.153)$$

and a similar expression holds for $\kappa_\downarrow^2$. The quantities $\kappa_\uparrow$ and $\kappa_\downarrow$ are referred to as the inverse up-spin and down-spin screening lengths.

Hint: Take the derivative with respect to x on both sides of equations (12.137) and use the continuity equations (12.133) and Poisson's equation.

For degenerate populations of carriers, the conductivities of the up-spin and down-spin electrons are related to their density of states and diffusion constant by Einstein's relations [37]

$$\sigma_\uparrow = e^2 g_{3D,\uparrow} D_\uparrow, \tag{12.154}$$

$$\sigma_\downarrow = e^2 g_{3D,\downarrow} D_\downarrow, \tag{12.155}$$

where $g_{3D,\uparrow}$ and $g_{3D,\downarrow}$ are the density of states in the up-spin and down-spin channels. In a non-magnetic metal, $g_{3D,\uparrow} = g_{3D,\downarrow} = \frac{1}{2} g_{3D}(E_F)$, where $g_{3D}(E_F)$ is given by Equation (12.8) in the Sommerfeld model of a metal. In a ferromagnetic contact in which the spin subbands are modeled using the Stoner-Wohlfarth model (see Section 11.2.1)

$$g_{3D,\uparrow} = \frac{m_0}{2\pi^2\hbar^3}\sqrt{2m_0 E_F}, \tag{12.156}$$

and

$$g_{3D,\downarrow} = \frac{m_0}{2\pi^2\hbar^3}\sqrt{2m_0\left(E_F - \Delta\right)}, \tag{12.157}$$

where E_F and Δ are the Fermi and exchange splitting energy of the ferromagnet, respectively.

• **Problem 12.6**

Following Hershfield and Zhao [36], we introduce the electrochemical potentials $\mu_\uparrow(x)$ and $\mu_\downarrow(x)$ for the up-spin and down-spin electrons such that their corresponding spin concentrations are given by

$$n_\uparrow(x) = \rho_{3D,\uparrow}\left(E_F\right)\left[\mu_\uparrow(x) + eV(x)\right], \tag{12.158}$$

and

$$n_\downarrow(x) = \rho_{3D,\downarrow}\left(E_F\right)\left[\mu_\downarrow(x) + eV(x)\right], \tag{12.159}$$

where $V(x)$ is the electrostatic potential.

Using the last two equations, show that the total current densities in Equations (12.135) can be written in the more compact form

$$J_\uparrow(x) = \frac{\sigma_\uparrow}{e}\frac{d\mu_\uparrow}{dx}, \tag{12.160}$$

and

$$J_\downarrow(x) = \frac{\sigma_\downarrow}{e}\frac{d\mu_\downarrow}{dx}. \tag{12.161}$$

Solution: Starting with Equation (12.159), we get

$$\frac{\mu_\uparrow(x)}{-e} = V(x) + \frac{n_\uparrow(x)}{eg_{3D,\uparrow}(E_F)} \tag{12.162}$$

and a similar expression for $\mu_\downarrow(x)/-e$.

Taking the derivative with respect to x on both sides of the last equation, we get

$$\frac{d}{dx}\left(\frac{\mu_\uparrow(x)}{-e}\right) = \frac{dV(x)}{dx} + \frac{1}{eg_{3D,\uparrow}(E_F)}\frac{dn_\uparrow(x)}{dx}. \tag{12.163}$$

Multiplying the last equation on both sides by $-\sigma_\uparrow$ and taking into account that the electric field $E(x) = -dV/dx$ and Einstein's relations (12.154 and 12.155), we get

$$J_\uparrow = -\sigma_\uparrow\frac{d}{dx}\left(\frac{\mu_\uparrow(x)}{-e}\right) = \sigma_\uparrow E + qD_\uparrow\frac{dn_\uparrow}{dx}, \tag{12.164}$$

and a similar expression for $J_\downarrow$.

The spin-dependent electrochemical potentials are useful quantities since they allow a direct calculation of the up-spin and down-spin electron concentrations, and simple spatial derivatives give the associated spin current densities.

- **Problem 12.7**

  Starting with the results of the previous problem, show that the electrochemical potentials obey the following set of coupled equations

  $$\frac{d^2}{dx^2}\begin{pmatrix}\mu_\uparrow \\ \mu_\downarrow\end{pmatrix} = \begin{pmatrix}+\frac{1}{D_\uparrow\tau_{\uparrow\downarrow}} & -\frac{1}{D_\uparrow\tau_{\uparrow\downarrow}} \\ -\frac{1}{D_\downarrow\tau_{\downarrow\uparrow}} & +\frac{1}{D_\downarrow\tau_{\downarrow\uparrow}}\end{pmatrix}\begin{pmatrix}\mu_\uparrow \\ \mu_\downarrow\end{pmatrix}. \tag{12.165}$$

  If we assume $\tau_{\downarrow\uparrow} = \tau_{\uparrow\downarrow} = \tau_s$, the spin flip scattering rate, show that the difference of the spin-dependent electrochemical potential $\mu_\uparrow - \mu_\downarrow$ obeys the diffusion equation

  $$\frac{d^2}{dx^2}(\mu_\uparrow - \mu_\downarrow) = \frac{\mu_\uparrow - \mu_\downarrow}{L_{sf}^2}, \tag{12.166}$$

  where the spin decay length is given by $L_{sf} = \sqrt{D_{eff}\tau_s}$ and D_{eff} is the effective diffusion coefficient and is equal to

  $$D_{eff} = \frac{D_\uparrow D_\downarrow}{D_\uparrow + D_\downarrow}. \tag{12.167}$$

12.10 References

[1] N. F. Mott, "The electrical conductivity of transitions metals", Proc. R. Soc. London, Ser. A, **156**, 699 (1936); "The resistance and thermo-electric properties of the transition metals", Proc. R. Soc. London, Ser. A, **156**, 368 (1936).

[2] A. Brataas, G. E. W. Bauer and P. J. Kelly, "Non-collinear magneto-electronics", Phys. Rep., **427**, 157 (2006).

[3] S. Srinivasan, V. Diep, B. Behin-Aein, A. Sarkar and S. Datta, "Modeling multi-magnet networks interacting via spin currents", arXiv:1304.0742.

[4] K. Y. Camsari, S. Ganguly and S. Datta, "Modular approach to spintronics", Sci. Rep., **5**, 10571 (2015).

[5] N. W. Ashcroft and N. D. Mermin, *Solid State Physics*, (Saunders College, Philadelphia, 1976).

[6] M. Jullière, "Tunneling between ferromagnetic films", Phys. Lett. A, **54**, 225 (1975).

[7] M. N. Baibich, J. M. Broto, A. Fert, F. Nguyen Van Dau, and F. Petroff, P. Eitenne, G. Creuzet, A. Friederich and J. Chazelas, "Giant magnetoresistance of (001)Fe/(001)Cr magnetic superlattices", Phys. Rev. Lett., **61**, 2472 (1988).

[8] G. Binasch, P. Grünberg, F. Saurenbach and W. Zinn, "Enhanced magnetoresistance in layered magnetic structures with antiferromagnetic interlayer exchange", Phys. Rev. B, **39**, 4828 (1989).

[9] G. Schmidt, D. Ferrand, L. W. Molenkamp, A. T. Filip and B. J. van Wees, "Fundamental obstacle for electrical spin injection from ferromagnetic metal into a diffusive semiconductor", Phys. Rev. B, **62**, R4790 (2000).

[10] E. I. Rashba, "Theory of electrical spin injection: Tunnel contacts as a solution of the conductivity mismatch problem", Phys. Rev. B, **62**, R16267 (2000).

[11] R. Feiderling, M. Klein, G. Reuscher, W. Ossau, G. Schmidt, A. Waag and L. Molenkamp, "Injection and detection of a spin-polarized current in a light-emitting diode", Nature (London), **402**, 787 (1999).

[12] G. Salis, R. Wang, X. Jiang, R. M. Shelby, S. S. P. Parkin, S. R. Bank and J. S. Harris, "Temperature independence of the spin injection efficiency of a MgO based tunnel spin injector", Appl. Phys. Lett., **87**, 262503 (2005).

[13] T. Santos, J. Lee, P. Migdal, I. Lekshmi, B. Satpati and J. S. Moodera, "Room temperature tunnel magnetoresistance and spin polarized tunneling through an organic semiconductor barrier", Phys. Rev. Lett., **98**, 016601 (2007).

[14] W. E. Pickett and J. S. Moodera, "Half metallic magnets", Physics Today, **54**, 39 (2001).

[15] P. A. Dowben and R. Skomski, "Are half metallic ferromagnets half metals?", J. Appl. Phys., **95**, 7453 (2004).

[16] M. S. Lundstrom, *Fundamentals of Carrier Transport*, Purdue Modular Series in Solid State Devices, Eds. G. W. Neudeck and R. F. Pierret, (Addison-Wesley, Reading, MA, 1990).

[17] A. D. Stone and A. Szafer, "What is measured when you measure a resistance? The Landauer formula revisited", IBM J. Res. Develop., **32**, 384 (1988).

[18] J. Bardeen, "Tunneling from a many-particle point of view", Phys. Rev. Lett., **6**, 57 (1961).

[19] S. J. Xie, K. H. Ahn, D. L. Smith, A. R. Bishop and A. Saxena, "Ground state properties of ferromagnetic metal/cojugated polymer interfaces", Phys. Rev. B, **67**, 125202 (2003).

[20] K. Hathaway and E. Dan Dahlberg, "Resource Letter STMN-1: Spin Transport in magnetic nanostructures", Am. J. Phys., **75**, 871 (2007).

[21] J. Mathon, "Phenomenological theory of giant magnetoresistance", Chapter 4 in *Spin Electronics*, edited by M. Ziese and M.J. Thornton, (Springer Verlag, Berlin, 2001).

[22] A. Barthélémy, A. Ferta, J.-P. Contour, M. Bowen, V. Cros, J. M. De Teresa,
A. Hamzic, J. C. Faini, J. M. George, J. Grollier, F. Montaigne, F. Pailleux, F. Petroff and C. Vouille, "Magnetoresistance and spin electronics", Journal of Magn. and Magn. Mater., Vol. 242-245, 68 (2002).

[23] M. A. M. Gijs and G. E. W. Bauer, "Perpendicular giant magnetoresistance of magnetic multilayers", Adv. Phys., **46**, 285 (1997).

[24] J. M. Daughton, "Magnetic tunneling applied to memory", J. Appl. Phys., **81**, 3758 (1997).

[25] J. M. Daughton, A. V. Pohm, R. T. Fayfield and C. H. Smith, "Applications of spin dependent transport materials", J. Phys. D, **32**, R169-R177 (1999).

[26] J. M. Daughton, J. Brown, E. Chen, R. Beech, A. V. Pohm and W. Kude, "Magnetic field sensors using GMR multilayer", IEEE Trans. Mag., **30**, 4608 (1994).

[27] G. A. Prinz, "Magnetoelectronics", Science, **282**, 1660 (1998).

[28] J. Hayakawa, S. Ikeda, Y. M. Lee, F. Matsukura and H. Ohno, "Effect of high annealing temperature on giant tunneling magnetoresistance ratio of CoFeB/MgO/CoFeB magnetic tunnel junctions", Appl. Phys. Lett., **89**, 232510 (2006).

[29] World magnetic sensor components. Frost & Sullivan (2005).

[30] A. Fert, J.-L. Duvail and T. Valet, "Spin relaxation effects in the perpendicular magnetoresistance of magnetic layers", Phys. Rev. B, **52**, 6513 (1995).

[31] P. C. van Son, N. van Kampen and P. Wyder, "Boundary resistance of the ferromagnetic-nonferromagnetic metal interface", Phys. Rev. Lett., **58**, 2271 (1987).

[32] In Equations (99) - (102), we introduce a negative sign so that the current densities associated with the flow of electrons from left to right in Fig. 11.22 are positive. The same convention is used in Equations (113), (124), (126) and (128).

[33] Spin diffusion length approaching 100 μm have been reported by J. M. Kikkawa and D. D. Awschalom, "Lateral drag of spin coherence in gallium arsenide", Nature (London), **397** (1999).

[34] Y. V. Pershin and M. Di Ventra, "Spin blockade at semiconductor/ferromagnet junctions", Phys. Rev. B, **75**, 193301 (2007).

[35] Y. V. Pershin and M. Di Ventra, "Current-voltage characteristics of semiconductor/ferromagnet junctions in the spin blockade regime", arXiv:0707.4475v1 (2007).

[36] S. Hershfield and H. L. Zhao, "Charge and spin transport through a metallic ferromagnetic-paramagnetic-ferromagnetic junction", Phys. Rev. B, **56**, 3296 (1997).

[37] S. Datta, *Electronic Transport in Mesoscopic Systems*, (Cambridge University Press, New York, 1995), p. 41.

13

Active Devices Based on Spin and Charge

In Chapter 12, we discussed passive spintronic devices. In this chapter, we introduce and discuss *active* spin-based devices such as spin transistors, that are capable of amplifying a signal. They can be used in signal processing and logic circuits.

13.1 Spin-based transistors

The basic definition of a "transistor" is a three-terminal device where the current flowing between two of the terminals can be modulated by a current or voltage applied to the third terminal. Device engineers would cringe at this simplistic definition of a transistor, since they will usually recognize that the transistor is much more than just a three-terminal device with current modulation. The current modulation is a necessary, but *not* a sufficient, condition for a device to qualify as a transistor. A true transistor must have other attributes as well, such as "gain" and "isolation between the input and output terminals." The "gain" (or amplification) requirement means that a small change in the current or potential at the third terminal (the input signal) should produce a much larger change in the current flowing between the other two terminals or the potential difference appearing between those terminals (the output signal). "Isolation" is an even more important property. It means that any change in current or potential at the input terminals should bring about a change in the current or potential at the output terminal, but this should be a *one-way* interaction. In other words, any change caused at the output terminal by external means should not induce a change in the input quantity. Stated otherwise, signal should be communicated "unidirectionally" from the input to the output and not the other way around. Both "gain" and "unidirectionality" (or "isolation") are required for digital logic circuits [1]. As we embark on a discussion of spin transistors, the reader will do well to keep these ideas in mind for reasons that will become clear later.

In the following sections, we will discuss two genres of spin-based transistors: Spin Field Effect Transistors and Spin Bipolar Transistors.

13.2   Spin field effect transistors (SPINFET)

In many ways, the first proposal for a Spin Field Effect Transistor (SPINFET) [2] – now called the *Datta-Das transistor* after the two proponents of this device – was a watershed event in the field of spintronics. It was the first time anyone proposed using the spin degree of freedom of a charge carrier to realize an active device which can process information in a tractable way. The SPIN-FET device proposed in [2] looks exactly like a conventional metal-insulator-semiconductor field effect transistor (MISFET) except that the source and drain contacts are ferromagnetic. Figure 13.1 shows a schematic of this device. To keep the ensuing discussion focused on the essential elements, we will assume that the channel is strictly one dimensional (a quantum wire) whose width is smaller than the Fermi wavelength of carriers in the channel, so that only the lowest subband is occupied by electrons.

Both source and drain contacts are magnetized in such a way that their magnetic moments are parallel to each other and point along the direction of current flow ($+x$-axis). When a potential difference is imposed between these contacts, the ferromagnetic source injects carriers into the channel with their spins polarized in the $+x$-direction. These are the majority spins in the ferromagnetic source contact. We will now make four idealized assumptions:

1. The ferromagnetic source injects only the majority spins and no minority spins at all. Moreover, there should be no loss of spin polarization at the interface due to spin flip scattering. In other words, the spin injection efficiency at the source end, defined as

$$\zeta = \frac{I_{maj} - I_{min}}{I_{maj} + I_{min}}, \tag{13.1}$$

is 100%, where I_{maj} is the current in the channel due to majority spins (in the source) and I_{min} is the current due to minority spins.

2. We will also assume that the drain is an idealized spin filtering ferromagnet which only transmits the majority spins and completely blocks the minority spins.

3. When the gate voltage is zero, there are no stray symmetry breaking electric fields in the ($y - z$) plane that induce any Rashba interaction. Note that the source-to-drain electric field, which drives the current in the channel, is in the x-direction which is the same direction as the carrier velocity; hence, it does not cause a Rashba spin-orbit interaction. Only when the gate voltage is non-zero can there be a Rashba interaction in the channel due to the electric field caused by the gate voltage. There is also no Dresselhaus spin-orbit interaction [3], and no

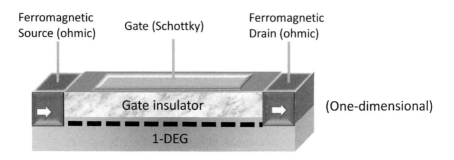

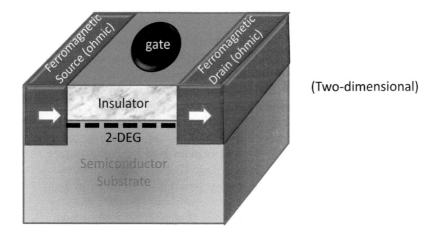

FIGURE 13.1

Schematic representation of a Spin Field Effect Transistor. The arrows indicate the direction of magnetization of the source and drain contacts.

stray magnetic fields in the channel (we ignore the magnetic field caused by the ferromagnetic contacts).

4. There is no spin relaxation in the channel. There can be momentum and energy relaxing collisions in the channel – transport need not be ballistic – but these scatterings must not relax spin. This is a realistic assumption since we know (see Chapter 8) that in a strictly one dimensional channel with a single transverse subband occupied, there is no D'yakonov-Perel' spin relaxation in space. There is also no Elliott-Yafet spin relaxation if there is no magnetic field in the channel. This is because if the only spin-orbit interaction is the Rashba interaction, then each spin-split band has a fixed (wavevector-independent) spin quantization axis (see Section 7.2.1.). In that case, intraband scattering does not change spin polarization and interband scattering (due to a non-magnetic scatterer) is forbidden because the spin eigenstates in two different bands are orthogonal. Therefore, there is no Elliott-Yafet relaxation.

5. We will ignore effects due to multiple reflections of an electron between the source and drain contacts. These effects have been considered elsewhere [4, 5].

We will explain the *switching action* of the strictly one-dimensional Datta-Das transistor by first invoking a *particle picture* as opposed to the *wave picture* used by Datta and Das in [2]. We do this purposely to dispel any notion that the Datta-Das switching transistor is a spin interferometer that requires *coherent* interference of spin. In fact, the switch is a purely classical device and can be explained in purely classical terms using a particle based picture. No phase coherent transport or "interference" is required for the switching action. With the particle picture, we will explain how the conductance assumes a maximum value at a certain gate voltage V_{ON} and a minimum value at another gate voltage V_{OFF}. The value of the minimum conductance at V_{OFF} is exactly zero if all of the above four idealized conditions are met.

However, there are certain things that we cannot deduce from the particle viewpoint. For example, we cannot tell what the conductance will be at any arbitrary gate voltage between V_{ON} and V_{OFF}. For that, we need the wave picture that Datta and Das used. We will present both pictures for the reader.

13.2.1   Particle viewpoint

In the particle viewpoint, we will consider electrons as particles with spin. Right after injection from the source, all electrons in the channel of the Datta-Das transistor will have their spins polarized along the direction of current flow, i.e., the $+x$-direction since the source is magnetized in the $+x$-direction and injects spin with 100% efficiency. When the gate voltage is turned on, it causes a transverse electric field in the y-direction, which causes Rashba spin-orbit interaction in the channel. This interaction will act as an effective magnetic field in the z direction which will be given by (see Equation (8.23))

$$B_{Rashba} = \left[\frac{2m^* a_{46}}{g\mu_B \hbar} \mathcal{E}_y v_x \right] \hat{z},\qquad(13.2)$$

where $\mathcal{E}_y$ is the y-directed gate electric field and v_x is the x-directed velocity of carriers in the channel. Note that we have used the spin-independent field which is proportional to velocity and not wavevector. This will allow us to work with a spin-independent precession frequency Ω (see Problem 8.5).

The pseudo-magnetic field B_{Rashba} is directed along the z-axis. It will cause the electrons entering the channel with x-polarized spins to execute Larmor precession in the x-y plane as they travel toward the drain. The angular frequency of this precession is the Larmor frequency

$$\Omega = \frac{d\phi}{dt} = \frac{g\mu_B B_{Rashba}}{\hbar} = \frac{2a_{46}m^*}{\hbar^2} \mathcal{E}_y v_x,\qquad(13.3)$$

which is **spin-independent** since we have used the spin-independent B_{Rashba}.

Therefore, the *spatial rate* of spin precession will be given by

$$\frac{d\phi}{dx} = \frac{d\phi}{dt}\frac{dt}{dx} = \frac{d\phi}{dt}\frac{1}{v_x} = \frac{2a_{46}m^*}{\hbar^2} \mathcal{E}_y.\qquad(13.4)$$

Needless to say, this is also spin-independent.

It is important to note from the above equation that the spatial rate of spin precession depends on the gate voltage (or $\mathcal{E}_y$) but is *independent* of the carrier velocity along the channel. This means that the spin of every electron, regardless of its velocity (and hence kinetic energy), precesses by exactly the *same* angle as it travels from source to drain. This angle is given by

$$\Phi = \left[\frac{2a_{46}m^*}{\hbar^2} \mathcal{E}_y \right] L,\qquad(13.5)$$

where L is the length of the channel (or source-to-drain separation).

An electron may suffer numerous momentum and energy randomizing collisions and arrive at the drain with arbitrary velocity, but it does *not* matter*. When *any* electron arrives at the drain, with whatever velocity and whatever scattering history, its spin has precessed by exactly the same angle Φ as any other electron in traversing the channel[†] since Φ is independent of the electron's velocity. The angle Φ depends only on $\mathcal{E}_y$, or the gate voltage, which

---

*It will matter only if the scattering is *spin-dependent*, but that does not become obvious from the particle picture. We will need the wave picture, presented next, to understand this.

[†]The reader should appreciate the fact that it is absolutely necessary that B_{Rashba} be proportional to the carrier velocity and *not* the carrier wavevector (which is not proportional to the velocity) for this discussion to be valid. This is a matter of consistency. If we assume

is the same for every electron. If the gate voltage is of such magnitude that $\Phi = (2n + 1)\pi$ [n = an integer], then *every* electron arriving at the drain will have its spin polarized *anti-parallel* to the drain's magnetization (i.e., in the x-direction), since they all entered the channel with spins polarized parallel to the drain's magnetization (in the $+x$-direction). These electrons are completely blocked by the drain since the drain transmits only those electrons whose spins are polarized in the $+x$-direction and reflects those whose spins are polarized in the x-direction. Therefore, the source-to-drain current falls to zero. Without a gate voltage, the spins do not precess ($\Phi = 0$ when $\mathcal{E}_y = 0$) so that every electron arriving at the drain has its spin polarized *parallel* to the drain's magnetization. These electrons are all transmitted by the drain so that the source-to-drain current is non-zero. Thus, when the gate voltage is zero, the conductance is maximum and when the gate voltage is V_{OFF}, corresponding to $\Phi = (2n + 1)\pi$, the conductance is minimum and ideally zero. Therefore, the gate voltage changes the conductance between a maximum and minimum value using spin precession and realizes basic *switching* transistor action. Note that the switching action has been explained by invoking a classical particle picture (Larmor precession of a particle about a pseudo magnetic field), and no wave interference was necesssary. This underscores the fact that the switch is a purely classical device.

The wave picture, however, becomes necessary if we want to find the conductance of the device at any arbitrary gate voltage. Datta and Das [2] viewed the operation of the device as interference of two "waves" between which a phase shift is introduced by the gate voltage. This phase shift is exactly the same as the angle of precession Φ given by Equation (13.5). Therefore, when the gate voltage is such that $\Phi = 2n\pi$, the conductance is maximum since the waves interfere constructively and when the gate voltage is such that $\Phi = (2n+1)\pi$, the conductance is zero since the waves interfere destructively. We present this picture in the next subsection.

13.2.2   Wave viewpoint

The energy dispersion relations in the one-dimensional channel of the Datta-Das transistor (with no Dresselhaus interaction and no channel magnetic field) are two horizontally displaced parabolas, as shown in Fig. 8.1. We have assumed that only the lowest subband in the quantum wire is occupied by electrons and all higher subbands are unoccupied. An electron with energy E has wavevectors k_1 and k_2 in the two spin resolved bands. The corresponding

---

a spin-independent precession frequency Ω, as we have done here, we must also invoke a spin-independent spin-orbit field B_{Rashba} since Ω is proportional to B_{Rashba}. We have seen earlier that if B_{Rashba} is proportional to v then it is spin-independent, but if it is proportional to k then it is not. Hence, the B_{Rashba} used in this discussion must be proportional to v.

eigenspinors are the +z and -z polarized states $[1\ 0]^\dagger$ and $[0\ 1]^\dagger$ (recall Chapter 7).

The source contact is magnetized in the $+x$-direction and hence injects only $+x$-polarized spins in the channel. Consider an electron injected with energy E. Its spin state at the source end can be written as a superposition of the channel eigenspinors:

$$\Psi_{source} = \frac{1}{\sqrt{2}}\begin{bmatrix} 1 \\ 1 \end{bmatrix} = \frac{1}{\sqrt{2}}\begin{bmatrix} 1 \\ 0 \end{bmatrix} + \frac{1}{\sqrt{2}}\begin{bmatrix} 0 \\ 1 \end{bmatrix}. \tag{13.6}$$

At the drain end, the spin state of the electron is

$$\Psi_{drain} = \frac{1}{\sqrt{2}}\begin{bmatrix} 1 \\ 0 \end{bmatrix} e^{ik_1 L} + \frac{1}{\sqrt{2}}\begin{bmatrix} 0 \\ 1 \end{bmatrix} e^{ik_2 L} = \frac{1}{\sqrt{2}}\begin{bmatrix} e^{ik_1 L} \\ e^{ik_2 L} \end{bmatrix}, \tag{13.7}$$

where L is the channel length.

The eigenspinor in the drain contact, which is also magnetized in the $+x$-direction, is $(1/\sqrt{2})[1\ 1]^\dagger$. Therefore, if we neglect multiple reflections within the channel, then the transmission amplitude into the drain is given by the projection of the Ψ_{drain} on the eigenspinor in the drain contact:

$$t(E) = \frac{1}{\sqrt{2}}[1\ 1]\frac{1}{\sqrt{2}}\begin{bmatrix} e^{ik_1 L} \\ e^{ik_2 L} \end{bmatrix} = \frac{1}{2}\left(e^{ik_1 L} + e^{ik_2 L}\right). \tag{13.8}$$

Therefore, the transmission probability is

$$T(E) = |t(E)|^2 = \frac{1}{4}\left|1 + e^{i(k_2-k_1)L}\right|^2 = \cos^2\left(\frac{\Phi}{2}\right), \tag{13.9}$$

since $k_2 - k_1 = 2m^*\eta/\hbar^2 = 2m^*a_{46}\mathcal{E}_y/\hbar^2$.

According to the Landauer formula for linear response conductance, the channel conductance will be given by

$$\begin{aligned} G_{SD} &= \frac{e^2}{h}\frac{1}{4kT}\int_0^\infty dE\, T(E)\mathrm{sech}^2\left(\frac{E-E_F}{2kT}\right) \\ &= \frac{e^2}{h}\frac{1}{4kT}\int_0^\infty dE\cos^2\left(\frac{\Phi}{2}\right)\mathrm{sech}^2\left(\frac{E-E_F}{2kT}\right), \end{aligned} \tag{13.10}$$

where E_F is the chemical potential (or Fermi energy) in the channel and kT is the thermal energy. Therefore, we can write the channel conductance as

$$G_{SD} = G_0\cos^2\left(\frac{\Phi}{2}\right), \tag{13.11}$$

where

$$G_0 = \frac{e^2}{h}\frac{1}{4kT}\int_0^\infty dE\,\mathrm{sech}^2\left(\frac{E-E_F}{2kT}\right) = \frac{e^2}{2h}\left[1 + \tanh\left(\frac{E_F}{2kT}\right)\right]. \tag{13.12}$$

Equation (13.11) can tell us what the conductance will be at any arbitrary gate voltage, since it relates the conductance to any arbitrary value of Φ, not just $\Phi = 2n\pi$ or $\Phi = (2n+1)\pi$. This is the advantage of the wave picture. However, the wave picture may also convey the false impression that the device is a wave interference device that requires phase coherent transport and cannot tolerate phase randomizing (inelastic) collisions even if they do not flip spin. This impression is false since we have already shown that the operation of the switching action can be explained with the particle picture alone. Nonetheless, it behooves us to critically examine what role collisions or scatterings play, if any. It turns out that only *spin-dependent* scattering can have a deleterious effect, but spin-independent scattering, regardless of whether it is elastic (phase conserving) or inelastic (phase randomizing), has no deleterious effect at all. This makes the Datta-Das device remarkably robust, unlike wave interference devices.

13.2.3   Effect of scattering on the Datta–Das SPINFET

Let us examine what happens if there is scattering in the channel of the Datta-Das device even if it does not flip spin. First, we will consider elastic scattering caused by a time-invariant scatterer such as an impurity or surface roughness. We will assume, for the sake of generality, that the scattering is *spin-dependent*. This type of scattering will introduce an additional (time-independent) phase shift so that we should replace $e^{ik_1 L}$ with $e^{ik_1 L + \theta_1}$ and $e^{ik_2 L}$ with $e^{ik_2 L + \theta_2}$ in Equation (13.8). Because of the spin-dependent nature of scattering, $\theta_1 \neq \theta_2$.

It is easy to see that Equation (13.9) will be still valid provided we replace Φ with $\Phi + \Delta\theta$ where $\Delta\theta = \theta_2 - \theta_1$. Now, if $\Delta\theta$ is *independent of energy*, then Equation (13.11) will also be valid if we replace Φ with $\Phi + \Delta\theta$. However, if $\Delta\theta$ depends on energy, then we cannot pull the squared cosine term in Equation (13.10) outside the integral since the argument $\Phi + \Delta\theta(E)$ depends on energy E. In that case, ensemble averaging, represented by the integral over energy in Equation (13.10), will have a deleterious effect, and the off-conductance G_{OFF} will not be zero so that the conductance modulation ratio, defined as $(G_{ON} - G_{OFF})/(G_{ON} + G_{OFF})$, will be less than 100%. Thus elastic scattering has a deleterious effect only if (1) it is spin-dependent, and (2) the differential phase shift $\Delta\theta$ depends on the electron's energy.

Inelastic scattering, on the other hand is caused by a time-varying scatterer, such as a phonon. In this case, as long as the scattering is spin-dependent, we have $\Delta\theta(t) \neq 0$. Since the differential phase shift depends on time t, different electrons passing through the device at different times will experience different differential phase shifts. Ensemble averaging will then invariably make $G_{OFF} \neq 0$ and the conductance modulation ratio will once again be less than 100%.

Therefore, the only type of scattering that adversely affects the one dimensional Datta–Das device is *spin-dependent* scattering.[‡] Elastic spin-dependent scattering is more forgiving since it degrades the device performance only if the differential phase shift is energy dependent, while inelastic spin-dependent scattering is less forgiving and degrades the device performance even if the differential phase shift is not energy dependent. If the scattering is spin-independent, then the one-dimensional Datta–Das transistor is not affected by it at all. Therefore, the one-dimensional device is surprisingly robust.

13.2.4   Transfer characteristic of the Datta–Das transistor

In Fig. 13.2 we plot the transfer characteristic (source-to-drain current I_{SD} versus gate voltage V_G) of an ideal SPINFET with no phase randomizing collisions in the channel. Basically, we have used Equation (13.11) with the further assumption that Φ (or $\mathcal{E}_y$) is linearly related to the gate voltage V_G. If we assume that there is no transverse electric field in the channel inducing the Rashba effect when $V_G = 0$, then obviously $V_{ON} = 0$. In that case, this device is "normally on," meaning that the current I_{SD} is maximum when the gate voltage V_G is zero and a gate voltage V_{OFF} is required to turn the device off. The transfer characteristic is *oscillatory*. Therefore, the transconductance $\partial I_{SD}/\partial V_G$) can be either positive or negative depending on the dc bias applied to the gate terminal. If two devices – one with positive transconductance and one with negative transconductance – are connected in series, then we will realize a device that acts like a complementary metal oxide semiconductor (CMOS) inverter. Oscillatory transfer characteristics have other unusual applications. Suppose we apply a sinusoidal voltage of frequency f to the gate and the amplitude of this voltage is n times V_{OFF}. Then the source-to-drain current I_{SD} will oscillate with a frequency nf. Thus, we have realized a single stage frequency multiplier.

It should be obvious that, in this device, although current modulation is achieved through spin precession, spin itself plays no direct role in information handling. Information is still encoded in charge which carries the current from the source to the drain. The transistor is still switched between the on and off states by changing the current or the amount of charge transmitting through the device. The role of spin is only to provide an alternate means of changing the current. Thus, this device is a quintessential hybrid spintronic device, where "spin" merely augments the role of "charge."

Significant effort has been made to demonstrate this device experimentally, but at the time of writing this textbook, no effort has borne fruit. The primary obstacles to realizing this device is that spin injection and detection efficiencies at ferromagnet/semiconductor interfaces are much smaller than 100%, contrary to what assumptions 1 and 2 assumed. If this technological barrier

---

[‡]If the scattering is spin-independent, then it is easy to see that $\Delta\theta$ is identically zero in all cases. Spin-independent scattering – both elastic and inelastic – therefore does not degrade the conductance modulation ratio at all, which remains at 100%.

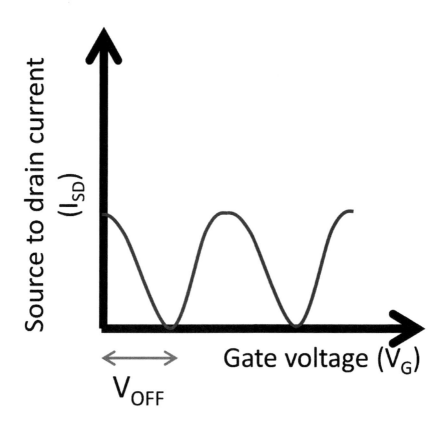

FIGURE 13.2

The transfer characteristic (source-to-drain current versus gate voltage) of an ideal SPINFET. The gate voltage required to turn the transistor from "on" to "off" is V_{OFF}. Note that the device is "normally-on" since the source-to-drain current is maximum when the gate voltage is zero.

can be overcome, the Datta-Das SPINFET might indeed be demonstrated some day. Whether it will lead to a better device, superior to traditional transistors, is another issue that we discuss later in this chapter.

13.2.5   Sub-threshold slope

One way the Datta-Das device is different from a conventional field effect transistor (FET) is that, in principle, the average carrier concentration in the channel need not change at all when the device is switched from "on" (conducting) to "off" (non-conducting), or vice versa.§ In conventional n-channel FETs, the current flowing between the source and drain is changed by altering the electron concentration in the channel with the gate voltage. If the FET is a so-called "enhancement mode" (or "normally off") device, then without any applied gate voltage, there should be no (or very few) electrons in the channel. When a positive voltage is applied to the gate, it will draw electrons in the channel from the source contact by Coulomb attraction, thereby increasing the channel conductance and the channel current for a fixed source-to-drain voltage. By contrast, the channel concentration in a SPINFET need not change; it is only the transmittivity of the carriers that changes since the transmittivity depends on the spin polarization. This mode of switching, where the channel carrier concentration does not change, may have some advantages. Consider the conventional FET in the "off" state. The conduction band energy profile along the length of the channel is shown in Figure 13.3. The channel is depleted of electrons, as it should be in the "off" state, and the Fermi level (determined by the Fermi level in the metallic contacts) is well below the conduction band edge, consistent with the fact that the channel is empty of carriers. When a positive gate voltage is applied to turn the transistor "on," it lowers the conduction band edge to a position below the Fermi level. The barrier at the source-channel interface is also made *thinner*, so that electrons get into the channel by tunneling through this barrier from the source. Since the conduction band edge is below the Fermi level, the channel becomes populated.

In the "off" state, the barrier is too thick to tunnel through, but some carriers can still get into the channel by thermionic emission over the barrier. These electrons cause a leakage current in the "off" state that leads to unwanted standby power dissipation in the device. Assuming Boltzmann statistics for the electrons, the population in the channel will fall off as $exp[-\phi_b/kT]$ where ϕ_b is the effective barrier height. We will assume that ϕ_b is proportional to eV_G, where V_G is the gate voltage. Therefore, the leakage current will fall off with a slope of $(kT/e)ln(10)$ per decade of gate voltage (assuming an en-

---

§In reality, the gate voltage will always cause some change in the channel carrier concentration since the charges on the gate will attract or repel charge carriers in the channel. However, this change could be very small if a very small gate voltage is required to turn the device on or off, or if the gate capacitance is very small.

hancement mode device). This slope is called the sub-threshold slope and has a minimum value of $(kT/e)ln(10)$/decade = 60 mV/decade at room temperature. It is the *minimum* value because we have completely neglected tunneling in the off state.

It should be obvious that the 60 mV/decade (at room temperature) limitation comes about because we switch the transistor on and off by populating and de-populating the channel. Instead, if we can switch the device by changing the carrier transmittivity with a gate voltage, without changing the carrier concentration, then the 60 mV/decade limit need not apply. This could have been one advantage of the SPINFET.[¶] However, we will show in the next subsection that, realistically speaking, a SPINFET will probably have a much larger leakage current than a conventional FET for entirely different reasons.

One last issue that we need to address before we embark on a realistic assessment of the device potential of SPINFETs is the issue of "normally on" versus "normally off." It is obvious that the SPINFET, as described, is "normally on", meaning that it is in the conducting state ("on") when the gate voltage is zero. This is not desirable. If we want to build the more desirable "normally off" device, we will simply have to magnetize the source and drain ferromagnets in the anti-parallel configuration so that the majority spins injected by the source are minority spins in the drain, and vice versa. We need not actually magnetize the two contacts in opposite directions so that their magnetic moments are pointing against each other (the north pole of one magnet facing the north pole of the other). Such a configuration will be unstable and both source and drain ferromagnets will demagnetize quickly. A better alternative is to choose two different materials, such as iron and cobalt, which have opposite signs of spin polarization of carriers at the Fermi level [6] so that majority spins in one are already minority spins in the other when both are magnetized in the same direction. A SPINFET with iron and cobalt contacts will be normally off. "Normally off" devices are always preferred to avoid standby power dissipation in circuits.

[¶]In the 1980s, a device known as *Velocity Modulation Transistor* became popular (H. Sakaki, Jpn. J. Appl. Phys., Pt. 2, Vol. 21, L381 (1982) and C. Hamaguchi, Jpn. J. Appl. Phys., Pt. 2, Vol. 23, L132 (1984)). The device structure was that of a MOSFET and the channel was a quantum well with a fixed carrier concentration. A positive gate voltage pulled the electrons toward an interface of the quantum well, thereby increasing interface roughness scattering in the channel and decreasing carrier mobility. This decreased the conductance of the channel, realizing transistor action. Just like the SPINFET, this device is not subject to the $(kT/e)ln(10)$/decade sub-threshold slope limitation. In fact, any device where the carrier concentration is not modulated for transistor action will not be subjected to this limitation. It turned out later that the on-off conductance ratio of the Velocity Modulation Transistor was too small for mainstream applications since mobility could not be changed by several orders of magnitude using this technique. As a result, the Velocity Modulation Transistor idea never caught on. We will show later that SPINFETs will also tend to have a very small on-to-off conductance ratio.

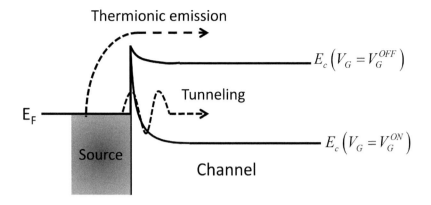

FIGURE 13.3
Conduction band energy profile in a MISFET along the length of the channel (axis joining source and drain). When the gate voltage is such that the transistor is OFF, the barrier at the source/channel interface is wide. That prevents carriers from tunneling into the channel. When the gate voltage turns the transistor ON, the conduction band edge is lowered and barrier made thin. This allows carriers to tunnel into and populate the channel. Thermionic emission can inject carriers into the channel even when the device is OFF.

13.2.6   Effect of non-idealities

Spin injection efficiency: Let us now reexamine some of the assumptions that we made when we presented the theory of the ideal SPINFET in Section 12.2. The first assumption was that the *spin injection efficiency* at the source-channel interface and the *spin detection efficiency* at the drain-channel interface are both 100%. At the time of writing this chapter, the highest electrical spin injection efficiency demonstrated at a ferromagnet/paramagnet interface was only 70% at or near room temperature [7]. Perfect 100% spin injection efficiency would require that the injecting ferromagnet is an ideal half metal with 100% spin polarization [8]. Likewise, perfect 100% spin detection efficiency would require that the detecting ferromagnet is an ideal ferromagnet with 100% spin polarization. It has been argued that ideal half metals cannot exist at any temperature above 0 K because of magnons and phonons, and even at 0 K, surfaces and inhomogeneities will reduce the spin polarization to less than 100% [9]. Thus, the spin injection and detection efficiencies are probably going to be less than 100%.

We will show later that the ratio of "on-conductance" to "off-conductance" of the SPINFET has a maximum value of $G_{on}/G_{off} = (1+\zeta_S\zeta_D)/(1-\zeta_S\zeta_D)$, where ζ_S is the spin injection efficiency at the source and ζ_D is the spin detection efficiency at the drain. If $\zeta_S = \zeta_D = 70\%$, then this ratio has a value of 2.92, which is clearly inadequate for any mainstream application. This small ratio means that the leakage current in the off state is huge; it is at least one-third of the on-current, if spin injection/detection efficiencies are 70%. In order to achieve a conductance on-off ratio of 10^5, required of modern transistors, $\zeta_S = \zeta_D = 99.9995\%$, which is a very tall order. The *primary impediment* to realization of usable SPINFETs at this time is inadequate spin injection efficiency. We will re-visit this later.

Channel magnetic field: Another questionable assumption is the absence of stray magnetic fields in the channel. Invariably, there will be a magnetic field along the channel because of the magnetized ferromagnetic contacts. This will cause problems. The first problem is that the spin precession rate in space will no longer be given by Equation (13.4) and will no longer be independent of the carrier velocity (or energy). This is because the spins will precess about the net magnetic field, and this net field is

$$\vec{B}_{total} = \vec{B}_{Rashba} + \vec{B}_{channel} \tag{13.13}$$

where the first term on the right hand side is linearly proportional to the carrier velocity v_x, but the second term is independent of velocity. Therefore, the spatial rate of spin precession, which is $d\phi/dx = g\mu_B B_{total}/(\hbar v_x)$, is no longer independent of carrier velocity v_x. Consequently, the angle by which an electron's spin has precessed when it arrives at the drain contact (Φ) will depend on the electron's velocity and scattering history. At a finite temperature, and in the presence of momentum randomizing scattering in the channel, the spread in the electron velocity will make different electrons suffer different amounts of spin precession Φ, so that when they arrive at the drain end, not all of them will have the same spin polarization. Those which have their spin polarizations anti-parallel to the drain's magnetization will be blocked, but the others will transmit to varying degrees and cause a leakage current in the "off" state. This is another source of leakage current, although it is less important than that due to imperfect spin injection and detection at the source and drain contacts. A more serious problem is that the channel magnetic field makes the eigenspinors wavevector dependent (recall Chapter 7). Without the magnetic field, the Rashba interaction lifts the spin degeneracy at any non-zero wavevector, but each spin-split band still has a fixed spin quantization axis that is parallel or anti-parallel to B_{Rashba}. This means that the spin polarization in each band is always the same and independent of wavevector. Furthermore, the spin polarizations in the two bands are anti-parallel to each other. Thus, there can be no scattering between the two bands because their eigenspinors are orthogonal. Electrons can scatter elastically or inelastically only within the same spin band, but not between two different spin bands. Intraband scattering does not alter the spin polarization since every state in the same band has exactly the same spin polarization. Therefore, scattering caused by any non-magnetic entity (phonons, impurities, etc.) does *not* relax spin, if there is no magnetic field in the channel. In other words, Elliott-Yafet type spin relaxation is not possible in a quantum wire channel if there is no channel magnetic field. We also know that there can be no D'yakonov-Perel' spin relaxation in a quantum wire with a single subband occupied (recall Problem 8.2). Therefore, there is virtually no spin relaxation in a quantum wire channel if there is no channel magnetic field.

If a channel magnetic field is present, then the spin polarizations in both bands become wavevector dependent and neither subband has a fixed spin quantization axis. Thus, two states in two spin bands, with different wavevectors, have different spin polarizations that are not completely orthogonal. A scattering event (due to a non-magnetic impurity, phonon, etc.) can cause a transition between these states because the matrix element for scattering between them is non-zero. Any such scattering will rotate the spin since the initial and final states have different spin polarizations. Therefore, ordinary non-magnetic scatterers can *relax spin*, and the Elliott-Yafet mechanism is revived in a quantum wire channel [10]. Any such spin relaxation will randomize the spin polarizations of electrons arriving at the drain and cause additional leakage current.

There is a way to reduce (or nearly eliminate) the leakage current due to the channel magnetic field. In addition to the Rashba spin-orbit interaction, there is also the Dresselhaus spin-orbit interaction in any semiconductor that lacks crystalline inversion symmetry. The ideal SPINFET analysis of Section 12.2 ignored this interaction, but it is there. In a one-dimensional SPINFET, the pseudo magnetic field due to the Dresselhaus interaction $B_{Dresselhaus}$ is directed along the channel and since it depends on the carrier velocity v_x (see Equation (8.23)), it can be altered by altering the average velocity of carriers in the channel. Therefore, it can be manipulated to partially offset the channel magnetic field caused by the ferromagnetic contacts. This can reduce the leakage current, as was shown in [11].

The total magnetic field that an electron will see in a quantum wire SPIN-FET is

$$\vec{B}_{total} = B_{Rashba}\hat{z} + B_{channel}\hat{x} + B_{Dresselhaus}\hat{x}. \qquad (13.14)$$

The last two terms are magnetic fields directed along the channel, in the x-direction. These two fields will cancel each other if

$$B_{Dresselhaus} = \frac{2m^*a_{42}}{g\mu_B\hbar}\left[\left(\frac{m\pi}{W_z}\right)^2 - \left(\frac{n\pi}{W_y}\right)^2\right]v_x = -B_{channel}, \qquad (13.15)$$

where a_{42} is a material constant, W_y, W_z are the transverse dimensions of the channel, v_x is the carrier velocity, and we have assumed the quasi one-dimensional channel to be in the [100] crystallographic direction. It is obvious that perfect cancellation is possible only for the carrier velocity v_x satisfying the condition

$$v_x = \left| \frac{B_{channel} g \mu_B \hbar}{2m^* a_{42} \left[\left(\frac{m\pi}{W_z} \right)^2 - \left(\frac{n\pi}{W_y} \right)^2 \right]} \right|. \tag{13.16}$$

Since different carriers will have different velocities at any non-zero temperature and in the presence of scattering, we can never completely eliminate leakage current by playing off the Dresselhaus interaction against the channel magnetic field. However, with a back gate terminal, we can adjust the channel carrier concentration so that the *average velocity* of the carriers satisfies the above equation. That will result in the minimum leakage current.

Structural modification to reduce the channel magnetic field: It should be obvious to the reader that the SPINFET works just as well if the contacts are magnetized in the +y direction instead of the +x direction. Electrons will enter the channel from the source with their spins polarized in the +y direction, and these spins will precess in the x-y plane when the gate voltage is turned on. Therefore, there is no difference between having the contacts magnetized in the *y*-direction versus having them magnetized in the *x*-direction, as long as the gate electric field is in the *y*-direction and the current flow is in the *x*-direction so that the pseudo magnetic field associated with the Rashba interaction is in the *z*-direction. The advantage of having the contacts magnetized in the *y*-direction is that the channel magnetic field will be much weaker now since it will only be a fringing field. This reduces leakage current.

13.2.7 The quantum well SPINFET

The original SPINFET proposed in [2] was not a quantum wire SPINFET that we have described here, but a quantum well (two-dimensional) SPINFET. For the quantum well SPINFET, the Hamiltonian describing an electron in the channel is

$$H = H_0 + H_R = \frac{\hbar^2}{2m^*} \left(k_x^2 + k_z^2 \right) + \epsilon_n + \eta \left(\sigma_z k_x - \sigma_x k_z \right)$$

$$= \begin{pmatrix} \frac{\hbar^2}{2m^*} \left(k_x^2 + k_z^2 \right) + \epsilon_n + \eta k_x & -\eta k_z \\ -\eta k_z & \frac{\hbar^2}{2m^*} \left(k_x^2 + k_z^2 \right) + \epsilon_n - \eta k_x \end{pmatrix}, \tag{13.17}$$

where we assume that the quantum well is in the x-z plane and ϵ_n are the transverse subband bottom energies due to confinement in the *y*-direction. The reader can easily deduce this Hamiltonian from the material presented in Chapter 7.

If $k_z = 0$, then the eigenspinors of the above Hamiltonian are $[1 \ 0]^\dagger$ and $[0 \ 1]^\dagger$, which are the +z-polarized and -z-polarized spins. This means that B_{Rashba} is directed along the z-axis. But if $k_x = 0$, the eigenspinors are

$(1/\sqrt{2})[1\ 1]^\dagger$ and $(1/\sqrt{2})[1\ -1]^\dagger$, which are the $+x$-polarized and $-x$-polarized spins. In that case, B_{Rashba} is directed along the x-axis. The pseudo magnetic field due to the Rashba interaction is always directed in the direction that is mutually perpendicular to the gate electric field and the direction of carrier velocity.

If the source injects x-polarized spins, then electrons which have no component of momentum along the channel ($k_x = 0$), do not undergo any spin precession at all since the original spin polarization and the pseudo magnetic field are both in the same direction (x-direction). In this case

$$\frac{d\vec{S}}{dt} = \vec{\Omega} \times \vec{S} = \frac{g\mu_B \vec{B}_{Rashba}}{\hbar} \times \vec{S} = 0 \qquad (13.18)$$

so that spin does not precess. But this does not matter so much since these electrons have no motion along the channel and therefore do not contribute to current, in any case. On the other hand, electrons that have some component of momentum along the channel ($k_x \neq 0$) do undergo spin precession. However, the axis of spin precession depends on the ratio of k_x and k_z. This has a deleterious effect when one ensemble averages over k_x and k_y. The net effect is to give rise to a larger leakage current in the off-state.

This effect was recognized in ref. [2], which advocated the use of a narrow channel that approximates a quantum wire channel. A quantum wire SPINFET is superior to a quantum well SPINFET in producing lower leakage current, but on the flip side, the maximum conductance of a true quantum wire channel is the Landauer conductance e^2/h per spin. The maximum conductance of a quantum well channel, on the other hand, is Ne^2/h per spin, where N is the number of transverse subbands along the z-direction that are occupied by carriers. The number N can be quite large for a quantum well. For a channel width of 1 μm and a Fermi De-Broglie wavelength of 10 nm, $N = 200$. Therefore, a quantum well channel has a larger channel conductance, which can ultimately lead to a larger device transconductance. The transconductance determines important figures of merit, such as transistor gain and bandwidth. Therefore, a quantum well channel may be preferred over a quantum wire channel in such applications (mostly analog applications) where the ratio of on to off conductance is not the main consideration, but the transistor gain and bandwidth are more important.

13.3 Analysis of the two-dimensional SPINFET

Although in their original proposal [2] Datta and Das proposed a transistor structure with a two-dimensional channel, they analyzed their device assuming that the electron's wavevector component transverse to the direction of

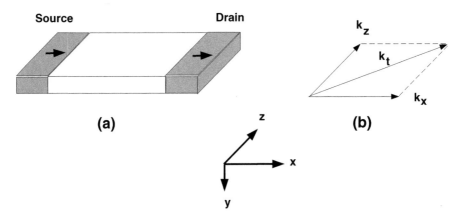

FIGURE 13.4

(a) A two-dimensional SPINFET channel, and (b) the wavevector components in the plane of the channel.

current flow is zero. Strictly speaking, this assumption corresponds to a one-dimensional channel since only there the transverse wavevector component will be zero. Datta and Das realized that the conductance modulation of the transistor will be reduced substantially if the channel is two-dimensional since then ensemble averaging over the transverse wavevector component will dilute the modulation. They did not derive an exact analytical expression for the source-to-drain current as a function of gate voltage for a two-dimensional channel. In this section we will complete this task by following [12]. This result was first derived by Pala et al. and later by others [12, 14].

Consider the two-dimensional channel of a Spin Field Effect Transistor (SPINFET) in the $(x - z)$ plane (shown in Fig. 13.4(a)), with current flowing in the x-direction. An electron's wavevector components in the channel are designated as k_x and k_z, while the total wavevector is designated as k_t. Note that $k_t^2 = k_x^2 + k_z^2$, as shown in Fig. 13.4(b).

The gate terminal induces an electric field in the y-direction which causes Rashba interaction. The Hamiltonian operator describing an electron in the channel is

$$H = \frac{p_x^2 + p_z^2}{2m^*} [\mathbf{I}] + \frac{\eta [V_G]}{\hbar} (\sigma_z p_x - \sigma_x p_z), \tag{13.19}$$

where the p-s are the momentum operators, the σ-s are the Pauli spin matrices, and $[\mathbf{I}]$ is the 2×2 identity matrix. Since this Hamiltonian is invariant in both x- and z-coordinates, the wavefunctions in the channel are plane wave states $e^{i(k_x x + k_z z)}$. Consequently, in the basis of these states, the Hamiltonian is

$$H = \begin{bmatrix} \frac{\hbar^2 k_t^2}{2m^*} + \eta [V_G] k_x & -\eta [V_G] k_z \\ -\eta [V_G] k_z & \frac{\hbar^2 k_t^2}{2m^*} - \eta [V_G] k_x \end{bmatrix}. \tag{13.20}$$

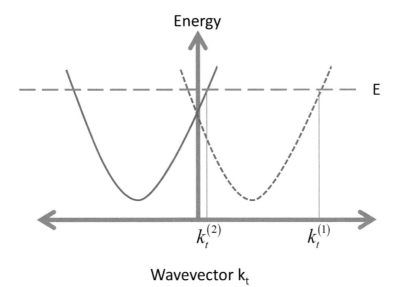

FIGURE 13.5
Schematic representation of the dispersion relations in the two spin split bands, under the influence of the gate voltage inducing Rashba interaction in the channel.

Diagonalization of this Hamiltonian yields the eigenenergies and the eigenspinors in the two spin-split bands in the two-dimensional channel:

$$E_l = \frac{\hbar^2 k_t^2}{2m^*} - \eta \left[V_G\right] k_t \text{ (lower band)}; \quad E_u = \frac{\hbar^2 k_t^2}{2m^*} + \eta \left[V_G\right] k_t \text{ (upper band)}.$$

$$(13.21)$$

$$[\Psi]_l = \begin{bmatrix} sin\theta \\ cos\theta \end{bmatrix} \text{ (lower band)}; \quad [\Psi]_u = \begin{bmatrix} -cos\theta \\ sin\theta \end{bmatrix} \text{ (upper band)}, \quad (13.22)$$

where $\theta = (1/2)arctan\,(k_z/k_x)$. The energy dispersion relations in the two bands (one broken and the other solid) are plotted in Fig. 13.5. Note that an electron of energy E has two different wavevectors in the two bands given by $k_t^{(1)}$ and $k_t^{(2)}$.

We will assume that the source contact of the SPINFET is polarized in the $+x$-direction and injects $+x$-polarized spins into the channel under a source-to-drain bias. We also assume that the spin injection efficiency at the source is 100%, so that only $+x$-polarized spins are injected to the complete exclusion of $-x$-polarized spins. An injected spin will couple into the two spin eigenstates in the channel. It is as if the x-polarized beam splits into two beams, each corresponding to one of the channel eigenspinors. This will yield

$$\frac{1}{\sqrt{2}} \begin{bmatrix} 1 \\ 1 \end{bmatrix} = C_1 \begin{bmatrix} sin\theta \\ cos\theta \end{bmatrix} + C_2 \begin{bmatrix} -cos\theta \\ sin\theta \end{bmatrix},$$
$$+x - polarized \tag{13.23}$$

where the coupling coefficients C_1 and C_2 are found by solving Equation (13.23). The result is

$$C_1 = C_1(k_x, k_z) = sin(\theta + \pi/4)$$
$$C_2 = C_2(k_x, k_z) = -cos(\theta + \pi/4) \tag{13.24}$$

Note that the coupling coefficients depend on k_x and k_z.

At the drain end, the two beams recombine and interfere to yield the spinor of the electron impinging on the drain. Since the two beams have the same energy E and transverse wavevector k_z (these are good quantum numbers in ballistic transport), they must have different longitudinal wavevectors $k_x^{(1)}$ and $k_x^{(2)}$ since $k_t^{(1)} \neq k_t^{(2)}$.

Hence, the spinor at the drain end will be

$$[\Psi]_{drain} = C_1 \begin{bmatrix} sin\theta \\ cos\theta \end{bmatrix} e^{i(k_x^{(1)}L + k_z W)} + C_2 \begin{bmatrix} -cos\theta \\ sin\theta \end{bmatrix} e^{i(k_x^{(2)}L + k_z W)}$$

$$= e^{ik_z W} \left\{ sin(\theta + \pi/4) \begin{bmatrix} sin\theta \\ cos\theta \end{bmatrix} e^{ik_x^{(1)}L} - cos(\theta + \pi/4) \begin{bmatrix} -cos\theta \\ sin\theta \end{bmatrix} e^{ik_x^{(2)}L} \right\}$$

$$= e^{ik_z W} \begin{bmatrix} sin(\theta + \pi/4) sin\theta e^{ik_x^{(1)}L} + cos(\theta + \pi/4) cos\theta e^{ik_x^{(2)}L} \\ sin(\theta + \pi/4) cos\theta e^{ik_x^{(1)}L} - cos(\theta + \pi/4) sin\theta e^{ik_x^{(2)}L} \end{bmatrix},$$
$$\tag{13.25}$$

where L is the channel length (distance between source and drain contacts) and W is the transverse displacement of the electron as it traverses the channel.

Since the drain is polarized in the same orientation as the source, it transmits only $+x$-polarized spins, so that spin filtering at the drain will yield a transmission probability $|T|^2$ where T is the projection of the impinging spinor on the eigenspinor of the drain. It is given by

$$T = \frac{1}{\sqrt{2}} [1\ 1] \begin{bmatrix} sin(\theta + \pi/4) sin\theta e^{ik_x^{(1)}L} + cos(\theta + \pi/4) cos\theta e^{ik_x^{(2)}L} \\ sin(\theta + \pi/4) cos\theta e^{ik_x^{(1)}L} - cos(\theta + \pi/4) sin\theta e^{ik_x^{(2)}L} \end{bmatrix} e^{ik_z W}$$

$$= \frac{1}{\sqrt{2}} e^{ik_z W} \left\{ sin(\theta + \pi/4)[sin\theta + cos\theta] e^{ik_x^{(1)}L} + cos(\theta + \pi/4)[cos\theta - sin\theta] e^{ik_x^{(2)}} \right.$$

$$= e^{ik_z W} \left[sin^2(\theta + \pi/4) e^{ik_x^{(1)}L} + cos^2(\theta + \pi/4) e^{ik_x^{(2)}L} \right].$$
$$\tag{13.}$$

Here, we have assumed 100% spin filtering efficiency.

Therefore,

$$
\begin{aligned}
|T|^2 &= \cos^4\left(\theta + \pi/4\right)\left|1 + \tan^2\left(\theta + \pi/4\right)e^{i\left[k_x^{(1)} - k_x^{(2)}\right]L}\right|^2 \\
&= \cos^4\left(\theta + \pi/4\right)\left[1 + \tan^4\left(\theta + \pi/4\right) + 2\tan^2\left(\theta + \pi/4\right)\cos\left(\Theta L\right)\right] \\
&= \cos^4\left(\theta + \pi/4\right) + \sin^4\left(\theta + \pi/4\right) + \frac{1}{2}\cos^2(2\theta)\cos(\Theta L),
\end{aligned}
$$
(13.27)

where $\Theta = k_x^{(1)} - k_x^{(2)}$.

The above expression can be further simplified to

$$
|T|^2 = \cos^2\left(\frac{\Theta L}{2}\right) + \sin^2\left(\frac{\Theta L}{2}\right)\sin^2(2\theta).
$$
(13.28)

From Equation (13.21), we get that $k_t^{(1)} - k_t^{(2)} = 2m^*\eta\left[V_G\right]/\hbar^2$. Expressing the wavevectors in terms of their x- and z-components, we obtain

$$
\sqrt{\left[k_x^{(1)}\right]^2 + k_z^2} - \sqrt{\left[k_x^{(2)}\right]^2 + k_z^2} = 2m^*\eta\left[V_G\right]/\hbar^2,
$$
(13.29)

which yields

$$
\Theta = k_x^{(1)} - k_x^{(2)} = \frac{m^*\eta\left[V_G\right]k_t^{(2)}/\hbar^2 + 2m^{*2}\eta^2\left[V_G\right]/\hbar^4}{\left[k_x^{(1)} + k_x^{(2)}\right]/2}.
$$
(13.30)

From Equation (13.21), we also get that when $\eta\left[V_G\right]$ is small,

$$
k_t^{(2)} = -\frac{m^*\eta\left[V_G\right]}{\hbar^2} \pm \sqrt{\left(\frac{m^*\eta\left[V_G\right]}{\hbar^2}\right)^2 + k_0^2} \approx k_0 - \frac{m^*\eta\left[V_G\right]}{\hbar^2},
$$
(13.31)

where $k_0 = \sqrt{2m^*E}/\hbar$ (the negative root is extraneous).

Now, since $\eta\left[V_G\right]$ is small, $\left[k_x^{(1)} + k_x^{(2)}\right]/2 \approx \sqrt{k_0^2 - k_z^2}$. Substituting these results in Equation (13.30), we get

$$
\begin{aligned}
\Theta &= \frac{\left(m^*\eta\left[V_G\right]/\hbar^2\right)k_0 + m^{*2}\eta^2\left[V_G\right]/\hbar^4}{\sqrt{k_0^2 - k_z^2}} \\
&= \frac{-\left(2m^*\eta\left[V_G\right]/\hbar^2\right)\sqrt{2m^*E}/\hbar - m^{*2}\eta^2\left[V_G\right]/\hbar^4}{\sqrt{2m^*E/\hbar^2 - k_z^2}}.
\end{aligned}
$$
(13.32)

The current density in the channel of the SPINFET (assuming ballistic transport) is given by the Tsu-Esaki formula:

$$
J = \frac{q}{W_y}\int_0^\infty \frac{1}{h}dE \int_0^{\sqrt{2m^*E}/\hbar} \frac{dk_z}{\pi}|T|^2\left(E, k_z\right)\left[f(E) - f(E + qV_{SD})\right],
$$
(13.33)

where q is the electronic charge, W_y is the thickness of the channel (in the y-direction), V_{SD} is the source-to-drain bias voltage, and $f(\epsilon)$ is the electron occupation probability at energy ϵ in the contacts. Since the contacts are at local thermodynamic equilibrium, these probabilities are given by the Fermi-Dirac factor.

In the linear response regime when $V_{SD} \to 0$, the above expression reduces to

$$J = \frac{q^2 V_{SD}}{W_y} \int_0^\infty \frac{1}{h} dE \int_0^{\sqrt{2m^* E/\hbar}} \frac{dk_z}{\pi} |T|^2 (E, k_z) \left[-\frac{\partial f(E)}{\partial E} \right]. \qquad (13.34)$$

This yields that the channel conductance G is

$$G = \frac{I_{SD}}{V_{SD}} = \frac{J W_y W_z}{V_{SD}} = \frac{q^2 W_z}{\pi h} \int_0^\infty dE \int_0^{\sqrt{2m^* E/\hbar}} dk_z |T|^2 (E, k_z) \left[-\frac{\partial f(E)}{\partial E} \right], \qquad (13.35)$$

where I_{SD} is the source-to-drain current and W_z is the channel width.

Using Equation (13.28), we finally get that the channel conductance is

$$G = \frac{q^2 W_z}{\pi h} \int_0^\infty dE \int_0^{\sqrt{2m^* E/\hbar}} dk_z \left\{ \left[cos^2 \left(\frac{\Theta L}{2} \right) + sin^2 \left(\frac{\Theta L}{2} \right) sin^2(2\theta) \right] \right.$$
$$\left. \left[-\frac{\partial f(E)}{\partial E} \right] \right\}$$
$$= \frac{q^2 W_z}{\pi h} \int_0^\infty dE \int_0^{\sqrt{2m^* E/\hbar}} dk_z \left\{ \left[\left(1 - \frac{\hbar^2 k_z^2}{2m^* E} \right) cos^2(\Theta L/2) + \frac{\hbar^2 k_z^2}{2m^* E} \right] \right.$$
$$\left. \left[-\frac{\partial f(E)}{\partial E} \right] \right\},$$
$$(13.36)$$

If the temperature is low so that $-\frac{\partial f(E)}{\partial E} \approx \delta(E - E_F)$ (where E_F is the Fermi energy), then the last equation reduces to

$$G = \frac{q^2 W_z}{\pi h} \int_0^{\sqrt{2m^* E_F/\hbar}} dk_z \left[\left(1 - \frac{\hbar^2 k_z^2}{2m^* E_F} \right) cos^2(\Theta L/2) + \frac{\hbar^2 k_z^2}{2m^* E_F} \right]. \qquad (13.37)$$

The last equation is very instructive. Clearly, the conductance modulations in the one- and the two-dimensional cases are very *different*. Particularly, the conductance modulation in a one-dimensional channel could reach 100%, whereas the last equation shows that the conductance modulation in a two-dimensional channel will never reach 100% because ensemble averaging represented by the integration over the transverse wavevector component k_z will dilute the modulation considerably.

Also note that the minimum source-to-drain current, or off-current, in the two-dimensional SPINFET is

$$I_{off}^{2-D} = \frac{q^2 W_z}{\pi h} \int_0^{\sqrt{2m^* E_F/\hbar}} dk_z \frac{k_z^2}{k_F^2} = \frac{q^2 W_z k_F}{3\pi h}, \qquad (13.38)$$

while the maximum source-to-drain current, or on-current, is

$$I_{on}^{2-D} = \frac{q^2 W_z}{\pi h} \int_0^{\sqrt{2m^* E_F/\hbar}} dk_z = \frac{q^2 W_z k_F}{\pi h}, \qquad (13.39)$$

where k_F is the Fermi wavevector $\left(k_F = \sqrt{2m^* E_F/\hbar}\right)$. Therefore, the best possible conductance on/off ratio of a 2D-SPINFET (even assuming 100% spin injection and detection efficiencies) is a paltry 3:1. This small on/off ratio makes the 2-D SPINFET unsuitable for most digital applications.

13.3.1 SPINFET based on the Dresselhaus spin–orbit interaction

The SPINFET of [2] was based on the Rashba interaction. The Dresselhaus spin-orbit interaction can also be gainfully employed to realize a different kind of SPINFET [15]. In a one dimensional channel with crystallographic inversion asymmetry, the strength of the pseudo magnetic field due to the Dresselhaus interaction depends on the physical width of the channel W_z (see Equation (13.15)). If we define the one-dimensional channel by a split-gate, then we can vary the potential applied on the split-gate to change the channel width W_z and therefore the strength of the pseudo magnetic field $B_{Dresselhaus}$ associated with the Dresselhaus interaction. This field also causes spins to precess in space at a rate independent of the carrier velocity.[||] By varying the split-gate voltage, one can change $B_{Dresselhaus}$ and therefore the precession rate. As a result, the gate voltage modulates the angle by which the spins precess as they traverse the channel from source to drain. This, in turn, modulates the source-to-drain current. The source and drains are magnetized either in the y-direction or z-direction since the Dresselhaus field $B_{Dresselhaus}$ is in the x-direction if we have the channel in the [100] crystallographic direction. Since the split-gate voltage can be used to modulate the source-to-drain current, we can realize transistor action. The schematic of a Dresselhaus type SPINFET is shown in Figure 13.6. There is no magnetic field along the channel (except weak fringing fields) since the ferromagnetic contacts are magnetized in the $(y - z)$ plane, which is transverse to the axis of the channel. This reduces the

---

[||]The velocity independence should be obvious since the pseudo magnetic field due to the Dresselhaus interaction is proportional to the carrier velocity (see Equation (13.15) just like the pseudo magnetic field due to the Rashba interaction.

leakage current. The SPINFET based on Dresselhaus interaction has been discussed in [15]. Here, we present the basic ingredients.

Consider a one-dimensional channel along the [100] crystallographic direction that has Dresselhaus interaction but no Rashba interaction and no magnetic field. This channel is defined by split-gates on a two-dimensional electron gas in the $(x - z)$ plane, so that the confining potential in the z-direction is parabolic. We will assume that the curvature of this parabolic confining potential is ω, so that the effective width of the channel in the z-direction is $\sqrt{\hbar/(2m^*\omega)}$. Current flows in the x-direction. We will make the same assumptions that were implicit in the Datta-Das SPINFET, namely, that the source injects spin with 100% efficiency, the drain detects them with 100% efficiency, and finally there is no spin relaxation in the channel, which is much shorter than the spin relaxation length.

The Hamiltonian describing this channel is

$$H = \epsilon + \frac{\hbar^2 k_x^2}{2m^*} + 2a_{42}\sigma_x k_x \left[\frac{m^*\omega}{2\hbar} - \left(\frac{\pi}{W_y} \right)^2 \right], \tag{13.40}$$

where ϵ is the lowest subband energy in the channel and W_y is the width of the two-dimensional electron gas (in the y-direction). We are assuming that only the lowest subband is filled with carriers.

Diagonalization of the Hamiltonian in Equation (13.40) shows that the eigenspinors in the channel are $(1/\sqrt{2})[1\ 1]^\dagger$ and $(1/\sqrt{2})[1\ -1]^\dagger$, which are $+x$-polarized and $-x$-polarized states. They have eigenenergies that differ by $2\beta k_x$ where $\beta = 2a_{42}[m^*\omega/(2\hbar) - (\pi/W_y)^2]$. Therefore, the energies of $+x$ polarized and $-x$-polarized states will be

$$E(+x\_pol) = \epsilon + \frac{\hbar^2 k_x^2}{2m^*} + \beta k_x,$$

$$E(-x\_pol) = \epsilon + \frac{\hbar^2 k_x^2}{2m^*} - \beta k_x. \tag{13.41}$$

An electron entering the channel with an energy E will have two different wavevectors k_{x+} and k_{x-} for $+x$-polarized and $-x$-polarized spins. If we inject only $+z$-polarized spins from the ferromagnetic source contact (magnetized in the $+z$ direction), then the incident electron will transmit equally into $+x$ and $-x$ polarized states since

$$\begin{bmatrix} 1 \\ 0 \end{bmatrix} = \frac{1}{\sqrt{2}} \frac{1}{\sqrt{2}} \begin{bmatrix} 1 \\ 1 \end{bmatrix} + \frac{1}{\sqrt{2}} \frac{1}{\sqrt{2}} \begin{bmatrix} 1 \\ -1 \end{bmatrix}. \tag{13.42}$$

The eigenspinor describing the electron's spin at the drain end will be

$$[\Psi] = \frac{1}{2} \begin{bmatrix} e^{ik_{x+}L} + e^{ik_{x-}L} \\ e^{ik_{x+}L} - e^{ik_{x-}L} \end{bmatrix}, \tag{13.43}$$

where L is the channel length.

If the drain is also magnetized in the +z direction and is an ideal half metallic ferromagnet that transmits only the majority spin (i.e., an ideal spin detector), then the transmission probability of the electron (and hence the linear response source-to-drain conductance G_{SD}) will be given by

$$G_{SD} \propto \left| [1 \ 0] \begin{bmatrix} e^{ik_{x+}L} + e^{ik_{x-}L} \\ e^{ik_{x+}L} - e^{ik_{x-}L} \end{bmatrix} \right|^2$$

$$\propto \cos^2 \left(\frac{(k_{x+} - k_{x-})L}{2} \right)$$

$$= \cos^2 \left(m^* \beta L / \hbar^2 \right)$$

$$= \cos^2 \left(\frac{\Phi}{2} \right), \tag{13.44}$$

where we have used Equation (13.41) to come up with the last two equalities. The angle

$$\Phi = \frac{2m^* \beta L}{\hbar^2} = \frac{4m^* a_{42} L}{\hbar^2} \left[\frac{m^* \omega}{2\hbar} - \left(\frac{\pi}{W_y} \right)^2 \right]. \tag{13.45}$$

We can modulate the curvature of the confining potential ω in the z-direction with split-gate potentials, which will change Φ and therefore the source-to-drain conductance G_{SD}. This realizes transistor action.

In order to precess the spin by π radians and switch the transistor from "on" to "off," we need

$$\Delta\Phi = \frac{4m^* a_{42} L}{\hbar^2} \left[\frac{m^* \Delta\omega}{2\hbar} \right] = \pi, \tag{13.46}$$

which mandates that

$$\Delta\omega = \frac{\pi \hbar^3}{2m^{*2} a_{42} L}, \tag{13.47}$$

where $\Delta\omega$ is the amount by which we have to change the curvature of the parabolic confining potential in the z-direction with the split-gate voltage.

Just as in the case of the Datta-Das SPINFET, we will benefit from a stronger spin-orbit interaction strength a_{42} and a long channel length L, since these would reduce the split-gate voltage required to turn the transistor from on to off.

13.4 Device performance of SPINFETs

In the world of electronics, the universally accepted benchmark for the transistor device is the celebrated metal-insulator-semiconductor-field effect-transistor

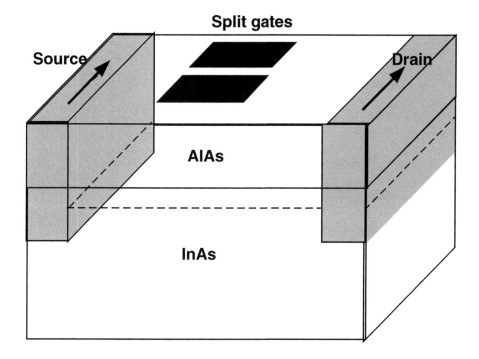

FIGURE 13.6
Schematic of a Spin Field Effect Transistor based on the Dresselhaus interaction. The 2-DEG is shown as the sheet enclosed by the broken lines.

(MISFET), which has been, and remains, the workhorse of electronics. Therefore, the SPINFET must be compared with an equivalent MISFET to determine if there are any advantages to utilizing spin. Surprisingly, in spite of many papers extolling perceived merits of SPINFETs, this elementary exercise was not carried out until recently [16]. The results were quite illuminating.

Bandyopadhyay and Cahay [16] compared a strictly one-dimensional (quantum wire) MISFET with a strictly one-dimensional SPINFET at very low temperatures and showed that the SPINFET is actually *inferior* (requires a larger gate voltage to switch states) unless its channel is several micrometers long. Here, we extend that analysis to a quasi two-dimensional channel with narrow width and at arbitrary temperatures.

We will compare primarily the switching voltage of a two-dimensional MISFET with that of a two-dimensional SPINFET (i.e., the voltage required to turn the device on or off) since that determines the dynamic energy dissipated during switching. The dissipated energy is $(1/2)C_g V_{switch}^2$ where C_g is the gate capacitance.** Therefore, a lower V_{switch} translates to less energy dissipated during switching. This can be a significant advantage in high density circuits where the primary obstacle to continuing device miniaturization is excessive energy dissipated during switching.

Let us focus on a narrow GaAs channel SPINFET where the width of the channel is sufficiently small that the component of an electron's velocity along the width is much smaller than the component along the length ($v_x \gg v_z$ or $k_x \gg k_z$). In that case, Equation (13.5) approximately holds and the electric field $\mathcal{E}_y$ required to turn off the SPINFET is found by setting $\Phi = \pi$ and solving for $\mathcal{E}_y$. This value is

$$\mathcal{E}_y^{off} = \frac{\pi \hbar^2}{2m^* a_{46} L}. \tag{13.48}$$

In GaAs, the value of $a_{46} \approx 10^{-38}$ C-m^2 and m^* is 0.067 times the free electron mass. Therefore, if the channel length $L = 1$ μm, then $\mathcal{E}_y^{off} = 280$ kV/cm. For a 0.1 μm long channel, $\mathcal{E}_y^{off} = 2.8$ MV/cm, which is beginning to approach the breakdown field in GaAs. The field would have been smaller if the value of a_{46} were larger. Unfortunately, there is no known material where the value of a_{46} is significantly (a few orders of magnitude) higher than what it is in GaAs. Therefore, the practicality of deep sub-micron n-channel SPINFETs is currently dubious since the value of $\mathcal{E}_y^{off}$ in such short channel SPINFETs may exceed the breakdown field in the channel.

There is, however, a small caveat. It is well known that spin-orbit interactions in many semiconductors are much stronger in the valence band than in the conduction band. This, of course, immediately brings up the possibility

---

**The energy dissipated can be much smaller than $(1/2)C_g V_{switch}^2$ if the gate voltage is applied in small steps. This mode of switching, known as "adiabatic switching," has been discussed by some authors [17], but is slow and unreliable.

that p-channel SPINFETs might require a smaller electric field to turn off, and therefore may produce better SPINFETs. However, spin transport in the valence band is complicated, and it is not really clear whether there will be any significant advantage with p-channel devices. This remains an issue for further research.

13.4.1   Comparison between MISFET and SPINFET

The SPINFET, as proposed in [2], is a "normally-on" device. Therefore, it is best implemented with a standard MISFET (with ferromagnetic contacts) in "accumulation mode", where a conducting channel is already formed when the gate voltage is zero. In standard MOS theory based on the so-called delta depletion approximation [18], the gate voltage V_G applied on the gate of a MISFET that is in accumulation mode will be dropped entirely across that gate insulator, with practically no drop on the semiconductor channel, since the accumulation layer charge will shield the semiconductor underneath from the electric field. Therefore,

$$V_G \approx \mathcal{E}_i d , \tag{13.49}$$

where $\mathcal{E}_i$ is the electric field in the gate insulator (assumed uniform since there are no fixed charges in an ideal gate insulator). Next, Poisson's equation dictates that

$$\kappa_s \mathcal{E}_y = \kappa_i \mathcal{E}_i , \tag{13.50}$$

where κ_s and κ_i are the dielectric constants of the semiconductor and insulator, respectively, and $\mathcal{E}_y$ is the transverse electric field in the semiconductor channel at the interface with the gate insulator. This field induces the Rashba interaction. Here, we have assumed that there is no sheet charge at the insulator-semiconductor interface due to such things as interface states.

Combining the two preceding equations, we obtain

$$V_G = \frac{\kappa_s}{\kappa_i} \mathcal{E}_y d. \tag{13.51}$$

Therefore, using Equation (13.48) in Equation (13.51), we find that the gate voltage required to turn off a SPINFET will be

$$V_{off}^{SPINFET} = V_{switch} = \frac{\kappa_s}{\kappa_i} \mathcal{E}_y^{off} d = \frac{\kappa_s}{\kappa_i} \frac{\pi \hbar^2}{2m^* a_{46}} \frac{d}{L}. \tag{13.52}$$

If we assume that the gate insulator is AlGaAs, then $\kappa_s \approx \kappa_i$. Therefore, if the gate insulator thickness is 10 nm, then for a 1 μm channel length device, $V_{off}^{SPINFET} = 284$ mV. For a 0.1 μm channel length device, $V_{off}^{SPINFET} = 2.84$ V. Remember that in the 0.1 μm device, the transverse electric field required to turn the device off is 2.8 MV/cm, which is close to the breakdown electric field in GaAs. Therefore, GaAs SPINFETs with channel lengths much smaller than 0.1 μm are not very practical. One should ruminate over this, knowing that silicon MISFETs with gate lengths of 22 nm are currently in production.

Now, if the same SPINFET structure were used as a traditional normally-on MISFET, then what would have been the switching voltage that would have turned the device off? We would simply have to put enough negative charges on the gate to drive all the electrons out of the channel in order to turn the MISFET off. That means

$$C_g V_{off}^{MISFET} = -en_s A , \qquad (13.53)$$

where e is the electronic charge, n_s is the average surface concentration of electrons in the channel, and A is the cross sectional area of the gate (which, we assume, covers the entire channel). Since $C_g = \kappa_i A/d$, we get

$$V_{off}^{MISFET} = V_{switch} = -\frac{en_s d}{\kappa_i}. \qquad (13.54)$$

If $n_s = 10^{11}$ cm^{-2} (typically obtained in GaAs channels) and $d = 10$ nm, then $V_{off}^{MISFET} = 14$ mV, independent of the channel length. From Equation (13.52), we see that this is also the $V_{off}^{SPINFET}$ of a SPINFET with 20 μm channel length. Therefore, if a typical GaAs accumulation mode MISFET structure is used either in the SPINFET mode or in the traditional MISFET mode, the SPINFET mode will have a lower switching voltage if the channel length is larger than 20 μm. Consequently, the SPINFET beats the MISFET only if its channel length is huge – larger than 20 μm! This comparison is valid at any arbitrary temperature.

If we use different material systems (e.g., InAs/AlAs heterostructures instead of AlGaAs/GaAs heterostructures), then the numbers will change slightly, perhaps in favor of the SPINFET, but the basic conclusion will not change. When all is said and done, the spin-orbit interaction in technologically important semiconductors is too weak to produce sufficient spin precession rate (in space) at small enough gate voltages. Therefore, not enough spin precession is produced to rotate the spin by π radians if the channel length is short and the gate voltage is not too high. In order to produce a rotation by π, one needs either very high gate voltages or very long channels. That is why the switching voltage of a SPINFET is typically larger than that of a comparable MISFET unless the channel of the SPINFET is unnaturally long. This conclusion is consistent with that reached in [16].

13.4.2 Comparison between HEMT and SPINFET

If a high electron mobility transistor (HEMT) [also known as a modulation doped field effect transistor, or MODFET] is used instead of a MISFET structure to implement a SPINFET, then the analysis is slightly different. The HEMT is a normally-on device, just like an accumulation mode MISFET. Here, however, the gate insulator is doped so that there are fixed charges in the insulator. The average electric field in the gate insulator is typically negligible compared to the electric field in the semiconductor right at the interface

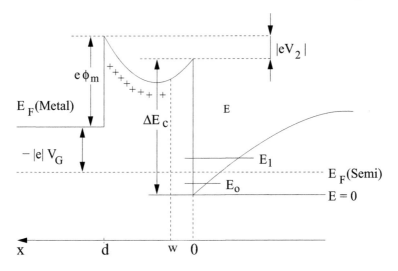

FIGURE 13.7
Energy band diagram in the direction perpendicular to the heterointerface
in a HEMT, when a gate voltage V_G is applied. The conduction band offset
between the gate insulator material and the semiconductor substrate material
is ΔE_c and ϕ_m is the metal-insulator barrier. The Fermi level is denoted by
E_F. The conduction band edge in the semiconductor substrate, adjoining the
interface, has a triangular shape (see Chapter 18 for a discussion of triangular
potential wells). The lowest two bound state energies in the triangular well
are E_0 and E_1.

with the insulator. The energy band diagram of a HEMT as a function of
distance in the direction perpendicular to the heterointerface is shown in Fig.
13.7.

Using Poisson's equation, and neglecting the electric field in the insulator
in comparison with the electric field in the semiconductor at the insulator-
semiconductor interface, we can write

$$\mathcal{E}_y \approx \frac{en_s}{\kappa_s}, \tag{13.55}$$

where n_s is the surface concentration of carriers in the HEMT channel at zero
gate voltage. Following [19], we can further stipulate that

$$n_s = C_g(V_G - V_t)/(eA)$$
$$C_g = A\kappa_i/(d + \Delta d), \tag{13.56}$$

where d is the gate insulator thickness, V_t is the threshold voltage (it is a
negative quantity), and Δd is the thickness of the two-dimensional electron

gas in the channel. Combining the previous two equations, we get

$$\mathcal{E}_y = \frac{\kappa_i}{\kappa_s} \frac{V_G - V_t}{d + \Delta d}. \tag{13.57}$$

The quantity V_t is given by [19]

$$V_t = \phi_m - \Delta E_c - qN_D(d - w)^2/(2\kappa_s) \tag{13.58}$$

and is usually a negative quantity. Here, ϕ_m is the Schottky barrier height between the gate metal and the gate insulator, ΔE_c is the conduction band offset between the insulator and the semiconductor, N_D is the doping density in the insulator, and w is the width of the spacer layer separating the doped region of the insulator from the semiconductor channel.

Note from Equation (13.57) that, in the HEMT, there is a large transverse electric field $\mathcal{E}_y$ in the channel. Setting this electric field equal to $\mathcal{E}_y^{off}$ given by Equation (13.48), we obtain

$$V'_{off}^{SPINFET} = \frac{\kappa_s}{\kappa_i} \frac{\pi\hbar^2}{2m^*a_{46}} \frac{d + \Delta d}{L} - |V_t|. \tag{13.59}$$

Typically $V_t \sim$ -1 V. Therefore, if we assume $d = 10$ nm, $\Delta d = 8$ nm and the channel material is GaAs, then $V'_{off}^{SPINFET} = 400$ mV if the channel length is 1 μm and 4 V if the channel length is 0.1 μm. In comparison, the switching voltage of the HEMT itself is $|V_t| = 1$ V. Therefore, the SPINFET does not produce any advantage over the HEMT in terms of a lower switching voltage or lower dynamic power dissipation during switching if the channel length is smaller than 0.25 μm. Once again, the reason for this is that spin-orbit interactions are too weak to produce sufficient spin precession within a short channel with small gate voltage. In order for a 10 nm channel length SPINFET to have the same switching voltage as a 10 nm channel length HEMT, the strength of Rashba interaction (a_{46}) has to be 25 times larger than what it is.

13.5 Power dissipation estimates

The dynamic energy dissipated during switching a transistor is

$$P_{dynamic} = (1/2)C_g V_{switch}^2. \tag{13.60}$$

We will consider 100-nm channel length SPINFET devices (because 22-nm gate length MISFETs are currently in production), for which $V_{off}^{SPINFET} = 2.8$ V if implemented in the MISFET configuration and 4 V if implemented in the HEMT configuration. Assume also that the channel width is 50 nm. In that case, the gate capacitance C_g is 57 aF in the MISFET configuration and 32 aF in the HEMT configuration. Therefore, the dynamic energy dissipated during switching the transistor is 0.22 fJ in the MISFET configuration and 0.26 fJ in the HEMT configuration. If the switching is carried out in ~ 10 ps, then the power dissipated is 22 μW and 26 μW respectively. With a device density of 10^8 transistors/cm^2 (comparable to the Pentium IV chip), the power density will be 22 kW/cm^2 and 26 kW/cm^2, respectively, which are too high (the Pentium IV dissipates less than 100 W/cm^2). Therefore, the SPINFET does not reduce power dissipation unless its channel is at least a few μm long. Having such a large SPINFET is not acceptable since it reduces device density (severalfold from what we currently have in microprocessor chips) and goes against the grain of Moore's law, which foresees doubling of the device density on a chip every 18 months.

13.6   Other types of SPINFETs

Slightly different types of SPINFET ideas have also appeared in the literature. They go by such names as "Non-ballistic SPINFET" [20, 21] or the "Spin Lifetime Transistor" [22].

13.6.1   Non-ballistic SPINFET

This device has a two-dimensional electron gas like an ordinary MISFET and has both Dresselhaus and Rashba spin orbit interactions. The latter can be changed with a gate voltage. Fig. 13.8 shows such a device. Since the channel is two-dimensional, the spin-split bands generally do not have a fixed spin quantization axis even if there is no magnetic field in the channel [meaning that the spin eigenstates are wavevector-dependent; recall Section 7.1.1 with $B_x = B_z = 0$]. The only exception to this situation is when the Rashba and Dresselhaus interactions in the channel have exactly the same strength. In that case, each band has a fixed spin quantization axis, and the spin eigenstate in either band is wavevector-independent because of the formation of the persistent spin helix state. In fact, these eigenspinors are $[-sin(\pi/8), cos(\pi/8)]^T$ and $[cos(\pi/8), sin(\pi/8)]^T$. Therefore, the eigenspin lies in the plane of the 2-DEG (x-z plane) and subtends an angle of 45° or 135° with the x-axis.

In the non-ballistic SPINFET, the Rashba interaction is first tuned with the gate voltage to make it exactly equal to the Dresselhaus interaction, which is independent of the gate voltage. This then makes the spin eigenstates wavevector independent, as we saw in Chapter 7. Electrons are injected into the channel of the transistor from a ferromagnetic source with a polarization

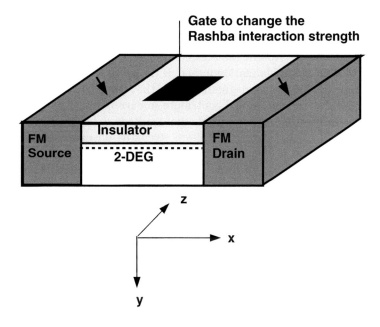

FIGURE 13.8

The non-ballistic SPINFET. The current flows in the x-direction and the symmetry breaking electric field is applied along the y-direction to modulate the strength of the Rashba interaction. When the strengths of the Rashba and Dresselhaus interactions are equal, the wavevector-independent eigenspin polarizations lie in the x-z plane and subtend an angle of 45° or 135° with the x-axis. The contacts are magnetized in the x-z plane subtending an angle of 45° with the x-axis so that they inject and detect spins polarized in one of the two eigenspin states.

that corresponds to the spin eigenstate in one of the spin-split bands. All carriers enter this band. Since the spin eigenstate is wavevector-independent, any momentum relaxing scattering in the channel, which will change the electron's wavevector, will *not alter* the spin polarization. Scattering can couple two states within the same band, but not in two different bands (with anti-parallel spin quantization axis) since the eigenspinors in these two bands are orthogonal. Therefore, regardless of how much momentum relaxing scattering takes place in the channel, there will be no spin relaxation. The only agents that can cause spin relaxation are hyperfine interaction with nuclear spins, magnetic impurities, and perhaps the Bir-Aronov-Pikus mechanims if transport is bipolar. Absent these mechanisms, the carriers will arrive at the drain with their spin polarization intact. If the drain ferromagnetic contact is magnetized parallel to the magnetization of the source contact, then all these carriers will exit the device and contribute to current.

In order to change the current, the gate voltage is changed, which makes the Rashba and Dresselhaus interaction strengths unequal, thereby making the spin eigenstates in the channel wavevector dependent (see, again, Equations (7.30) and (7.31)). In that case, if the electrons suffer momentum relaxing collisions due to impurities, defects, surface roughness, etc., their spin polarizations will also rotate, resulting in spin relaxation. Thus, the carriers that arrive at the drain no longer have all their spins aligned along the drain's magnetization. Consequently, the overall transmission probability of the electrons decreases, and the current drops. This is how the gate voltage changes the source-to-drain current and produces transistor action. Gate-voltage modulation of spin relaxation length in quantum wells, based on this kind of mechanism, has been demonstrated experimentally [23].

One simple way of viewing the transistor action is that when the Rashba and Dresselhaus interactions are balanced, the channel current is 100% spin polarized (all carriers arriving at the drain have exactly the same spin, namely the original spin), but when the two interactions are unbalanced, then the spin polarization of the current decreases owing to spin relaxation. The spin polarization can decrease to zero, but *no less*, meaning that, on average, 50% of the spins will be aligned and 50% anti-aligned with the drain's magnetization. The aligned component in the current will transmit and the anti-aligned component will be blocked. Thus, the minimum current (off-current) of this transistor is no less than one-half of the maximum current (on-current). Since the maximum ratio of the on-to-off conductance is only 2, this device is clearly unsuitable for most, if not all, mainstream applications. A recent simulation has shown that the on-to-off conductance ratio is actually only about 1.2 for realistic device parameters [24].

The situation can be improved dramatically if the source and drain contacts have anti-parallel magnetizations, instead of parallel. In that case, when the Rashba and Dresselhaus interactions are balanced, the transmitted current will be exactly zero, but when they are unbalanced, it is non-zero. Therefore, the on-to-off conductance becomes infinity. However, there is a caveat. This argument presupposes that the ferromagnetic contacts can inject and detect spins with 100% efficiency, meaning that only the majority spins are injected and transmitted by the ferromagnetic source and drain contacts, respectively. If, on the other hand, the spin injection efficiency is only about 70% [7] (which is the record at room temperature at the time of writing this book), so that

$$\frac{I_{maj} - I_{min}}{I_{maj} + I_{min}} = 0.7 , \tag{13.61}$$

then 15% of the injected current is due to minority spins. These minority spins will transmit through the drain even when the Rashba and Dresselhaus interactions are balanced. Thus the off-current is 15% of the total injected current $(I_{maj} + I_{min})$, whereas the on-current is still at best 50% of the total injected current. Therefore, the on-to-off ratio of the conductance is 0.5/0.15 = 3.3, which is not a whole lot better than 2. Consequently, achieving a large conductance ratio is very difficult, particularly at room temperature, when the spin injection efficiency tends to be small. Furthermore, the drain will also not be a perfect spin filter. The reader can easily show that the ratio of on-to-off conductance of this transistor will be

$$\frac{G_{on}}{G_{off}} = \frac{1}{1 - \zeta_S \zeta_D}, \tag{13.62}$$

where ζ_S is the spin injection efficiency at the source and ζ_D is the spin detection efficiency at the drain. Therefore, in order to make the conductance ratio $\sim 10^5$, which is what today's transistors have, the spin injection and detection efficiencies have to be 99.9995%, which appears impractical at this time.

13.6.2   Spin relaxation transistor

A device very similar to the non-ballistic SPINFET was proposed in [22]. We will call it the Spin Relaxation Transistor (SRT) since it realizes transistor action by modulating the spin relaxation rate in the channel with a gate voltage. With zero gate voltage, the Rashba spin-orbit interaction in the channel is weak or non-existent, which makes the spin relaxation time very long. When the gate voltage is turned on, the spin-orbit interaction strength increases, which makes the spin relaxation time short. This phenomenon has been demonstrated experimentally in quantum wells [25].

The source and drain ferromagnets are magnetized in the anti-parallel configuration. The source injects its majority spins in the channel and they arrive with their spin polarizations nearly intact at the drain when the gate voltage is zero, because of negligible spin relaxation in the channel. The drain dutifully blocks them since they are minority spins in the drain contact. Accordingly, the source-to-drain current is nearly zero when the gate voltage is zero.

When the gate voltage is turned on, spins flip in the channel, so that a significant fraction (but at most 50%) of the electrons arrive at the drain with their spins flipped. These electrons are transmitted by the drain because their spins are majority spins in the drain. Therefore, the source-to-drain current increases when the gate voltage is turned on. This device is a normally-off device.

Since the drain transmits only the flipped spins, [22] proposed the following equation for the source-to-drain current I_{SD}:

$$I_{SD} = \frac{Aen_s}{2\tau_{transit}} \left(1 - e^{-\tau_{transit}/T_1}\right) , \qquad (13.63)$$

where A is the cross-sectional area of the channel, e is the electronic charge, and n_s is the two-dimensional electron density in the channel. The quantities $\tau_{transit}$ and T_1 are the transit times through the channel and the spin relaxation time. Since, for short channel devices, $\tau_{transit} \ll T_1$ always, the above equation simplifies to

$$I_{SD} \approx \frac{Aen_s}{2T_1}. \qquad (13.64)$$

Based on the above, [22] states that the current (or conductance) on-off ratio will be $T_1(V_G = 0)/T_1(V_G = V_t)$, where V_t is the "threshold voltage" that turns the device fully on and causes the maximum source-to-drain current to flow through the channel. The authors of [22] further assume that $T_1(V_G = 0)$ = 1 μsec and $T_1(V_G = V_t)$ = 10 psec, so that the conductance on-off ratio is 10^5.

The bane of "spin injection efficiency": A little reflection will convince the reader that this analysis tacitly assumes that the spin injection efficiency at the source and the spin filtering efficiency at the drain are both 100%. That assumption is actually implicit in Equation (13.63). If we do not make that unrealistic assumption, then of course this device cannot be any better than the non-ballistic SPINFET with anti-parallel magnetizations of the source and drain contacts. In fact, Equation (13.62) applies for this device as well (see Problem 13.4), so that, in order to achieve a conductance on-off ratio of 10^5, the required value of ζ_S or ζ_D is 99.9995%, which is not practical in the near term. If the spin injection and detection efficiencies are 70% each, then the maximum conductance on-off ratio of this device is a mere 1.96, which is a far cry from the claimed value of 10^5. The purpose of pointing this out is not to be overly critical of the device design, but to highlight the critical role of spin injection and detection efficiencies in all these devices. Without an extremely high spin injection/detection efficiency, these devices have a scant chance of being useful for mainstream applications. Therefore, the bulk of research effort in this area should be directed toward improving spin injection and detection efficiencies in these structures.

Equation (13.63) is actually not the best representation for the current. Since the transit time for different electrons transiting the device will be different owing to their having different velocities, but they all traverse the same channel length L, it is more appropriate to write the current as

$$I_{SD} = \frac{Aen_s v_d}{4L} \left[1 + \zeta_S \zeta_D - 2\zeta_S \zeta_D e^{-L/L_s} \right] , \qquad (13.65)$$

where v_d is the drift velocity of the electrons and L_s is the spin relaxation length. Therefore, the conductance on/off ratio is

$$\frac{G_{on}}{G_{off}} = \frac{1 + \zeta_S \zeta_D - 2\zeta_S \zeta_D e^{-L/L_{V_G=0}}}{1 + \zeta_S \zeta_D - 2\zeta_S \zeta_D e^{-L/L_{V_G=V_T}}}. \qquad (13.66)$$

In the event $L \ll L_{V_G=V_T}, L_{V_G=0}$, this reduces to

$$\frac{G_{on}}{G_{off}} = \frac{1 - \zeta_S \zeta_D + 2\zeta_S \zeta_D (L/L_{V_G=0})}{1 - \zeta_S \zeta_D + 2\zeta_S \zeta_D (L/L_{V_G=V_T})}. \qquad (13.67)$$

The authors of [22] claimed that their SRT device will have a very low value of the threshold voltage V_t, of the order of 0.1 V. The quantity $1/T_1$ is proportional to the electric field $\mathcal{E}_y^2$. Assuming $n_s = 10^{12}$ cm^{-2}, [22] claimed that $\mathcal{E}_y = 50$ kV/cm is sufficient to reduce T_1 to 10 psec (from 1 μs when $\mathcal{E}_y = 0$) and turn the device on. Using Equation (13.51), we find that the voltage required to turn a GaAs based SRT on will be 0.05 V if the gate insulator thickness d is 10 nm and $\mathcal{E}_y^{off} = 50$ kV/cm ([22] assumed that effectively $d = 20$ nm, which explains the factor of 2 difference). Here, we have assumed, as always, that the SRT is implemented with a MISFET structure in accumulation. If this structure were used in the MISFET mode instead of the SRT mode, then the MISFET would be normally-on and to turn it off would require a gate voltage of 0.14 V (see Equation (13.54)). A high channel carrier concentration favors the SRT and a low channel carrier concentration favors the MISFET. If the channel concentration were 10^{11} cm^{-2}, then the MISFET would have had a switching voltage of 14 mV, and the SRT would have had switching voltage in excess of 50 mV. However, the disadvantage of having a high carrier concentration in the SRT is that it tends to degrade the spin injection efficiency even further, leading to an even lower conductance on-off ratio and higher leakage current in the off-state. Of course, the MISFET will always have a much higher conductance on-off ratio than the SRT unless extremely high spin injection efficiencies are achieved; thus, the SRT is currently not competitive with a MISFET in any case.

HEMT implementation: If a HEMT structure were used to implement the SRT, the device would be normally-on with anti-parallel magnetizations of the source and drain contacts (since there is already a strong built-in transverse channel electric field in a HEMT when the gate voltage is zero). With parallel magnetizations, the device will be normally-off. This device can be switched "off" (in the case of anti-parallel magnetizations) or "on" (in the case of parallel magnetizations) by applying a negative gate voltage to null the built-in electric field. Since this built-in field has a magnitude of $(\kappa_i/\kappa_s)(|V_t|/(d +$

Δd)) (see Equation (13.57)), it is obvious that the gate voltage we need to apply to null the electric field is V_t. Therefore, both the HEMT and the SRT will have exactly the same switching voltage. The advantage of using the HEMT structure to implement the SRT is that the leakage current in the off-state will be close to zero (even if the spin injection and detection efficiencies are poor) since application of a gate voltage equal to V_t will not only null the electric field in the channel, but it will also deplete the channel of mobile carriers so that there is no carrier left to carry a leakage current. However, the reduction in the leakage current in this case is not due to any spintronic principle, but rather due to the basic MODFET principle, which switches the transistor off by making the channel carrier concentration go to zero. To sum up, the SRT has no advantage over a conventional HEMT in terms of switching voltage, if implemented with a HEMT structure. However, it may have a *disadvantage*. Since at best only 50% of the injected carriers are transmitted in the SRT when it is in the on-state, while 100% of the injected carriers are transmitted in a ballistic HEMT, the on-current of the HEMT will be at least twice that of the SRT. Since their switching voltages are the same, on average, the transconductance of the HEMT will be roughly twice that of the SRT. Thus, the SRT is not competitive with a traditional HEMT as a transistor, if implemented with a HEMT-like structure.

13.7   Importance of spin injection efficiency

It should be obvious to the reader that in all of the SPINFETs discussed so far, the spin injection efficiency is critical. Without a very high spin injection efficiency, approaching 100%, none of the SPINFETs will have a high enough on-to-off conductance ratio to be of much use in anything. Below, we carry out a systematic analysis of how spin injection efficiency affects the transfer characteristics (source-to-drain current versus gate voltage) of SPINFETs, on-off ratio of the conductance, and the maximum transconductance.

Consider the basic SPINFET proposed by Datta and Das [2]. Since the contacts are magnetized along the direction of current flow (say, the $+x$-direction), majority spins that are injected from the source into the channel are $+x$-polarized spins with eigenspinor $(1/\sqrt{2})[1 \ 1]^\dagger$ and the minority spins are $-x$-polarized spins with eigenspinors $(1/\sqrt{2})[1 \ -1]^\dagger$. When an electron that had the majority spin in the source ultimately arrives at the drain, the spinor describing this electron will be

$$\Psi_{maj} = \frac{1}{\sqrt{2}} \begin{bmatrix} 1 \\ e^{i\Phi} \end{bmatrix} , \tag{13.68}$$

where Φ is the angle (remember that it is velocity independent) by which the spin has precessed in the channel under the influence of the gate potential. This electron is detected by the majority spin band in the drain with efficiency $(1 + \zeta_D)/2$, where ζ_D is the spin detection efficiency at the drain. The probability of this detection is

$$P_{maj}^{maj} = \frac{1 + \zeta_D}{2} \left| \frac{1}{\sqrt{2}} [1 \ 1] \frac{1}{\sqrt{2}} \begin{bmatrix} 1 \\ e^{i\Phi} \end{bmatrix} \right|^2 = \frac{1 + \zeta_D}{2} cos^2 \left(\frac{\Phi}{2} \right). \qquad (13.69)$$

The minority spin band also detects this electron with a probability

$$P_{maj}^{min} = \frac{1 - \zeta_D}{2} \left| \frac{1}{\sqrt{2}} [1 \ -1] \frac{1}{\sqrt{2}} \begin{bmatrix} 1 \\ e^{i\Phi} \end{bmatrix} \right|^2 = \frac{1 - \zeta_D}{2} sin^2 \left(\frac{\Phi}{2} \right). \qquad (13.70)$$

But the source injects majority spins with an efficiency $(1 + \zeta_S)/2$, where ζ_S is the spin injection efficiency at the source. Consequently, the component of the source-to-drain current due to majority spin injection from the source is

$$I_{SD}^{maj} = \frac{1 + \zeta_S}{4} I_0 \left[(1 + \zeta_D) cos^2 \left(\frac{\Phi}{2} \right) + (1 - \zeta_D) sin^2 \left(\frac{\Phi}{2} \right) \right]$$

$$= \frac{1 + \zeta_S}{4} I_0 (1 + \zeta_D cos \Phi). \qquad (13.71)$$

Similarly, the component of the source-to-drain current due to minority spin injection from the source is

$$I_{SD}^{min} = \frac{1 - \zeta_S}{4} I_0 \left[(1 - \zeta_D) cos^2 \left(\frac{\Phi}{2} \right) + (1 + \zeta_D) sin^2 \left(\frac{\Phi}{2} \right) \right]$$

$$= \frac{1 - \zeta_S}{4} I_0 (1 - \zeta_D cos \Phi). \qquad (13.72)$$

Since the majority and minority spins are orthogonal in the contacts, the total source-to-drain current is

$$I_{SD} = I_{SD}^{maj} + I_{SD}^{min} = \frac{1}{2} I_0 \left(1 + \zeta_S \zeta_D cos \Phi \right). \qquad (13.73)$$

Note from the above equation that if either the spin injection or the spin detection efficiency is zero, then there is no current modulation, since the device does not discriminate between spin polarizations. In that case, the source-to-drain current is constantly $I_0/2$ irrespective of the gate voltage. On the other hand, if $\zeta_S = \zeta_D = 100\%$, then the current oscillates between I_0 and 0, with the average value of $I_0/2$.

The on-current corresponds to $\Phi = 0$ and the off-current to $\Phi = \pi$. Therefore, the ratio of on-to-off conductance is

$$\frac{G_{on}}{G_{off}} = \frac{1 + \zeta_S \zeta_D}{1 - \zeta_S \zeta_D}. \qquad (13.74)$$

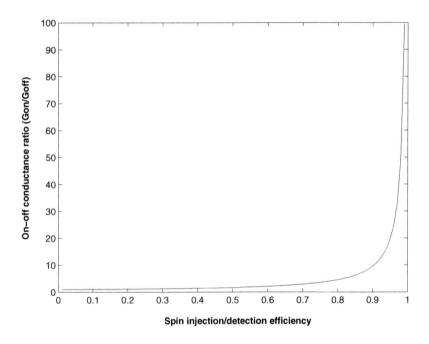

FIGURE 13.9

The ratio of on-to-off conductance of a SPINFET as a function of spin injection
and detection efficiency $\zeta = \zeta_S = \zeta_D$.

This ratio is infinity if $\zeta_S = \zeta_D = 100\%$, but drops precipitously to only 9.5 if $\zeta_S = \zeta_D = 90\%$. Fig. 13.9 shows how the conductance on-to-off ratio falls off rapidly with decreasing ζ_D or ζ_S.

The transconductance of the transistor also depends on ζ_S and ζ_D. It is given by

$$g_m = \frac{\partial I_{SD}}{\partial V_G} = -I_0\zeta_S\zeta_D\frac{m^*a_{46}}{\hbar^2}\frac{\kappa_i}{\kappa_s}sin\Phi \times \left\{ \begin{array}{ll} L/d & (MISFET\ mode) \\ L/(d+\Delta d) & (HEMT\ mode) \end{array} \right\}$$

$$g_m^{max} = \frac{\partial I_{SD}}{\partial V_G} = -I_0\zeta_S\zeta_D\frac{m^*a_{46}}{\hbar^2}\frac{\kappa_i}{\kappa_s} \times \left\{ \begin{array}{ll} L/d & (MISFET\ mode) \\ L/(d+\Delta d) & (HEMT\ mode). \end{array} \right\}$$

(13.75)

The transconductance is proportional to the product of the spin injection and detection efficiencies.

Spin injection and detection efficiencies are of paramount importance in determining SPINFET performance. Flatté and Hall [26] have suggested two possible routes to achieving $\sim 100\%$ spin injection efficiency: (i) the use of 100% spin polarized half metallic ferromagnets as spin injectors, and (ii) the use of spin selective barriers. Unfortunately, there are no 100% spin polarized half metals at any temperature above 0 K because of phonons and magnons [9] and even at 0 K, the 100% spin polarization may be destroyed by surfaces and inhomogeneities [9]. On the other hand, the best spin selective barriers are resonant tunneling devices [27] that have spin-resolved resonant levels. When the carrier energy is resonant with one of these levels, only the corresponding spin transits through the barrier. However, at any temperature exceeding 0 K, these levels are thermally broadened and therefore there is always a non-zero probability of both spins transmitting (albeit unequally), which makes the spin injection efficiency considerably less than 100%. There are other methods for achieving very high spin injection efficiency [28], but they too are ineffective at room temperature. Therefore, it is very difficult to achieve nearly 100% spin injection efficiency at room temperature. As a result, all of the SPINFETs discussed in this chapter are very likely to have either a low conductance on-off ratio or be constrained to work at very low temperatures, or both. Unfortunately, this does not make them competitive with the workhorse of modern electronics, the silicon MISFET.

13.8 Transconductance, gain, bandwidth, and isolation

At the beginning of this chapter, we mentioned that a real transistor must have sufficiently high gain and isolation between input and output terminals. The small signal equivalent circuit [29] of the SPINFET is shown in Fig. 13.10. The feedback impedance from the drain node to the gate node is negligible if

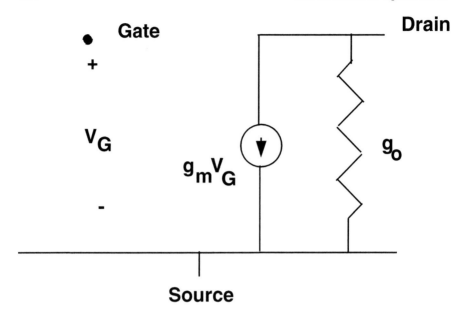

FIGURE 13.10
The small signal equivalent circuit model of a SPINFET.

the gate-to-drain capacitance C_{gd} is small, as it typically is. This indicates that any influence at the output does not back propagate to the input. In other words, the isolation between input and output is good as long as C_{gd} is small.

The transconductance g_m of the device determines the gain and bandwidth. In an accumulation mode transistor, the source-to-drain current does not saturate with increasing source-to-drain voltage. Instead, it continues to increase linearly with voltage (Ohm's law) so that the average source to drain current $I_0/2 = g_0 V_{SD}$. Using this result in Equation (13.75), we obtain

$$\frac{g_m^{max}}{g_0} = \frac{\partial I_{SD}}{\partial V_G} = -2 V_{SD} \zeta_s \zeta_D \frac{m^* a_{46}}{\hbar^2} \frac{\kappa_i}{\kappa_s} \times \left\{ \begin{array}{ll} L/d & (MISFET \ mode) \\ L/(d + \Delta d) & (HEMT \ mode). \end{array} \right\}$$

(13.76)

The left hand side of the above equation is the voltage gain of the transistor in the common gate or common source configuration. If we substitute realistic values for the quantities in the right hand side (i.e., use the same values that we used before) and assume 100% spin injection and detection efficiencies, we find that, in order to produce a gain larger than unity, V_{SD} must be at least 0.9 V. If $V_{SD} = 5$ V (the normal power supply voltage today), then a GaAs SPINFET's maximum voltage gain is 5.5, if $L = 0.1$ μm and $d = 10$ nm. This is better than that of single electron transistors (SETs) and may be adequate for some applications, but it is certainly not competitive with that of 0.1 μm length Si MISFETs.

The unity gain frequency (bandwidth) is given by

$$f_T = \frac{1}{2\pi} \frac{g_m}{C_g}. \tag{13.77}$$

In a quantum wire SPINFET, the maximum value of g_0 is $2e^2/h = 78$ μSiemens. Therefore, the maximum transconductance of a GaAs SPINFET with a 0.1 μm long channel and a 10 nm thick gate insulator is 430 μSiemens if $V_{SD} = 5$ V (normal power supply voltage). Since we estimated that $C_g \sim 50$ aF, the unity gain frequency is 10^{13} Hz. This high value, however, is due to the small gate capacitance. It is not a consequence of SPINFET principles.

13.8.1 Silicon SPINFETs

Lately, there has been a spate of activity in silicon based SPINFETs [30, 31] following the successful demonstration of spin injection in silicon [32]. These transistors are a little different from the Datta-Das construct. The Datta-Das device of [2] relies on modulating the strength of Rashba interaction in the channel with a gate voltage to alter the spin precession rate and thus realize transistor action. This will not be very effective in silicon which has weak spin-orbit interaction. Therefore, Appelbaum and Monsma [31] decided on a different approach. They have an external static magnetic field B_{dc}, and spins in the channel precess about this field as they travel from the ferromagnetic source to ferromagnetic drain. The precession rate is the Larmor frequency $g\mu_B B_{dc}/\hbar$. Therefore, the angle by which a spin precesses in traversing the device is

$$\theta = \frac{g\mu_B B_{dc}\tau_t}{\hbar} = \frac{g\mu_B B_{dc}L}{v\hbar}, \tag{13.78}$$

where $\tau_t = L/v$ is the transit time, v is the electron velocity, and L is the distance between the two ferromagnetic contacts.

The angle θ is varied by varying the electron velocity v. That, in turn, is varied by changing the voltage across the channel, which changes the accelerating electric field. The drain is a spin selective filter that passes spins aligned parallel to its own magnetization and blocks the anti-parallel spins. Thus, by changing θ with a voltage, the drain current can be modulated and a SPINFET is realized. This device has a complicated structure and the interested reader is referred to [31] for a complete description. Huang et al. [30] have claimed to have demonstrated the operation of this device recently.

One disadvantage of this device is that θ depends on v (unlike in the case of the Datta-Das device) and therefore different electrons with different velocities will precess by different angles. Consequently, ensemble averaging over electrons will dilute the current modulation.

Huang et al. [30] use an interesting approach for spin injection in this device. They use a hot electron emitter that relies on the difference between the scattering rates of majority and minority spins to selectively inject majority spins. For further details of this approach, the reader is referred to [30, 31, 32].

13.9 Spin Bipolar Junction Transistors (SBJT)

The Spin Bipolar Junction Transistor (SBJT) is identical with the normal Bipolar Junction Transistor, except that the base is ferromagnetic and has a non-zero spin polarization. The conduction energy band diagram of a hetero-junction npn SBJT is shown in Fig. 13.11. The expressions for the collector (I_C), emitter (I_E) and base (I_B) currents were derived in [33] and are

$$
\begin{aligned}
I_C = {} & qA\frac{D_{nb}}{L_{nb}}\frac{1}{sinh(W/L_{nb})}n_{be}\left[e^{qV_{EB}/kT}-1\right] \\
& -qA\frac{D_{nb}}{L_{nb}}coth(W/L_{nb})n_{bc}\left[e^{qV_{CB}/kT}-1\right] \\
& -qA\frac{D_{pc}}{L_{pc}}coth(W_c/L_{pc})p_{oc}\left[e^{qV_{CB}/kT}-1\right] ,
\end{aligned}
\tag{13.79}
$$

$$
\begin{aligned}
I_E = {} & qA\frac{D_{nb}}{L_{nb}}coth(W/L_{nb})n_{be}\left[e^{qV_{EB}/kT}-1\right] \\
& -qA\frac{D_{nb}}{L_{nb}}\frac{1}{sinh(W/L_{nb})}n_{bc})\left[e^{qV_{CB}/kT}-1\right] \\
& +qA\frac{D_{pe}}{L_{pe}}coth(W_e/L_{pe})p_{oe}\left[e^{qV_{EB}/kT}-1\right] ,
\end{aligned}
\tag{13.80}
$$

$$
I_E + I_B + I_C = 0,
\tag{13.81}
$$

where A is the cross-sectional area of the transistor, W_c is the width of the collector, W_e is the width of the emitter, D_{nb} (L_{nb}) is the minority carrier diffusion constant (length) for electrons in the base, D_{pc} (L_{pc}) is the minority carrier diffusion constant (length) for holes in the collector, D_{pe} (L_{pe}) is the minority carrier diffusion constant (length) for holes in the emitter, $n_{be} = (n_i^2/N_{AB})(1+\alpha_e\alpha_{0b})/\sqrt{1-\alpha_{0b}^2}$, $n_{bc} = (n_i^2/N_{AB})(1+\alpha_c\alpha_{0b})/\sqrt{1-\alpha_{0b}^2}$, $p_{oc} = (n_i^2/N_{DC})$, $p_{oe} = (n_i^2/N_{DE})$, n_i is the intrinsic carrier concentration in the material, N_{AB} is the acceptor dopant concentration in the base, N_{DC} is the donor dopant concentration in the collector, N_{DE} is the donor dopant concentration in the emitter, α_e and α_c are the non-equilibrium spin polarizations

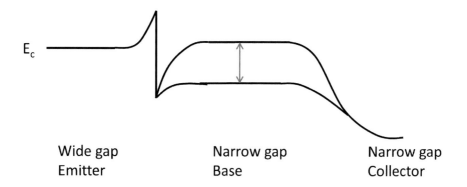

FIGURE 13.11
Conduction band energy profile showing the emitter, base, and collector regions of a heterojunction SBJT with a wide-bandgap emitter and a narrow bandgap base and collector. The base is spin-polarized or ferromagnetic.

in the emitter and collector, α_{0b} $(= tanh(\Delta/kT))$ is the equilibrium spin polarization in the base, and 2Δ is the magnitude of energy splitting between the majority and minority spin in the base.

Based on a small signal analysis, it has been shown that the voltage and current gains of the SBJT are about the same as those of a conventional BJT [34], but the short circuit current gain β $(= \partial I_C/\partial I_B)$ has a dependence on the degree of spin polarization in the base (or, equivalently, the spin-splitting energy Δ), which can be altered with an external magnetic field using the Zeeman effect. Thus, the external magnetic field can act as a *fourth terminal*. This can lead to non-linear circuits such as mixers/modulators. For example, if the ac base current is a sinusoid with a frequency w_1 and the magnetic field is an ac field which is another sinusoid with frequency w_2, then the collector current will contain frequency components $w_1 \pm w_2$. This is one example where spin augments the role of charge, making the SBJT more powerful than the traditional BJT (3-terminal versus 4-terminal).

13.10 GMR-based transistors

The concept of giant magnetoresistance (GMR) was explored in Chapter 13. Its application has been mostly in *passive* devices, such as read-heads. However, it has also spawned a number of proposals for *active* devices (transistors). We will discuss a few of them here.

13.10.1   All-metal spin transistor

In [35, 36], Johnson proposed and analyzed an all-metal spin-based transistor to capitalize on the first successful reports of GMR measurements in the current-perpendicular-to-plane (CPP) geometry by Pratt et al. [37] at low temperatures using superconducting electronics and later extended to room temperature by Gijs et al. using microstructure samples [38]. As depicted in Fig. 13.12, Johnson's all-metal spin transistor consists of a spin valve structure with a third terminal connected to the non-magnetic spacer layer. The transistor requires that the thickness of the layers be comparable to or smaller than the spin diffusion length in the material, which is on the order of a few nm in impure metallic samples [39]. An elementary description of the basic principle of operation of this device has been given by Johnson in [35, 36]. Preliminary experimental results were reported by Johnson himself [40], leading to only small voltage output swing and no power or current gain. If such a device could generate a current gain, it could potentially be used to make logic devices. It can also serve as an MRAM. If logic and memory functions can be embedded in the same device, it will yield unprecedented advantages since it obviates the need for communication between logic and memory, which is the primary bottleneck in computation. Moreover, since the electrical characteristics of this purely ohmic device are magnetically tunable, it can potentially be used as a field sensor. Thus, it is a "multifunctional" device with many potential roles that make it interesting.

13.10.2   Spin valve transistor

Soon after the proposal by Johnson, Monsma et al. [41] fabricated the first hybrid spintronic device in which ferromagnets and semiconductors were closely integrated (see Fig. 13.13). This device is now usually referred to as the *spin valve transistor* (SVT) and is basically a reincarnation of the well-known metal base transistor proposed in the 1960s [42]. First fabricated in 1995, the SVT consists of the typical emitter/base/collector configuration of a bipolar junction transistor in which the thin base region is formed of a spin CPP multilayer containing at least two magnetic layers separated by a normal metal spacer. The two magnetic layers act as a spin polarizer and a spin analyzer for electrons injected from the n-type emitter. The relative orientation of the magnetic layers in the base region, which can be altered with a magnetic field in the plane parallel to the emitter-base and base-collector interfaces, determines the amount of injected current transmitted across the base and reaching the collector contact. In the active mode of operation, the collector Schottky barrier is reverse-biased and the emitter Schottky is forward-biased (see Fig. 13.14). As a result, in the SVT, the collector current depends strongly on the magnetic configuration of the base region.

In the first SVT report by Monsma et al. [41], a 215% change in the collector current was observed with an applied magnetic field of 500 Oe at

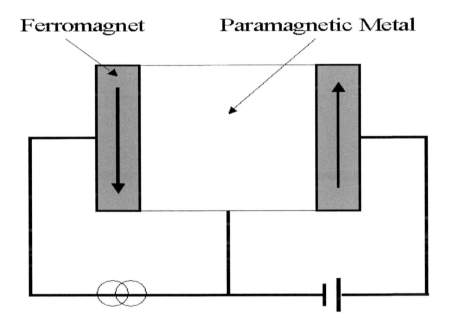

FIGURE 13.12
Schematic of the Johnson spin transistor (after [35]).

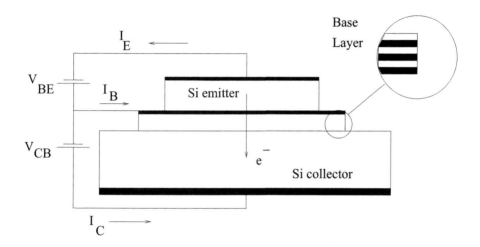

FIGURE 13.13

Schematic cross section of the spin valve transistor proposed by Monsma et al. [41]. The base region is composed of a short superlattice of alternating ferromagnetic and paramagnetic metal layers similar to those used in the demonstration of the giant magnetoresistance effect. The transistor is biased in the common base configuration showing electron injection from the emitter to collector region. Reproduced with permission from D. J. Monsma, et al., Phys. Rev. Lett., **74**, 5260 (1995). Copyright (1995) American Physical Society. http://link.aps.org/abstract/PRL/v74/p5260.

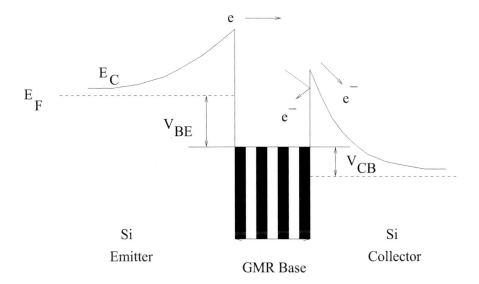

FIGURE 13.14
Schematic energy band diagram from emitter to collector of a spin valve transistor biased in the forward active mode showing the Schottky barriers at the emitter-base and base-collector contacts. Reprinted with permission from D. J. Monsma, et al., Phys. Rev. Lett., **74**, 5260 (1995). Copyright (1995) American Physical Society. http://link.aps.org/abstract/PRL/v74/p5260.

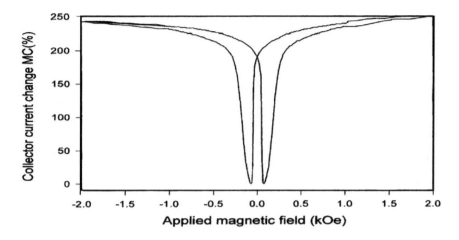

FIGURE 13.15

Magnetocurrent of the first spin valve transistor reported by Monsma et al. [41] with a base composed of four (Cu: 2 nm)/(Co: 1.5 nm) bilayers. The measurements were carried out at 77 K. There is a large (390 %) change in the collector current at fairly small value of the external magnetic field. Reprinted with permission from D. J. Monsma, et al., Phys. Rev. Lett., **74**, 5260 (1995). Copyright (1995) American Physical Society. http://link.aps.org/abstract/PRL/v74/p5260.

low temperature (77 K), with the typical hysteresis characteristic of the spin-valve effect (see Fig. 13.15).

As illustrated in Fig. 13.14, the SVT is a hot electron device as a result of the injection of highly energetic (hot) electrons from the emitter into the base. Even if the total base thickness is small (a few hundred angstroms), the hot electrons will lose some energy when crossing the base, depending on the bulk (or volume) scattering in each layer and the amount of scattering at the various interfaces bordering the base. This hot electron scattering depends strongly on the magnetic configuration of the multilayers. In the case of anti-ferromagnetic alignment (where magnetizations in adjacent layers are anti-parallel) in the base, both up-spin and down-spin carriers experience heavy inelastic scattering in one of the magnetic layers. As a result, the average kinetic energy of both types of spin injected from the emitter drops exponentially with distance in the base region. On the other hand, for the case of parallel alignment of adjacent ferromagnetic layers in the base, only one type of spin will get scattered heavily whereas the other will reach the base-collector interface without much scattering. In the SVT, the number of hot electrons retaining enough energy to surmount the collector Schottky barrier and contributing to the collector current is therefore strongly dependent on the magnetic configuration of the base. In the case of parallel magnetic alignment, more carriers will be emitted over the collector barrier and the collected current will be larger.

The challenge of SVT technology is to design the transistor in such a way that the magneto-current ratio γ_{MC} defined as

$$\gamma_{MC} = (I_P - I_{AP})/I_{AP} \qquad (13.82)$$

(where I_P and I_{AP} refer to the collector currents for the parallel (P) and anti-parallel (AP) configurations in the base) is the largest for a given magnetic field applied to switch between the two configurations. Efforts to meet this challenge have led to various modifications of the emitter-base junction design so as to increase the hot electron emitter injection efficiency, and the collector-base junction design to increase the amount of hot carrier collection.

In the original SVT proposed by Monsma et al. [41], the injection energy of hot electrons incident from the emitter is fixed by the height of the Schottky barrier between the emitter and the base. In a related device based on a metal(emitter)/thin insulator/metal(GMR base)/semiconductor(collector) structure proposed by Mizushima et al. [43] (referred to as the magnetic tunnel transistor or MTT), this limitation was overcome by using a tunnel junction as the emitter. The thin insulator acts as a tunnel barrier, injecting hot electrons into the metal base at an energy given by the tunnel bias voltage between the metal emitter and base. This provides additional tunability of the hot electron energy, allowing more detailed spectroscopic studies of spin transport across the base.

One major point of concern in MTT structures is the reliability of the ultra-thin insulator (oxide barrier) which is subjected to a large bias (1–2 V) and large current density. To improve reliability, LeMinh et al. [44] proposed an MTT device in which the metal emitter is replaced by silicon. As a result, the voltage drop across the emitter-base region is dropped partly across the semiconductor emitter depletion region and partly across the tunnel oxide, thus reducing the voltage stress across the latter. However, the energy of the hot electrons injected into the base is still tunable, as the portion of the emitter-base bias dropped across the tunnel insulator raises the conduction band minimum at the semiconductor/tunnel barrier interface with respect to the Fermi level of the base metal. Compared to a magnetic tunnel transistor with a metal emitter, the voltage drop over the thin tunnel oxide is reduced, enabling stable device operation at higher bias voltages. LeMinh et al. recently fabricated such a device with a magneto-current ratio up to 166% and a steep enhanced transfer ratio reaching 6×10^{-4} at an emitter current of 200 mA [44].

In SVT technology, the poor transfer ratio has been the limiting factor leading to a current transfer ratio (ratio of the collector current divided by the emitter current) much smaller than unity. This was a major drawback encountered in the early development of the metal base transistor. Despite this, the Monsma transistor represents a very important step in the evolution of spintronics because it is characterized by electrical characteristics that are magnetically tunable. As pointed out in the recent review article by Jansen on SVT technology [45], the most influential aspect of the Monsma transistor (and its derivatives) is that they showed the value of functional integration of semiconductor and ferromagnetic materials into hybrid electronic devices. Room temperature magneto-current ratios as large as 400% in small magnetic fields (typically a few tens of Oe) are now routinely obtained in SVT technology [46].

More recently, the investigation of MTT structures has been successfully extended to GaAs based devices [7]. Much progress has been made in optimizing the device performance, focusing on the output current level and noise sources. Further improvements are still required in order to capitalize on the huge magnetic sensitivity of these structures which make them very attractive candidates for future generations of magnetic sensors or magnetic memories. In that context, MTT structures, scaled to submicron dimensions, will be relevant [45].

SVT and MTT technologies have opened up a new route to the systematic study of the fundamental physics of spin-dependent transport of hot electron at energies on the order of 1 eV. New insights into the fundamental physics of spin-dependent hot-electron scattering have been obtained, including the dominance of the volume effects over interface effects in the spin-dependent transmission across the base, the effects of thermal spin waves, and the surprisingly important role of elastic scattering processes. For instance, it has been realized that hot-electron spin-filtering may have some attractive features for spin-injection into a semiconductor, in particular the ability to reach a spin injection efficiency close to 100 % with conventional ferromagnets. Furthermore, the development of SVT and MTT technologies has led to a detailed study of hot-electron spin-transport through half-metallic ferromagnets and oxides, which will lead to the proposal of new types of hybrid electronic devices combining ferromagnets and semiconductors. The status of SVT research was recently reviewed by Jansen [45].

13.11   Concluding remarks

In concluding this chapter, we would like to leave the reader with the impression that hybrid spin devices – such as SPINFETs and SBJTs (including the GMR based SBJTs) – are interesting constructs with interesting physics. They were the pioneers and standard bearers in the field of spintronics and were primarily responsible for generating interest in this field within the device community. Their performance may not be stellar and they may not be competitive (currently) with standard semiconductor CMOS or BiCMOS technology for mainstream applications. Nonetheless, interest in these devices persists, not necessarily for killer applications, but more for the wonderful and fascinating physics that they continue to reveal.

13.12   Problems

- **Problem 13.1**

    Consider the one-dimensional Datta-Das SPINFET. Show that the maximum transconductance of this device is

    $$|g_m^{max}| = \left|\frac{\partial I_{SD}}{\partial V_G}\right| = \frac{m^* e^2}{2\pi \hbar^3}\frac{\partial \eta}{\partial V_G}LV_{SD}, \tag{13.83}$$

    where I_{SD} is the source-to-drain current, V_G is the gate voltage, V_{SD} is the source-to-drain voltage. and L is the channel length.

    In the common-source or common-gate configuration, the voltage gain of the transistor is $g_m/g_o = g_m/G_{SD}$, where G_{SD} is the (linear-response) channel conductance. Show that the voltage gain of the transistor in either of these configurations (when the transconductance maximizes so that the unity gain frequency is maximum) is

    $$a_v^{max} = \frac{2m^*}{\hbar^2}L\frac{\partial \eta}{\partial V_G}V_{SD}. \tag{13.84}$$

    In deriving the above, use the fact that the transconductance is maximum at a gate voltage where the drain current is about one-half of its maximum value.

    Bandyopadhyay and Cahay [16] found that for realistic device configurations, $\frac{\partial \eta}{\partial V_G} \approx 5 \times 10^{-29}$ Coulomb-meter. Using this value, and assuming the effective mass of InAs ($m^* = 0.03m_0$), show that the maximum voltage gain of a one dimensional SPINFET, with an InAs channel of length 50 nm, exceeds unity if V_{SD} exceeds $\sim$ 81 mV. Also, show that for $V_{SD} = 1$ V, the maximum voltage gain is greater than 12.

- **Problem 13.2**

    The transconductance g_m of the SBJT is defined as the quantity $\partial I_C/\partial V_{BE}$, the output conductance g_o is defined as $\partial I_C/\partial V_{CE}$, and the feedback conductance g_μ is defined as $\partial I_B/\partial V_{CB}$. These three quantities are the so-called small signal parameters that characterize the small signal model of any bipolar junction transistor. These parameters determine quantities such as input/output ac impedance, voltage amplification, and current amplification.

Using these definitions and the expression for the collector current, show that

$$g_m \approx \frac{q^2 A}{2kT} \left[\frac{D_{nb}}{L_{nb}} \frac{1}{sinh(W/L_{nb})} \frac{n_i^2}{N_{AB}} (1 + \alpha_e tanh(\Delta/kT)) \right] e^{q(V_{EB}+\Delta)/kT}$$

$$g_o \approx \frac{q^2 A}{2kT} \left[\frac{D_{nb}}{L_{nb}} coth \left(\frac{W}{L_{nb}} \right) \frac{n_i^2}{N_{AB}} (1 + \alpha_c tanh(\Delta/kT)) \right] e^{q(V_{CB}+\Delta)/kT}$$

$$g_\mu \approx \frac{q^2 A}{2kT} \left[\frac{D_{nb}}{L_{nb}} \frac{W}{2L_{nb}} \frac{n_i^2}{N_{AB}} (1 + \alpha_c tanh(\Delta/kT)) \right] e^{q(V_{CB}+\Delta)/kT}$$

if $\Delta > kT$.

- **Problem 13.3**

Consider the ideal SPINFET realized with a quantum wire with a single subband occupied. The axis of the quantum wire is along the x-direction. Show that if B_{Rashba} were proportional to k_x instead of v_x, then the SPINFET will have significant leakage current in the off-state because Φ will be different for different electrons with different velocities. In that case, ensemble averaging will result in a leakage current in the off-state since Φ will never be exactly π radians for every electron in the ensemble. However, since B_{Rashba} is proportional to v_x and not k_x, the leakage current in the off-state is exactly zero since Φ is exactly π radians for every electron in the ensemble regardless of their velocities.

Hint: Remember Equation (7.62). In the presence of spin-orbit interaction, $v_x \neq \hbar k_x/m^*$.

- **Problem 13.4**

Show that the conductance on-to-off ratio of the spin relaxation transistor of Section 13.5.2 is

$$\frac{G_{on}}{G_{off}} = \frac{1}{1 - \zeta_S \zeta_D}, \tag{13.85}$$

where ζ_S and ζ_D are the spin injection efficiency at the source/channel interface and spin detection efficiency at the drain/channel interface, respectively.

- **Problem 13.5**

Derive Equation (13.65).

- **Problem 13.6**

Derive the form of Equation 13.65 when the two ferromagntic contacts of the spin relaxation transistor have parallel magnetizations as opposed to anti-parallel magnetizations. In this case, the device is normally-on.

13.13 References

[1] D. A. Hodges and H. G. Jackson, *Analysis and Design of Digital Integrated Circuits*, 2nd edition, (McGraw Hill, New York, 1988), Chapter 1, p. 2.

[2] S. Datta and B. Das, "Electronic analog of electro-optic modulator", Appl. Phys. Lett., 56, 665 (1990).

[3] The effect of Dresselhaus interaction on the performance of the Datta-Das transistor has been examined in A. Lusakowski, J. Wróbel and T. Dietl, "Effect of bulk inversion asymmetry on the Datta-Das transistor", Phys. Rev. B., **68**, 081201(R), (2003). The conclusion was that certain crystallographic directions are better; for example, if the channel is aligned along the [110] crytstallographic direction in certain materials, it may offer superior performance than if the channel was aligned along another crystallographic direction, say, [111].

[4] H-W Lee, S. Caliskan and H. Park, "Mesoscopic effects in a single-mode Datta-Das spin field effect transistor", Phys. Rev. B., **72**, 153305 (2005).

[5] L. Xu, X-Q Li and Q-f Sun, "Revisit the spin-FET: Multiple reflection, inelastic scattering, and lateral size effects", Sci. Rep., **4**, 7527 (2014).

[6] E. Y. Tsymbal, O. N. Mryasov and P. R. LeClair, "Spin dependent tunneling in magnetic tunnel junction", J. Phys.: Condens. Matt., **15**, R109 (2003).

[7] G. Salis, R. Wang, X. Jiang, R. M. Shelby, S. S. P. Parkin, S. R. Bank and J. S. Harris, "Temperature independence of the spin injection efficiency of a MgO based tunnel spin injector", Appl. Phys. Lett., **87**, 262503 (2005).

[8] R. Skomski and J. M. D. Coey, *Permanent Magnetism*, (Taylor and Francis, New York, 1999).

[9] P. A. Dowben and R. Skomski, "Are half metallic ferromagnets half metals?", J. Appl. Phys., **95**, 7453 (2004).

[10] M. Cahay and S. Bandyopadhyay, "Phase coherent quantum mechanical spin transport in a weakly disordered quasi one-dimensional channel", Phys. Rev. B, **69**, 045303 (2004).

[11] S. Bandyopadhyay and M. Cahay, "A spin field effect transistor for low leakage current", Physica E, **25**, 399 (2005).

[12] P. Agnihotri and S. Bandyopadhyay, "Analysis of the two-dimensional Datta-Das spin field-effect transistor", Physica E, **42**, 1736 (2010).

[13] M. G. Pala, M. Governale, J. Knig and U. Zlicke, "Universal Rashba spin precession of two-dimensional electrons and holes", Europhys. Lett., **65**, 850 (2004).

[14] A. N. M. Zainuddin, S. Hong, L. Siddiqui, S. Srinivasan and S. Datta, "Voltage-controlled spin precession", Phys. Rev. B, **84**, 165306 (2011).

[15] S. Bandyopadhyay and M. Cahay, "Alternate spintronic analog of the electro-optic modulator", Appl. Phys. Lett., **85**, 1814 (2004).

[16] S. Bandyopadhyay and M. Cahay, "Reexamination of some spintronic field effect device concepts", Appl. Phys. Lett., **85**, 1433 (2004).

[17] R. K. Cavin, V. V. Zhirnov, J. A. Hutchby and G. I. Bourianoff, "Energy barriers, demons and minimum energy operation of electronic devices", Fluctuation and Noise Letters, **5**, C29 (2005).

[18] R. F. Pierret, *Field Effect Devices*, 2nd edition, Modular Series on Solid State Devices, Eds. G. W. Neudeck and R. F. Pierret, (Addison-Wesley, Reading, MA, 1990).

[19] M. Shur, *Physics of Semiconductor Devices*, (Prentice Hall, Englewood Cliffs, New Jersey, 1990).

[20] J. Schliemann, J. C. Egues and D. Loss, "Non ballistic spin field effect transistor", Phys. Rev. Lett., **90**, 146801 (2003).

[21] X. Cartoixá, D. Z-Y Ting and Y-C Chang, "A resonant spin lifetime transistor", Appl. Phys. Lett., **83**, 1462 (2003).

[22] K. C. Hall and M. E. Flatté, "Performance of spin based insulated gate field effect transistor", Appl. Phys. Lett., **88**, 162503 (2006).

[23] G. Wang, et al., "Gate control of the electron spin-diffusion length in semiconductor quantum wells", Nature Commun., **4**, 2372. doi: 10.1038/ncomms3372.

[24] E. Safir, M. Shen and S. Saikin, "Modulation of spin dynamics in the channel of a non-ballistic spin field effect transistor", Phys. Rev. B, **70**, 241302 (2004).

[25] S. Iba, S. Koh and H. Kawaguchi, "Room temperature gate modulation of electron spin relaxation time in (110)-oriented GaAs/AlGaAs quantum wells", Appl. Phys. Lett., **97**, 202102 (2010).

[26] M. E. Flatté and K. C Hall, www.arXiv.org/cond-mat/0604532.

[27] T. Koga, J. Nitta, H. Takayanagi and S. Datta, "Spin filter device based on the Rashba effect using a non-magnetic resonant tunneling diode", Phys. Rev. Lett., **88**, 126601 (2002).

[28] J. Wan, M. Cahay and S. Bandyopadhyay, "Can a non-ideal ferromagnet inject spin into a semiconductor with 100% efficiency without a tunnel barrier?", J. Nanoelectron. Optoelectron., **1**, 60 (2006).

[29] A. S. Sedra and K. C. Smith, *Microelectronic Circuits*, 3rd edition, (Oxford University Press, New York, 1991).

[30] B. Huang, D. J. Monsma and I. Appelbaum, "Experimental realization of a silicon spin field effect transistor", Appl. Phys. Lett., **91**, 072501 (2007).

[31] I. Appelbaum and D. J. Monsma, "Transit time field effect transistor", Appl. Phys. Lett., **90**, 262501 (2007).

[32] I. Appelbaum, B. Huang and D. J. Monsma, "Electronic measurement and control of spin transport in silicon", Nature (London), **447**, 295 (2007).

[33] J. Fabian and I. Zutic, www.arXiv.org/cond-mat/0409196.

[34] S. Bandyopadhyay and M. Cahay, "Are spin junction transistors suitable for signal processing?", Appl. Phys. Lett., **86**, 133502 (2005).

[35] M. Johnson, "Bipolar Spin Switch", Science, **260**, 320 (1993).

[36] M. Johnson, "The all-metal spin transistor", IEEE Spectrum, **31** (5), pp.47-51 (1994).

[37] W. P. Pratt, S.-F. Lee, J.M. Slaughter, R. Loloee, P. A. Schroeder and J. Bass, "Perpendicular giant magnetoresistance of Ag/Co multilayers", Phys. Rev. Lett., **66**, 3060 (1991).

[38] M. A. M. Gijs, S.K.J. Lenczowski and J. B. Giesbers, "Perpendicular giant magnetoresistance of microstructured Fe/Cr magnetic multilayers from 4.2 to 300 K, Phys. Rev. Lett., **70**, 3343 (1993).

[39] I. Zutic, J. Fabian and S. Das Sarma, "Spintronics: Fundamentals and applications", Rev. Mod. Phys., **76**, 351 (2004).

[40] M. Johnson, "Spin polarization of gold films via transport", J. Appl. Phys., **75**, 6714 (1994).

[41] D. J. Monsma, J. C. Lodder, Th. J. A. Popma, and B. Dieny, "Perpendicular hot electron spin valve effect in a new magnetic field sensor: The Spin Valve Transistor", Phys. Rev. Lett., **74**, 5260 (1995).

[42] S. M. Sze, *Physics of Semiconductor Devices*, (Wiley Interscience, New York, 1969), Chapter 11.

[43] K. Mizushima, T. Kinno, T. Yamauchi and K. Tanaka, "Energy dependent hot electron transport across a spin valve", IEEE Trans. Magn., **33** 3500 (1997).

[44] P. LeMinh, H. Gokcan, J. C. Lodder, and R. Jansen, "Magnetic tunnel transistor with a silicon hot electron emitter", J. Appl. Phys., **98**, 076111 (2005).

[45] For a recent review of the basic science and technology of the spin-valve transistor, see R. Jansen, "Spin valve transistor: Review and outlook", J. Phys. D: Appl. Phys., **36**, R289-R308 (2003).

[46] R. Jansen, H. Gokcan, O. M. J. van't Erve, F. M. Postma and J. C. Lodder, "Spin valve transistors with high magnetocurrent and 40 μA output current", J. Appl. Phys., **95**, 6927 (2004).

14

All-Electric spintronics with Quantum Point Contacts

As emphasized in the previous chapter, the central challenge in most spintronic devices (SPINFETs, sensors, etc.) is to find ways to inject, manipulate, and detect the spin of an electron with high efficiency while employing purely electrical means. To achieve this goal, quantum nanostructures, including quantum point contacts (QPCs) and quantum dots (QDs), have been pressed into action over the last decade [1, 2, 3]. This chapter describes some of the recent breakthroughs toward the realization of all electric spin-based devices with the aid of QPCs, resulting in high spin injection and detection efficiencies.

14.1 Quantum point contacts

The interest in point contacts dates back to the pioneering work of Sharvin who successfully demonstrated the injection and detection of a beam of electrons in a metal by means of point contacts much smaller than the mean free path [4]. In 1988, two groups, one in Cambridge [5] and another in Delft [6], first observed and measured the conductance of a semiconductor QPC formed in a 2-DEG. The main advantage of semiconductor QPCs over metal point contacts is that the electron density in a semiconductor structure can be varied by means of side gates, as described next.

A QPC is a very short quantum wire or constriction. It is usually made by depositing a pair of surface gates (Schottky) over a 2-DEG (e.g., the one that exists at the interface of a semiconductor heterostructure such as an AlGaAs/GaAs). Figures 14.1 (a) and (b) are schematic illustrations of a QPC with top gates [5, 6, 7].

A negative bias electrostatically depletes the 2-DEG directly underneath and in the vicinity of the gates, creating a narrow constriction and resulting in the formation of a quasi one dimensional electron gas (1-DEG) between the gates. The width of this channel can be controlled by appropriately biasing the gates. This approach allows an electrostatic tuning of the aspect ratio W/L (channel width/channel length) of the QPC.

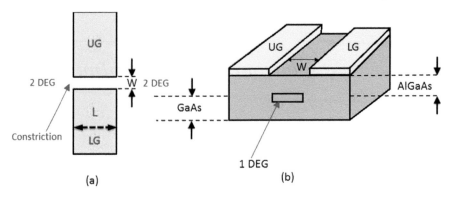

FIGURE 14.1

(a) A QPC created by two gates (UG and LG) deposited on the top surface of an AlGaAs/GaAs semiconductor heterostructure; (b) The 2-DEG formed at the AlGaAs/GaAs interface is depleted by applying a negative potential on the top gates, resulting in the formation of a quasi one-dimensional channel or electron gas (1-DEG) in the constriction between the two gates. Typical width and length of the narrow portion of the QPC are a few hundred nm. [7]

The conductance G of the QPC is given by [8]

$$G = \frac{2e^2}{h} \sum_{n,m=1}^{N} |t_{mn}|^2, \tag{14.1}$$

where N is the total number of electron subbands (see Figure 14.2) participating in conduction through the narrow portion of the QPC, i.e., whose bottom is below the Fermi level. The sum in Equation (14.1) includes an increasing number of terms as the Fermi level in the contacts is swept through upper energy subbands as the width of the QPC channel is adjusted with the gate biases. At sufficiently low temperature, the QPC dimensions can be made smaller than both the elastic and the inelastic scattering lengths. In that case, transport through the device is ballistic and the transmission coefficients $|t_{nn}|^2| = 1$ for $m = n$ and $|t_{mn}|^2 = 0$ for $m \neq n$. As a result, in the case of ballistic transport, the QPC conductance is quantized in units of G_0, as originally demonstrated by Wharam et al. [5] and van Wees et al. [6].

The factor of 2 in Equation (14.1) accounts for the spin degeneracy of the subbands. When the spin degeneracy is lifted with an external magnetic field, the conductance has been found to be quantized in integral values of $0.5\, G_0$ as long as the thermal energy $k_B T$ is small compared to the separation between the Zeeman spin-split energy levels. This typically requires working at very low temperature when using QPCs with top gates, as illustrated in Figure 14.3. These results imply that an experimental signature of a completely spin-

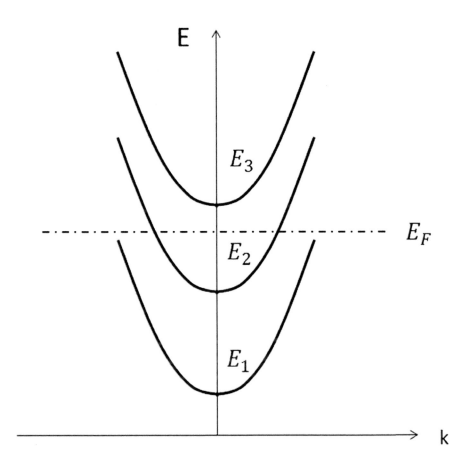

FIGURE 14.2

Schematic illustration of the energy subbands participating in electron conduction through the 1-DEG formed in the narrow constriction of the QPC. The bottoms of the energy subbands are located at energies E_1, E_2, E_3, ... Typically, a small bias is applied ($< 100\text{mV}$) between the source and drain contacts to maintain linear response transport and avoid electron heating. Here, E_F is the Fermi level in the source, and k is the wavevector in the direction of current flow in the narrow portion of the QPC.

polarized electric current is the occurrence of a 0.5 G_0 plateau in a conductance plot, albeit it is an indirect evidence.

The observation of a 0.5 G_0 plateau at 4.2 K was recently achieved by Debray et al. without the use of an external magnetic field [9]. They successfully demonstrated a novel way to create almost completely spin-polarized current by purely electrical means with the use of QPCs with in-plane SGs instead of surface gates [9, 10, 11, 12]. Figure 14.4 shows a three-dimensional rendition of a side gated QPC showing the different contacts for electrical measurements. Just as in the case of a QPC with top gates, the channel width of the side gated device can be controlled by the gate voltages on the SGs, allowing transport through several energy subbands formed in the narrow portion of the QPC. When the Fermi level is between the first and second subbands, the transport is strictly single-channeled. In this scenario, the electron density in the narrow portion of the QPC is very low, which in turn gives rise to strong electron-electron (e-e) interactions. This is one of the fundamental ingredients for the observation of 0.5 G_0 plateau in side-gated QPCs, as will be discussed in Sections 14.9 and 14.10.

14.2 Recent experimental results with QPCs and QDs

QPCs and QDs have played a very significant role in spintronics. In 2012, Kohda et al. used a QPC to realize a nanoscale version of the Stern-Gerlach experiment without the use of an external magnetic field. In their device, the electron spin separation in space was achieved via local tuning of the Rashba spin-orbit (SO) interaction [13]. Otsuka et al. recently demonstrated a new technique to probe the existence of local spin polarization in semiconductor microdevices in low and zero magnetic fields [14]. By connecting a single-lead QD to the device and monitoring electron tunneling into singlet and triplet states in the QD, they were able to detect the local spin polarization of the target device. These experiments have raised hopes of being able to inject and/or detect spins with high efficiencies.

Following the pioneering work of Thomas et al. [15], there have been many experimental reports of anomalies in the quantized conductance of QPCs appearing at non-integer multiples of $G_0 = 2e^2/h$, the unit of quantum conductance, including 0.25 G_0, 0.5 G_0, and 0.7 G_0 [16, 17, 18]. In their original work, Thomas et al. found that the 0.7 G_0 conductance anomaly showed up at zero magnetic field and evolved into a 0.5 G_0 conductance plateau when a strong enough magnetic field was applied to lift the spin degeneracy. From this result, they concluded that the 0.7 G_0 anomaly was indirect evidence of a spontaneous ferromagnetic spin polarization in the narrow portion of the QPC. There is now mounting evidence that the number and location of these

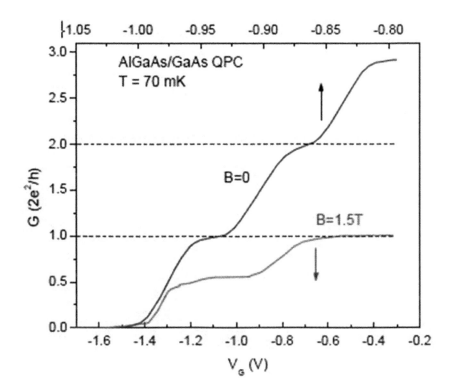

FIGURE 14.3
Experimental demonstration of the Zeeman spin splitting of the first quantized conductance plateau in the presence of an external magnetic field B applied perpendicular to the plane of the 2-DEG. Also shown for comparison is the conductance plot as a function of the potential V_G applied to the top gates of an AlGaAs/GaAs QPC. For $B = 1.5$ Tesla, a 0.5 G_0 conductance plateau appears indicating that the spin degeneracy has been lifted. Reproduced from [9] with permission of MacMillan Publishers.

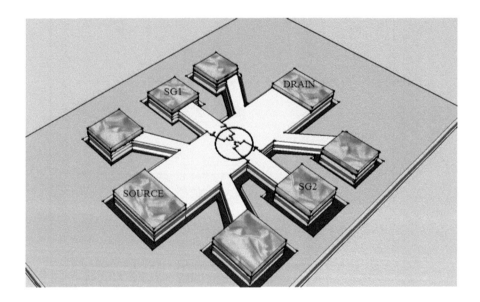

FIGURE 14.4

Schematic of a QPC with in-plane SGs. The dark lines represent the mesa etched regions defining the QPC. The circle indicates the location of the QPC. The unmarked pads are used for four probe conductance measurements. A spontaneous spin polarization can be created in the narrow portion of the QPC for a specific range of an asymmetric bias between the two SGs.

conductance anomalies can be further tuned by deliberately introducing a broken symmetry in the QPCs electrostatic confining potential [19, 20, 21, 22].

Recent experiments report that an asymmetry in the potential applied between the two side gates of a QPC can be used to create strongly spin polarized currents using both top and in-plane side gates (SGs), different materials (GaAs and InAs based heterostructures), two-dimensional electron gases (2-DEG) with different electron mobilities, and a wide variety of heterostructure design and QPC dimensions [9, 10, 11, 12, 21, 23]. As will be discussed in Section 14.5, this all-electric control of spin polarized current by tuning the lateral electric field between two SGs results from the electrostatic control of the SO interaction due to the lateral confinement between the two SGs. The latter is referred to as lateral SO interaction (LSOC). Building on this approach, Chuang et al. have recently demonstrated the operation of an all-electric spin valve by using asymmetrically biased QPCs spin injectors and detectors with near 100% efficiency [24, 25]. Additionally, with the use of middle gates, Chuang et al. achieved electrostatic control of the spin precession in the semiconducting channel between the two QPCs while maintaining ballistic transport through that region to minimize the effects of spin decoherence. They observed an oscillatory modulation of the current I through the spin valve with variations, $(I_{ON} - I_{OFF})/I_{OFF}$, as high as 500% at a temperature of 300 mK. Though this ON/OFF ratio is still a far cry from values in state of the art field effect transistors at room temperature [26], it is hoped that further improvements in the design of the device will allow the realization of an all-electric version of the Datta-Das SpinFET in the not too distant future.

The experimental breakthroughs listed above are described in more detail in later sections. In this chapter, the importance of the Rashba SO interaction is revisited in the context of QPC physics. We also discuss the physics of LSOC and its influence on the onset of spin polarization in asymmetrically biased QPCs. The use of asymmetrically biased QPCs to realize an all-electric spin valve is also addressed.

14.3   Spin–orbit coupling

The single electron Hamiltonian due to SOC in semiconductors was discussed in Chapter 6. This Hamiltonian is sometimes written in the form [27]

$$H_{SOC} = \lambda \vec{\sigma} . (\vec{k} \times \vec{\nabla} U), \tag{14.2}$$

where λ is a parameter characterizing the strength of SOC, $\vec{\sigma} = (\sigma_x, \sigma_y, \sigma_z)$ are the Pauli spin matrices, and U is the potential energy profile. This Hamiltonian can have many different physical origins, e.g., an internal potential gradient or an externally applied electric field [28].

14.4   Rashba spin–orbit coupling (RSOC)

The Rashba SOC was originally studied in a 2-DEG formed in quantum structures with asymmetric wells [28]. This SOC can be intrinsic or extrinsic, depending on its source. It arises from structural inversion asymmetry which can be tailored by introducing either internal (through asymmetric confinement of the 2-DEG) or external (applying voltages through top gates) electric fields. In triangular wells formed at the interface between narrow and wide band gap semiconductors or in quantum wells with different barrier heights on both sides, a gradient in the potential energy profile resulting from a difference in the conduction band edges and space-charge effects will lead to an internal electric field. The latter can further be controlled with a gate voltage. An electron moving in the 2-DEG in these heterostructures will experience a spatially varying electric field which will amount to an effective magnetic field lying in the plane of the 2-DEG and perpendicular to both the direction of the electric field and the momentum of the electron. This effective magnetic field creates SOC, lifts the degeneracy of the subbands in the 2-DEG, and is responsible for spin splitting. This SOC is of course the Rasbha interaction or Rashba SO Coupling (RSOC). It is described by the following Hamiltonian and the corresponding spin-splitting energy dispersion relations [28], as discussed in Chapters 6 and 7:

$$H_{RSOC} = \frac{\alpha}{\hbar}(\vec{k} \times \hat{z}).\vec{\sigma}, \tag{14.3}$$

and

$$E_{\vec{k}} = \frac{\hbar^2 k^2}{2m^*} \pm \alpha k, \tag{14.4}$$

where α is the Rashba coupling constant which is proportional to the average electric field in the 2DEG characterizing the strength of the RSOC and k is the magnitude of the wavevector of the electron in the 2-DEG; α is typically tuned with an external gate voltage [29] and is of the order of $(5-10) \times 10^{-12}$ eV-m in many narrow gap compound semiconductors like InAs or InSb. If a 2-DEG is further confined to a 1-D channel, the RSOC leads to energy dispersion relationships for spin-up and spin-down electrons, as shown in Figure 14.5. In that case, k is the wavevector component in the direction of current flow through the 1-D channel. The parabolic dispersion curves for opposite spins are shifted along k, as we saw in Chapter 7.

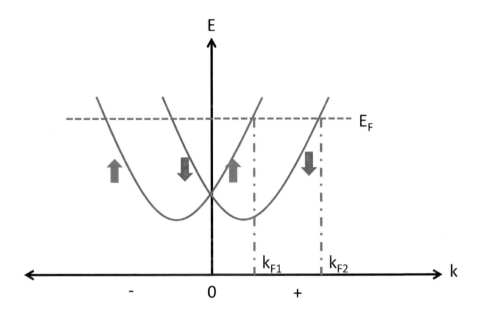

FIGURE 14.5
Lowest energy subbands in the quasi 1D channel formed in the narrow portion
of the QPC due to the top or SGs. The number of subbands participating in
conduction through the device depends on the location of the Fermi level E_F,
which can be adjusted with the potential on the gates.

14.5   Lateral spin–orbit coupling (LSOC)

LSOC was theoretically predicted independently by K. Hattori et al. [30], Y. Xing et al. [31] and Y. Jiang et al. [32]. Hereafter, we consider LSOC in a side-gated QPC schematically shown in Figure 14.6. The white region represents the mesa etched quantum channel and its surroundings, while the gray areas represent the etched isolation trenches defining the dimensions of the QPC. There are four contacts connected to the QPC device, source, drain, and two SGs. Symmetric and asymmetric voltages (V_{sg1} and V_{sg2}) can be applied to the two SGs.

   If the QPC is made from a nominally symmetric quantum well (QW), spatial inversion asymmetry can be assumed to be negligible along the growth axis (z-axis) of the QW and the corresponding RSOC can be neglected. For simplicity, the Dresselhaus SO coupling due to the bulk inversion asymmetry in the direction of current flow is also neglected. The only SO interaction is then the LSOC due to the lateral confinement of the QPC channel, provided by the isolation trenches and the bias voltages of the SGs. The effect of LSOC on the conductance of the QPC can be understood starting with the single-particle Hamiltonian in the narrow portion of the QPC [33, 34]:

$$H = H_0 + H_{SO}, \tag{14.5}$$

where

$$H_0 = \frac{1}{2m^*}({p_x}^2 + {p_y}^2) + U(x,y), \tag{14.6}$$

and

$$H_{SO} = \beta \vec{\sigma}.(\vec{k} \times \vec{\nabla}U) = \vec{\sigma}.\vec{B}_{SO}, \tag{14.7}$$

where β is the intrinsic SOC parameter characterizing the strength of LSOC, $\vec{\sigma}$ is the vector of Pauli spin matrices, and B_{SO} is the effective magnetic field, which is induced by the LSOC. The 2-DEG is assumed to be located in the $(x - y)$ plane, x being the direction of current flow from source to drain and y the direction of confinement of the channel. The quantity $U(x,y)$ is the confinement potential.

   Figure 14.7 (top) is a schematic representation of the confining potential along the y-direction when the same side-gate voltage is applied to both SGs (symmetric configuration). In that case, the effective magnetic field has exactly the same magnitude, but opposite directions, at the opposite sides of the QPC. Electrons moving through the narrow portion of the QPC with opposite spins experience opposite SOC that leads to an accumulation of opposite spins on opposite sides of the QPC. "Spin-up" is the majority spin species on edge I of the QPC and the minority species on edge II. The difference in spin density is anti-symmetric about $y = w_1/2$, resulting in zero net polarization (Figure 7 14.7 (bottom)).

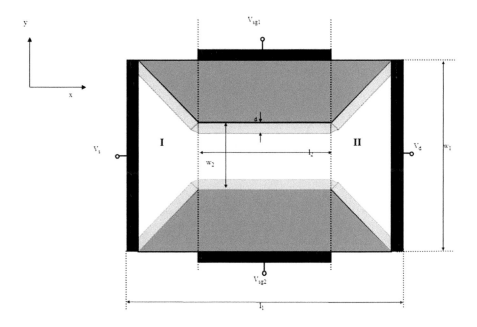

FIGURE 14.6

Schematic of a QPC with in-plane SGs. In the light gray areas, the conduction band energy profile changes abruptly from the bottom of the conduction band inside the semiconducting channel to the vacuum level in the isolation trenches defined by wet chemical etching (this abrupt change occurs in the dark gray areas). In the narrow portion of the QPC (of width w_2 and length l_2) the sharp potential discontinuity on the sidewalls leads to LSOC. A bias asymmetry between the SGs can lead to spin polarization in the narrow channel of the QPCs. Also shown are the source and drain contacts. The current flows in the x-direction. Reproduced with permission from J. Wan, M. Cahay, P. Debray, and R.S. Newrock, "Possible origin of the 0.5 plateau in the ballistic conductance of quantum point contacts", Physical Review B 80, 155440 (2009). Copyright [2009], American Physical Society.

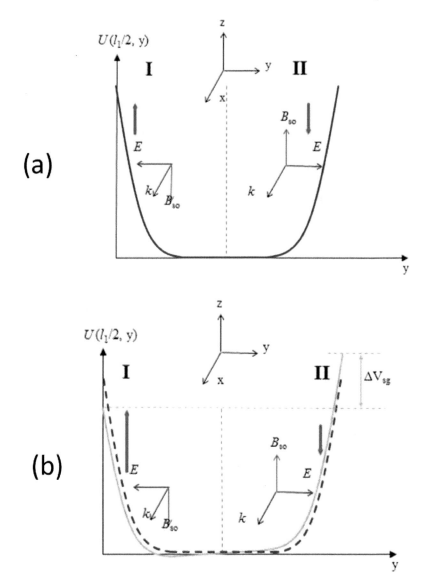

FIGURE 14.7
Schematic representation of the confining potential along the y-direction half
way through the narrow portion of the QPC shown in Fig. 14.6 when (top) a
symmetric and (bottom) an asymmetric bias is applied to the two SGs, respectively. Reproduced with permission from J. Wan, M. Cahay, P. Debray, and
R.S. Newrock, "Possible origin of the 0.5 plateau in the ballistic conductance
of quantum point contacts", Phys. Rev. B, **80**, 155440 (2009). Copyright
[2009], American Physical Society.

When asymmetric SG voltages (i.e., $V_{sg1} \neq V_{sg2}$ in Figure 14.6) are applied on the QPC, the potential profile changes from the symmetric dashed line to the asymmetric solid line as shown in Figure 14.7 (bottom). The spin-up population on the left edge I exceeds the spin-down one on the right edge II. This results in a net spin-up polarization due to the initial imbalance between spin-up and spin-down electrons induced by the asymmetric LSOC. When the asymmetry between V_{sg1} and V_{sg2} is reversed, the direction of spin polarization is reversed as well. Non-equilibrium Green's function (NEGF) simulations show that the presence of a strong repulsive Coulomb electron-electron interaction in the narrow portion of the QPC enhances the initial spin imbalance created by the asymmetry in the SG voltages [33]. As a result, the spontaneous spin polarization can reach nearly 100% when electrons occupy only the lowest quantum confined subband in the constriction and transport is sigle-channeled. When that happens, a 0.5 G_0 conductance plateau can appear. Specific experimental and theoretical examples of the conductance of QPC with an asymmetric potential applied between their SGs are discussed in Sections 14.9 and 14.10, respectively.

14.6 Stern–Gerlach type spatial spin separation in a QPC structure

As discussed in Chapter 1, Stern and Gerlach [35] demonstrated in 1922 the quantization of spin angular momentum by showing the spatial separation of uncharged silver particles deflected in a region with an inhomogeneous magnetic field. Spin-up and spin-down angular momenta were accelerated in opposite directions as a result of the spatial modulation of the Zeeman field. In the case of charged particles (such as electrons), the Lorentz force modulates the orbital motion of the particles and hinders the spin separation of the electrons. This is the reason why the Stern-Gerlach experiment has been limited to atomic beams.

In a 2-DEG, although the electron orbital motion is restricted in-plane due to the strong confining potential, external magnetic fields or micromagnets produce either insufficient or uncontrollable field gradients, making spin separation very difficult. As discussed in Chapter 9, the Spin Hall effect provides a method for generating spin current in semiconductors. However, the efficiency of the spin current generation is represented by the Spin Hall angle θ_{SH}, the ratio between the longitudinal conductivity and the Spin Hall conductivity. For conventional semiconductor materials, θ_{SH} is typically around 0.0001 [36] which is too small for the realization of a practical spintronic device based on the Spin Hall effect. However, the recent advent of giant spin Hall effect based on materials like β-tantalum and topological insulators (as discussed in

Chapter 9) can change all that.

Recently, Kohda et al. [13] demonstrated a nanoscale version of the Stern-Gerlach experiment with electrons utilizing spatial modulation of the RSOC existing in a QPC formed in the 2-DEG of an InGaAsP/InGaAs-based heterostructure. The key ingredient was the combination of a strong Rashba RSOC system with a QPC to induce spatial modulation of the effective magnetic field in the narrow portion of the QPC. As illustrated in Figure 14.8, a spin-dependent force is generated by the lateral potential confinement, which separates the spin-up and -down electrons, and the selective filtering of the spin components at the QPC potential results in a spin-split single channel as schematically shown in Figure 14.8. Channel conductance is controlled by the trench-type side gates shown in the scanning electron microscope (SEM) image in Figure 14.8. Kohda et al. have shown that the non-uniform spin-orbit interaction (SOI) induces spin-dependent deflection before the electron is transmitted through the QPC constriction, as is the case in the Stern-Gerlach spin separation, and the scattering due to the potential of the QPC constriction results in spin-dependent transmission/reflection, which in effect produces a spin-polarized electric current. As a result of the spin separation, a spin-resolved plateau with a half integer in units of $2\,e^2/h$ appears.

Referring to Figure 14.8, when the strength $\alpha(y)$ of the RSOC depends on the y-direction owing to the lateral confinement, the effective force acting on the electrons in the vicinity of the QPC is given by

$$F_y = -\frac{\partial}{\partial y}\alpha(y)\sigma_y k_x, \tag{14.8}$$

where σ_y is y-component of the Pauli spin matrix and k_x is the conduction electron momentum in the x-direction (the direction of current flow through the QPC). This force results in spin separation in the y-direction when electrons are propagated in the x-direction, as shown in Figure 14.8. In the original Stern-Gerlach experiment, a relatively large (meter scale) apparatus was necessary to observe a measurable separation of up- and down-spin beams owing to the small magnetic field gradient on the order of $10^3 - 10^4$ Tesla/m present in the apparatus. In their nanoscale version of the Stern-Gerlach experiment, Kohda et al. have shown that the effective magnetic field gradients are much larger (on the order of 10^8 Tesla/m), which allows resolving the spin-up and spin-down beams in a miniaturized (nanoscale) set-up.

14.7   Detection of spin polarization

Otsuka et al. [14] demonstrated a new method to probe local spin polarization of conduction electrons in semiconductor microdevices at low and zero

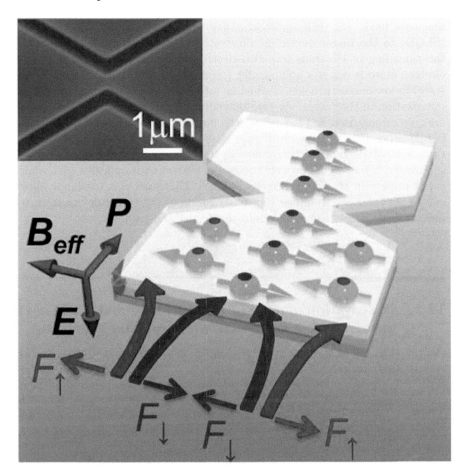

FIGURE 14.8

Schematic of a QPC device fabricated using an InGaAs/InGaAsP 2-DEG with strong SO interaction. The three black arrows represent the electron momentum **P**, the internal electric field **E** (perpendicular to the plane of the 2-DEG), and the effective magnetic field **B**$_{\mathrm{eff}}$. Also shown pointing away (towards) the QPC channel are arrows labeled $F_\uparrow$ ($F_\downarrow$) which represent the directions of the spin-dependent force acting on up- and down-spins. The thick curved arrows pointing toward the walls and inside of the QPC represent the deflected electron trajectories. Also shown are the directions of the electron spins associated with the specific electron trajectories. The inset is an SEM micrograph of a fabricated QPC channel. The scale bar is 1 μm [13]. Reprinted by permission from MacMillan Publishers Ltd: Nature Communications, M. Kohda, S. Nakamura, Y. Nishihara, K. Kobayashi, T. Ono, J.I. Ohe, Y. Tokura, T. Minemo and J. Nitta, "Spin-Orbit Induced Electronics Spin Separation in Semiconductor Nanostructures", **3**, 1082 (2012). Copyright [2012].

magnetic fields. In their approach, they couple a single-lead quantum dot (SLQD) to the target device, as illustrated in Figure 14.9(a), and measure the tunneling of electrons into two-electron states in the SLQD. The two-electron state is either a spin singlet (ground state) or triplet state (excited state) in low magnetic fields. Tunneling into these states reflects electron spin polarization in the target. At zero magnetic field, the electrons in the target are spin unpolarized and tunneling into both singlet and triplet states in the SLQD is possible, as shown in Figure 14.9(b). On the other hand, if the electrons are polarized in the target device, the tunneling into the singlet state is suppressed because this process needs an electron with opposite spin, which is not present in the target [Figure 14.9(c)]. Therefore, this technique allows one to get information of the spin polarization of the target device from the measurement of tunneling rates into the singlet and triplet states.

14.8   Observation of a 0.5 G_0 conductance plateau in asymmetrically biased QPCs with in-plane side gates

Debray et al. [9] have fabricated QPCs with in-plane SGs starting with heterostructures containing a low band gap semiconductor InAs quantum well (QW). A schematic of the heterostructure used is shown in Figure 14.10. In this structure, the 2-DEG is confined to a 3.5 nm thick InAs QW. The 2-DEG was characterized by Shubnikov-de Haas and quantum Hall measurements yielding electron density and mobility of 1.2×10^{12} cm^{-2} and 5×10^4 cm^2/V-s, respectively.

The details of the device fabrication and conductance measurements are given in [9]. Figure 14.11(a) is an SEM picture of a typical InAs QPC device. The QPC channel was created by negatively biasing the two SGs, G_1 and G_2. Figure 14.11(b) shows a conductance plot of the QPC at 4.2 K with $V_{G_1} - V_{G_2}$ equal to 7.5 V as a function of common-mode bias V_G applied to both SGs and swept in the forward bias direction, i.e., from -3.5 V to 0 V. A small conductance at 0.5 G_0 was observed in the absence of any applied magnetic field, an indirect signature of complete spin polarization in the channel of the QPC. The 0.5 structure in Figure 14.11(b) was observed only when the transverse confining potential of the QPC was made asymmetric by appropriately adjusting the SG voltages. The 0.5 G_0 plateau was also observed when the confinement asymmetry was reversed between the two SGs, i.e, $V_{G_1} - V_{G_2}$ set to –7.5 V. In addition, no 0.5 G_0 conductance plateau was found when the SGs were held at the same potential (symmetric case) [9].

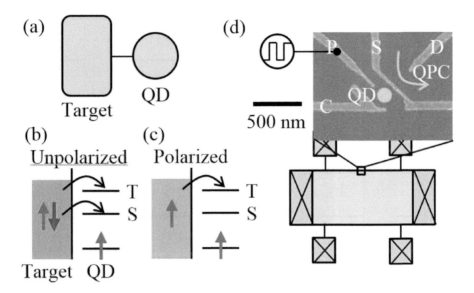

FIGURE 14.9

(a) Schematic of the single-lead quantum dot (SLQD) [14]. A QD is coupled to the target through a single tunneling barrier. Energy diagrams of (b) the unpolarized case and (c) the polarized case. In the unpolarized case, electrons tunnel into singlet and triplet states in the QD. In the polarized case, i.e., in the presence of a small external magnetic field (1.5 T in the experiments) tunneling into singlet state is suppressed. (d) Schematic of the overall device structure. At the edge of a Hall bar, a SLQD is fabricated. Reproduced with permission from T. Otsuka, Y. Sugihara, J. Yoneda, S. Katsumoto and S. Tarucha, "Detection of spin polarization utilizing singlet and triplet states in a single-lead quantum dot", Phys. Rev. B, **86**, 081308(R) (2012). Copyright [2012], American Physical Society.

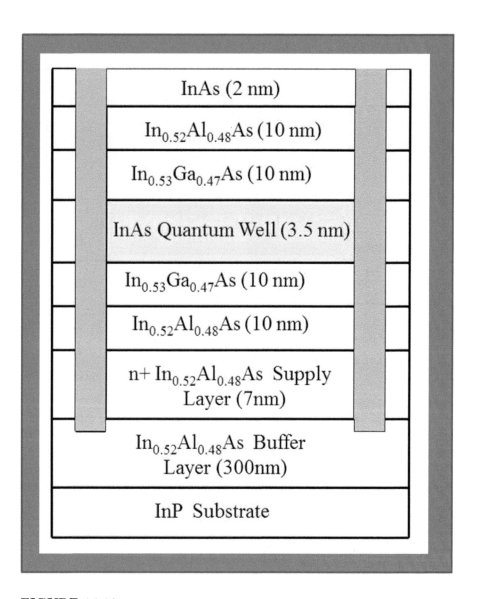

FIGURE 14.10
Schematic of the InAs heterostructure used to build QPCs with in-plane SGs
defined by etching the heterostructure through the 2DGE. The InAs QW is
surrounded by identical InGaAs layers to minimize the effects of RSOC.

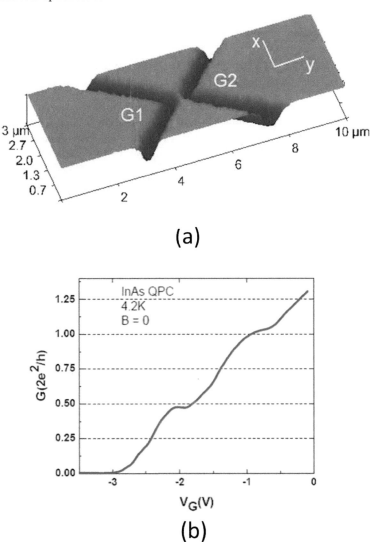

FIGURE 14.11

(a) Scanning electron micrograph (SEM) of a typical side-gated InAs QPC, showing the two SGs (G_1 and G_2). The trenches defined by wet chemical etching can be clearly seen. Reproduced with permission from P. P. Das, K. Chetry, N. Bhandari, J. Wan, M. Cahay, R. S. Newrock and S. T. Herbert, "Evolution of the anomalous conductance plateau in an asymmetrically biased InAs/InAlAs quantum point contact", Appl. Phys. Lett., **99**, 122105 (2011). Copyright [2011], AIP Publishing LLC. (b) Observation of a 0.5 G_0 conductance plateau in the presence of an asymmetric potential between the two SGs of an InAs based QPC device. The measurements were made at T = 4.2 K. [9].

14.9   Prospect for generation of spin-polarized current at higher temperatures

The semiconductor InAs, which has high intrinsic SOC, would normally be preferred for generating spin polarized current, but it has a short spin coherence length – about a micron at 4.2 K [30]. This would reduce to a few nanometers at room temperature. That makes InAs, or any other semiconductor with large intrinsic SOC, unsuitable for making practical devices operational at room temperature. An exception occurs if the devices are strictly one-dimensional, meaning that only the lowest subband is occupied by electrons and transport is single-channeled. In the latter case, D'yakonov-Perel' spin relaxation will be suppresssed (see Problems 8.3 and 8.4) and the spin relaxation length may be sufficiently large, even at room temperature [37]. A spin relaxation length of ~ 40 nm has been reported in InSb nanowires at room temperature [38].

QPCs can be made to act as strictly one-dimensional quantum wires so that the room temperature spin relaxation length can easily exceed the length of the constriction in the QPC. This raises hopes for generating spin-polarized current at room temperature with QPCs.

Alternately, one can use materials that have weak SOC and hence long spin coherence lengths at room temperature. Even a very weak SOC can cause significant spin polarization provided the electron-electron interaction is strong enough. This means that QPCs made from a material like GaAs, which has a weak intrinsic SOC, can also be used to generate spin-polarized current by purely electrical means, as was shown by Bhandari et al. [12]. GaAs may have a long spin coherence length of perhaps several hundred nm in quantum wires at room temperature [see simulations for 77 K in Figure 11.4], as compared to a few nanometers for InAs. It is also possible to grow GaAs samples with very low electron concentration, which ensures a strong e-e interaction. Since GaAs is a mainstream material with a mature and well-established processing technology with the added advantage of a large Schottky barrier with metals, making it relatively easy to deposit surface gates, it is an ideal potential candidate for developing all-electric spin devices that could be operational at a temperature of perhaps tens of Kelvin, if not the liquid nitrogen temperature of 77 K.

Operation of QPC based spin polarizers at even higher temperatures (maybe even room temperature) could be achieved with the use of antimonide-based heterostructures [39, 40, 41]. InSb has the advantage of both a low electron effective mass and a strong SO interaction [42]. This would lead to a larger spacing between the subbands in the QPC and better control of spin injection and detection over a larger range of gate biases. Furthermore, a combination of etching techniques and the use of additional gates could make the transport through the channel between the QPCs truly single-moded. Disorder effects could be reduced by fabricating samples using a wafer with higher electron mobility, or using an undoped heterostructure and electrostatically inducing the 2-DEG. There are already experimental reports of 40 nm spin relaxation length in single-moded InSb nanowires (50 nm diameter) at room temperature in very disordered samples [38], and a significant reduction in the disorder can increase the relaxation length to perhaps hundreds of nm.

14.10   Prospect for an all-electric SpinFET

Recently, Chuang et al. have demonstrated the operation of an all-electric spin valve by fabricating spin injectors and detectors with an efficiency near 100% using asymmetrically biased QPCs [24]. Furthermore, they have shown electrostatic control of the spin precession in the semiconducting channel between the two QPCs by maintaining ballistic transport, which minimizes the effects of spin decoherence in that region.

A schematic of the device investigated by Chuang et al. is shown in Figure 14.12(a). Each QPC was formed by depositing a pair of top gates over a two-dimensional electron gas that exists at the interface of an InAlAs/InGaAs semiconductor heterostructure (Figure 14.12(b). The high efficiency of the QPCs as spin injector and detector was obtained by applying different biases to the two top gates in both the left (V_{G1}, V_{G2}) and right (V_{G3}, V_{G4}) QPC. The bias difference at each QPC creates a local electric field in the y-direction (Figure 14.12(a)) that, as a result of the SO coupling, acts on the electrons as an effective magnetic field $\vec{B_1}$ in the z-direction. Because the SO interaction is due to the lateral confinement in the y-direction, it is referred to as lateral spin-orbit coupling. The direction of $\vec{B_1}$ can be reversed by flipping the sign of the potential difference between the gates, which changes the polarity (up or down) of the injected spins. The injection of spin-up or spin-down electrons with near 100% efficiency through the left QPC is achieved when the Fermi level E_F in the source contact is below the crossing point of the lowest energy subbands for up and down spins in the QPC (Figure 14.12(c)).

As discussed in Section 14.9, in the presence of LSOC, an asymmetrically-biased QPC with in-plane SGs has been shown to create a highly spin-polarized current [9]. Three ingredients are essential to generate a strong spin polarization: lateral spin-orbit coupling, an asymmetric lateral confinement, and a strong e-e interaction [33]. The effects of the electron-electron interaction are hard to assess in the QPCs used by Chen and colleagues. However, they provided convincing evidence for the spin filtering action of each QPC using magnetic focusing experiments, which showed near 100% spin injection efficiency when an asymmetric bias was applied between the top gates of the QPC.

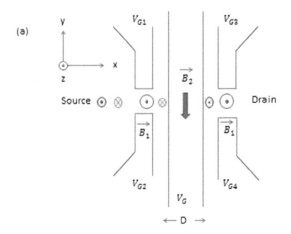

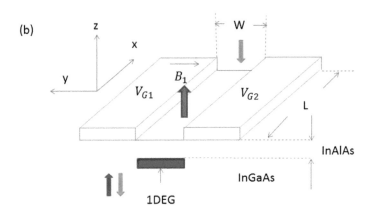

FIGURE 14.12
(a) Schematic of the all-electric spin valve demonstrated by Chuang et al.
[24, 25]. An asymmetric bias ($V_{G1} - V_{G2} \neq 0$) between the two top side gates
of the left QPC injects spin-down in the constriction between the two QPCs.
The field $\vec{B}_1$ represents the effective magnetic field in each QPC (pointing
up in both QPCs). The spin precession under the middle gate (region D) is
due to the effective magnetic field $\vec{B}_2$ resulting from the RSOC in that region
and modulates the drain current. The spin precession in the middle channel
rotates the spins by $\sim180°$, causing them to be blocked by the right QPC.
This causes the minimum in the drain current. (b) Cross-sectional view of
a QPC formed with two negatively biased top gates (V_{G1}, V_{G2}). A 1-DEG
is formed in the constriction by the electrostatic confinement of the 2-DEG
at the InAlAs/InGaAs interface. With a carefully selected asymmetrical bias
between the top gates, only one type of spin (down-spin) is injected through
the left QPC.

If the length of the region between the QPCs (D in Figure 14.12(a) is much shorter than the spin decoherence length in the 2-DEG, the injected spins move ballistically under the middle gate and undergo the same precession, as a result of RSOC in that region. The latter originates from the electric field along the z-axis due to the inversion asymmetry at the InAlAs/GaAs interface, and it gives rise to the magnetic field $\vec{B}_2$ that induced spin precession around the y-direction in D (Figure 14.12(a)). The field $\vec{B}_2$ can be tuned by varying the potential on the middle gate, V_G. As a result, the spin exiting the middle region D can be made either parallel or anti-parallel to the spin polarization in the right QPC. The parallel and anti-parallel scenarios lead to a maximum (ON condition) or minimum (OFF condition) current collected at the drain, respectively, when $V_{G1} - V_{G2} = V_{G3} - V_{G4} \neq 0$.

The ON/OFF ratio should be infinite for the spin valve to act as a perfect all-electric SpinFET. Chuang et al. observed an oscillatory switching of the current (I) through the spin valve with variations, $(I_{ON} - I_{OFF})/I_{OFF}$, as high as 500 % at a temperature of 300 mK. Though this ON/OFF ratio is still a far cry from values in state of the art field effect transistors at room temperature [26], it is hoped that further improvements in the design of the device will allow the realization of an all-electric Datta-Das SpinFET in the near future.

14.11 Conclusion

The full potential of spin-based devices can only be realized if injection, manipulation, and detection of the electron spin can be performed by purely electrical means [43, 44, 45]. This chapter has described some of the recent breakthroughs toward the realization of all electric spin-based devices.

14.12 Problems

This set of problems studies the condition for the existence of spin polarized edge states in a quantum wire in the presence of LSOC due to the lateral confinement, as discussed in detail in [46]. Consider a heterostucture with a 2-DEG confined inside a symmetric quantum well of width d grown along the z-axis, as shown in Figure 14.13. The temperature is assumed to be small enough that transport in the 2-DEG occurs in the lowest bound state with energy ϵ_z resulting from the spatial confinement in the z-direction.

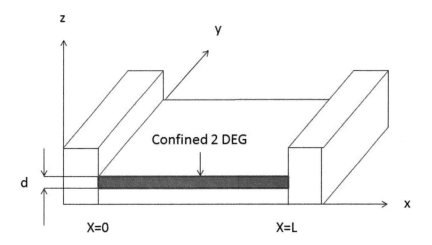

FIGURE 14.13

Schematic representation of a 2-DEG formed in a symmetric QW of width d where z is the axis of growth of the heterostructure. The electrons are constricted along the x axis as a result of lateral confinement, leading to the formation of a quantum wire with edges located at $x = 0$ and $x = W$. Electrons move freely along the y-axis. For symmetric structures, the quantum wire has inversion symmetry along the x-axis. The confinement at the edges leads to a local inversion asymmetry accompanied by lateral (or edge) spin-orbit couplings, which are opposite in sign at the opposite edges of the channel.

If transport through the well is further limited by potential barriers at $x = 0$ and $x = W$, the resulting step-like edge potentials will force electrons to move in a quantum wire extending along the y-axis. Hereafter, we analyze the formation of edge states at $x = 0$ and $x = L$ as a result of the lateral confinement in the x-direction and the accompanying LSOC. The lateral confinement along the x-axis could result from a wet or dry etching process through the heterostructure, leaving sharp barriers on both sides of the quantum wire. The height of the channel/barrier interface could also be controlled via regrowth of a different material with a wider bandgap than the channel in the trenches following the etching process.

If we model the difference in electron affinity barrier/channel interface with a hardwall potential, the edge potential can be approximated as

$$V_{edge}(x) = V_0\theta(-x) - V_0\theta(x - W), \qquad (14.9)$$

where the strength V_0 can be several eV if the edges of the quantum wire are formed by etching of trenches through the 2-DEG. In this case, the edges are due to a channel/vacuum interface, like in the work of Debray et al. [9].

Associated with this hardwall potential, there will be a SO contribution to the Hamiltonian of the form [27]

$$H_{SO} = \alpha \vec{\sigma} . [\vec{k} \times \vec{\Delta} V_{edge}], \qquad (14.10)$$

where α characterizes the strength of the lateral SO coupling, $\vec{\sigma}$ are the Pauli matrices, and $\hbar \vec{k}$ the electron momentum operator.

First we analyze the formation of edge states near $x = 0$, assuming that the width W is large enough that the overlap of the wave functions localized on opposite edges of the quantum wire can be neglected. Taking into account the difference in effective masses between the channel and the cladding barriers, the effective Schrödinger equation in the quantum wire near $x = 0$ is given by

$$H_{2D} \phi_\sigma(x, y) = E_\sigma \phi_\sigma(x, y), \qquad (14.11)$$

where

$$H_{2D} = p_x [\frac{1}{m^*(x)} p_x] + \frac{p_y{}^2}{2m^*(x)} + V_0 \theta(-x) + \epsilon_z + \frac{\alpha}{\hbar} \delta(x) p_y \sigma_z, \qquad (14.12)$$

and E_σ are the eigenenergies in the 2-DEG measured with respect to the bottom of the conduction band in the quantum wire. In the Hamiltonian of Equation (14.12), $p_x (p_y)$ is the x-component (y-component) of the electron momentum, σ_z the Pauli matrix, and $\delta(x)$ and $\theta(x)$ the Dirac and unit step functions, respectively. The effective mass is set equal to $m_b{}^*$ and $m_{QW}{}^*$ in the regions $x < 0$ and $x > 0$, respectively.

According to the effective Hamiltonian in Equation (14.12), both the y component of the momentum and the z-component of the spin are good quantum numbers. The eigenfunctions of the Hamiltonian can therefore be written as:

$$\phi_\sigma(x, y) = \frac{1}{\sqrt{L}} e^{ik_y y} \Phi_x \Xi_\sigma, \qquad (14.13)$$

where L is the length over which the eigenfunctions are normalized in the y-direction, Ξ_σ are the eigenspinors of σ_z (σ +1 and -1, for up- and down-spins, respectively).

- **Problem 14.1**

  Starting with the effective 2D Hamiltonian in Equation (14.12) and the ansatz (14.13) for its eigenfunctions, show that the wave functions $\phi_\sigma(x)$ must satisfy the effective mass Schrödinger equation

$$[p_x (\frac{1}{m^*(x)} + V_\sigma(x)] \Phi_\sigma(x) = \epsilon_\sigma \psi_\sigma(x), \qquad (14.14)$$

  where

$$\epsilon_\sigma = E_\sigma - (\epsilon_y + \epsilon_z), \qquad (14.15)$$

and $V_\sigma(x)$ is a spin dependent effective potential

$$V_\sigma(x) = V_{1D}\theta(-x) + \alpha\sigma k_y \delta(x), \tag{14.16}$$

with

$$V_{1D} = V_0 - (1 - r_m)\epsilon_y, \tag{14.17}$$

where we have introduced the effective mass ratio $r_m = m_{QW}^*/m_b^*$.
The eigenergies ϵ_σ are measured with respect to the value $\epsilon_y + \epsilon_z$,where
$\epsilon_y = \frac{\hbar^2 k_y{}^2}{2m_{QW}^*}$.

- **Problem 14.2**

Assuming that $\alpha > 0$ in Equation (14.16), draw qualitatively the effective edge potentials for both spin-up and spin-down electrons for both $k_y > 0$ and $k_y < 0$.

- **Problem 14.3**

Using the results of the previous problem, show that for the case of attractive edge potential at the interface, the existence of bound states near the interface $x = 0$ requires solutions of Equation (14.14) such that

$$\epsilon_\sigma = -\frac{\hbar^2 k_\sigma^2}{2m_{QW}^*}, \tag{14.18}$$

with corresponding eigenfunctions

$$\Psi_\sigma(x) = \sqrt{\frac{2\kappa_\sigma k_\sigma}{\kappa_\sigma + k_\sigma}} e^{\kappa_\sigma x}, \tag{14.19}$$

for $x < 0$, and

$$\Psi_\sigma(x) = \sqrt{\frac{2\kappa_\sigma k_\sigma}{\kappa_\sigma + k_\sigma}} e^{-\kappa_\sigma x}, \tag{14.20}$$

for $x > 0$, where

$$\kappa_\sigma = \sqrt{(k_\sigma{}^2 + q_0{}^2)/r_m}, \tag{14.21}$$

and

$$q_0 = \frac{1}{\hbar}\sqrt{2m_{QW}^* V_{1D}}. \tag{14.22}$$

- **Problem 14.4**

Show that the wavevector k_σ obeys the relation

$$k_\sigma + \sqrt{r_m(k_\sigma{}^2 + q_0{}^2)} + \sigma\lambda_{SO}k_y = 0, \tag{14.23}$$

where $\lambda_{SO} = 2m_{QW}^*\alpha/\hbar^2$ is a dimensionless parameter characterizing the strength of LSOC.

- **Problem 14.5**

Assuming $m_b^* > m_{QW}^*$ and q_0 is real, the eigenfunctions $\Psi_\sigma(x)$ in Equation (14.23) represent true bound states if k_σ is real. Starting with Equation (14.20), show that the k_σ associated with true bound states is given by

$$k_\sigma = \frac{-\sigma\lambda_{SO}k_y - \sqrt{r_m\lambda_{SO}^2 k_y^2 + r_m(1 - r_m)q_0^2}}{1 - r_m}. \qquad (14.24)$$

- **Problem 14.6**

Starting with the results of the previous problem, show that a necessary (but not sufficient) condition for the existence of bound states is $k_y\sigma < 0$. In other words, LSOC-induced edge states are of two kinds, representing two spin-polarized counterpropagating channels along the y-axis with opposite spins. Starting with Equation (14.24), show that edge bound states do not exist for arbitrary k_y.

(a) Show that, if $\lambda_{SO}^2 < 1 - r_m$, the range of k_y values for which edge states exist is given by

$$\frac{b}{(1 - r_m) + \frac{\lambda_{SO}^2}{r_m}} < (k_y d)^2 < \frac{b}{(1 - r_m) - \frac{\lambda_{SO}^2}{1 - r_m}}, \qquad (14.25)$$

where $b = \frac{V_0}{\epsilon_0}$ and $\epsilon_0 = \frac{\hbar^2}{2m_{QW}^*}\frac{1}{d^2}$.

(b) Show that, if $\lambda_{SO}^2 > 1 - r_m$, edge states are only possible if

$$(k_y d)^2 > \frac{b}{(1 - r_m) + \frac{\lambda_{SO}^2}{r_m}}. \qquad (14.26)$$

- **Problem 14.7**

Edge bound states will exist within the bandgap of the quantum well if the LSOC strength is such that the total energy of these edge states E_σ is below ϵ_z. This requires that the following relation holds:

$$\epsilon_y + \epsilon_z < 0. \qquad (14.27)$$

or $k_y^2 < k_\sigma^2$.

Starting with Equations (14.24) and (14.27), show that the necessary condition for the existence of edge states within the bandgap of the quantum well material is

$$|\lambda_{SO}| > 1 + r_m. \qquad (14.28)$$

- **Problem 14.8**

  Starting with Equation (14.28), show that in the extreme case where $r_m \ll 1$, the energy dispersion relation of the edge states within the bandgap of the 2-DEG is given by

  $$E_\sigma\left(k_y\right) = \epsilon_z + \frac{\hbar^2 k_y{}^2}{2\mu}, \qquad (14.29)$$

  where μ is the LSOC-dependent effective mass and is given

  $$\frac{1}{\mu} = \frac{1}{m^*_{QW}}\left(1 - \lambda^2_{SO}\right). \qquad (14.30)$$

  Note that, if $\lambda_{SO} > 1$, μ changes sign and the energy dispersion will be located within the bandgap region of the material forming the quantum well.

14.13   References

[1] M. Eto, T. Hayashi and Y. Kurotani, "Spin Polarization at Semiconductor Point Contacts in Absence of Magnetic Field", J. Phys. Soc. Jpn., **74**, 1934 (2005).

[2] S. Nadj-Perge, S. M. Frolov, E. P. A. M. Bakkers and L. P. Kouwenhoven, "Spin-orbit qubit in a semiconductor nanowire", Nature, **468**, 1084 (2010).

[3] Y. Kanai, R. S. Deacon, S. Takahashi, A. Oiwa, K. Yoshida, K. Shibata, K. Hirakawa, Y. Tokura and S. Tarucha, "Electrically tuned spin-orbit interaction in an InAs self-assembled quantum dot", Nature Nanotechnol., **6**, 511 (2011).

[4] Yu. V. Sharvin, "A possible method for studying Fermi surfaces", Sov. Phys. JETP, **21**, 655 (1965).

[5] D. A. Wharam, T. J .Thornton, R. Newbury, M. Pepper, H. Ahmed, J. E. F. Frost, D. G. Hasko, D. C. Peacock, D. A. Ritchie and G. A. C. Jones, "One-dimensional transport and the quantisation of the ballistic resistance", J. Phys. C, **21**, L209 (1988).

[6] B. J. van Wees, H. van Houten, C. W. J. Beenakker, J. G. Williamson, L. P. Kouwenhoven, D. van der Marel and C. T. Foxon, "Quantized Conductance of Point Contacts in a Two-Dimensional Electron Gas", Phys. Rev. Lett., **60**, 848 (1988).

[7] M. Cahay, "Quantum Transport: Immune to Local Heating", Nature Nanotechnol., **9**, 97 (2014).

[8] S. Datta, *Electronic Transport in Mesoscopic Systems* (Cambridge University Press, Cambridge, 1995).

[9] P. Debray, S. M. S. Rahman, J. Wan, R. S. Newrock, M. Cahay, A. T. Ngo, S. E. Ulloa, S. T. Herbert, M. Muhammad and M. Johnson, "All Electrical Quantum Point Contact Spin Valves", Nature Nanotechnol., **4**, 759 (2009).

[10] P. P. Das, K. Chetry, N. Bhandari, J. Wan, M. Cahay, R. S. Newrock and S. T. Herbert, "Understanding the anomalous conductance in asymmetrically biased InAs/InAlAs quantum point contacts - a step towards a tunable all electric spin valve", Appl. Phys. Lett., **99**, 122105 (2011).

[11] P. P. Das, N. Bhandari, J. Wan, J. Charles, M. Cahay, K. B. Chetry, R. S Newrock and S. T. Herbert, "Influence of surface scattering on the anomalous plateaus in an asymmetrically biased InAs/InAlAs quantum point contact", Nanotechnology, **23**, 215201 (2012).

[12] N. Bhandari, P. P. Das, M. Cahay, R. S. Newrock and S. T. Herbert, "Observation of a 0.5 conductance plateau in asymmetrically biased GaAs quantum point contact", Appl. Phys. Lett., **101**, 10241 (2012).

[13] M. Kohda, S. Nakamura, Y. Nishihara, K. Kobayashi, T. Ono, J.I. Ohe, Y. Tokura, T. Minemo and J. Nitta, "Spin-Orbit Induced Electronics Spin Separation in Semiconductor Nanostructures", Nature Communications, **3**, 1082 (2012).

[14] T. Otsuka, Y. Sugihara, J. Yoneda, S. Katsumoto and S. Tarucha, "Detection of spin polarization utilizing singlet and triplet states in a single-lead quantum dot", Phys. Rev. B, **86**, 081308(R) (2012).

[15] K. J. Thomas, J. T. Nicholls, M. Y. Simmons, M. Pepper, D. R. Mace and D. A. Ritchie, "Possible Spin Polarization in a One-Dimensional Electron Gas", Phys. Rev. Lett., **77**, 135 (1996).

[16] M. Pepper and J. Bird , "The 0.7 feature and interactions in one-dimensional systems", J. Phys: Condens. Matter, **20**, 16301 (2008).

[17] A. P. Micolich, "What lurks below the last plateau: experimental studies of the 0.7 2e2/h conductance anomaly in one-dimensional systems", J. Phys.: Condens. Matter, **23**, 443201 (2011).

[18] K.-F. Berggren and M. Pepper, "Electrons in one dimension", Philos. Trans. R. Soc. London, Ser. A, **368**, 1141 (2010).

[19] A. Shailos, A. Shok, J. P. Bird, R. Akis, D. K. Ferry, S. M. Goodnick, M.P. Lilly, J.L. Reno and J.A. Simmons, "Linear conductance of

quantum point contacts with deliberately broken symmetry", J. Phys. Condens. Matter, **18**, 1715 (2006).

[20] J. C. Chen, Y. Lin, K. T. Lin, T. Ueda and S. Koniyma, "Effects of impurity scattering on the quantized conductance of a quasi-one-dimensional quantum wire", Appl. Phys. Lett., **94**, 01205 (2009).

[21] M. Seo and Y. Chung, "Anomalous Conductance quantization in a quantum point contact with an asymmetric confinement potential", J. of the Korea Phys. Society, **60**, 1907 (2012).

[22] K. M. Liu, C. H. Juang, V. Umansky and S. Y. Hsu, "Effect of impurity scattering on the linear and nonlinear conductances of quasi-one-dimensional disordered quantum wires by asymmetrically lateral confinement", J. Phys.: Condens. Matter, **22**, 395303 (2010).

[23] S. Kim, Y. Hashimoto, Y. Iye and S. Katsumoto, "Evidence of Spin-Filtering in Quantum Constrictions with Spin-Orbit Interaction", J. Phys. Soc. Jpn., **81**, 054706 (2012).

[24] P. Chuang, S-C Ho, L. W. Smith, F. Sfigakis, M. Pepper, C-H. Chen, J-C. Fan, J. P. Griffiths, I. Farrer, H. E. Beere, G. A. C. Jones, D. A. Ritchie and T-M. Chen, "All Electric All-Semiconductor Spin Field Effect Transistors", Nature Nanotechnol., **10**, 35 (2015).

[25] M. Cahay, "Spin Transistors: Closer to an All-Electric Device", Nature Nanotechnol., **10**, 21 (2015).

[26] The International Roadmap for Semiconductors, available at http://www.itrs.net.

[27] H.-A. Engel, E.I. Rashba and B. Halperin, "Theory of Spin Hall Effects in Semiconductors", in *Handbook of Magnetism and Advanced Magnetic Materials*, Vol.5 (John Wiley & Sons, 2007).

[28] Y. A. Bychkov and E. I. Rashba, "Oscillatory effects and the magnetic susceptibility of carriers in inversion layers", J. Phys. C, **17**, 6039 (1984).

[29] M. Kohda and J. Nitta, "Enhancement of spin-orbit interaction and the effect of interface diffusion in quaternary InGaAsP/InGaAs heterostructures", Phys. Rev. B, **81**, 115118 (2010).

[30] K. Hattori and H. Okamoto, "Spin separation and spin Hall effect in quantum wires due to lateral-confinement-induced spin-orbit-coupling", Phys. Rev. B, **74**, 155321 (2006).

[31] Y. Xing, Q-F. Sun, L. Tang and J. Hu, "Accumulation of opposite spins on the transverse edges of a two-dimensional electron gas in a longitudinal electric field", Phys. Rev. B, **74**, 155313 (2006).

[32] Y. Jiang and L. Hu, "Kinetic magnetoelectric effect in a two-dimensional semiconductor strip due to boundary-confinement-induced spin-orbit coupling", Phys. Rev. B, **74**, 075302 (2006).

[33] J. Wan, M. Cahay, P. Debray and R. S. Newrock, "On the physical original of the 0.5 plateau in the conductance of quantum point contacts", Phys. Rev. B, **80**, 155440 (2009).

[34] J. Wan, M. Cahay, P. Debray and R. S. Newrock, "Spin texture of conductance anomalies in quantum point contacts", J. Nanoelectron. Optoelectron., **6**, 95 (2011).

[35] W. Gerlach and O. Stern, "Der experimentalle nachweis der richtungsquantelung in magnetic field", Z Phys., **9**, pp. 349-352 (1922).

[36] K. Ando and E. Saitoh, "Observation of the inverse spin Hall effect in silicon", Nature Communications, **3**, 629 (2012).

[37] S. Bandyopadhyay, Md. I. Hossain, H. Ahmad, J. Atulasimha and S. Bandyopadhyay, "Coherent spin transport and suppression of spin relaxation in InSb nanowires with single subband occupancy at room temperature", Small, **10**, 4379 (2014).

[38] Md. I. Hossain, S. Bandyopadhyay, J. Atulasimha and S. Bandyopadhyay, "Modulation of D'yakonov-Perel' spin realxation in InSb nanowires with infrared illumination at room temperature", unpublished.

[39] G. Tuttle, H. Kroemer and J. H. English, "Electron concentration and mobilities in AlSb/InAs/AlSb quantum wells", J. Appl. Phys., **65**, 5239 (1989).

[40] S.J. Koester, C.R. Bolognesi, E.J. Hu and H. Kroemer, "Design and analysis of InAs/AlSb ballistic constrictions for high temperature operation and low gate leakage", J.Vac. Sci. Technol. B, **11**, 2528 (1993).

[41] S. J. Koester, B. Brar, C. R. Bolognesi, E. J. Caine, A. Patlach, E. L. Hu, H. Kroemer and M. J. Rooks, "Length dependence of quantized conductance in ballistic constrictions fabricated on InAs/AlSb quantum wells", Phys. Rev. B, **53**, 13063 (1996).

[42] I. van Weperen, S. R. Plissard, E. P. A. Bakkers, S. M. Frolov and L. P. Kouwenhouven, "Quantized conductance in an InSb nanowire", Nano Letters, **13**, 387 (2013).

[43] D. D. Awschalom and M. E. Flatté, "Challenges for semiconductor spintronics", Nature Physics, **3**, 153 (2007).

[44] D. Awaschalom and N. Samarth, "Spintronics without magnetism", Physics, **2**, 50 (2009).

[45] M. Flatté, "Solid-state physics: Silicon spintronics warms up", Nature, **462**, 419 (2009).

[46] A. Matos-Abiague, "Helical edge states induced by lateral spin-orbit coupling", Phys. Rev. B, **87**, 155306(2013).

15

Single Spin Processors

15.1   Single spintronics

In the last chapter, we saw that Spin Transistors - whether they are of the "field effect" type, or the "bipolar junction" type - do not really yield significant advantages over their traditional charge based counterparts. Spin transistors are hybrid spin devices where charge still plays the dominant role and spin merely augments the role of charge. Digital information (i.e., binary bit 0 or 1) is ultimately encoded by charge. For example, in the case of the SPINFET, when charge flows through the channel of the transistor causing a source-to-drain current, the device is "on" and could encode the binary bit 1. When no current flows and the device is "off," the encoded bit could be 0. Switching between logic bits is therefore associated with turning on or off a current, which involves *physical motion of charges*. This physical motion consumes considerable energy, which is ultimately dissipated as heat.

Heat dissipation is an extremely serious issue in electronics since it is the primary threat to the fabled "Moore's law." That "law" is an empirical law which predicts that transistor density on a chip will roughly double every 18 months. Moore's law has driven the commercial juggernaut that we call the electronics industry, and it is sacrosanct. Anything that threatens this law is a looming catastrophe which must be eliminated. If heat dissipation cannot be reduced dramatically by switching from electronics to spintronics, and thereby perpetuating Moore's law, then perhaps spintronics has little chance of displacing the silicon juggernaut.

Fortunately, spintronics can, *in principle*, reduce heat dissipation significantly. With charge based electronics, there is always a *fundamental* limitation as far as energy dissipation is concerned. Charge is a *scalar* quantity, which only has a magnitude. Thus, logic levels can be demarcated solely by a difference in the *magnitude* of charge, or by the presence and absence of charge. Consequently, to switch from one bit to another, we must change the magnitude of charge in the active region of the device, or move charge around in space. That invariably causes a current (I) flow and an associated $I^2 R$ dissipation (R is the resistance in the path of the current). This dissipation cannot be avoided.

Spin, unlike charge, is not a scalar quantity. It is a pseudo vector, with a fixed magnitude of $\hbar/2$, but a variable "polarization". We can make the polarization "bistable" by placing an electron in a static magnetic field, so that the polarization has only two allowed states – parallel to the field and anti-parallel. No other polarization is an eigenstate. Spin polarization therefore becomes a *binary* variable. We can encode bits 0 and 1 in these two polarizations. For example, the polarization parallel to the field could encode 1 and the anti-parallel polarization could encode 0, or vice versa. In that case, switching can be accomplished simply by flipping the spin, *without causing physical motion of charges*, or a current flow. This may result in considerable energy saving. There is still some energy dissipated in flipping the spin, but it is of the order of $g\mu_B B$, where B is the flux density of the global magnetic field that we need to keep the spin polarization bistable. This could be made arbitrarily small by making $B \to 0$. A smaller B, of course, causes more random bit flips, but bit flip errors can be handled with software or error correcting codes up to a point. In fact, errors occurring with a probability as high as 3% can be handled by the most sophisticated codes that are available today [1]. We will later show that, at any temperature T, $g\mu_B B$ must be kept larger than $kT ln(1/p)$ where p is the error probability (associated with random bit flips) that we can tolerate or handle with error correcting codes. This is true as long as the spins are in equilibrium with their thermal environment [2]. Therefore, a smaller B results in a higher error probability, since $p > \exp[-g\mu_B B/(kT)]$. The point to note here is that we need some minimum energy dissipation $g\mu_B B \geq kT ln(1/p)$ *not* because the switching mechanism demands it, but because we have to keep the error probability manageable.

Since B is required only to make the spin polarization bistable, we ask: why does spin polarization have to be *bistable* in this paradigm? In charge based electronics, logic states are ultimately encoded by voltage or current, which are *continuous* (not discrete) variables. They are certainly not bistable. So, why does spin polarization have to be discrete and bistable? The answer is that, in charge based electronics, there are voltage or current amplifiers with *non-linear* transfer characteristics that restore strayed logic levels at circuit nodes to one of two values: 0 and 1. A discussion of this can be found in [3]. There is no equivalent device in spintronics to restore logic levels encoded in spin polarization. Therefore, we must ensure that only two spin polarizations are allowed and intermediate polarizations are not stable (or eigenstates). Hence, the need for bistability.

15.1.1 Bit stability and fidelity

In order to encode logic bits reliably, spin polarization must be robust and should not flip (spontaneously) too easily. Spontaneous bit flip is a problem that afflicts charge based electronic memory as well. There, the stored voltage level can decay with time owing to charge leakage causing loss of stored data. This is usually countermanded by using a clock to refresh the charge or voltage. The same can be done with spin. We can periodically read the spin polarization of target electrons using a variety of methods. If any spin is found to have flipped spontaneously (as determined by error detecting algorithms), then we can rewrite the errant spin by suitable methods. This is equivalent to "refreshing".

Of course, we do not want to have to refresh spin bits too often since this can be overwhelming and consumes energy. Therefore, it is imperative that the spin relaxation time is long. If the electron hosting the spin is localized in a quantum dot, then indeed the spin relaxation time can be quite long. In InP quantum dots, the single electron spin flip time has been reported to exceed 100 μs at 2 K [4]. More recently, several experiments have been reported claiming spin flip times (or so-called longitudinal relaxation time T_1) of several milliseconds, culminating in a recent report of 1 second in a GaAs quantum dot at low temperature [5]. Organics are even better because of the weak spin-orbit interaction in these materials. The spin flip time in the π-conjugated molecule tris(8-hydroxyquinolinolato aluminum), popularly known as *Alq3*, is reported to possibly exceed 1 second at 100 K [6]. Let us assume conservatively that the spin flip time is 10 msec at the operating temperature. If the clock frequency for refresh cycles is 50 GHz, then the clock period is 20 ps, which is a factor of 5×10^8 smaller than the spin flip time. Therefore, the probability that an unintentional spin flip will occur between two successive refresh clock pulses is $1 - e^{-2 \times 10^{-9}} = 2 \times 10^{-9}$, which can be easily handled even in ultra large scale integrated circuits.

15.2 Reading and writing single spin

The next issue to address is how can one align the spin polarization of a single electron in a quantum dot to write data, and how does one read data by measuring a single electron's spin polarization? We can host single electrons in single quantum dots and orient (write) the spin polarization in a target quantum dot by a variety of methods (see, for example, [7]). The easiest method is to generate a local magnetic field with inductors (something routinely done in magnetic random access memory chips) which will orient the local spin parallel to the field and write a bit. Reading is slightly more difficult. There are, however, three different successful experimental demonstrations of the reading of a single spin in a solid [8, 9, 10]. Among them, the technique of [8] is conceptually the simplest. Here, highly precise magnetic resonance force microscopy is utilized for ascertaining the spin polarization of a single electron in a solid.

15.3   Single spin logic

The *Single spin logic* (SSL) paradigm [11], proposed in 1994, uses the idea
of encoding binary bits in the bistable spin polarization of single electrons
housed in quantum dots that are placed in a global magnetic field. Exchange
interaction between spins in nearest neighbor dots elicits computational activity. This is an esoteric idea, perhaps too futuristic, but it is instructive, which
is why we discuss it in this textbook. More important, it is the progenitor of
spin-based quantum computing, which we discuss in the next chapter. This
idea, however, is relevant to classical computing, namely, Boolean logic.

A Boolean logic gate, or an entire combinational/sequential logic circuit,
can be implemented with a two-dimensional array of single-electron quantum
dots with nearest neighbor exchange interaction. Certain dots in the array are
designated as input ports, and the spins of the single electrons in those dots
are oriented and held fixed with some external device so that they conform
to the input bits. That is how the input stream is fed to the SSL circuit.
The array is then allowed to relax to the thermodynamic ground state by
emitting phonons. Exchange interactions between nearest neighbor quantum
dots and the states of the input spins determine the ground state spin polarizations in every dot of the array. By arranging quantum dots in appropriate
two-dimensional patterns on a wafer (which engineers the exchange interaction between nearest neighbors), we can ensure that the spin polarizations
in quantum dots that are designated as output spins conform to the desired
output(s) in accordance with the truth table of the logic gate or the logic
circuit that is being implemented. This is best explained through examples,
and we will provide that later in this chapter.

Different logic circuits are implemented using different layouts of the quantum dots. We illustrate this idea with the example of a NAND gate in the
following section. Since this gate is universal, any combinational or sequential
logic circuit can, in principle, be built with the SSL NAND gate.

15.3.1   SSL NAND gate

The NAND gate is realized with a linear chain of three electrons in three
quantum dots, as shown in Fig. 15.1. We will assume that the dots are small
enough that the energy cost to accommodate more than one electron in a dot
is prohibitive. Hence, each dot contains a single electron. There is a global dc
magnetic field of flux density B directed in an arbitrary direction to define the
global spin quantization axis, i.e., the stable/metastable spin polarization of
the electron in any dot is either parallel or anti-parallel to this field. We will
assume that only nearest neighbor electrons interact via exchange since their
wavefunctions overlap. Second nearest neighbor interactions are negligible

since exchange interaction strength decays exponentially with separation.*

Let us now regard the two outside spins in Fig. 15.1 (dots A and C) as the two inputs to the logic gate and the spin in the middle (dot B) as the output. Assume that the down-spin state ($\downarrow$), which is parallel to the global magnetic field, represents logic bit 1, and the up-spin state ($\uparrow$), which is anti-parallel to the global magnetic field, is logic bit 0. We will show the following result: If we apply logic inputs to the peripheral dots A and C by aligning their spins either parallel or anti-parallel to the global magnetic field, and then let the system relax to the thermodynamic ground state, the logic output in dot B will always settle to a polarization that will be the correct output conforming to the truth table of a NAND gate.

TABLE 15.1
Truth Table of a Boolean
NAND gate

Input 1	Input 2	Output
1	1	0
0	0	1
0	1	1
1	0	1

Once the NAND gate is realized, any arbitrary circuit can be implemented. Sarkar et al. have implemented a large array of combinational and sequential logic circuits using this approach [12]. In the next section, we show that the ground state configurations indeed conform to the rows in Fig. 15.1 which represent the truth table of the NAND gate.

15.3.2   Input-dependent ground states of the NAND gate

In this section, we will prove the NAND operation rigorously using quantum mechanics.

Consider a linear array of three quantum dots designated A, B, and C, as in Fig. 15.1. A and C are the input dots and B is the output dot. One can describe this system with the so-called Hubbard Hamiltonian [13]. The Hubbard Hamiltonian is a convenient form of the Hamiltonian for many electron systems. We have not discussed it in this textbook, but excellent descriptions are found in many solid state physics and quantum mechanics textbooks, such

---

*The exchange interaction strength between two electrons depends on the spatial overlap of their wavefunctions. If the electrons are confined in two different quantum dots, the overlap decreases exponentially with increasing separation between the dots.

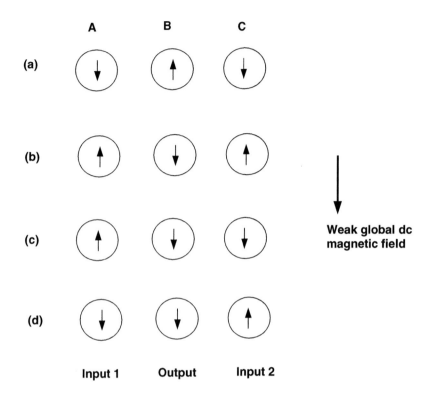

FIGURE 15.1

A 3-dot array with each dot containing a single conduction band electron. Only nearest neighbors are exchange coupled, i.e., there is exchange coupling between A and B and between B and C, but not between A and C. A global magnetic field is applied to define the global spin quantization axis, i.e., the stable/metastable spin orientations in every dot are parallel and anti-parallel to this field. The peripheral dots A and C are the "input ports" in which the spin polarizations are aligned with external agents to conform to input bit 0 or 1. The central dot B is the "output port" and the spin polarization of the electron in this dot encodes the output bit produced in response to the input bits. Figures (a) – (d) show the ground state configurations when the inputs are aligned as indicated. In all cases, the output bit is the NAND function of the two input bits (in accordance with the truth table of a NAND gate), showing that this array implements a NAND gate.

as [13]. For our 3-electron system, the Hubbard Hamiltonian is

$$H_{Hubbard} = \sum_{i\sigma} \epsilon_0 n_{i\sigma} + \sum_{input\ dots,\ \sigma} g\mu_B B_i sign(\sigma) + \sum_{<ij>} \left[c_{i\sigma}^\dagger c_{j\sigma} + h.c. \right]$$
$$+ \sum_i U_i n_{i\uparrow} n_{i\downarrow} + \sum_{<ij>\alpha\beta} J_{ij} c_{i\alpha}^\dagger c_{i\beta} c_{j\beta}^\dagger c_{j\alpha} + B_z \sum_{i\sigma} g\mu_B n_{i\sigma} sign(\sigma) ,$$

$$(15.1)$$

where *h.c.* stands for "Hermitian conjugate". Here the first term represents the site energy (kinetic + potential) in the i-th dot, the second term is the Zeeman splitting energy associated with applying a local magnetic field to the i-th dot to selectively align the spin in that dot along the direction of the applied local field (providing input data), the third term is the tunneling between nearest neighbor dots (the angular braket denotes "nearest neighbors"), the fourth term is the charging energy of the dot, which is the energy cost associated with putting a second electron in the same dot (Pauli exclusion principle ordains that the second electron must have a spin anti-parallel to that of the first, since we assume that each dot has only one bound state), the fifth term is the exchange interaction between nearest neighbor dots, and the last term is the splitting energy due to the global magnetic field, applied over the entire array, in the z-direction, to define the spin quantization axes. Here, we have used the second quantization operators – the so-called creation and annihilation operators. Readers not familiar with these operators can refer to any advanced quantum mechanics textbook where they are discussed. However, for what follows, knowledge and understanding of second quantization operators are not necessary.

If the charging energy (intradot Coulomb repulsion) is sufficiently strong (meaning that the energy cost to accommodate more than one electron in a dot is prohibitive), the Hubbard model at half filling (one electron per dot) reduces to the so called Heisenberg model, which was briefly discussed in Chapter 10. This model is much simpler to handle than the full Hubbard model, and yet captures all the essential physics. This was the approach adopted by Molotkov and Nazin in analyzing SSL gates [14, 15]. Using this approach, the Hamiltonian reduces to

$$H_{Heisenberg} = \sum_{<ij>} J_{ij}^{\parallel} \sigma_{zi} \sigma_{zj} + \sum_{<ij>} J_{ij}^{\perp} (\sigma_{xi} \sigma_{xj} + \sigma_{yi} \sigma_{yj})$$
$$+ \sum_{input\ dots} \sigma_{zi} h_{zi}^{input} + \sum_i \sigma_{zi} h_{zi}^{global} , \qquad (15.2)$$

where we have assumed that the input data are provided to the input dots by z-oriented local magnetic fields. We will assume the isotropic case whereby $|J_{ij}^{\parallel}| = |J_{ij}^{\perp}| = J$. Here J is the exchange energy that is non-zero if the wavefunctions in dots i and j overlap in space. Roughly speaking, it is one-half of the energy difference between the triplet state (when spins in dots i

and j are parallel) and the singlet state (when the spins in dots i and j are anti-parallel). In the above equation, the σ-s are of course the Pauli spin matrices.

Now, the spins in dots A, B, and C can be either $+z$-polarized (anti-parallel to the global magnetic field) or $-z$-polarized (parallel to the global magnetic field), which we call "up-spin" and "down-spin," respectively. Remember that the "up-spin" state is logic 0 and the "down-spin" state is logic 1. Therefore, the 3-spin basis states that span the entire Hilbert space are $|\downarrow_A\downarrow_B\downarrow_C>$, $|\downarrow_A\downarrow_B\uparrow_C>$, $|\downarrow_A\uparrow_B\downarrow_C>$, $|\downarrow_A\uparrow_B\uparrow_C>$, $|\uparrow_A\downarrow_B\downarrow_C>$, $|\uparrow_A\downarrow_B\uparrow_C>$, $|\uparrow_A\uparrow_B\downarrow_C>$, $|\uparrow_A\uparrow_B\uparrow_C>$. For the sake of brevity, we will write them as $|\downarrow\downarrow\downarrow>$, $|\downarrow\downarrow\uparrow>$, $|\downarrow\uparrow\downarrow>$, $|\downarrow\uparrow\uparrow>$, $|\uparrow\downarrow\downarrow>$, $|\uparrow\downarrow\uparrow>$, $|\uparrow\uparrow\downarrow>$, $|\uparrow\uparrow\uparrow>$. These eight basis functions form a complete orthonormal set. The matrix elements $<\phi_\alpha|H_{Heisenberg}|\phi_\beta>$ are summarized in the matrix below [16]. Here the ϕ-s are the 3-electron basis states enumerated above.

$$\begin{pmatrix} D_1 & 0 & 0 & 0 & 0 & 0 & 0 & 0 \\ 0 & D_2 & 2J & 0 & 0 & 0 & 0 & 0 \\ 0 & 2J & D_3 & 0 & 2J & 0 & 0 & 0 \\ 0 & 0 & 0 & D_4 & 0 & 2J & 0 & 0 \\ 0 & 0 & 2J & 0 & D_5 & 0 & 0 & 0 \\ 0 & 0 & 0 & 2J & 0 & D_6 & 2J & 0 \\ 0 & 0 & 0 & 0 & 0 & 2J & D_7 & 0 \\ 0 & 0 & 0 & 0 & 0 & 0 & 0 & D_8 \end{pmatrix},$$

where

$$D_1 = 2J - h_A - h_C - 3Z$$
$$D_2 = -h_A + h_C - Z$$
$$D_3 = -2J - h_A - h_C - Z$$
$$D_4 = -h_A + h_C + Z$$
$$D_5 = h_A - h_C - Z$$
$$D_6 = -2J + h_A + h_C + Z$$
$$D_7 = h_A - h_C + Z$$
$$D_8 = 2J + h_A + h_C + 3Z. \tag{15.3}$$

In the above matrix, Z is one-half of the Zeeman splitting energy associated with the global magnetic field, while h_A and h_C are the energies supplied at the inputs. If the local magnetic field providing input to an isolated dot aligns the spin in the direction of the global magnetic field (down-spin direction), then the corresponding h is positive. Otherwise, it is negative. The quantity J in Equation (15.3) is always positive.[†]

[†]If we evaluate J from the overlap of the wavefunctions in neighboring dots, it will turn

We will evaluate the 8 eigenenergies E_n $(n = 1....8)$ by finding the eigenvalues of the 8×8 matrix above and the corresponding eigenstates

$$\psi_n = c_1^n |\downarrow\downarrow\downarrow\rangle + c_2^n |\downarrow\downarrow\uparrow\rangle + c_3^n |\downarrow\uparrow\downarrow\rangle + c_4^n |\downarrow\uparrow\uparrow\rangle$$
$$+ c_5^n |\uparrow\downarrow\downarrow\rangle + c_6^n |\uparrow\downarrow\uparrow\rangle + c_7^n |\uparrow\uparrow\downarrow\rangle + c_8^n |\uparrow\uparrow\uparrow\rangle$$
$$= [c_1^n, c_2^n, c_3^n, c_4^n, c_5^n, c_6^n, c_7^n, c_8^n]. \tag{15.4}$$

We will repeat this for four cases: $h_A = \pm h$ and $h_C = \pm h$. These four cases correspond to the four entries in the truth table of the NAND gate.

Case I: $h_A = h_C = h > 0$: This is the case when input bits [1] and [1] are written in dots A and C by the local magnetic fields. In this case, the eigenenergies and eigenstates are obtained by diagonalizing the matrix in Equation (15.3). They are given in Table 15.2.

TABLE 15.2
Eigenenergies and eigenstates when inputs are $[1, 1]$

Eigenenergy	Eigenstate
$-J - h - Z - \Delta_1$	$[0, 2/\Upsilon_1, \Lambda_1/(J\Upsilon_1), 0, 2/\Upsilon_1, 0, 0, 0]$
$2J - 2h - 3Z$	$[1, 0, 0, 0, 0, 0, 0, 0]$
$-J + h + Z - \Delta_2$	$[0, 0, 0, 2/\Upsilon_3, 0, -\Lambda_3/(J\Upsilon_3), 2/\Upsilon_3, 0]$
$-Z$	$[0, 1/\sqrt{2}, 0, 0, -1/\sqrt{2}, 0, 0, 0]$
Z	$[0, 0, 0, -1/\sqrt{2}, 0, 0, 1/\sqrt{2}, 0]$
$-J - h - Z + \Delta_1$	$[0, 2/\Upsilon_6, \Lambda_6/(J\Upsilon_6), 0, 2/\Upsilon_6, 0, 0, 0]$
$-J + h + Z + \Delta_2$	$[0, 0, 0, 2/\Upsilon_7, 0, -\Lambda_7/(J\Upsilon_7), 2/\Upsilon_7, 0]$
$2J + 2h + 3Z$	$[0, 0, 0, 0, 0, 0, 0, 1]$

where

$$\Delta_1 = \sqrt{(h + J)^2 + 8J^2},$$
$$\Delta_2 = \sqrt{(h - J)^2 + 8J^2},$$
$$\Lambda_1 = -J - h - \Delta_1,$$
$$\Lambda_3 = J - h + \Delta_2,$$
$$\Lambda_6 = -J - h + \Delta_1,$$
$$\Lambda_7 = J - h - \Delta_2$$
$$\Upsilon_n = \left[8 + (\Lambda_n/J)^2\right]^{1/2}. \tag{15.5}$$

---

out to be negative (see Chapter 10). However, refer back to Equation (10.85), where the Heisenberg Hamiltonian was written as $\widehat{H} = -\frac{1}{2} \sum_{i<j} J_{ij} \vec{\sigma}_i \cdot \vec{\sigma}_j$. Since we have written the Hamiltonian in Equation (15.2) in the form $+ \sum_{i<j} J_{ij} \vec{\sigma}_i \cdot \vec{\sigma}_j$, our J has to be positive for consistency. A positive J makes the singlet state lower in energy than the triplet state, which is the correct result.

In the above table, the eigenenergies are arranged in ascending order (i.e., the first entry is the ground state and the last entry is the highest excited state), *provided $h \gg J$ and $J > Z/2$*. The last inequality merely ensures that the first excited state has a higher energy than the ground state. The reason for the first inequality will become clear shortly.

Note that the ground state wavefunction is the entangled state.[‡]

$$\psi^{11}_{ground} = \frac{2}{\Upsilon_1}| \downarrow\downarrow\uparrow> + \frac{\Lambda_1}{J\Upsilon_1}| \downarrow\uparrow\downarrow> + \frac{2}{\Upsilon_1}| \uparrow\downarrow\downarrow> . \tag{15.6}$$

However, when the two inputs are [1, 1], we want the output to be [0], since this is the situation shown in Fig. 15.1(a) and conforms to the truth table of a NAND gate. Therefore, the *desired* ground state is the unentangled state

$$\psi^{11}_{desired} = | \downarrow\uparrow\downarrow>, \tag{15.7}$$

which corresponds to $A = 1$, $C = 1$ and $B = 0$, as shown in Fig. 15.1(a).

From Equation (15.6) it is obvious that we can make $\psi^{11}_{ground} \approx \psi^{11}_{desired}$ if

$$\left|\frac{\Lambda_1}{2J}\right| = \frac{h + J + \sqrt{(h + J)^2 + 8J^2}}{2J} \gg 1, \tag{15.8}$$

i.e., if $h \gg J$. In other words, *the ground state will approximate* the state in Fig. 15.1(a) if we make $h \gg J$. That means if we apply the inputs [1, 1] to the input dots A and C and let the system relax thermodynamically to the ground state (by emitting phonons, etc.), it will reach the state in Fig. 15.1(a) where the output bit (in dot B) will be [0] and we will have realized the first entry in the truth table of the NAND gate. All that is required is $h \gg J$.

Case II: $h_A = h_C = -h < 0$: This is the case when input bits [0 0] are written in dots A and C by the local magnetic fields. In this case, the eigenenergies and eigenstates are given in Table 15.3.

[‡]An entangled state is one which cannot be written as the tensor product of constituents. The famous example is the so-called singlet state of two electron spins $(1/\sqrt{2})[| \uparrow\downarrow> -| \downarrow\uparrow>]$ which cannot be written in terms of the tensor product $| \uparrow> \otimes | \downarrow>$ or $| \downarrow> \otimes | \uparrow>$. In plain English, the singlet state indicates that one spin is "up" and the other is "down," but it does not tell us which is up and which is down. Each has a 50% probability of being up or down. The tensor product, on the other hand, tells us that the first spin is up and the second spin is down, or the other way around.

Entangled states remain entangled forever. If one of the two electrons went to New York and the other to Tokyo, and a measurement at Tokyo told us that the spin measured there was down, then it immediately tells us that the spin in New York is up. Information about the New York spin travels to Tokyo instantaneously with infinite speed, apparently violating the tenets of Einstein's relativity. This was the basis of the famous Einstein-Podolsky-Rosen paradox. It has led to deep questions about quantum non-locality and is the basis of Bell's theorem addressing such issues as hidden variables in quantum mechanics. The entangled state is actually used in quantum teleportation. We shall revisit this matter in the next chapter.

TABLE 15.3

Eigenenergies and eigenstates when inputs are $[0, 0]$

Eigenenergy	Eigenstate
$-J - h + Z - \Delta_1$	$[0, 0, 0, 2/\Upsilon_1, 0, \Lambda_1/(J\Upsilon_1), 2/\Upsilon_1, 0]$
$2J - 2h + 3Z$	$[0, 0, 0, 0, 0, 0, 0, 1]$
$-J + h - Z - \Delta_2$	$[0, 2/\Upsilon_3, -\Lambda_3/(J\Upsilon_3), 0, 2/\Upsilon_3, 0, 0, 0]$
$-Z$	$[0, -1/\sqrt{2}, 0, 0, 1/\sqrt{2}, 0, 0, 0]$
Z	$[0, 0, 0, -1/\sqrt{2}, 0, 0, 1/\sqrt{2}, 0]$
$-J - h + Z + \Delta_1$	$[0, 0, 0, 2/\Upsilon_6, 0, \Lambda_6/(J\Upsilon_6), 2/\Upsilon_6, 0]$
$-J + h - Z + \Delta_2$	$[0, 2/\Upsilon_7, -\Lambda_7/(J\Upsilon_7), 0, 2/\Upsilon_7, 0, 0, 0]$
$2J + 2h - 3Z$	$[1, 0, 0, 0, 0, 0, 0, 0]$

It is obvious that Table 15.3 can be derived from Table 15.2 by simply replacing h with $-h$.

The ground state wavefunction is the entangled state

$$\psi^{00}_{ground} = \frac{2}{\Upsilon_1}| \downarrow\uparrow\uparrow> + \frac{\Lambda_1}{J\Upsilon_1}| \uparrow\downarrow\uparrow> + \frac{2}{\Upsilon_1}| \uparrow\uparrow\downarrow>, \qquad (15.9)$$

whereas the desired state shown in Fig. 15.1(b) is

$$\psi^{00}_{desired} = | \uparrow\downarrow\uparrow> . \qquad (15.10)$$

Once again, we can make $\psi^{00}_{desired} \approx \psi^{00}_{ground}$, if we make $h >> J$. Therefore, if we apply inputs $[0\ 0]$ to dots A and C and let the system relax to the ground state, dot B will have output $[1]$ corresponding to Fig. 15.1(b), and we will have realized the second entry in the truth table of the NAND gate, provided $h >> J$.

Case III: $-h_A = h_C = h > 0$: This is the case when input bits 0 and 1 are written in dots A and C by the local magnetic fields. In this case, the eigenenergies and eigenstates are given in Table 15.4

TABLE 15.4

Eigenenergies and eigenstates when inputs are [0, 1]

Eigenenergy	Eigenstate
$-\Theta_4/2 - 2J/3 - Z + \sqrt{3}i/2\Theta_3$	$[0, \Pi_3^{(1)}/(J^2\Pi_4^{(1)}), 2\Pi_1^{(1)}/(J\Pi_4^{(1)}), 0, 4/\Pi_4^{(1)}, 0, 0, 0]$
$-\Theta_4/2 - 2J/3 + Z + \sqrt{3}i/2\Theta_3$	$[0, 0, 0, \Pi_3^{(2)}/(J^2\Pi_4^{(2)}), 0, 2\Pi_1^{(2)}/(J\Pi_4^{(2)}), 4/\Pi_4^{(2)}, 0]$
$-\Theta_4/2 - 2J/3 - Z - \sqrt{3}i/2\Theta_3$	$[0, \Pi_3^{(3)}/(J^2\Pi_4^{(3)}), 2\Pi_1^{(3)}/(J\Pi_4^{(3)}), 0, 4/\Pi_4^{(3)}, 0, 0, 0]$
$-\Theta_4/2 - 2J/3 + Z - \sqrt{3}i/2\Theta_3$	$[0, 0, 0, \Pi_3^{(4)}/(J^2\Pi_4^{(4)}), 0, 2\Pi_1^{(4)}/(J\Pi_4^{(4)}), 4/\Pi_4^{(4)}, 0]$
$2J - 3Z$	$[1, 0, 0, 0, 0, 0, 0, 0]$
$2J + 3Z$	$[0, 0, 0, 0, 0, 0, 0, 1]$
$\Theta_4 - 2J/3 - Z$	$[0, \Pi_3^{(7)}/(J^2\Pi_4^{(7)}), 2\Pi_1^{(7)}/(J\Pi_4^{(7)}), 0, 4/\Pi_4^{(7)}, 0, 0, 0]$
$\Theta_4 - 2J/3 + Z$	$[0, 0, 0, \Pi_3^{(8)}/(J^2\Pi_4^{(8)}), 0, 2\Pi_1^{(8)}/(J\Pi_4^{(8)}), 4/\Pi_4^{(8)}, 0]$

where

$$\Theta_1 = J\left[9(h/J)^2 - 10 + 3i\sqrt{3(h/J)^6 + 12(h/J)^4 + 69(h/J)^2 + 27}\right]^{1/3},$$

$$\Theta_2 = -\frac{4J^2}{3\Theta_1}\left[\left(\frac{h}{J}\right)^2 + \frac{7}{3}\right],$$

$$\Theta_3 = \frac{2\Theta_1}{3} + \frac{3\Theta_2}{2} = 2iIm\left(\frac{2\Theta_1}{3}\right),$$

$$\Theta_4 = \frac{2\Theta_1}{3} - \frac{3\Theta_2}{2} = 2Re\left(\frac{2\Theta_1}{3}\right),$$

$$\Pi_1^{(1)} = -\Theta_4/2 - 2J/3 + 2h + (\sqrt{3}i/2)\Theta_3,$$

$$\Pi_2^{(1)} = -\Theta_4/2 - 2J/3 - Z + (\sqrt{3}i/2)\Theta_3,$$

$$\Pi_3^{(1)} = \left[\Pi_2^{(1)}\right]^2 + 2\Pi_2^{(1)}(Z + J + h) + 4Jh + 2JZ - 4J^2 + Z^2 + 2hZ,$$

$$\Pi_4^{(1)} = \left[\frac{[\Pi_3^{(1)}]^2}{J^4} + \frac{4[\Pi_1^{(1)}]^2}{J^2} + 16\right]^{1/2},$$

$$\Pi_1^{(2)} = \Pi_1^{(1)},$$

$$\Pi_2^{(2)} = \Pi_2^{(1)} + 2Z,$$

$$\Pi_3^{(2)} = \left[\Pi_2^{(2)}\right]^2 + 2\Pi_2^{(1)}(-Z + J + h) + 4Jh - 2JZ - 4J^2 + Z^2 - 2hZ,$$

$$\Pi_4^{(2)} = \left[\frac{[\Pi_3^{(2)}]^2}{J^4} + \frac{4[\Pi_1^{(2)}]^2}{J^2} + 16\right]^{1/2},$$

$$\Pi_1^{(3)} = \Pi_1^{(1)} - \sqrt{3}i\Theta_3,$$

$$\Pi_2^{(3)} = \Pi_2^{(1)} - \sqrt{3}i\Theta_3,$$

$$\Pi_3^{(3)} = \left[\Pi_2^{(3)}\right]^2 + 2\Pi_2^{(3)}(Z + J + h) + 4Jh + 2JZ - 4J^2 + Z^2 + 2hZ,$$

$$\Pi_4^{(3)} = \left[\frac{[\Pi_3^{(3)}]^2}{J^4} + \frac{4[\Pi_1^{(3)}]^2}{J^2} + 16 \right]^{1/2},$$

$$\Pi_1^{(4)} = \Pi_1^{(2)} - \sqrt{3}i\Theta_3,$$

$$\Pi_2^{(4)} = \Pi_2^{(2)} - \sqrt{3}i\Theta_3$$

$$\Pi_3^{(4)} = \left[\Pi_2^{(4)} \right]^2 + 2\Pi_2^{(4)}(-Z + J + h) + 4Jh - 2JZ - 4J^2 + Z^2 - 2hZ,$$

$$\Pi_4^{(4)} = \left[\frac{[\Pi_3^{(4)}]^2}{J^4} + \frac{4[\Pi_1^{(4)}]^2}{J^2} + 16 \right]^{1/2},$$

$$\Pi_1^{(7)} = \Theta_4 - 2J/3 + 2h,$$

$$\Pi_2^{(7)} = \Theta_4 - 2J/3 - Z,$$

$$\Pi_3^{(7)} = \left[\Pi_2^{(7)} \right]^2 + 2\Pi_2^{(7)}(Z + J + h) + 4Jh + 2JZ - 4J^2 + Z^2 + 2hZ,$$

$$\Pi_4^{(7)} = \left[\frac{[\Pi_3^{(7)}]^2}{J^4} + \frac{4[\Pi_1^{(7)}]^2}{J^2} + 16 \right]^{1/2},$$

$$\Pi_1^{(8)} = \Pi_1^{(7)},$$

$$\Pi_2^{(8)} = \Pi_2^{(7)} + 2Z,$$

$$\Pi_3^{(8)} = \left[\Pi_2^{(8)} \right]^2 + 2\Pi_2^{(8)}(-Z + J + h) + 4Jh - 2JZ - 4J^2 + Z^2 - 2hZ,$$

$$\Pi_4^{(8)} = \left[\frac{[\Pi_3^{(8)}]^2}{J^4} + \frac{4[\Pi_1^{(8)}]^2}{J^2} + 16 \right]^{1/2},$$

where Im stands for the imaginary part, Re stands for the real part, and i is the imaginary square root of –1.

It can be verified that even though the Θ-s are complex, the energy eigenvalues in Table 15.4 are all real, as they must be since they are eigenvalues of a Hermitian matrix.

The ground state wavefunction is the entangled state

$$\psi_{ground}^{01} = \Pi_3^{(1)}/(J^2\Pi_4^{(1)})| \downarrow\downarrow\uparrow> +2\Pi_1^{(1)}/(J\Pi_4^{(1)})| \downarrow\uparrow\downarrow> +4/\Pi_4^{(1)}| \uparrow\downarrow\downarrow>, \tag{15.11}$$

whereas the desired state shown in Fig. 15.1(c) is

$$\psi_{desired}^{01} = | \uparrow\downarrow\downarrow> . \tag{15.12}$$

Once again, we can make $\psi_{desired}^{01} \approx \psi_{ground}^{01}$, if we make $h >> J$. Therefore, if we apply inputs [0 1] to dots A and C and let the system relax to the ground state, dot B will have output [1] corresponding to Fig. 15.1(c), and we will have realized the third entry in the truth table of the NAND gate. As before, all we need for this to happen is $h >> J$.

Case IV: $-h_A = h_C = -h < 0$: This is the case when input bits 1 and 0 are written in dots A and C by the local magnetic fields. In this case, the eigenenergies and eigenstates are given by Table 15.4 with h replaced with $-h$. The eigenenergies do not change since they depend on h^2 and are therefore insensitive to the sign of h. However, the eigenstates are sensitive to the sign of h, and change.

The ground state wavefunction is

$$\psi_{ground}^{10} = \hat{\Pi}_3^{(1)}/(J^2\hat{\Pi}_4^{(1)})|\downarrow\downarrow\uparrow> +2\hat{\Pi}_1^{(1)}/(J\hat{\Pi}_4^{(1)})|\downarrow\uparrow\downarrow> +4/\hat{\Pi}_4^{(1)}|\uparrow\downarrow\downarrow>, \tag{15.13}$$

where we have used the "hats" for distinction, i.e., to remind the reader that $\hat{\Pi} \neq \Pi$ since the Π-s are sensitive to the sign of h.

The desired state shown in Fig. 15.1(d) is

$$\psi_{desired}^{10} = |\downarrow\downarrow\uparrow> . \tag{15.14}$$

Once again, we can make $\psi_{desired}^{10} \approx \psi_{ground}^{10}$, if we make $h >> J$. Therefore, if we apply inputs [1 0] to dots A and C and let the system relax to the ground state, dot B will have output [1] corresponding to Fig. 15.1(d), and we will have realized the fourth and final entry in the truth table of the NAND gate. All we need is $h >> J$.

15.3.3   Ground state computing with spins

What we have shown in the previous section is that, if we provide binary inputs to the two peripheral spins in the 3-spin array using local magnetic fields applied selectively to the input dots, and then let the array relax to the ground state thermodynamically (e.g., by emitting phonons), then we will realize the operation of the NAND gate. This is an important result since the NAND gate is *universal*, meaning that any Boolean logic circuit can be implemented with NAND gates alone. This idea of relaxation to the ground state has some connection with the idea of artificial neural networks, where a system is allowed to relax to ground state to compute.

The ground state computing notion appears in many contexts. One particularly relevant scheme was proposed by Pradeep Bakshi and co-workers in 1991 [17], where they considered an array of "quantum dashes", each with a single electron. This is shown in Fig. 15.2. Because of Coulomb repulsion, electrons in neighboring dashes are pushed towards opposite corners in antipodal positions. Therefore, the ground state of the array has spontaneous polarization and assumes an anti-ferroelectric configuration, just like our arrays tend to assume an anti-ferromagnetic configuration. Any three dashes will realize a NAND gate [18] in a way very similar to our construct. Two peripheral dashes are input ports and the central dash is the output port. Obviously, when the inputs are [0 0] and [1 1], the output is [1] and [0], respectively. When the inputs are [0 1] or [1 0], there is a tie, which can be resolved in favor of the output bit being [1], by applying a weak global biasing dc electric field, much like our global dc magnetic field.

The ground state computing scheme, however, has some problems. We now discuss some of them.

Error probability

It is the natural tendency of any physical system to relax to the ground state, which is the basis of ground state computing. However, once a system relaxes to ground state, it need not stay there forever. If it gets out of the ground state, and it does [19], it will produce wrong results and cause errors in computation.

Consider a system which is thermodynamically coupled to its environment (phonon bath), which allows it to relax to the ground state, after it is provided with inputs. Once the computation is over and the system has attained equilibrium with the environment, the probability of finding the system in any state (including the ground state) with energy ϵ_n, at a temperature T, is given by the Fermi-Dirac occupation probability $1/\exp[(\epsilon_n - E_F)/kT + 1]$ where E_F is the Fermi level. This probability is *not* unity for the ground state. If we approximate the Fermi-Dirac statistics with the Boltzmann statistics, then the relative probability of being in two states E_p and E_q is $\exp[-(E_p - E_q)/kT]$. If we designate the ground state energy as E_{ground} and an excited state energy as $E_{excited}$, then the ratio of the probabilities of being in the excited state and the ground state is $\exp[-(E_{excited} - E_{ground})/kT]$. We can call this the *intrinsic error probability* P_{error} associated with being in the excited state, since straying from the ground state into any excited state causes an error in the result. The total error probability is the sum of P_{error} carried out over all excited states.

Case I: inputs are [1 1]: Here $E_1 - E_{ground} \approx 4J - 2Z$ and $E_2 - E_{ground} \approx 2h + 2Z + 2J \approx 2h$ if we assume that $h >> J, Z$. Because of the last inequality, we only have to worry about the first excited state E_1, since the second excited state E_2 is so far above in energy than E_1 that the probability of straying into the second excited state is negligible compared to the probability of straying into the first excited state.

Case II: inputs are [0 0]: Here $E_1 - E_{ground} \approx 4J + 2Z$ and $E_2 - E_{ground} \approx 2h - 2Z + 2J \approx 2h$ if we assume that $h >> J, Z$. Therefore, the same considerations as Case I apply.

Case III and IV: inputs are [1 0] or [0 1]: In these cases also, we need to worry only about the first excited states, since the second excited states are far above in energy than the first excited states. Here, $E_1 - E_{ground} = 2Z$.

The total intrinsic error probability is $P_{error} = p \approx \exp[-(E_1 - E_{ground})/kT]$. Therefore, $E_1 - E_{ground} = kT ln(1/p)$.

Considering the four cases above, it is obvious that the following conditions must be fulfilled: (i) in cases III and IV, $Z = (1/2)kT ln\left(1/p^{III}\right)$ and $Z =$

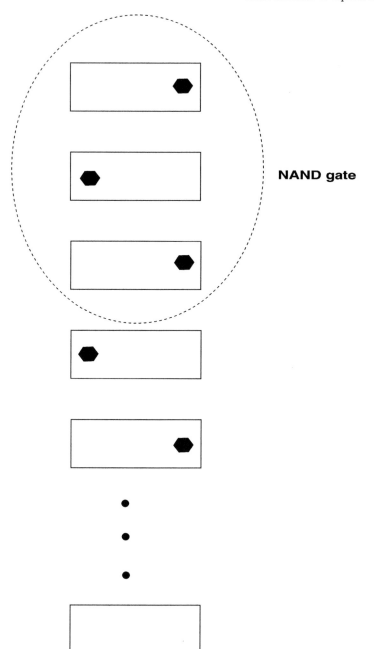

FIGURE 15.2

An array of quantum dashes with nearest neighbor Coulomb coupling. Each dash has a single electron. Coulomb repulsion pushes electrons in neighboring dashes to opposite corners so that the array assumes an anti-ferroelectric order. Three consecutive dashes form a NAND gate. After [17].

$(1/2)kT ln \left(1/p^{IV}\right)$, respectively. Here, p^{III} is the error probability in case III and p^{IV} is the error probability in case IV. This result shows that $p^{III} = p^{IV} = p$; (ii) in case I, $4J - 2Z = 4J - kT ln(1/p) = kT ln \left(1/p^{I}\right)$. Here, p^{I} is the error probability in case I; and finally (iii) in case II, $4J + 2Z = 4J + kT ln(1/p) = kT ln \left(1/p^{II}\right)$. Here, p^{II} is the error probability in case II.

If we want the same error probability in cases I, III, and IV, then $p^{I} = p^{III} = p^{IV} = p$. Condition (ii) then yields that $J = (1/2)kT ln(1/p)$, which, along with condition (i), implies that $J = Z$ if the error probabilities in cases I, III, and IV are to be the same. Condition (iii) then yields that $3 ln(1/p) = ln(1/p^{II})$, or $p^{II} = p^3$, which means that the error probability in case II is much smaller than in cases I, III, or IV. The condition $J = Z = (1/2)kT ln(1/p)$ determines the value of J or Z required to maintain a given error probability p at a temperature T.

Let us say that we can tolerate an error probability p of 3% since it is possible to correct for this error rate using elaborate and sophisticated error correction schemes used in quantum computing [1]. Therefore, $J/(kT) = Z/(kT) = (1/2) ln(1/.03) = 1.75$. Present technology can produce $J \sim 1$ meV with semiconductor quantum dots [20]. Therefore, $T \approx 6.5$ K. Molecular systems can produce $J \sim 6$ meV [21], which raises the temperature of operation to 39 K. In order to operate at room temperature, we will need $J = 46$ meV, which is extremely challenging and probably unattainable in the near term.

Handling an error probability of 3% is possible, but requires immense resources. If we cannot afford that and can only tolerate an error probability of 10^{-9} typical of ultra large scale integrated circuits today, then $T \approx 1.1$ K, as long as $J \sim 1$ meV. Operating at the liquid helium-4 temperature of 4.2 K will require increasing J to about 4 meV, which is possible with molecular systems.

The Issue of Unidirectionality

There is, however, another serious issue in ground state computing which may not be apparent at first. One of the requirements for logic circuits is "unidirectionality," i.e., logic signals must propagate unidirectionally from the input port to the output port *and not the other way around* [3]. Device engineers call this property "isolation" between the input and output. Traditional transistors have inherent isolation between input and output,§ but in SSL, there is unfortunately no isolation between the input dot and output dot of the logic gate since exchange interaction is bidirectional. Consider just two exchange coupled spins in two neighboring quantum dots. Their ground state will be the singlet state (recall the discussion of parahydrogen in Chapter 10)

---

§For example, in SPINFETs, if we change the gate voltage, it will alter the source-to-drain current, but any change in the source-to-drain current does not affect the gate voltage. The interaction is one-way, meaning that there is isolation between the input (gate) terminal and the output (drain) terminal.

and therefore this system will act as a natural NOT gate if one spin is the input and the other is the output [the output is always the logic complement of the input]. However, exchange interaction, being bidirectional, cannot distinguish between which spin is the input bit and which is the output. The input will influence the output just as much as the output influences the input. If something spurious or extraneous (such as noise) changes the output, it will change the input! The master-slave relation between input and output is lost. Since the input and output are indistinguishable, it becomes ultimately impossible for logic signal to flow unidirectionally from an input stage to an output stage and not the other way around.

Of course, one can forcibly impose unidirectionality in some (but not all) cases by holding the input spin in a fixed orientation until the desired output spin orientation is produced in the output dot. Once the *final* output has been harvested, it does not matter if things go astray. We can release the input once the final output has been produced. In that case, the input signal itself enforces unidirectionality because it is a symmetry-breaking influence. This approach was actually used to demonstrate a device called a "magnetic cellular automaton" where the input enforced unidirectionality and produced the correct final output [22]. However, there are problems. First, this approach can only work for a small number of bits before the influence of the input dies out. Second, and more important, the input cannot be changed until the final output has been produced since otherwise the correct output may not be produced at all. That makes such architectures *non-pipelined* and therefore unacceptably slow. There may also be additional problems associated with random errors when this approach is employed. Because of the lack of unidirectionality, the system can get stuck in metastable states [23], which could prevent relaxation to the ground state. These issues have been discussed in [19].

In our case, we have to hold the inputs in dots A and C at their intended states until the output in dot B has been produced. In other words, we cannot let go of the inputs in one particular gate until the output in *that* gate has been produced, but we do not have to hold the inputs in place until the outputs in all *succeeding* gates (or, in other words, the final outputs) have been produced. Therefore, pipelining is possible in this approach, if we use some clever scheme to impose unidirectionality. This is discussed next.

Unidirectionality in time: Clocking

If unidirectionality cannot be imposed in *space* (because exchange interaction is bidirectional), then it must be imposed in *time*. This is accom-

---

¶Pipelined architectures are those where a new input can be provided to the computing machinery before the output in response to a previous input has been generated. One does not have to wait for the output to be generated before a new input can be provided. This increases the processing speed vastly. All modern computing architectures allow pipelining.

plished by using clocking to activate successive stages sequentially in time [24]. This strategy is routinely adopted in bucket brigade devices, such as charge-coupled-device (CCD) shift registers [25], where a push clock and a drop clock are used to lower and raise barriers between neighboring charge coupled devices and thus steer a charge packet unidirectionally from one device to the next. In single spin circuits, we can do the same. We will delineate a gate pad between every two neighboring quantum dots and apply a clock signal to this pad. This is shown in Fig. 15.3. During the positive clock cycle, a positive potential will appear over the potential barrier isolating two neighboring quantum dots which will lower the barrier temporarily to make the wavefunctions overlap and exchange couple these two spins. Then during the negative clock cycle, the barrier is raised again to decouple the two spins. In this way, pairs of spin bits can be coupled sequentially in time and logic information transferred unidirectionally from one dot to the next in a bucket brigade fashion. However, there are some subtle issues. Bandyopadhyay. [26] showed that a single phase clock does not work; instead, a 3-phase clock is required to do this job.

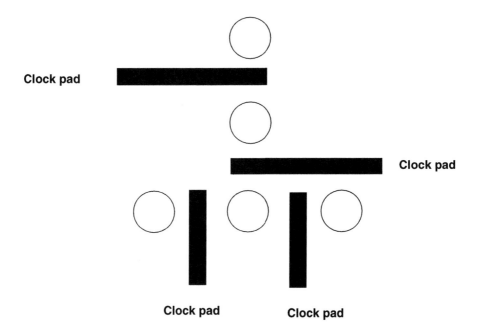

Clock pad

Clock pad

Clock pad   **Clock pad**

FIGURE 15.3

A clock pad is inserted between neighboring dots. A positive potential appearing at a clock pad will lower the potential barrier between the two flanking dots and cause exchange coupling. A 3-phase clock can propagate signals unidirectionally.

15.4 Energy dissipation issues

15.4.1 Energy dissipated in the gate during switching

Consider the ground state energies in Cases I – IV that we considered. These energies are

$$E^I_{ground} = -J - h - Z - \Delta_1,$$
$$E^{II}_{ground} = -J - h + Z - \Delta_1,$$
$$E^{III}_{ground} = E^{IV}_{ground} = -Re\left(\frac{2\Theta_1}{3}\right) - \frac{2J}{3} - Z - \sqrt{3}Im\left(\frac{2\Theta_1}{3}\right),$$

$$(15.15)$$

where Re and Im stand for real and imaginary parts.

When $h >> J, Z$,

$$E^I_{ground} = -2J - 2h - Z,$$
$$E^{II}_{ground} = -2J - 2h + Z,$$
$$E^{III}_{ground} = E^{IV}_{ground} = -2h - 2J/3 - Z. \qquad (15.16)$$

Therefore, when we switch between any two of the four states in Fig. 15.1, the maximum energy that can be dissipated is $2Z$. We saw that this energy is equal to $kTln(1/p)$. Therefore, the maximum energy dissipated in switching the NAND gate is $kTln(1/p)$. With $p = 10^{-9}$, it is ~ 21 kT, whereas modern electronic transistors dissipate several thousands of kT during switching. At a temperature of 1 K, the energy dissipated is 3×10^{-22} J. Even with a superfast 50 GHz clock, the corresponding dissipated power is about 15 pW/gate. This will allow us to continue to shrink electronic devices well into the next few decades, i.e., continue the fabled Moore's law,[||] without having to worry about excessive power dissipation and thermal management on the chip.

The dissipated energy of $kTln(1/p)$ is actually the minimum energy that needs to be dissipated to switch a NAND gate if we do not wish to exceed an error probability of p. This is the famous *Landauer-Shannon result* [27], which stipulates that in a *logically irreversible* operation like the NAND operation, the energy dissipated is $kTln(1/p)$, where p is the error probability. If $p > 0.5$, then the distinguishability between bits 0 and 1 is lost. Therefore, the

---

[||]Moore's law is an empirical law postulated by Gordon Moore, one of the founders of Intel Corporation, which states that the device density on a chip will roughly double every 18 months. In order for this to remain feasible, the energy dissipation per device should also reduce by approximately 50% every 18 months. Otherwise, the power dissipation (per unit area) on the chip will continue to increase and stretch the limits of heat sinking technology.

absolute minimum energy that must be dissipated to switch any irreversible logic gate is $kTln2$.

Landauer also showed that it is possible to switch *logically reversible* gates without dissipating any energy at all, but at the expense of losing error correction capability [27]. Logically reversible gates are those where the inputs can be inferred from a knowledge of the outputs, meaning that the logic function implemented by the gate has a unique inverse. An inverter is logically reversible, since the input is always the logic complement of the output. Therefore, knowledge of the output is sufficient to tell us unambiguously what the input was. However, a NAND gate is logically irreversible. When the output is [0], we can tell precisely what the inputs were since they must have been [1 1]. However, when the output is [1], we cannot tell with certainty what the inputs were. They could have been [0 0], or [0 1], or [1 0]. Therefore, a NAND gate is logically irreversible. There is no way to switch such a gate without dissipating energy.

Exercise: A simple proof of the Landauer-Shannon result

The energy dissipated in changing the state of a system which is in equilibrium with the environment (phonon bath) is

$$\Delta E = kT\Delta S + S\Delta(kT), \tag{15.17}$$

where S is the entropy.

According to the second law of thermodynamics, $\Delta(kT) \geq 0$, so that

$$\Delta E_{min} = kT\Delta S. \tag{15.18}$$

Shannon's definition of entropy is

$$S = \sum_{i}^{N} p_i ln(p_i), \tag{15.19}$$

where the system in question can exist in N states and p_i is the probability of being in the i-th state.

A binary system has only two states $(i = 0, 1)$ and $N = 2$. Therefore,

$$S = p_{[0]} ln\left(p_{[0]}\right) + p_{[1]} ln\left(p_{[1]}\right). \tag{15.20}$$

After switching the system, let us say that we find it is in state [1] so that $p_{[1]}^{final} = 1$ and $p_{[0]}^{final} = 0$. Therefore,

$$S_{final} = 0ln(0) + 1ln(1) = 0. \tag{15.21}$$

If the gate is *logically irreversible* and we could not infer the input state from knowledge of the output, then $p_{[0]}^{initial} = p_{[1]}^{initial} = 0.5$. In that case,

$$S_{final} = 0.5ln(0.5) + 0.5ln(0.5) = -ln(2). \tag{15.22}$$

Therefore, $\Delta S = S_{final} - S_{initial} = ln(2)$. Substituting this in Equation (15.18), we get the Landauer-Shannon result:

$$E_{min} = kTln2. \tag{15.23}$$

If the gate were *logically reversible*, i.e., we could infer the input from the output, then the situation would be different. Consider an inverter. If the output were bit 1, then input must have been bit 0. In that case, $S_{initial} = 1ln(1) + 0ln(0) = 0$, and $S_{final} = 0ln(0) + 0ln(0) = 0$. Therefore, $\Delta S = S_{final} - S_{initial} = 0$, and $\Delta E_{min} = 0$, meaning dissipationless switching is possible. In other words, a logically reversible operation can, in principle, be physically reversible and dissipate no energy at all. However, finding the physically reversible implementation may be extremely challenging and whether such an implementation can have any error-resilience at all is the subject of intense debate.

Physical interpretation of the Landauer-Shannon result

A bistable element used to store and process binary logic bits must have two stable (or metastable) states. Viewed from the perspective of an energy landscape, these two states will form two potential minima, as shown in Fig. 15.4. In order to prevent spontaneous transitions between these two states, which will cause bit errors, there must be an energy barrier separating these two minima. When the logic element is to be switched, the potential barrier must be lowered by some external agency so that one state becomes accessible from the other via elastic transitions.

Many authors have argued that since no energy need to be dissipated in lowering the barrier adiabatically, it should be possible to switch with zero dissipation. There are schemes involving time-modulated potentials that allow dissipationless switching following this notion, but ghey require exquisite synchronization in time. If the synchronization fails, the switch malfunctions. These schemes, therefore, have virtually no error-resilience and will be extremely sluggish because the barrier has to be lowered very slowly for the sake of adiabaticity. On the other hand, if we insist on abrupt and non-adiabatic switching, then we can lower the barrier quickly by making the double-well potential profile asymmetric in such a way that one well is lower than the other in energy by an amount at least equal to the barrier height. In this case, switching is both fast and certain because the system will inevitably migrate to the lower minimum, but the minimum energy dissipated in this process is the difference in the energy between the two minima, which is at least equal to th barrier height. Therefore, the minimum energy dissipated in non-adiabatic switching is the minimum barrier height that we have to maintain in order to suppress spontaneous transitions between the two potential wells. These transitions are caused by thermionic emission over the barrier. The probability of thermionic emission over a barrier of height E_b is $\exp[-E_b/kT]$ assuming

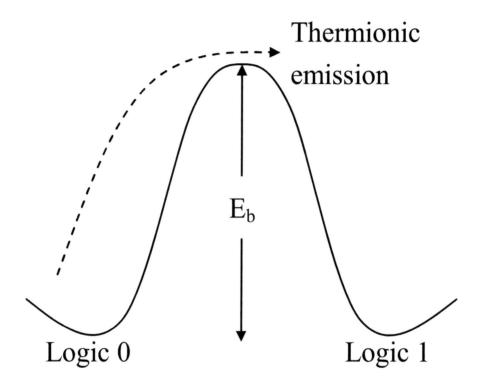

FIGURE 15.4
The potential energy landscape of a bistable device. An energy barrier E_b
separates the two potential minima which 'store' the two states. The barrier
is high and/or wide enough to suppress tunneling. Thermionic emission over
the barrier can cause spontaneous transitions between the two states resulting
in bit errors.

that the carriers are distributed in a Boltzmann distribution characterized by a carrier temperature T. If this probability approaches 0.5, then it is as likely that the logic element is in one state as the other, meaning that there is no preference to be in either state. The distinguishability of the two states will then be completely lost. Therefore, we need

$$e^{-\frac{E_b}{kT}} \leq 0.5 \tag{15.24}$$

or $E_b > kTln2$. This is the physical basis of the Landauer-Shannon limit on the minimum energy dissipated during switching.

It is obvious that there are many assumptions implicit in Equation (15.24). For example, we assumed that the carriers are in a Boltzmann distribution. They need not be. If they are driven far out of equilibrium, then they could be in a non-thermal distribution and Equation (15.24) will not apply. This situation has been examined in ref. [28], which reached the conclusion that it may be possible to perform irreversible logic operations while dissipating less than $kTln2$ amount of energy per bit flip, using non-equilibrium systems.

15.4.2 Energy dissipated in the clocking circuit

The clocking circuit that one needs to impose unidirectionality in SSL introduces additional dissipation. The energy dissipated in the clock pad is $(1/2)CV^2$ if the clock pulse is applied abruptly. It can be much less if applied in small steps (adiabatically) [29], but we will not consider that scenario here since it requires extreme precision. Appropriate pulse shaping can also reduce this energy dissipation, but usually at the expense of slower speed. Since we want to be conservative, we will assume the maximum dissipation of $(1/2)CV^2$. Here, C is the capacitance of the gate pad and V is the potential applied to the pad to lower the potential barrier between neighboring dots to exchange couple neighboring spins. The value of V may be as small as 10 mV.

15.5 Comparison between spin transistors and single-spin-processors

For the sake of comparison, we will assume that the capacitance of a gate pad in SSL is about the same as the gate capacitance of a SPINFET, which was estimated in the previous chapter to be 50 aF. We will carry out a comparison at 4.2 K, since that is the order of temperature required to achieve high spin injection efficiency in a SPINFET.

The energy dissipated in switching a single gate in SSL is $kTln(1/p)$. With an error probability of $p = 10^{-9}$ (achieved in modern integrated circuits), this energy is 7.5 meV at 4.2 K. We require that J, Z is one half of this, which is 3.75 meV. It is not impossible to achieve this magnitude of J [20]. In order to make $Z = 3.75$ meV, we need a magnetic field of only 0.07 Tesla if we use materials with giant g-factors (e.g., $InSb_xN_{1-x}$, which is reported to have a g-factor of 900 [30]). Of course, these materials may not be optimum for long spin relaxation times. Even if we can find materials with g-factors of ~ 10 that have long spin relaxation times, the required magnetic field is about 6 Tesla, which is easily achievable.

The energy dissipated in the clock is 2.5×10^{-21} J, while the energy dissipated in the gate is 7.5 meV $= 1.2 \times 10^{-21}$ J. Therefore, the total energy dissipated in an SSL gate during switching is $\sim 4 \times 10^{-21}$ J which is ~ 70 kT at 4.2 K temperature. While these numbers depend somewhat on structural and material parameters that were assumed, it is safe to assert that SSL – an example of single-spin processing – will be an extremely energy-efficient computing paradigm if implemented. Its drawbacks are that it may be able to work only at very low temperatures, may be unacceptably error-prone because of random spin flipping due to coupling with the environment, and the actual implementation is immensely challenging. There have been some proposals for implementing SSL with graphene nanoflakes with the hope that this would increase the temperature of operation and in the process make the architecture robust [31], but at the time of writing this chapter, there has been no hint of any experimental demonstration of SSL or a similar paradigm.

15.6   Concluding remarks

In the preceding material, we always calculated the excited and ground states of many-spin configurations using the Heisenberg model. We showed that the ground state spin configurations under different inputs always obey the truth table of the NAND gate. What we did not show *rigorously* is that when we switch the inputs, the output switches to the new ground state. We know that this should happen since any physical system tends to relax to the ground state, although the probability of remaining in the ground state is not unity and is given by Fermi-Dirac statistics. Examining the switching dynamics rigorously is a difficult proposition since it requires studying a quantum system in the presence of dissipation (coupling to phonons). Sophisticated theoretical techniques exist for the purpose, but are outside the scope of this book.

In the end, the real question is how practical is SSL? Recent advances in controlling and manipulating single electrons in quantum dots [32, 33, 34, 35, 36, 37, 38, 39] have now brought us to the point where a paradigm like SSL does not appear completely far-fetched. The drawback is the temperature of operation; it may have to be much lower than room temperature, which makes it unappealing. Low error-resilience is another possible shortcoming, but it could be mitigated with sophisticated (software-based) error correction schemes, just as in the case of spin-based quantum computing paradigms. In the end, if we ever reach the point where we must operate at the limit of lowest possible energy dissipation, SSL may emerge as an important approach. Until then, it will remain a theoretical curiosity. In Chapter 17, we will discuss a scaled-up version of SSL, where a single spin is essentially replaced with a single-domain ferromagnet acting as a giant single classical spin. It is holding out promise for an extremely energy-efficient logic paradigm that works at room temperature.

15.7   Problems

- **Problem 15.1**

  There are two sources of bit flip errors: (i) spins flipping spontaneously within a clock period due to spin relaxation mechanisms discussed in Chapter 8. This error probability is $p_1 = 1 - e^{-T/T_1}$, where T is the refresh clock period and T_1 is the spin relaxation time, and (ii) the system getting out of the ground state to the first excited state (we ignore the second and higher excited states since they are too far away in energy). That probability is $p_2 = \exp[-(E_1 - E_{ground})/kT]$. At what clock frequency will p_1 be equal to the maximum value of p_2 if the spin relaxation time is 10 msec, the global magnetic flux density is 4 Tesla and the quantum dot material has a g-factor of 20? Assume that the temperature is 4.2 K.

- **Problem 15.2**

  Let us say that the up-spin state is the $+z$-polarized state $[1\ 0]^{\dagger}$ and the down-spin state is the $-z$-polarized state $[0\ 1]^{\dagger}$. Show that

$$\sigma_z | \uparrow > = | \uparrow >,$$
$$\sigma_z | \downarrow > = -| \downarrow >,$$
$$J\sigma_{zA}\sigma_{zB} | \uparrow_A \downarrow_B > = -J | \uparrow_A \downarrow_B >,$$
$$J\sigma_{zA}\sigma_{zB} | \uparrow_A \uparrow_B > = J | \uparrow_A \uparrow_B >,$$
$$\sigma_x | \uparrow > = | \downarrow >,$$
$$\sigma_x | \downarrow > = | \uparrow >,$$
$$\sigma_y | \uparrow > = i | \downarrow >,$$
$$\sigma_y | \downarrow > = -i | \uparrow >,$$
$$J\sigma_{xA}\sigma_{xB} | \uparrow_A \downarrow_B > = J | \downarrow_A \uparrow_B >,$$
$$J\sigma_{yA}\sigma_{yB} | \uparrow_A \downarrow_B > = J | \downarrow_A \uparrow_B >,$$
$$J\sigma_{xA}\sigma_{xB} | \uparrow_A \uparrow_B > = J | \downarrow_A \downarrow_B >,$$
$$J\sigma_{yA}\sigma_{yB} | \uparrow_A \uparrow_B > = -J | \downarrow_A \downarrow_B > . \qquad (15.25)$$

- **Problem 15.3**

Derive the matrix in Equation (15.3) starting with the Heisenberg Hamiltonian.

- **Problem 15.4**

If we changed the designation so that the down-spin state became logic 0 and the up-spin state logic 1, will we have realized the AND gate? Explain.

- **Problem 15.5**

We showed that the NAND gate functionality is realized if $h \gg J$. The question is how much larger should h be compared to J. Show that it is sufficient that $h \geq 10J$.

Hint

Consider the case when the inputs are $[0\ 0]$ or $[1\ 1]$. The requirement that the ground state wavefunction becomes the desired wavefunction is that $|\Lambda_1/(J\Upsilon_1)|^2 \to 1$ and $|2/\Upsilon_1|^2 \to 0$. Plot these two quantities as a function of h/J and show that $|\Lambda_1/(J\Upsilon_1)|^2$ begins to saturate toward 1 and $|2/\Upsilon_1|^2$ begins to saturate toward 0 when h/J exceeds 10.

- **Problem 15.6**

Show that when the inputs are $[1\ 0]$ or $[0\ 1]$, it is still sufficient that $h \geq 10J$ in order to yield NAND gate functionality.

- **Problem 15.7**

We showed that $Z = g\mu_B B_{global}/2 = (1/2)kTln(1/p)$ where B_{global} is the flux density of the global magnetic field. Let us say that we are operating at $T = 1$ K and $p = 10^{-9}$. What is the value of B_{global} required if the g-factor of the quantum dots hosting the spins is 50?

What would be the value of B_{global} if we could operate at room temperature with an error probability of 3%?

Are these reasonable values?

- **Problem 15.8**

  An SSL circuit is operating at $T = 1$ K and with bit error probability of $p = 10^{-9}$. The quantum dot material has a g-factor of –50. What is the minimum magnetic field B_{input} required to write input data, i.e., orient the input spins in the desired direction?

  Solution

  We know $J = Z = (1/2)kTln(1/p)$. Therefore, $J = Z = 0.9$ meV, which is quite achievable with present day quantum dot technology. We also saw in Problems 15.5 and 15.6 that the minimum value of h that we require for NAND gate functionality is $10J$. Therefore, $h = |g|\mu_B B_{input} \geq 9$ meV, which yields that the minimum value of $B_{input} = 3.1$ Tesla.

15.8   References

[1] E. Knill, "Quantum computing with realistically noisy devices", Nature (London), **434**, 39 (2005).

[2] S. Salahuddin and S. Datta, "Interacting systems for self correcting low power switching", Appl. Phys. Lett., **90**, 093503 (2007).

[3] D. A. Hodges and H. G. Jackson, *Analysis and Design of Digital Integrated Circuits*, 2nd edition, McGraw Hill, New York, 1988, Chapter 1, p. 2.

[4] M. Ikezawa, B. Pal, Y. Masumoto, I. V. Ignatiev, S. Yu. Verbin and I. Ya. Gerlovin, "Sub millisecond electron spin relaxation in InP quantum dots", Phys. Rev. B, **72**, 153302 (2005).

[5] S. Amasha, K. MacLean, Iuliana Radu, D. M. Zümbuhl, M. A. Kastner, M. P. Hanson and A. C. Gossard, "Electrical control of spin relaxation in a quantum dot", arXiv:0707.1656.

[6] S. Pramanik, C-G Stefanita, S. Patibandla, S. Bandyopadhyay, K. Garre, N. Harth and M. Cahay, "Observation of extremely long spin relaxation time in an organic nanowire spin valve", Nature Nanotech, **2**, 216 (2007).

[7] S. Bandyopadhyay, "Power dissipation in spintronic devices: A general perspective", J. Nanosci. Nanotech., **7**, 168 (2007).

[8] D. Rugar, R. Budakian, H. J. Mamin and B. H. Chui, "Single spin detection by magnetic resonance force microscopy", Nature (London), **430**, 329 (2004).

[9] J. M. Elzerman, R. Hanson, L. H. Willems van Beveren, B. Witkamp, L. M. K. Vandersypen and L. P. Kouwenhoven, "Single shot read out of an electron spin in a quantum dot", Nature (London), **430**, 431 (2004).

[10] M. Xiao, I. Martin, E. Yablonovitch and H. W. Jiang, "Electrical detection of the spin resonance of a single electron in a silicon field effect transistor", Nature (London), **430**, 435 (2004).

[11] S. Bandyopadhyay, B. Das and A. E. Miller, "Supercomputing with spin polarized single electrons in a quantum coupled architecture", Nanotechnology, **5**, 113 (1994).

[12] S. K. Sarkar, T. Bose and S. Bandyopadhyay, "Single spin logic circuits", Phys. Low Dim. Struct., **2**, 69 (2006).

[13] N. W. Ashcroft and N. D. Mermin, *Solid State Physics*, Saunders College, Philadelphia, 1976.

[14] S. N. Molotkov and S. S. Nazin, "Single electron spin with quantum dot logical gates with ferromagnetic chains", Phys. Low Dim. Struct., **10**, 85 (1997).

[15] S. N. Molotkov and S. S. Nazin, "Single electron spin logical gates", JETP Lett., **62**, 256 (1995); S. N. Molotkov and S. S. Nazin, "Single electron computing: Quantum dot logic gates", Zh. Eksp. Teor. Fiz. **110**, 1439 (1996); [Sov. Phys.: JETP, **83**, 794 (1996)].

[16] H. Agarwal, S. Pramanik and S. Bandyopadhyay, "Single spin universal Boolean logic gate", New J. Phys., **10**, 015001 (2008).

[17] P. Bakshi, D. Broido and K. Kempa, "Spontaneous polarization of electrons in quantum dashes", J. Appl. Phys., **70**, 5150 (1991); K. Kempa, D. A. Broido and P. Bakshi, "Spontaneous polarization in quantum dot systems", Phys. Rev. B, **43**, 9343 (1991).

[18] P. Bakshi, private communication, 1992.

[19] M. Anantram and V. P. Roychowdhury, "Metastable states and information propagation in a one-dimensional array of locally coupled bistable cells", J. Appl. Phys., **85**, 1622 (1999).

[20] D. V. Melnikov and J-P Leburton, "Single particle state mixing in two electron double quantum dots", Phys. Rev. B, **73**, 155301 (2006); J-P Leburton, private communication.

[21] C. F. Hirjibehedin, C. P. Lutz and A. G. Heinrich, "Spin coupling in engineered atomic structures", Science, **312**, 1021 (2006).

[22] R. P. Cowburn and M. E. Welland, "Room temperature magnetic quantum cellular automata", Science, **287**, 1466 (2000).

[23] R. Landauer, "Is quantum mechanics useful?", Philos. Trans. Royal Soc. London, Ser. A, **353**, 367 (1995).

[24] S. Bandyopadhyay and V. P. Roychowdhury, "Computational paradigms in nanoelectronics: Quantum coupled single electron logic and neuromorphic networks", Jpn. J. Appl. Phys., **35**, Part 1, 3350 (1996).

[25] D. K. Schroeder, *Advanced MOS Devices*, Modular Series in Solid State Devices, Eds. G. W. Neudeck and R. F. Pierret, (Addison-Wesley, Reading, MA, 1987).

[26] S. Bandyopadhyay, "Computing with spins: From classical to quantum computing", Superlat. Microstruct., **37**, 77 (2005).

[27] R. W. Keyes and R. Landauer, "Minimal energy dissipation in logic", IBM J. Res. Develop., **14**, 152 (1970); R. Landauer, "Uncertainty principle and minimal energy dissipation in the computer", Int. J. Theor. Phys., **21**, 283 (1982).

[28] R. K. Cavin, V. V. Zhirnov, G. I. Bourianoff, J. A. Hutchby, D. J. C. Herr, H. H. Hosack, W. H. Joyner and T. A. Woolridge, "A long term view of research targets in nanoelectronics", J. Nanoparticle Res., **7**, 573 (2005); V. V. Zhirnov, R. K. Cavin, J. A. Hutchby and G. I. Bourianoff, "Limits to binary logic switch scaling - A Gedanken model", Proc. IEEE, **91**, 1934 (2003).

[29] R. K. Cavin, V. V. Zhirnov, J. A. Hutchby and G. I. Bourianoff, "Energy barriers, demons and minimum energy operations of electronic devices", Fluctuation and Noise Letters, **5**, C29 (2005).

[30] X. W. Zhang, W. J. Fan, S. S. Li and J. B. Xia, "Giant and zero g-factors of dilute nitride semiconductor nanowires", Appl. Phys. Lett., **90**, 193111 (2007).

[31] W. L. Wang, O. V. Yazyev, S. Meng and E. Kaxiras, "Topological frustration in graphene nanoflakes: Magnetic order and spin logic devices", Phys. Rev. Lett., **102**, 157201 (2009).

[32] M. Ciorga, A. S. Sachrajda, P. Hawrylak, C. Gould, P. Zawadzki, S. Jullian, Y. Feng and Z. Wasilewski, "Addition spectrum of a lateral quantum dot from Coulomb and spin blockade spectroscopy", Phys. Rev. B, **61**, R16315 (2000).

[33] M. Piero-Ladriere, M. Ciorga, J. Lapointe, P. Zawadzki, M. Korukusin-ski, P. Hawrylak and A. S. Sachrajda, "Spin blockade spectroscopy of a two level artificial molecule", Phys. Rev. Lett., **91**, 026803 (2003).

[34] C. Livermore, C. H. Crouch, R. M. Westervelt, K. L. Campman and A. C. Gossard, "The Coulomb blockade in coupled quantum dots", Science, **274**, 1332 (1996).

[35] T. H. Oosterkamp, T. Fujisawa, W. G. van der Wiel, K. Ishibashi, R. V. Hijman, S. Tarucha and L. P. Kouwenhoven, "Microwave spectroscopy of a quantum dot molecule", Nature (London), **395**, 873 (1998).

[36] A. W. Holleitner, R. H. Blick, A. K. Huttel, K. Eberl and J. P. Kotthaus, "Probing and controlling the bonds of an artificial molecule", Science, **297**, 70 (2001).

[37] N. J. Craig, J. M. Taylor, E. A. Lester, C. M. Marcus, M. P. Hanson and A. C. Gossard, "Tunable non-local spin control in a coupled quantum dot system", Science, **304**, 565 (2004).

[38] R. Hanson, B. Witkamp, L. M. K. Vandersypen, L. H. W. van Beveren, J. M. Elzerman and L. P. Kouwenhoven, "Zeeman energy and spin relaxation in a one electron quantum dot", Phys. Rev. Lett., **91**, 196802 (2003).

[39] J. R. Petta, A. C. Johnson, J. M. Taylor, E. A. Laird, A. Yacoby, M. D. Lukin, C. M. Marcus, M. P. Hanson and A. C. Gossard, "Coherent manipulation of single electron spins in coupled quantum dots", Science, **309**, 2180 (2005).

16

Quantum Computing with Spins

In the last chapter, we saw that it is theoretically possible to switch a *logically reversible* gate without dissipating any energy at all (recall the proof of the Landauer-Shannon result). If there is no dissipation, then no information is ever being discarded; therefore, if we know the state of the system completely at any instant of time, we should be able to infer its state at any previous instant. This is consistent with the equations of quantum mechanics (e.g., the Schrödinger equation, or the Pauli equation) which are reversible in time. This property of quantum mechanical systems, known as "time reversal symmetry", guarantees that if we know the state of a system at a time t_0, we can tell what the state was at any time $t < t_0$. Because of this fundamental connection between physical reversibility, logical reversibility, and the equations of quantum mechanics, we would expect logically reversible gates to behave as non-classical (quantum mechanical) computing machinery.[*] This will bring us to the topic of quantum computing and quantum logic gates, but before we get there, let us explore reversible logic in a little more detail.

16.1   Quantum inverter

The inverter, or the NOT gate, is a logically reversible gate since whenever we know the output of the gate, we can tell unambiguously what the input was. After all, the input is always the logic complement of the output. Since this gate is logically reversible, it should be possible to operate it without dissipating energy. Taking it one step further, we should be able to switch this gate in accordance with the laws of quantum mechanics. This brings us to the topic of a *quantum inverter*.

The quantum inverter is a NOT gate where the output is always the logic complement of the input. Therefore, if we change the input, the output should change accordingly to become the logic complement of the input. In a quantum inverter, this change should take place without dissipating energy and

---

[*]Not all reversible gates are quantum mechanical. Classical gates can also be reversible, as we will see in the example of the Toffoli-Fredkin gate.

in accordance with the laws of quantum mechanics, i.e., following a unitary evolution.

A quantum inverter can be implemented with two exchange coupled spins. One spin will represent the input bit and the other will be the output bit. Input-output relations will be determined by exchange coupling between the spins, in much the same way as the SSL-based NAND gate of the previous chapter, but with one major difference. The dynamics of the SSL-based NAND gate was dissipative. In fact, dissipation is required to make the system relax to the ground state and produce the correct result. In the case of the quantum inverter, there will be no dissipation. Therefore, we will have to rely entirely on the machinery of quantum mechanics to produce the correct result every time. We will show that quantum mechanics is up to the task, but we shall require extreme precision. We must allow the system to evolve according to quantum mechanical laws up to a certain time and then abruptly halt the system, at *precisely the right juncture*, if we want the correct result. If we overshoot or undershoot in time, we will not get the correct answer. This is the Achilles' heel of quantum gates. They demand perfect precision, which may be unrealistic in most situations. Operating flawlessly in the presence of imperfection will require in-built error correction, which will always engender some energy dissipation, but that is forbidden for quantum gates.[†] Therefore, if we have errors due to any imperfection, they will build up with time and ultimately render the gate useless. That is why precision is vital for quantum gates. Insofar as perfect precision is unrealistic, quantum gates are notoriously error-prone and massive error correction resources are usually required to make them viable. There is a school that believes that the immense resources needed for error correction will more than offset any advantage that a quantum gate may have over a classical gate.

The SSL-based NAND gate, which we discussed in the previous chapter, did not require this kind of precision. All we had to do after providing inputs to the system was to let the system relax to the ground state by dissipating energy. We did not have to bother how long it took or what route the system traced in order to reach the ground state. Those details were irrelevant. There was no issue of overshooting or undershooting. Dissipation forgave all our transgressions and made the gate work in spite of imprecisions and inaccuracies. Unfortunately, we do not have that luxury when we deal with dissipationless systems. Therefore, reversibility and absence of dissipation come with a price, namely, the need for extreme precision.

Consider the quantum inverter shown in Fig. 16.1. The spin polarization in one dot (say, the left dot) is the input bit, and the spin polarization in the other dot is the output bit. We will assume that there is only one size quantized level in each dot. A weak magnetic field B_z is applied globally to

---

[†]Errors can be corrected with "software", such as error correcting codes, but not with "hardware." There are quantum control techniques to minimize errors, but they are not exactly hardware-based approaches.

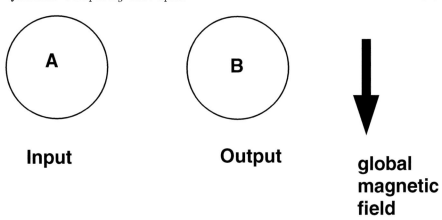

FIGURE 16.1
A quantum inverter realized with two exchange coupled spins in quantum dots. A is the input and B the output of the inverter.

define the spin quantization axes, meaning that the spin in any dot is always either parallel or anti-parallel to this field. We will assume that the field is in the z-direction.

The Hubbard Hamiltonian for this system is [1]

$$H_{Hubbard} = \sum_{i\sigma} \epsilon_0 n_{i\sigma} + \sum_{input\ dot,\ \sigma} g\mu_B B_i sign(\sigma) + \sum_{<ij>} \left[c_{i\sigma}^\dagger c_{j\sigma} + h.c. \right]$$
$$+ \sum_i U_i n_{i\uparrow} n_{i\downarrow} + \sum_{<ij>\alpha\beta} J_{ij} c_{i\alpha}^\dagger c_{i\beta} c_{j\beta}^\dagger c_{j\alpha} + B_z \sum_{i\sigma} g\mu_B n_{i\sigma} sign(\sigma) \,,$$
$$(16.1)$$

where $h.c.$ stands for "Hermitian conjugate". Here the first term represents the site energy (kinetic + potential) in the i-th dot, the second term is the Zeeman splitting energy associated with applying a local magnetic field to the i-th dot to selectively align the spin in that dot along the direction of the applied local field (providing input data), the third term is the tunneling between nearest neighbor dots (the angular braket denotes "nearest neighbors"), the fourth term is the charging energy of the dot, which is the energy cost associated with putting a second electron in the same dot (Pauli exclusion principle mandates that the second electron must have a spin anti-parallel to that of the first, since we assume that each dot has only one bound state), the fifth term is the exchange interaction between nearest neighbor dots, and the last term is the Zeeman energy due to the global magnetic field, applied over the entire 2-spin array, in the z-direction, to define the spin quantization axes.

Once again, we will simplify this to the Heisenberg model, which yields

$$\mathcal{H} = J \sum_{<ij>} \sigma_{zi}\sigma_{zj} + J \sum_{<ij>} [\sigma_{xi}\sigma_{xj} + \sigma_{yi}\sigma_{yj}] + \sum_{input\ dot} \sigma_{zi}h_{zi}^{input} + \sum_{i} \sigma_{zi}h_{zi}^{global}.$$
(16.2)

The quantity J is the exchange coupling energy, h_{zi}^{input} is the Zeeman splitting caused by the local magnetic field applied to the left dot (input) to orient the spin of its electron, and h_{zi}^{global} is the Zeeman splitting associated with the global magnetic field.

We will define "up-spin" as the polarization parallel to the global magnetic field and "down-spin" as anti-parallel. In the basis of two electron states, the Hamiltonian in Equation (16.2) can be written as

$$
\begin{array}{cccc}
|\downarrow\downarrow> & |\downarrow\uparrow> & |\uparrow\downarrow> & |\uparrow\uparrow>
\end{array}
$$
$$
\begin{pmatrix}
h_A + J + 2Z & 0 & 0 & 0 \\
0 & h_A - J & 2J & 0 \\
0 & 2J & -h_A - J & 0 \\
0 & 0 & 0 & -h_A + J - 2Z
\end{pmatrix}
\begin{array}{c}
|\downarrow\downarrow> \\
|\downarrow\uparrow> \\
|\uparrow\downarrow> \\
|\uparrow\uparrow>
\end{array}
$$
(16.3)

where $2h_A$ is the Zeeman splitting caused by the externally applied local magnetic field in input dot A and $2Z$ is the Zeeman splitting caused in any dot by the global magnetic field. If the local magnetic field providing input to dot A aligns the spin in the direction of the global magnetic field (up-spin direction), then the corresponding h is positive. Otherwise, it is negative. The quantity J is always positive to ensure that the singlet state has lower energy than the triplet state.

The eigenenergies E_n and eigenstates ϕ_n $(= c_1| \downarrow\downarrow> + c_2| \downarrow\uparrow> + c_3| \uparrow\downarrow> + c_4| \uparrow\uparrow>)$ of the above Hamiltonian are found by diagonalizing the above matrix:

where, once again, we have arranged the eigenstates in ascending order (ground state first and highest excited state last).

In the absence of any input ($h_A = 0$), the ground state energy is $-3J$ and the ground state wave function is $\frac{1}{\sqrt{2}}(| \uparrow\downarrow> - | \downarrow\uparrow>)$, which is an entangled state. In this state, neither the input dot nor the output dot has a definite spin polarization.

If the system is initially without any input and is in the ground state at time $t = 0$, then obviously $c_1(0) = c_4(0) = 0$, $c_2(0) = -1/\sqrt{2}$, and $c_3(0) = 1/\sqrt{2}$. Thereafter, if a local magnetic field is applied to the input dot A at time $t=0$ to align its spin polarization to the "down" state, then, in the absence of any coupling to the environment, the wavefunction of the system evolves unitarily according to

$$\psi(t) = \exp[-i\mathcal{H}t/\hbar]\psi(0),$$
(16.4)

TABLE 16.1

Eigenenergies and eigenstates of the quantum inverter

Eigenenergies	Eigenstates
$-J - \sqrt{h_A^2 + 4J^2}$	$\sqrt{\frac{1}{2}\left(1 - \frac{h_A}{\sqrt{h_A^2+4J^2}}\right)}\lvert \uparrow\downarrow> - \sqrt{\frac{1}{2}\left(1 + \frac{h_A}{\sqrt{h_A^2+4J^2}}\right)}\lvert \downarrow\uparrow>$
$-h_A + J$	$\lvert \uparrow\uparrow>$
$-J + \sqrt{h_A^2 + 4J^2}$	$\sqrt{\frac{1}{2}\left(1 + \frac{h_A}{\sqrt{h_A^2+4J^2}}\right)}\lvert \uparrow\downarrow> + \sqrt{\frac{1}{2}\left(1 - \frac{h_A}{\sqrt{h_A^2+4J^2}}\right)}\lvert \downarrow\uparrow>$
$h_A + J + 2Z$	$\lvert \downarrow\downarrow>$

where $\psi(0) = \frac{1}{\sqrt{2}}(\lvert \uparrow\downarrow> - \lvert \downarrow\uparrow>)$.

Since the initial state wavefunction does not contain the states $\lvert \uparrow\uparrow>$ or $\lvert \downarrow\downarrow>$, wavefunction at any arbitrary time t will be a mixture of states $\lvert \downarrow\uparrow>$ and $\lvert \uparrow\downarrow>$ only. This can be shown rigorously by solving Equation (16.4) with the Hamiltonian H given by Equation (16.3). Therefore, the wavefunction at time t is given by

$$\psi(t) = c_2(t)\lvert \downarrow\uparrow> + c_3(t)\lvert \uparrow\downarrow>, \qquad (16.5)$$

where

$$c_2(t) = \frac{e^{iJt/\hbar}}{\sqrt{2}}\left[\cos(\omega t) - i\left(\frac{h_A}{\hbar\omega} - \sqrt{1 - \frac{h_A^2}{\hbar^2\omega^2}}\right)\sin(\omega t)\right],$$

$$c_3(t) = -\frac{e^{iJt/\hbar}}{\sqrt{2}}\left[\cos(\omega t) + i\left(\frac{h_A}{\hbar\omega} + \sqrt{1 - \frac{h_A^2}{\hbar^2\omega^2}}\right)\sin(\omega t)\right], \qquad (16.6)$$

and $\hbar\omega = \sqrt{h_A^2 + 4J^2}$.

If the system were to act as an inverter, it must, at some point in time, reach the unentangled state $\lvert \downarrow\uparrow>$. In that state, the left dot (dot A) will be in the down-spin state because of the input provided to it, and the right dot will be in the up-spin state, thereby satisfying the requirement of the inverter, namely, that the output is the logic complement of the input. This desired state corresponds to $\lvert c_2 \rvert = 1$ and $c_3 = 0$. It is not clear at the outset whether the system can *ever* reach the desired state by evolving unitarily according to Equation (16.4). It turns out (see Problem 14.1) that it can after a time delay

$$\tau_d = \frac{h}{4\sqrt{h_A^2 + 4J^2}} \qquad (16.7)$$

provided $h_A = 2J$.

This desired state, however, is not an eigenstate of the system (see Table 16.1). Consequently, the system will continue to evolve to a different state unless a "read" operation collapses the wavefunction as soon as the desired state is reached. There is no margin of error allowed. We must "halt" the system as soon as the duration τ_d elapses. If we undershoot or overshoot, we will incur an error that will build up with time. These errors can only be corrected with "software", since there are no "hardware" antidotes. Amazing advances, however, have been made in "software", and as mentioned in the previous chapter, error correction codes exist that can correct errors occurring with a probability of up to 3% in every computational step [2].

In this system, there is no dissipation whatsoever except during the read operation. Therefore, the product of dissipated energy (ignoring the read operation) and switching delay τ_d is exactly zero. On the other hand, the product of applied energy and switching delay is

$$h_A \tau_d = \frac{h h_A}{4\sqrt{h_A^2 + 4J^2}}. \tag{16.8}$$

We immediately see that any energy-time uncertainty limit that might have been expected is non-existent, since

$$h_A \tau_d < \frac{\hbar}{2} \quad if \; h_A < \frac{2J}{\sqrt{\pi^2 - 1}}. \tag{16.9}$$

It should be emphasized that h_A is the energy applied to switch the inverter and does not have to be *dissipated*. Even if it were dissipated, the above equation clearly shows that there is no energy-time uncertainty limit on (dissipated) energy-delay product, contrary to the popular view espoused in [3, 4]. In a paper that is now a classic, Rolf Landauer showed that there is no lower limit imposed by the uncertainty principle on dissipated energy [5]. In fact, concrete and detailed classical models of dissipationless computation have been provided by several authors, and numerous quantum mechanical models of dissipationless computation have also been put forward, starting with the early work of Benioff [6, 7, 8]. These models require no dissipated energy, but usually do require input energy to switch. What we see in the current example is that there is no energy-time uncertainty limit associated with the *applied energy* either. Furthermore, computation can proceed by applying arbitrarily small energy to initiate the process. The quantity h_A can be arbitrarily small, but the smaller we make h_A, the longer will be τ_d according to Equation (16.7). Therefore, the price we pay for being miserly with energy is long switching delay.

16.2   Can the NAND gate be switched without dissipating energy?

The reader may wonder whether the NAND gate of the previous chapter could have been switched in a similar fashion without dissipating energy. Looking at any of the tables listing the eigenenergies and the eigenstates of the NAND gate in Chapter 15, it becomes obvious that, if there are no input signals applied to the input dots ($h_A = h_C = 0$), then the ground state of the NAND gate is

$$\psi_{NAND} = \frac{2}{\sqrt{6}} | \downarrow\uparrow\downarrow> - \frac{1}{\sqrt{6}} | \downarrow\downarrow\uparrow> - \frac{1}{\sqrt{6}} | \uparrow\downarrow\downarrow> . \qquad (16.10)$$

This is an entangled state where none of the three dots has any net spin polarization.[‡]

If we start with this ground state (as in the quantum inverter), apply inputs h_A and h_C, and then allow the system to evolve unitarily in accordance with Equation (16.4), will we realize NAND gate operation? In other words, will the correct outputs be produced? This is a profound question since its answer will tell us whether the NAND gate can be switched coherently without dissipating energy. The answer is actually surprising. When the inputs are [1 1], and the local magnetic fields writing input bits are in the same direction as the global magnetic field (down-spin direction), the answer is YES. With these specific inputs, the NAND gate can switch coherently without dissipation, in accordance with Equation (16.4), so that at some point in time, the 3-body wavefunction will be the unentangled state $| \downarrow\uparrow\downarrow>$, which corresponds to Fig. 15.1(a). The time required to attain this state is approximately $0.32\hbar/J$ and the requirement on the input signal strength is $h_A = h_B = h = 3J$ (see Problem 14.2).

The general wavefunction describing the state of the NAND gate is given in Equation (15.4). The amplitude of the target state $| \downarrow\uparrow\downarrow>$ is $c_3(t)$. In Fig. 16.2, we plot $|c_3(t)|^2$ as a function of time t when $h/J = 3$. Note that it reaches unity when $t \approx 0.32\hbar/J$ provided $h = 3J$. This means that the pure target state is *attainable* through coherent dynamics.

When the input bits are [1 0] or [0 1], the target states are $| \downarrow\downarrow\uparrow>$ and $| \uparrow\downarrow\downarrow>$, shown in Figs. 15.1(c) and (d), respectively. They can only be approached, but never completely attained. The amplitudes of these states are c_2 and c_5 (see Equation (15.4)). In Figs. 16.3 and 16.4, we plot the probabilities $|c_2|^2$ and $|c_5|^2$ as a function of time when $h/J = 2.03$. It can be shown that the probabilities reach a maximum when $h/J = 2.03$. Note that they never reach unity (the maximum value is about 0.97), indicating that the NAND gate CANNOT be switched coherently, without dissipation, when the inputs are [1 0] or [0 1].

[‡]To find the spin polarization in any dot, algebraically add up the like spin amplitudes in that dot. For example, in dot A, the down-spin amplitude is $2/\sqrt{6} - 1/\sqrt{6} = 1/\sqrt{6}$, while the up-spin amplitude is $-1/\sqrt{6}$. The probability of dot A having up-spin is therefore $1/6$ and down-spin is also $1/6$. Since these probabilities are equal, the dot has no net spin polarization. If they were unequal, the dot would have had a net spin polarization. Do not think that the two probabilities should add up to 1. Since we are dealing with many-body states where the three spins act in unison and are connected, we cannot expect the probabilities in individual dots to add up to 1.

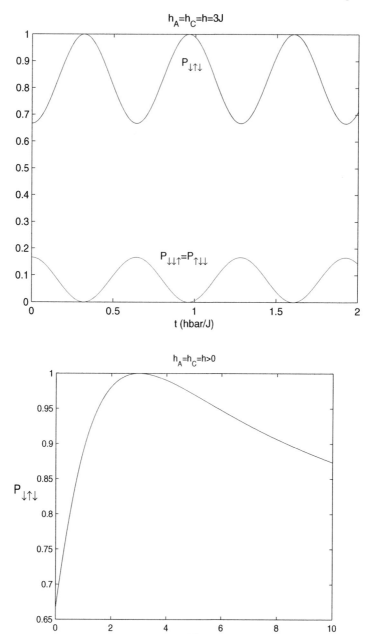

FIGURE 16.2
Probability $|c_3(t)|^2$ ($= P_{\downarrow\uparrow\downarrow}$) of the target state $|\downarrow\uparrow\downarrow\rangle$ as a function of time when $h/J = 3$. The input bits are [1 1]. The bottom figure shows the maximum value of $|c_3(t)|^2$ as a function of h/J.

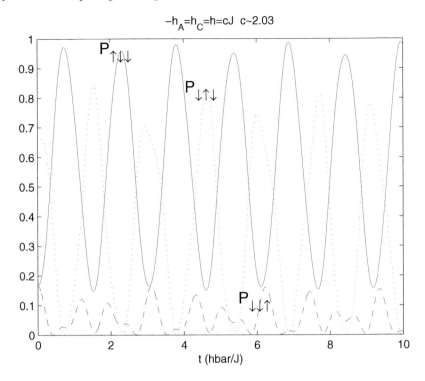

FIGURE 16.3

Probability $|c_2|^2$ $(=P_{\downarrow\downarrow\uparrow})$ as a function of time. The ratio $h/J = 2.03$. The inputs are $[1\ 0]$.

The state in Fig. 15.1(b) is even less attainable, since $|c_6|^2$ is always zero. This is not surprising since the initial state ψ_{NAND} does not contain the target state $|\uparrow\downarrow\uparrow\rangle$ conforming to Fig. 15.1(b), i.e., the desired state when the inputs are $[0\ 0]$.

In the end, the NAND gate CANNOT be switched coherently between any two arbitrary states without dissipation. Only one entry in the truth table can be produced through coherent evolution (without dissipation) but the other three entries cannot be produced. *It is intriguing to note that the only entry in the truth table that can be produced through coherent evolution happens to be the one that is logically reversible.* When the output is 0, we know unambiguously that both inputs must have been 1, meaning that this situation is logically reversible. In this case, the gate can be switched without dissipation. In the other three cases, which are logically irreversible since the inputs cannot be inferred from the output, the gate cannot be switched coherently. This reinforces the connection between physical reversibility and logical reversibility, as originally expounded by Landauer.

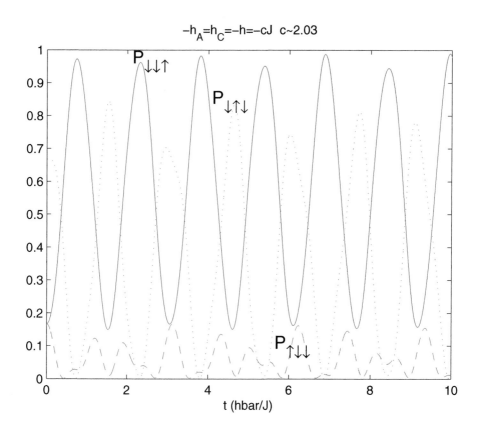

FIGURE 16.4
Probability $|c_5|^2$ $(=P_{\uparrow\downarrow\downarrow})$ as a function of time. The ratio $h/J = 2.03$. The inputs are [01].

16.3 Universal reversible gate: Toffoli–Fredkin gate

We have seen so far that the inverter is both logically and physically reversible, while the NAND gate is both logically and physically irreversible. The inverter, unfortunately, is not a universal gate in the sense that any arbitrary Boolean logic circuit cannot be built with inverters alone. We therefore seek a universal reversible logic gate. Such a gate is the Toffoli-Fredkin (T-F) gate [9], which has three inputs (a, b, c) and three corresponding outputs (a', b', c').

The truth table of a T-F gate is

TABLE 16.2

Truth table of the
Toffoli-Fredkin gate

a	b	c	a'	b'	c'
0	0	0	0	0	0
0	0	1	0	0	1
0	1	0	0	1	0
0	1	1	0	1	1
1	0	0	1	0	0
1	0	1	1	0	1
1	1	0	1	1	1
1	1	1	1	1	0

Looking at the above truth table, it is obvious that the input-output relationship is

$$a' = a$$
$$b' = b$$
$$c' = c \oplus (a \cdot b), \tag{16.11}$$

where $\oplus$ denotes the exclusive OR Boolean function and the "dot-product" denotes the Boolean AND function. The variables a and b are called control bits and the variable c is the target bit. The control bits pass through the gate without change. The target bit is flipped only if both control bits are logical 1.

We can devise a spintronic implementation of the T-F gate using the same system used for the NAND gate, namely, three spins in three quantum dots (Fig. 16.5) with nearest neighbor exchange coupling. The entire array is placed in a global magnetic field which defines the spin quantization axes. In this case, the two peripheral spins are control bits a and b and the central spin

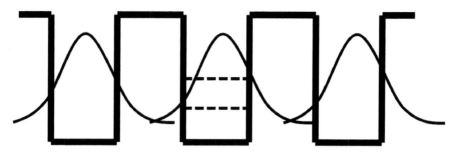

FIGURE 16.5
A Toffoli-Fredkin gate realized with three spins in quantum dots placed in a global magnetic field. Wavefunctions of nearest neighbor electrons overlap causing exchange interaction between them. Inputs are directly written in the control dots A and B with local magnetic fields that could be generated with inductors. An ac magnetic field of the right frequency flips the spin in the central dot C only when the spins in dots A and B are down, i.e., $a = b = 1$.

is the target bit c. Assume that the global magnetic field is in the direction of "down-spin" which encodes the logic bit 1. Inputs are provided with local magnetic fields to the peripheral dots **A** and **B** to align the spins in these dots such that they conform to the intended control bits.

Consider the case when the control bits $a = b = 1$. In this case, $h_A = h_B = h$ and the eigenenergies/eigenstates are obtained from Table 15.2. The two states of interest are $| \downarrow\uparrow\downarrow >$ and $| \downarrow\downarrow\downarrow >$ since they alone correspond to the situation when the two peripheral spins are "down" representing $a = b = 1$. It is easy to show (Problem 14.5) that, if $h >> J$, then the ground state is nearly the pure unentangled state $| \downarrow\uparrow\downarrow >$ while the first excited state is the pure unentangled state $| \downarrow\downarrow\downarrow >$. The energy difference between these two states is $4J - -2Z$, which is positive as long as $J > Z/2$. We will call this energy difference $\hbar\omega_{11}$ where the subscripts indicate the Boolean values of control bits a and b. Therefore,

$$\hbar\omega_{11} = 4J - 2Z. \tag{16.12}$$

Note that the above is the energy difference between states corresponding to the central spin being down and up respectively – or the target bit being 1 and 0 – when the control bits are both 1.

Similarly, when the control bits are $a = 0$, $b = 1$, i.e., $-h_A = h_B = h$, the eigenenergies/eigenstates are obtained from Table 15.4. This time, the two states of interest are $| \uparrow\uparrow\downarrow >$ and $| \uparrow\downarrow\downarrow >$ since they alone correspond to the situation when $a = 0$, $b = 1$. Once again, we can show that when $h >> J$, the first excited state is nearly the unentangled state $| \uparrow\uparrow\downarrow >$ and the ground state is nearly the unentangled state $| \uparrow\downarrow\downarrow >$. The energy difference between them is $2Z$. Therefore,

$$\hbar\omega_{01} = 2Z. \tag{16.13}$$

Proceeding in this fashion, and always ensuring that $h \gg J$, we can show that

$$\hbar\omega_{10} = 2Z \tag{16.14}$$

and

$$\hbar\omega_{00} = 4J + 2Z. \tag{16.15}$$

The purpose of all this was to show that

$$\omega_{11} \neq \omega_{00}, \omega_{10}, \omega_{01}. \tag{16.16}$$

16.3.1 Dynamics of the T-F gate

The T-F gate is realized if the spin in dot $\mathbf{C}$ toggles only if $a = b = 1$, and not otherwise. To achieve this, we utilize the Rabi oscillations of Chapter 4. An ac magnetic field, rotating in the $(x - y)$ plane, is turned on and has an angular frequency ω_{11}. This field will flip the spin in the central dot only when $a = b = 1$, by virtue of the inequality in the last equation. This realizes the T-F gate. Note that the spin in the central dot undergoes a transition by coherently emitting or absorbing a photon from the ac magnetic field. No dissipation takes place consistent with the reversible nature of the T-F gate.

The switching time is determined by the amplitude of the ac magnetic field. It is given by

$$t_s = \frac{h}{2g\mu_B B_{ac}}, \tag{16.17}$$

where B_{ac} is the amplitude of the flux density associated with the ac magnetic field.

We can make the time t_s quite short by using quantum dot materials that have large g-factors. For a reasonable value of $B_{ac} = 0.01$ Tesla and $g = 900$, the switching time is ~ 4 ps. A g-factor of 900 has been calculated for certain nitrides such as $InSb_{1-x}N_x$ [10].

The ac magnetic field is turned on only for the duration t_s given by the above equation. Such a pulse is called a π-pulse. This mode of implementing reversible dynamics has been proposed by a number of authors (see, for example, [11]).

16.4 A-matrix

When dealing with reversible logic gates, a useful concept is the so-called "A-matrix". Consider an inverter. An inverter converts a 0 to a 1 and vice versa.

Thus, the 2×2 matrix operator relating an input vector to the output vector (the vectors span the Hilbert spaces of the inputs and outputs) is given by the relation

$$\begin{bmatrix} 0 \\ 1 \end{bmatrix} = \begin{bmatrix} 0 & 1 \\ 1 & 0 \end{bmatrix} \begin{bmatrix} 1 \\ 0 \end{bmatrix}.$$

$$output \ A - matrix \ input \tag{16.18}$$

The matrix operator for a logic gate is called an *A-matrix*. For an inverter, or for that matter any logically reversible gate, the A-matrix must be invertible and its determinant must not vanish. It also stands to reason that a reversible gate must have as many outputs as inputs.

We wrote the A-matrix of the basic 1-bit inverter in Equation (16.18) using the computational basis states $|0>$ and $|1>$. We can write an A-matrix for the 3-bit T-F gate using the 3-bit computational basis states $|000>$, $|001>$, $|010>$, $|011>$, $|100>$, $|101>$, $|110>$ and $|111>$ (here the basis states are possible values of $|abc>$):

$$A_{Toffoli-Fredkin} = \begin{bmatrix} 1&0&0&0&0&0&0&0 \\ 0&1&0&0&0&0&0&0 \\ 0&0&1&0&0&0&0&0 \\ 0&0&0&1&0&0&0&0 \\ 0&0&0&0&1&0&0&0 \\ 0&0&0&0&0&1&0&0 \\ 0&0&0&0&0&0&0&1 \\ 0&0&0&0&0&0&1&0 \end{bmatrix}. \tag{16.19}$$

The A-matrix is not diagonal, but it is obviously unitary. The unitarity guarantees reversibility.

16.5 Quantum gates

The inverter and the T-F gate discussed so far are *reversible* gates and can be switched coherently in accordance of the laws of quantum mechanics. We loosely call them quantum gates, but they are not true quantum gates since they still deal with classical logic bits 0 and 1. We are now ready to explore true quantum gates.

16.5.1 The strange nature of true quantum gates: The "square root of NOT" gate

True quantum gates are strange entities that may not have any classical analog. Let us say that we want to invert a classical bit by operating with an ordinary NOT gate. We choose to do this by operating on the input bit with two strange gates in succession that we call "square-root-of-NOT" gates denoted by $\sqrt{NOT}$. The name comes about from the fact that the successive operation of two gates can be represented mathematically as the scalar product of their A-matrices. Thus, the logic operation that we are after can be written as

$$\sqrt{NOT} \cdot \sqrt{NOT} \equiv NOT. \tag{16.20}$$

What makes this gate "quantum" (or at least non-classical) is that it is impossible to have a single-input/single-output classical binary gate that works in this fashion [12]. If the $\sqrt{NOT}$ gate were classical, it would have an output of 0 or 1 for each possible input of 0 or 1. Suppose that we define the action of a $\sqrt{NOT}$ gate as the pair of transformations

$$\sqrt{NOT}_{classical}(0) = 0,$$
$$\sqrt{NOT}_{classical}(1) = 0. \tag{16.21}$$

Then, two consecutive applications of this gate will invert a 1 successfully, but not a 0, thereby violating Equation (16.20). We can try any other transformation:

$$\sqrt{NOT}_{classical}(0) = 0.$$
$$\sqrt{NOT}_{classical}(1) = 1. \tag{16.22}$$

This particular transformation does not invert any bit at all. In fact, it is impossible to define $\sqrt{NOT}$ classically such that two successive operations of this gate will reproduce the behavior of a NOT gate.

The A-matrix of a $\sqrt{NOT}$ gate can be defined as

$$A_{\sqrt{NOT}} = \begin{bmatrix} \frac{1+i}{2} & \frac{1-i}{2} \\ \frac{1-i}{2} & \frac{1+i}{2} \end{bmatrix}. \tag{16.23}$$

Note that

$$\begin{bmatrix} \frac{1+i}{2} & \frac{1-i}{2} \\ \frac{1-i}{2} & \frac{1+i}{2} \end{bmatrix} \cdot \begin{bmatrix} \frac{1+i}{2} & \frac{1-i}{2} \\ \frac{1-i}{2} & \frac{1+i}{2} \end{bmatrix} = \begin{bmatrix} 0 & 1 \\ 1 & 0 \end{bmatrix}, \tag{16.24}$$

so that $A_{\sqrt{NOT}} \cdot A_{\sqrt{NOT}} = A_{NOT}$, as required. Moreover, the matrix $A_{\sqrt{NOT}}$ is unitary, which it must be.

Note that, if a $\sqrt{NOT}$ gate is fed an input data string (0,1), it produces an output data string (α, β) given by

$$\alpha = \frac{1+i}{2}|0> + \frac{1-i}{2}|1>$$
$$\beta = \frac{1-i}{2}|0> + \frac{1+i}{2}|1>, \tag{16.25}$$

where the output bits are coherent superpositions of $|0>$ and $|1>$.[§] Such a bit is a "qubit". Since the $\sqrt{NOT}$ gate generates a qubit, it is a true quantum gate.

Consider now a different A-matrix:

$$A_{random} = \frac{1}{\sqrt{2}} \begin{bmatrix} 1 & 1 \\ -1 & 1 \end{bmatrix}, \tag{16.26}$$

which is unitary.

If this gate is fed an input data string (0,1), it produces an output data string (α', β') given by

$$\alpha' = \frac{1}{\sqrt{2}}|1> +\frac{1}{\sqrt{2}}|0>$$
$$\beta' = \frac{1}{\sqrt{2}}|1> -\frac{1}{\sqrt{2}}|0>, \tag{16.27}$$

which are also qubits.

Now a measurement of the output (either $|\alpha'>$ or $|\beta'>$) yields the answer 0 (state $|0>$) or 1 (state $|1>$) with equal probability 1/2, yielding a *perfectly random* generator of bits 0s and 1s. A single computation with a single gate yields a perfectly random bit! This is an extension of "quantum parallelism", which we will discuss later in this chapter. But here is a taste of the power of quantum computers. A classical algorithm will require more than one gate and many computational steps to generate a random number. More important, the number will be *never truly random*. Mathematically, there is *no function* that generates a true random number (only pseudo-random numbers can be generated using algorithms). Thus, a classical Turing machine can only generate pseudo random numbers; it cannot generate a true random number. This also shows that the classical Turing machine is not really universal since it cannot model the quantum mechanical process of generating a true random number. The gate that we have postulated here beats even the most powerful classical computer conceivable in generating random numbers.

16.6   Qubits

The "qubit" concept is at the heart of quantum computation and requires more introspection. We have seen in earlier chapters that a qubit can be written as

$$qubit = a_0|0> +a_1|1>, \tag{16.28}$$

[§]To show this, just operate $A_{\sqrt{NOT}}$ on the vector $[0, 1]^T$.

where $|0>$ denotes the state in which the qubit has a value of 0 and $|1>$ denotes the state in which the qubit has a value of 1. The coefficients a_0 and a_1 are complex quantities whose squared magnitudes denote the probability that, if a measurement is performed on the qubit, it will be found to have a value of 0 and 1, respectively. Since the measurement can yield a value of only 0 or 1, and nothing else, it follows that

$$|a_0|^2 + |a_1|^2 = 1. \qquad (16.29)$$

We emphasize that the above equation *does not* imply that the qubit is sometimes in state $|0>$ with probability $|a_0|^2$, and the rest of the time in state $|1>$ with probability $|a_1|^2$. It is only after measurement that the qubit collapses to a classical bit and assumes a definite value of 0 or 1. Prior to the measurement, it *does not* have a definite value; it is both 0 and 1, *all the time*. It is therefore called a superposition state.

While one can easily understand the meaning of "superposition" from the foregoing discussion, to understand the implication of a "coherent" superposition requires more reflection. Coherence means that the phase relationship between a_0 and a_1 (being complex quantities, they have an amplitude and a phase) must be preserved. It might appear that the phase is never relevant since a_0 and a_1 seemingly determine only the probabilities which, in turn, depend only on the squared magnitudes. This is actually not true. Consider a hypothetical bit-flip operator $\hat{O}_{flip}$ whose action is to flip a bit from 0 to 1 and vice versa (a classical realization of a bit flip operator is an inverter or NOT gate). Its action on a bit can be represented as

$$\hat{O}_{flip}|0>= 1|1>$$
$$\hat{O}_{flip}|1>= 1|0>. \qquad (16.30)$$

Note from the above that $0>$ and $|1>$ are not eigenstates of the operator. If we want to calculate the expected value of this operator for a system described by a qubit, we will get

$$
\begin{aligned}
expected\ value &= < qubit|\hat{O}_{flip}|qubit > \\
&= |a_0|^2 < 0|\hat{O}_{flip}|0 > +|a_1|^2 < 1|\hat{O}_{flip}|1 > \\
&\quad +a_0 * a_1 < 0|\hat{O}_{flip}|1 > +a_1 * a_0 < 1|\hat{O}_{flip}|0 > \\
&= a_0 * a_1 + a_1 * a_0 \\
&= 2|a_0||a_1|cos\theta, \qquad (16.31)
\end{aligned}
$$

where θ is the phase angle difference between a_0 and a_1. In deriving the last equation, we used Equations (16.29) and (16.30) and also the fact that $|0>$ and $|1>$ are orthonormal states by themselves. The fact that $|0>$ and $|1>$ are orthogonal to each other is obvious because any measurement outcome can produce either a 0 or a 1, and nothing in between. Thus the two possible outcomes are mutually exclusive, meaning that the two states are orthogonal to each other.

The above equation clearly shows that there are quantities that depend on the phases of a_0 and a_1. Thus maintaining the correct phase relations between these quantities, or "coherence," is required. Quantum computation depends critically on this coherence. This is where *spin has an advantage over charge when it comes to hosting qubits.* A qubit can be constructed with a coherent superposition of two anti-parallel spin states. The coherence time can be quite long. It has been measured as 100 nanoseconds in n-type GaAs at a temperature of 5 K [13], and it has been argued that the actual coherence time is much longer since the single spin decoherence time is longer than the ensemble averaged decoherence time that is usually measured in experiments [14]. de Sousa and Das Sarma [14] claim that the single spin decoherence time will be $1 - 100$ μsec. At the time of writing this chapter, there has been a report of observing ~ 0.6 s spin coherence time in nitrogen vacancy centers in diamond at 77 K! [15]. In contrast, the coherence time for charge in solids tends to saturate to about 1 nsec as the temperature of the solid is reduced to sub milliKelvins [16]. Consequently, spin-based quantum computers are much longer lived and more robust than charge-based quantum computers. To put all this into perspective, let us think in terms of what happens if the coherence is lost. In that event, the quantum computer will falter and generate wrong results. An error will occur if the qubit loses coherence during a clock cycle. Therefore, the error probability is roughly $1 - \exp[-T/T_2]$, where T is the clock period and T_2 is the time it takes for the qubit to decohere. In order to limit the error probability to 3%, which is the largest error probability that can be handled with error correcting codes at the time of writing this book [2], we need $T < 0.03T_2$. If $T_2 = 1$ nanosecond (charge), we need a clock frequency $1/T = 33$ GHz, whereas if $T_2 = 1$ microsecond (spin), we need a clock frequency of 33 MHz. Alternately, with a fixed clock period of 5 GHz, the error probability with spin ($T_2 = 1$ microsecond) is 2×10^{-4}, while the error probability with charge ($T_2 = 1$ nanosecond) is 0.18. Therefore, spintronics is considerably superior to electronics when it comes to quantum computing. This realization has provided a tremendous boost to spintronics.

16.7   Superposition states

We are now ready to elucidate how superposition endows a quantum computer with immense prowess. Consider a bistable spin (a single electron placed in a magnetic field). We can use the up-spin state to encode the binary bit 1 and the down-spin state to encode the binary bit 0. Classically, this electron can then store only one bit of information (whose value can be 0 or 1) in its spin degree of freedom because we have two states (up and down) and two possible values of the binary bit. Since each electron can store only one bit,

N electrons can store N binary bits classically. But now, if we can create a coherent superposition of the states of two electrons and *entangle their spins*, the corresponding qubit will be written as

$$qubit_{2\ electrons} = a_{00}|00> +a_{01}|01> +a_{10}|10> +a_{11}|11>, \qquad (16.32)$$

where the state $|ij>$ corresponds to the first electron's spin being in state $|i>$ and the second electron's spin in state $|j>$. This system can, in principle, store 2^2 bits of information corresponding to (i) both electrons being in the down-spin state, (ii) first in the down-spin state and the second in the up-spin state, (iii) the first in the up-spin state and the second in the down-spin state, and (iv) both in the up-spin state. Note that the classical case would correspond to only one of the coefficients in Equation (16.32) being non-zero. Extending this to N bistable spins, we can write the corresponding qubit as

$$qubit_{N-electron} = \sum_{x_1 x_2 ... x_N} a_{x_1 x_2 ... x_N}|x_1 x_2 ... x_N>, \qquad (16.33)$$

where the x-s can take the value 0 or 1. The above equation has 2^N terms. It means that while N bistable spins can store N binary bits classically, the same system can store 2^N classical bits quantum mechanically. Now ponder the following. If we want to store 2^{300} binary bits of information, it is impossible in the classical realm since the number 2^{300} is larger than the number of electrons in the known universe. However, in the quantum mechanical realm, just 300 electrons with entangled spins in a coherent superposition state can do the job!

Now the tempered vision. It is extremely difficult to entangle 300 qubits. So far, only a few qubits (less than 10) have been entangled. Furthermore, one might be misled to think that we can make a huge "memory" storing 2^N binary bits of information using just N entangled spins maintained in a coherent superposition state. This point of view would be *incorrect*. A reliable memory should be such that every time it is accessed, it returns the *same* data which is stored in it. For instance, if we have stored the binary string 11001, we should get this string *every time* the memory is read. But the quantum memory we are talking about does not satisfy this requirement. Every time we read the memory, the qubits collapse to classical bits and we only get N (not 2^N) bits out of it. More important, the values of these bits change with each measurement because of the probabilistic nature of quantum measurement. It is never the same bit string in two successive measurements! Surely, this cannot be judged a reliable memory.

16.8   Quantum parallelism

If the above is true, then how is the power of quantum computing ever harnessed? The answer is in something that has been termed "quantum parallelism." We discuss that next.

In the world of classical computers, parallelism refers to parallel (simultaneous) processing of different information in different processors. Quantum parallelism refers to simultaneous processing of different information or inputs in the *same* processor. This idea, due to Deutsch [17], refers to the notion of evaluating a function once on a superposition of all possible inputs to the function to produce a superposition of outputs. Thus, all outputs are produced in the time taken to calculate one output classically. Of course, not all of these outputs are accessible since a measurement on the superposition state of the output will produce only one output. However, it is possible to obtain certain joint properties [18] of the outputs, and that is a remarkable possibility.

Let us exemplify quantum parallelism more concretely. Consider the situation when N inputs x_1, x_2, ... x_N are provided to a computer and their functions $f(x_1)$, $f(x_2)$, ... $f(x_N)$ are to be computed. The results are then fed to another computer to calculate the functional $F(f(x_1), f(x_2), ...f(x_N))$.

With a classical computer, we will calculate $f(x_1)$, $f(x_2)$, ... $f(x_N)$ serially, one after the other. With a quantum computer, the story is different.

Prepare the initial state as a superposition of the inputs:

$$|I> = \frac{1}{\sqrt{N}}(|x_1 > +|x_2 > +... + |x_N >). \tag{16.34}$$

Let it evolve in time to produce the output

$$|O> = \frac{1}{\sqrt{N}}(|f(x_1) > +|f(x_2) > +... + |f(x_N) >). \tag{16.35}$$

Note that $|O>$ has been obtained in the time required to perform a single computation. Now, if $C = F(f(x_1), f(x_2), ...f(x_N))$ can be computed from $|O>$, then a quantum computer will be of great advantage. This is an example where "quantum parallelism" can be used to speed up the computation tremendously.

There are two questions now. First, can C be computed from a knowledge of the superposition of various $f(x_i)$-s and not the individual $f(x_i)$-s? The answer is "yes," but for a small class of problems. These are called the Deutsch-Josza class of problems, which can benefit from quantum parallelism. Second, can C be computed correctly with unit probability? The answer is "no." C cannot be computed with unit probability. However, if the first answer is wrong (hopefully, the computing entity can differentiate right from wrong answers), then the experiment or computation is repeated until the right answer is obtained. The probability of getting the right answer within k iterations is $(1 - p^k)$ where p is the probability of getting the wrong answer in any iteration. The mean number of times the experiment should be repeated is $N^2 - 2N - 2$.

The following is an example of the Deutsch-Josza class of problems. For integer $0 \leq x \leq 2L$, given that the function $f_k(x) \in [0,1]$ has one of two properties – (i) either $f_k(x)$ is independent of x, or (ii) one half of the numbers $f_k(0), f_k(1), ...f_k(2L-1)$ are zero – determine which type the function belongs to using the fewest computational steps.

The most efficient classical computer will require $L+1$ evaluations, whereas according to Deutsch and Josza, a quantum computer can solve this problem with just two iterations.

16.9 Universal quantum gates

The important question that concerns engineers and applied scientists is whether there exists a *universal* quantum gate. If it exists, we will focus on how to implement and synthesize this gadget, since it will ultimately lead to a quantum computer. The first universal quantum gate that was postulated is a 3-qubit gate due to Deutsch. It is fashioned after the universal classical gate of Toffoli and Fredkin. Using the computational basis states $|000 >$, $|001 >$, $|010 >$, $|011 >$, $|100 >$, $|101 >$, $|110 >$, and $|111 >$, the A-matrix of the Deutsch gate is given by

$$A_{Deutsch} = \begin{bmatrix} 1 & 0 & 0 & 0 & 0 & 0 & 0 & 0 \\ 0 & 1 & 0 & 0 & 0 & 0 & 0 & 0 \\ 0 & 0 & 1 & 0 & 0 & 0 & 0 & 0 \\ 0 & 0 & 0 & 1 & 0 & 0 & 0 & 0 \\ 0 & 0 & 0 & 0 & 1 & 0 & 0 & 0 \\ 0 & 0 & 0 & 0 & 0 & 1 & 0 & 0 \\ 0 & 0 & 0 & 0 & 0 & 0 & A & B \\ 0 & 0 & 0 & 0 & 0 & 0 & B & A \end{bmatrix}, \tag{16.36}$$

where

$$A = ie^{\frac{i\pi\alpha}{2}} (1 + e^{i\pi\alpha}) \text{ and } B = ie^{\frac{i\pi\alpha}{2}} (1 - e^{i\pi\alpha}) \tag{16.37}$$

and α may be an irrational number.

It is easy to check that $A_{Deutsch}$ is unitary. The T-F gate can be synthesized by connecting a number of Deutsch gates in series. The number of Deutsch gates required for this is N where N must satisfy the relation $[A_{Deutsch}]^N =$

$A_{Toffoli-Fredkin}$. Noting that

$$\begin{bmatrix} ie^{\frac{i\pi\alpha}{2}}(1 + e^{i\pi\alpha}) & ie^{\frac{i\pi\alpha}{2}}(1 - e^{i\pi\alpha}) \\ ie^{\frac{i\pi\alpha}{2}}(1 - e^{i\pi\alpha}) & ie^{\frac{i\pi\alpha}{2}}(1 + e^{i\pi\alpha}) \end{bmatrix}^N = i^N e^{\frac{iN\pi\alpha}{2}} \begin{bmatrix} ie^{\frac{i\pi\alpha}{2}}(1 + e^{i\pi\alpha}) & ie^{\frac{i\pi\alpha}{2}}(1 - e^{i\pi\alpha}) \\ ie^{\frac{i\pi\alpha}{2}}(1 - e^{i\pi\alpha}) & ie^{\frac{i\pi\alpha}{2}}(1 + e^{i\pi\alpha}) \end{bmatrix}$$

(16.38)

the quantity $e^{iN\pi\alpha}$ can be made arbitrarily close to any complex number of unit norm by changing N. Choosing N such that $e^{iN\pi\alpha} = -1$ yields the Toffoli-Fredkin gate [19].

16.9.1   Two-qubit universal quantum gates

Later, it was shown that there exist 2-qubit quantum gates which are universal. The proof of universality of these gates was given by DiVincenzo [20] using Lie algebra; however, it can also be shown that the Deutsch gate can be realized with these 2-qubit gates. This, in itself, is sufficient proof of universality. The Deutsch gate is not the most primitive universal quantum gate; the DiVincenzo gate is more primitive. Later, it was shown by Seth Lloyd that almost any 2-qubit gate is universal [21].

The computational basis of the 2-terminal (2 inputs and 2 outputs) universal quantum gate can be chosen as $|00 >, |01 >, |10 >, |11 >$. In this basis, the A-matrix of the universal 2-qubit gate is

$$A_{DiVincenzo} = \begin{bmatrix} 1 & 0 & 0 & 0 \\ 0 & 1 & 0 & 0 \\ 0 & 0 & A & B \\ 0 & 0 & B & A \end{bmatrix},$$

(16.39)

where A and B have been given by Equation 16.37.

16.10   A 2-qubit "spintronic" universal quantum gate

Today, most experimental effort in realizing semiconductor solid-state versions of universal 2-qubit gates is focused on the realization of a quantum controlled NOT or XOR gate. These are the two most popular gates and both are "universal." In these gates, one of the two qubits is called the target qubit and the other is the control qubit. The basic operation of the controlled NOT gate has two ingredients: (i) arbitrary rotation of the target qubit, and (ii) controlled rotation of the target qubit through specified angles if and only if the control qubit is in a certain state. The controlled XOR gate has similar dynamics: (i) arbitrary rotation of the target qubit, and (ii) a so-called "square-root-of-swap" operation where the quantum states of the target and control qubits are half-way interchanged [28].

Numerous authors have proposed various spintronic embodiments of the 2-qubit universal quantum gate. It is not possible to discuss all, or even a large subset of them, in a textbook. Therefore, we take a historic perspective and discuss some of the early ideas for universal quantum gate.

16.10.1 Silicon quantum computer based on nuclear spins

The first comprehensive solid state scalable realization of a universal 2-qubit spintronic quantum gate is probably the one due to Kane [29]. Qubits are encoded in the nuclear spin orientations of ^{31}P dopant atoms in silicon. These nuclear spin orientations are coherent superpositions of two anti-parallel polarizations. The coherence is long lived since nuclear spins have very long coherence times, possibly several milliseconds at 4.2 K temperature. Therefore, the qubits are robust.

To select a qubit for rotation, the spin-splitting energy of the target nucleus is first altered (using hyperfine interaction) by changing the wavefunction of the electron bound to that nucleus. This is achieved by applying a potential via a lithographically defined gate (called an "A-gate") placed precisely on top of that nucleus. The gate potential attracts or repels the electron bound to that nucleus, thereby changing its wavefunction, and hence the nucleus' spin-splitting energy via hyperfine interaction. The spin-splitting energy is made resonant with a global ac magnetic flux density of amplitude B_{ac} for specific duration (Rabi oscillation). If the duration is τ, then the angle by which the spin is rotated is

$$\theta(\tau) = \frac{g\mu_B B_{ac}}{\hbar}\tau. \tag{16.40}$$

This realizes single qubit rotation.

The 2-qubit operation is achieved by yet another set of lithographically defined gates (called "J gates") which raise or lower a potential barrier between two neighboring nuclei's electrons (one nucleus is the "target" and the other is the "control" qubit). As the barrier is lowered, the wavefunctions of the two electrons bound to those nuclei overlap, and the resulting exchange interaction lowers the energy of the singlet state with respect to the triplet state (as long as there is no magnetic field to induce a Zeeman splitting that exceeds the exchange splitting). As a result, the spin-splitting energy in any nucleus depends on the spin state of its neighbor. Thus, the rotation of the target qubit by the global ac magnetic field can be made conditional on the spin state of the control qubit. This realizes the quantum controlled NOT operation. When the gate operations are over, the nuclear spin is "read" by transducing the nuclear spin to electron spin, and then reading that electron spin with spin transistors or other devices that are capable of spin-to-charge transduction, and thus allowing a measurement of the spin.

A closely related proposal that hosts qubits in electronic spin, rather than nuclear spin, has been put forward by Vrijen et al. [30]. This model eliminates the need to transduce the qubit between nuclear and electron spin. Instead of using hyperfine interaction to change the spin-splitting energy in a target qubit, they utilize a Si/SiGe heterostructure. The ^{31}P dopant atom has to be placed exactly at the interface of these two materials and a lithographically delineated gate has to be aligned exactly on top of this buried dopant atom. The gate potential pushes the electron wavefunction into either the SiGe side or the Si side. Since these two materials have different g-factors, the spin-splitting energy in a constant magnetic field can be varied by the gate potential. This eliminates the need for hyperfine interaction (which is weak) to effect single qubit rotations. The downside is that extreme precision is required to place the buried dopant atom exactly underneath the gate (lateral precision) and also exactly at the interface of Si and SiGe (vertical precision). The lateral alignment may be achieved by ion-implanting the dopant through a mask [31], but the vertical alignment is much more difficult. One has to virtually eliminate "straggle" in ion implantation, and that is extremely difficult.

16.10.2   Quantum dot-based spintronic model of universal quantum gate

The idea of using the spin of a single electron in a quantum dot to serve as a vehicle for qubits has been around since at least 1996 [1, 32]. This has spawned many proposals for realizing universal quantum gates using the spin of a single electron in a quantum dot. We outline one such proposal [33].

Consider a penta-layered semiconductor quantum dot defined by split-gates on a heterostructure consisting of ferromagnet-insulator-semiconductor-insulator-ferromagnet layers, as shown in Fig. 16.6. We also show the conduction band diagram in the direction perpendicular to the heterointerfaces. The ferromagnets are magnetized and generate a permanent magnetic field over the semiconductor which spin-splits the quantum dot energy levels in the semiconductor because of Zeeman interaction. There may also be additional spin-splitting due to spin-orbit interaction. The Fermi level, which is determined by the metallic ferromagnetic contacts, is initially below the lower spin-split level, so that the quantum dot is depleted of all electrons (we assume very low temperature for this discussion).

Next, a positive potential is applied to the split-gates which effectively makes the quantum dot slightly larger and pulls the lower Zeeman spin-split level below the Fermi level. A spin polarized electron (whose spin is the majority spin in the ferromagnet) now tunnels into the semiconductor dot from the contacts. Only one electron can enter provided its spin is the majority spin in the ferromagnet because of the Pauli exclusion principle. Therefore, we have "written" and initialized a qubit.

The insulating layers provide a confining potential for the electron in the semiconductor dot. They also facilitate coherent spin injection from the ferromagnet because they increase the spin injection efficiency [34].

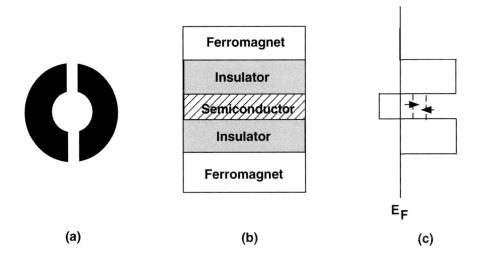

FIGURE 16.6
(a) Top view of the quantum dot defined by ring shaped split-gates. (b) A pentalayered heterostructure. The electron is confined in the semiconductor layer. (c) The conduction band energy diagram in the direction perpendicular to the heterointerfaces. The Fermi level is determined by the metallic ferromagnets. The broken lines show the initial spin-split energy levels in the semiconductor dot.

To select a qubit for rotation, we apply a differential potential between the two arms of the split-gate to cause a structural inversion asymmetry and an ensuing Rashba interaction. This modifies the total spin-splitting energy in the quantum dot (see Chapter 7). We apply just enough differential potential to make the total spin-splitting energy resonant with a global ac magnetic field. We hold the resonance for a finite duration τ. This rotates the spin in accordance with Equation (16.40). By controlling the time duration of the resonance, i.e., the quantity τ in Equation (16.40), we can rotate the spin through arbitrary angles θ. This satisfies the first requirement for a universal quantum gate, namely, arbitrary single qubit rotation.

There are, of course, many other ways of rotating a spin in a quantum dot resulting in single qubit rotation. Some of them do not use a magnetic field at all, but instead use an electric field. However, they typically need exotic structures and phenomena (e.g., electrically induced g-factor variation). Such rotations have been demonstrated experimentally [22, 23, 24, 25] and apparently can be carried out quite fast, in fractions of a nanosecond. Much faster spin rotation (on time scales of picoseconds or femtoseconds) is possible using optical manipulation [26, 27]. Thus, single qubit rotation is now an established art that has been demonstrated repeatedly with increasing sophistication.

The second and last ingredient is the 2-qubit controlled NOT or square-root-of-swap operation. For this purpose, we apply a positive potential to the barrier separating two adjacent quantum dots that lowers the barrier and makes the wavefunctions in these two dots overlap. That generates an exchange interaction. If the exchange interaction is allowed to take place for a duration $h/(16J)$, where J is the exchange interaction strength, we will have realized the "square-root-of-swap" operation [28]. The controlled NOT operation can also be realized since the spin-splitting energy in one dot (the difference between the up-spin and the down-spin energies) definitely depends on the spin state in its neighbor because of exchange interaction (recall the quantum inverter). Because of this, we can make the spin-splitting energy in the target dot resonant with a global magnetic field only if its neighbor's spin is up or down. Simply put, we can make the spin rotation in the target dot conditional on the spin state in the control dot. This realizes the controlled NOT operation, which is the second (and final) ingredient of a universal 2-qubit quantum gate.

16.11   Conclusion

In conclusion, we have provided the reader with a brief glimpse of spin-based quantum computing. We have omitted any discussion of opto-spintronic implementations of quantum gates since they are more complex and require the introduction of ideas and topics that are not in this textbook. These implementations are, however (see, for example, [35]), important and probably more numerous than the purely spintronic implementations.

16.12 Problems

- **Problem 16.1**

  Using Equation (16.6), show that $|c_2(\tau_d)| = 1$ where τ_d is given by Equation (16.7) and the following condition holds: $h_A = 2J$.

- **Problem 16.2**

  Consider a NAND gate with no inputs applied to the input dots. Assume that the gate is in the ground state whose wavefunction is given by Equation (16.10). Next, inputs [1 1] are applied to input dots **A** and **C**. Using Equation (16.4) and noting that the Hamiltonian H is given by Equation (15.3), show that the correct desired state is reached in time t_d, i.e., $\psi(t_d) = |\downarrow\uparrow\downarrow\rangle$, where $t_d \approx 0.32\hbar/J$, provided $h_A = h_C = h = 3J$. You can use software tools like MATLAB for symbolic manipulation. This example shows that the NAND gate, initially in the ground state without any inputs, can be switched coherently (without dissipating energy) to the correct state (corresponding to the truth table of a NAND gate) when inputs [1 1] are applied.

  Note that only for the input combination [1 1], the NAND gate is logically reversible since there is a unique input-output relation that allows one to infer the input bits from the output bit. This does not happen for any other combination of inputs such as [0 1], or [1 0] or [0 0].

- **Problem 16.3**

  Repeat the last problem when the input bits to a NAND gate are [0 0] to show that the amplitude of the "wrong" state $|\downarrow\uparrow\downarrow\rangle$ reaches zero at time $0.32\ \hbar/J$ when $h = 3J$. Can the correct state $|\uparrow\downarrow\uparrow\rangle$ be reached through coherent dynamics?

- **Problem 16.4**

  Repeat Problem 16.2 when the input bits to a NAND gate are [0 1] or [1 0] to show that the amplitude of the target state $|\uparrow\downarrow\downarrow\rangle$ or $|\downarrow\downarrow\uparrow\rangle$ reaches a maximum when $h \approx 2J$. Show also that, when this happens, the amplitude of the undesired state $|\downarrow\uparrow\downarrow\rangle$ vanishes.

- **Problem 16.5**

  Show that, when $h \gg J$, the ground state in Table 15.2 is nearly the pure unentangled state $| \downarrow\uparrow\downarrow >$. Hint: Evaluate the coefficients of the three different constituents of the ground state and show that the coefficient of the $| \downarrow\uparrow\downarrow >$ constituent reaches nearly unity when $h \gg J$.

- **Problem 16.6**

  The Fredkin swap gate is a universal classical reversible gate with three inputs A, B and C and three outputs P, Q, and R. The input-output relationship is the following: if $A = 0$, then $Q = B$ and $R = C$. However, if $A = 1$, then the outputs are swapped, namely, $Q = C$ and $R = B$.

  We will realize this gate with two quantum dots each hosting a single spin. The quantum dots are placed in a global magnetic field that defines the spin quantization axis, i.e., only polarizations parallel and anti-parallel to this magnetic field are stable or metastable. These two polarizations encode the logic bits 0 and 1.

  The potential barrier between these two dots is sufficiently high that the wavefunctions of the two electrons do not overlap in space and there is no exchange interaction between them. The potential barrier is, however, gated and when the gate potential is held high, the barrier is lowered to allow exchange coupling between the two spins.

  We will call the spin polarizations in the two dots at time $t = 0$ the bits B and C, while the spin polarizations at some later time τ will be bits Q and R. The gate potential encodes the bit A. When $A = 0$, the two spins are uncoupled, and, as long as τ is much smaller than the spin relaxation time, we will have the situation that $Q = B$ and $R = C$. When $A = 1$, the two spins are exhange coupled. Show that, if $\tau = h/8J$ (where J is the exchange interaction strength), then the Fredkin swap is realized, i.e., $Q = C$ and $R = B$.

  Hint: Consider all four cases when the spins in the two dots are $| \uparrow\uparrow >$, $| \uparrow\downarrow >$, $| \downarrow\uparrow >$, $| \downarrow\downarrow >$. Show that if the spins are exchange coupled for a duration $h/8J$, then

  $$| \uparrow\uparrow > \rightarrow | \uparrow\uparrow >$$
  $$| \uparrow\downarrow > \rightarrow | \downarrow\uparrow >$$
  $$| \downarrow\uparrow > \rightarrow | \uparrow\downarrow >$$
  $$| \downarrow\downarrow > \rightarrow | \downarrow\downarrow >$$

  which realizes the Fredkin swap operation.

16.13 References

[1] S. Bandyopadhyay and V. P. Roychowdhury, "Switching in a reversible spin logic gate", Proc. Superlat. Microstruct. Conf., Liege, Belgium, July 1996. (Superlat. Microstruct., **22**, 411 (1997)).

[2] E. Knill, "Quantum computing with realistically noisy devices", Nature (London), **434**, 39 (2005).

[3] R. T. Bate in *VLSI Microstructure Science and Engineering*, Vol. 4, Ed. N. G. Einspruch, (Academic Press, New York, 1981).

[4] C. Mead and L. Conway, *Introduction to VLSI Systems*, (Addison-Wesley, Reading, MA, 1980).

[5] R. Landauer, "Uncertainty principle and minimal energy dissipation in the computer", Intl. J. Theor. Phys., **21**, 283 (1982).

[6] P. Benioff, "The computer as a physical system- A microscopic quantum mechanical Hamiltonian model of computers as represented by Turing machines", J. Stat. Phys., **22**, 563 (1980).

[7] P. Benioff, "Quantum mechanical models of Turing machines that dissipate no energy", Phys. Rev. Lett., **48**, 1581 (1982).

[8] P. Benioff, "Quantum mechanical Hamiltonian models of discrete processes that erase their own histories: Applications to Turing machines", Intl. J. Theor. Phys., **21**, 177 (1982).

[9] E. Fredkin and T. Toffoli, "Conservative logic", Intl. J. Theor. Phys., **21**, 219 (1982).

[10] X. W. Zhang, W. J. Fan, S. S. Li and J. B. Xia, "Giant and zero electron g-factors of dilute nitride semiconductor nanowires ", Appl. Phys. Lett., **90**, 193111 (2007).

[11] S. Lloyd, "A potentially realizable quantum computer ", Science, **261**, 1569 (1993).

[12] C. P. Williams and S. H. Clearwater, *Explorations in Quantum Computing*, (Springer-Verlag, New York, 1998).

[13] J. M. Kikkawa and D. D. Awschalom, "Resonant spin amplification in n-type GaAs ", Phys. Rev. Lett., **80**, 4313 (1998).

[14] R. deSousa and S. Das Sarma, "Electron spin coherence in semiconductors: Considerations for a spin-based solid-state quantum computer architecture", Phys. Rev. B, **67**, 033301 (2003).

[15] N. Bar-Gill, L. M. Pham, A. Jarmola, D. Budker and R. L. Walsworth, Nature Commun., **4**, 1743 (2013).

[16] P. Mohanty, E. M. Q. Jariwalla and R. A. Webb, "Intrinsic decoherence in mesoscopic systems", Phys. Rev. Lett, **78**, 3366 (1997).

[17] D. Deutsch, "Quantum theory, the Church-Turing principle and the universal quantum computer ", Proc. Royal Soc. London A, **400**, 97 (1985).

[18] R. Jozsa, "Characterizing classes of functions computable by quantum parallelism", Proc. Royal Soc. London A, **435**, 563 (1991).

[19] V. P. Roychowdhury, private communication.

[20] D. P. DiVincenzo, "2 bit gates are universal for quantum computation", Phys. Rev. A., **51**, 1015 (1995).

[21] S. Lloyd, "Almost any quantum logic gate is universal ", Phys. Rev. Lett., **75**, 346 (1995).

[22] K. C. Nowack, F. H. L. Koppens, Yu. V. Nazarov and L. M. K. Vandersypen, "Coherent control of single electron spins with electric fields", Science, **318**, 1430 (2007).

[23] Y. Kato, R. C. Meyers, D. C. Driscoll, A. C. Gossard, J. Levy and D. D. Awschalom, "Gigahertz eelctron spin manipulation using voltage controlled g-tensor modulation", Science, **299**, 1201 (2003).

[24] F. H. L. Koppens, et al., "Driven coherent oscillations of a single electron spin in a quantum dot", Nature (London), **442**, 766 (2006).

[25] Y. Tokura, W. van der Wiel, T. Obata and S. Tarucha, "Coherent single electron spin control in a slanting Zeeman field", Phys. Rev. Lett., **96**, 047202 (2006).

[26] M. V. Gurudev Dutt, et al., "Ultrafast optical control of electron spin coherence in charged GaAs quantum dots", Phys. Rev. B., **74**, 125306 (2006).

[27] D. Press, T. D. Ladd, B. Zhang and Y. Yamamoto, "Complete quantum control of a single quantum dot spin using ultrafast optical pulses", Nature (London), **456**, 218 (2008).

[28] G. Burkard, D. Loss and D. P. DiVincenzo, "Coupled quantum dots as quantum gates", Phys. Rev. B, **59**, 2070 (1999).

[29] B. E. Kane, "A silicon based nuclear spin quantum computer ", Nature (London), **393**, 133 (1998).

[30] R. Vrijen, et al., "Electron spin resonance computers for quantum computing in silicon-germanium heterostructures", Phys. Rev. A, **62**, 12306 (2000).

[31] R. P. McKinnon, et al., "Nanofabrication processes for single ion implantation of silicon quantum computer devices", Smart Materials and Structures, **11**, 735 (2002).

[32] A. M. Bychkov, L. A. Openov and I. A. Semenihin, "Single electron computing without dissipation", JETP Lett., **66**, 298 (1997).

[33] S. Bandyopadhyay, "Self assembled nanoelectronic quantum computer based on the Rashba effect in quantum dots", Phys. Rev. B, **61**, 13813 (2000).

[34] E. I. Rashba, "Theory of electrical spin injection: Tunnel contacts as a solution of the conductivity mismatch problem", Phys. Rev. B, **62**, R16267 (2000).

[35] T. Calarco, A. Datta, P. Fedichev, E. Pazy and P. Zoller, "Spin based all optical quantum computation with quantum dots: Understanding and suppressing decoherence", Phys. Rev. A., **68**, 012310 (2003).

17

Nanomagnetic Logic: Computing with Giant Classical Spins

The idea of classical Boolean computing with single electron spins was described in Chapter 15. It is an interesting notion, but it is also somewhat academic since the operating temperature is likely to be a few Kelvin. There is however a room-temperature version of single spin logic where a single spin is replaced with a single-domain *shape anisotropic* ferromagnet, as shown in Fig. 17.1. The domain contains many spins, but, because of strong exchange interaction between the spins, they all point in the same direction. If a magnetic field or some other entity makes the spins reorient by rotating, all the spins will rotate in unison so that the entire ensemble of spins acts like one giant classical spin [1]. This has an important ramification. Magnets of certain shapes (e.g., the elliptical disk shown in Fig. 17.1) have only two stable magnetization orientations. In this case, the orientations are along the major axis of the ellipse ("up" and "down"), which can be used to encode logic bits "0" and "1" just as a single electron spin's two stable orientations in a magnetic field can be used to encode binary bits. Therefore, the single domain nanomagnet can act as a binary switch since it has two stable states. The difference with single electron spin in a magnetic field is that there the two stable states were energetically non-degenerate, but here they are degenerate and no magnetic field is needed to maintain the bistability. The bistability is entirely due to the anisotropic shape of the nanomagnet.

One can switch the magnet from one stable state to the other (which is equivalent to switching the stored bit from 0 to 1, or vice versa) in a variety of ways. Let us briefly compare the magnetic switch with a transistor switch at this point. A transistor (e.g., a MISFET) also has two stable conductance states ("on" and "off") and switching between them is accomplished by driving charges in and out of the transistor's channel. Each charge acts as a single degree of freedom and if N charges are involved in the switching action, then the minimum energy that will be dissipated in the switching process can be shown to be $NkTln(1/p)$ where p is the probability of static error (i.e the probability that a bit will switch spontaneously, resulting in error). However, in the case of the magnet switch, even if there are N spins, since they all rotate in concert, they act like a *single* degree of freedom. Hence the minimum energy that will be dissipated in switching a magnet is $kTln(1/p)$ [2]. Therefore, the magnet has an intrinsic energy advantage over the transistor –

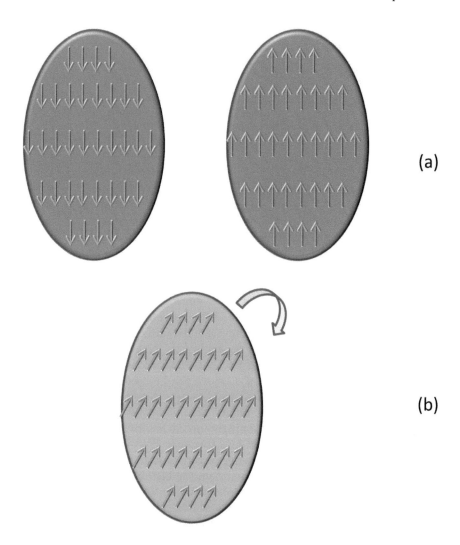

FIGURE 17.1

(a) An elliptical nanomagnet small enough to contain a single domain. In the nanomagnet, exchange interaction between the spins makes all of them point in the same direction as shown. Because of the elliptical shape of the magnet, which makes it shape-anisotropic, the spins tend to point along the major axis of the ellipse. The two directions – "up" and "down" along the major axis – are stable orientations for the magnetization vector of the magnet and they can be used to encode the logic bits 0 and 1. (b) When the magnetization of the magnet rotates under some external perturbation, all the spins rotate together in unison. That makes the entire single-domain nanomagnet behave like one giant classical spin.

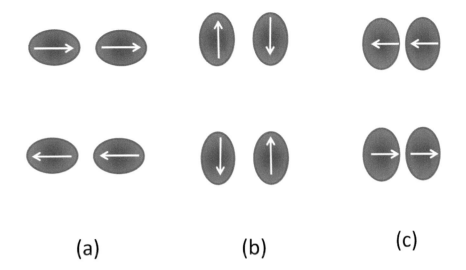

(a) (b) (c)

FIGURE 17.2
(a) Two elliptical nanomagnets with the line joining their centers aligned along their common major axis. The ordering of their magnetizations is ferromagnetic (mutually parallel). The magnetizations point along the major axis. (b) Two elliptical nanomagnets with the line joining their centers aligned along their common minor axis. The ordering is anti-ferromagnetic (mutually anti-parallel) and the magnetizations again point along the major axis. (c) Two elliptical nanomagnets with the line joining their centers aligned along their common minor axis but the magnets are so close to each other that the dipole coupling energy in the ferromagnetic state exceeds the shape anisotropy energy barrier in both magnets. The ordering of their magnetizations then becomes ferromagnetic, with the magnetizations pointing along the minor axes.

particularly when N is large.

Next, let us examine how an array of nanomagnets placed in certain patterns on a wafer can implement Boolean logic gate functionality in the manner of SSL. If two nanomagnets are placed close to each other, then they interact via dipole interaction just as two spins in close proximity interact via exchange interaction. Dipole interaction is much longer range than exchange interaction – exchange interaction decays exponentially with distance, while dipole interaction tends to decay as the inverse cube of the separation between two magnets.

If two elliptical nanomagnets are placed such that the line joining their centers is along the common major axis, then their magnetizations tend to be mutually parallel, as shown in Fig. 17.2(a) because of dipole interaction. If the line joining their centers is along the common minor axis, then the magneti-

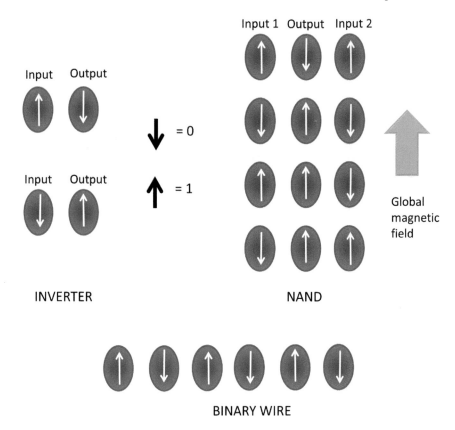

FIGURE 17.3
Nanomagnetic implementations of an inverter, a NAND gate, and a binary
wire.

zations will tend to be mutually anti-parallel, as long as the dipole interaction
energy is smaller than the energy barrier (called the "shape anisotropy energy barrier") that separates the two stable magnetization orientations along
the major axis of a single magnet. This is shown in Fig. 17.2(b). On the
other hand, if the magnets are placed so close to each other that the dipole
coupling energy in the parallel configuration exceeds the shape anisotropy energy barrier, then the magnetizations of both magnets tend to point along
the common minor axis as shown in Fig. 17.2(c). The configuration in Fig.
17.2(b) is useful for implementing Boolean logic.

17.1 Nanomagnetic logic and Bennett clocking

Just as in SSL, an inverter can be implemented with two weakly dipole-coupled nanomagnets and a NAND gate with three weakly dipole-coupled nanomagnets placed in a global bias magnetic field. This is shown in Fig. 17.3. Just as exchange coupling between spins enforces anti-ferromagnetic ordering, the dipole coupling between the nanomagnets will enforce anti-ferromagnetic ordering, thereby realizing the inverter and the NAND gate. An array of nanomagnets placed along a line coinciding with their minor axes will act as a "binary wire", whereby the bit encoded in the first nanomagnet will be replicated in every odd numbered magnet because of the dipole coupling enforcing anti-ferromagnetic ordering. This last system has been termed "magnetic quantum cellular automata" [3]. These magnetic systems have been discussed in [4].

The reader might have (correctly) guessed that neither any gate nor any binary wire would ever work without appropriate clocking. To understand why this is the case, consider the situation when the input bit to the inverter is flipped by an external agent and we expect the output bit to flip in response to carry out the inverter operation. This situation is depicted in Fig. 17.4(b). In order to flip, the output bit has to rotate by ~180° and in the process overcome the energy barrier that separates the two stable magnetization orientations along the major axis of the ellipse. This is called the shape anisotropy energy barrier, which arises owing to the shape anisotropy (eccentricity) of the ellipse and without it, we would not have had two stable orientations along the major axis. The magnetization may not be able to flip by itself in response to the input flipping. Therefore, the remedy is to provide additional energy to the output magnet in the form of a Bennett clock [5] to aid the output magnet's magnetization to overcome the shape anisotropy barrier, flip, and assume an orientation anti-parallel to that of the input magnet (successful inversion).

To understand what the clock must do, we should look at the potential energy profile of the magnet as a function of the magnetization's orientation in space. Let the polar and azimuthal angles of the magnetization vector be θ and ϕ, respectively, as shown in Fig. 17.5(a). If we take an isolated nanomagnet and plot its potential energy profile in the magnet's plane ($\phi = 90°, 270°$), then the profile will look like that in Fig. 17.5(a). Note that there are two degenerate global minima at $\theta = 180°$ and $\theta = 0°$, which correspond to the two orientations along the major axis of the ellipse. The major axis is therefore the "easy" axis and the magnetization's stable orientations are the two mutually anti-parallel directions along the major axis. Note also the energy barrier separating the two stable magnetization orientations. This is the "shape-anisotropy energy barrier." For the magnetization to go from one stable state to the other (i.e., for the magnetization to "flip"), the energy barrier has to be transcended. The orientation along the minor axis is the

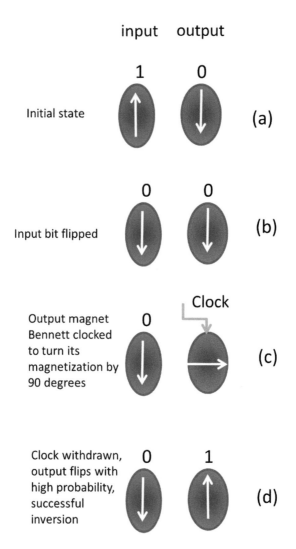

FIGURE 17.4

(a) Two dipole-coupled magnets in the ground state showing bit inversion.
(b) The input bit is flipped with an external agent, placing the two magnetizations temporarily in a parallel configuration. (c) The output magnet's
magnetization is rotated by ~90° with an external clocking agent. (d) The
clocking agent is removed and the output magnet's magnetization assumes an
orientation anti-parallel to that of the input magnet's with high probability
(because of dipole coupling with the neighbor), resulting in successful inverter
operation.

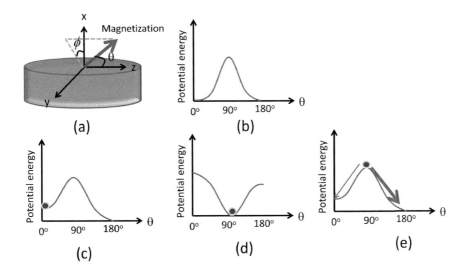

FIGURE 17.5

(a) Magnetization vector. (b) Potential energy profile in the plane of the magnet ($\phi = 90°, 270°$) as a function of the polar angle θ. (c) Potential energy profile of the output nanomagnet of the inverter dipole-coupled with the input nanomagnet. The purple circle with a yellow border denotes the location of the magnetization orientation (system state) and this figure corresponds to the state in Fig. 17.4(b). (d) Potential energy profile of the output nanomagnet and this figure corresponds to the state in Fig. 17.4(c). (e) Potential energy profile of the output nanomagnet after the clock has been withdrawn. There is a much higher chance of the system going to $\theta = 180°$ as opposed to $\theta = 0°$ since the former state is lower in energy, resulting in the configuration shown in Fig. 17.4(d).

maximum energy position and hence the minor axis is called the 'hard axis'.

If there are two dipole coupled nanomagnets, as shown in Fig. 17.4(b), with the fixed input magnet's magnetization pointing down (we will assume that the "down" orientation corresponds to $\theta = 0°$), then the energy profile of the output magnet will look like that in Fig. 17.5(c), with a slight preference for $\theta = 180°$ over $\theta = 0°$ because of the dipole-coupling preferring anti-ferromagnetic ordering. The coupling has made the energy profile asymmetric, resulting in a local minimum at $\theta = 0°$ and a global minimum at $\theta = 180°$.

The system should go to the global minimum and the magnetization of the output magnet should flip to assume the configuration in Fig. 17.4(d), except it cannot do so without overcoming the shape anisotropy energy barrier. Therefore, flipping can happen only if we temporarily *remove* the energy barrier with a clock. There are many ways to accomplish that. For example, one can apply a magnetic field along the minor axis [6], or pass a spin polarized current through the magnet with spins polarized along the minor axis, or apply uniaxial stress along the major axis [7]. All of these approaches can invert the potential barrier, resulting in a monostable profile shown in Fig. 17.5(d), whereby the energy minimum has moved to $\theta = 90°$, which is along the minor axis of the ellipse. Therefore, the magnetization will rotate by 90° and temporarily align along the minor axis or in-plane hard axis, as shown in Fig. 17.4(c). Next, when the clocking agent is removed, the magnetization gets perched on the maximum energy state, which is unstable. It will therefore decay to either the global ground state ($\theta = 180°$), resulting in successful inversion (as shown in Fig. 17.4(d)), or to the local ground state at $\theta = 0°$, resulting in unsuccessful inversion. Since the global ground state is lower in energy than the local ground state owing to dipole coupling, the probability of success is higher than the probability of failure, and the difference obviously increases with increased dipole coupling strength, which makes the global energy minimum increasingly lower in energy compared to the local minimum. Since, there is a limit to how strong the dipole coupling can be without upsetting the anti-ferromagnetic ordering, it is obvious that there is a lower bound on the error probability, which is unfortunately not zero. Therefore, nanomagnetic logic tends to be error-prone. This is the Achilles' heel of nanomagnetic logic, and a great deal of research is being devoted to overcome this impasse.

One obvious way to reduce error is to adjust the clock strength just so that the shape anisotropy energy barrier is eroded by the clock, but not inverted. This would result in the magnetization never being placed in the maximum energy state from which it has two final states (correct and wrong) available. Instead, now there is only one energy minimum, which is the global minimum, and the system should go there with very high probability. The corresponding potential energy profiles in the plane of the magnet are shown in Fig. 17.6. This remedy has been studied [8, 9, 10, 11] and seems to work, but it still does not necessarily reduce the error probability to acceptable limits at room temperature, where thermal noise can cause erratic magnetization dynamics.

There are many ways to rotate the magnetization of a nanomagnet through a chosen angle with an external agent. The external agent acts like a *torque* on the magnetization, which rotates the magnetization. The evolution of the magnetization vector under a torque is modeled by the Landau-Lifshitz-Gilbert (LLG) equation as long as the magnet has a single domain and all the spins in the domain rotate in unison. This is called the "macrospin model". In reality, all the spins may not rotate coherently in absolute unison, and in that case, there is no single equation that can be used to extract the temporal evolution of the magnetization vector under the torque. The magnetization dynamics is then obtained numerically by solving the local dynamics. There are many software packages available in the public domain for this purpose, some examples being OOMMF (Object Oriented Micro Magnetic Framework) [12], available from the National Institute of Standards and Technology in the

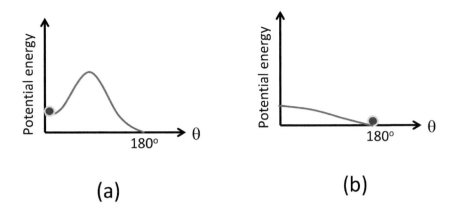

FIGURE 17.6
(a) Potential energy profile of the output nanomagnet of the inverter (in the plane of the nanomagnet). The output magnetization is in the state shown in Fig. 17.4(b), (b) Potential energy profile under critical clocking when the shape anisotropy energy barrier has been just eroded but not inverted. The profile is monostable and the system migrates to the only stable state (global energy minimum) if given sufficient time to settle.

United States and Magpar (Parallel Finite Element Micromagnetics Package) [13].

In the macrospin approximation, the temporal evolution of magnetization vector $\vec{M}(t)$ is described by the LLG equation [14, 15]:

$$\frac{d\vec{M}(t)}{dt} - \alpha \frac{\vec{M}(t)}{|\vec{M}(t)|} \times \frac{d\vec{M}(t)}{dt} + \gamma \vec{M}(t) \times \vec{H}_{eff}(t) = 0, \qquad (17.1)$$

where α is the dimensionless Gilbert damping factor due to dissipation within the magnet caused by such effects as spin-orbit interaction and γ is the gyromagnetic ratio given by $\gamma = 2\mu_B\mu_0/\hbar$ with μ_B being the Bohr magneton and μ_0 being the permeability of free space. The value of γ is a universal constant equal to $2.21{\times}10^5$ rad-m/A-s.

This equation is also sometimes written as

$$(1+\alpha^2)\frac{d\vec{M}(t)}{dt} + \alpha\gamma\frac{\vec{M}(t)}{|\vec{M}(t)|} \times \left(\vec{M}(t) \times \vec{H}_{eff}\right) + \gamma\vec{M}(t) \times \vec{H}_{eff}(t) = 0, \; (17.2)$$

and it is easy to show that the two forms are equivalent.

The field $\vec{H}_{eff}$ is a magnetic field due to the external agent. Note that, since the torque on the magnetization vector generated by the external agent is

$$\vec{T}(t) = \mu_0\Omega\vec{M}(t) \times \vec{H}_{eff}(t), \qquad (17.3)$$

where Ω is the magnet's volume, we can rewrite the LLG equation in Equation (17.1) in a more convenient form:

$$\frac{d\vec{M}(t)}{dt} - \alpha \frac{\vec{M}(t)}{M_s} \times \frac{d\vec{M}(t)}{dt} + \frac{\gamma}{\mu_0 \Omega} \vec{T}(t) = 0, \tag{17.4}$$

where we have used the fact that $\left|\vec{M}(t)\right| = M_s$ at all times. The quantity M_s is the saturation magnetization of the magnet and is like the magnetization moment per unit volume if the unit volume has been magnetized to saturation.

All that remains is to find the torque per unit volume due to various types of external agents. The usual external agents utilized to switch the magnetization of a magnet are an external magnetic field, a spin-polarized current generating a spin transfer torque (the spin polarized current can be generated by first passing the current through another magnet magnetized in the intended direction of switching), or by the Spin Hall effect [16, 17] and mechanical strain which generates a torque only in a magnetostrictive magnet via the Villari effect. These are shown in Fig. 17.7. There are, however, other ways to generate a torque, such as through spin-orbit interaction [18] (since we know that the latter acts as an effective magnetic field), exchange coupling with the surface of a topological insulator [19], etc. These latter methods are not yet mainstream (at the time of writing) and hence are not discussed.

In general, the torque per unit volume acting on the magnetization will require knowledge of $\vec{H}_{eff}(t)$. Since the total potential energy of the magnet is

$$U(t) = -\mu_0 \left[\vec{M}(t) \cdot \vec{H}_{eff}(t)\right] \Omega, \tag{17.5}$$

we find $\vec{H}_{eff}(t)$ by taking the derivative of $U(t)$ with respect to $\vec{M}(t)$:

$$\vec{H}_{eff}(t) = -\frac{1}{\mu_0 \Omega} \frac{\partial U(t)}{\partial \vec{M}(t)}. \tag{17.6}$$

The potential energy consists of the shape-anisotropy energy (in a magnet whose shape does not have spherical symmetry) and the energy due to the agent initiating the switching action (stress, magnetic field, etc.). Both of these contributors depend on the orientation of the magnetization vector. If the magnet has a single domain, then one can make the macrospin assumption whereby the *magnitude* of the magnetization vector is assumed to be spatially uniform, although its *direction* varies. The magnitude is equal to the saturation magnetization of the magnet times the permeability times the magnet's volume.

In that case, the shape-anisotropy energy term [as a function of time t] is given by

$$U_{sh}(t) = \frac{\mu_0}{2} M_s^2 \Omega N_d(t), \tag{17.7}$$

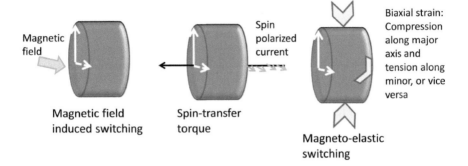

FIGURE 17.7

Rotating the magnetization of an elliptical nanomagnet through 90° with: (a) an external magnetic field directed in the intended direction, (b) a spin-polarized current passing through the magnet with the spins pointing in the intended direction, and (c) biaxial strain (compression along the major axis and tension along the minor axis, or vice versa). The latter mode works only for magnetostrictive magnets. If the magnet's magnetostriction coefficient is positive (e.g., Terfenol-D), then one should apply compression along the major axis and tension along the minor axis. If the magnetostriction coefficient is negative (e.g., cobalt), then one should interchange the compression and tension axes.

where $N_d(t)$ is the demagnetization factor given by

$$N_d(t) = N_{d-xx}\left(M_x(t)/M_s\right)^2 + N_{d-yy}\left(M_y(t)/M_s\right)^2 + N_{d-zz}\left(M_z(t)/M_s\right)^2,$$
(17.8)

where $(M_x(t), M_y(t), M_z(t))$ are the Cartesian components of the magnetization vector at the instant of time t. The quantities $N_{d-xx}, N_{d-yy}, N_{d-zz}$ depend on the geometric shape of the magnet (i.e., whether it is a cylinder, rectangular disk, elliptical disk, etc.).

If an external magnetic field $\vec{H}(t)$ is present, it contributes a potential energy term

$$U_{mag}(t) = -\mu_0 \left[\vec{M}(t) \cdot \vec{H}(t)\right] \Omega = -\mu_0 \left|\vec{M}(t)\right| \left|\vec{H}(t)\right| cos\theta(t)$$
$$= -\mu_0 M_s \Omega \left|\vec{H}(t)\right| cos\theta(t),$$
(17.9)

where $\theta(t)$ is the angle between the magnetic field and the magnet's magnetization at the instant of time t.

If an uniaxial stress $\sigma(t)$ is present, then the potential energy contribution due to stress will be given by

$$U_{st}(t) = -(3/2)\lambda_s\sigma(t)\Omega cos^2\nu(t),$$
(17.10)

where $\nu(t)$ is the angle between the stress axis and the magnetization vector at the instant of time t and $(3/2)\lambda_s$ is the magnetostriction coefficient.

From Equations (17.3) and (17.6), we can write the torque on the magnetization vector due to external agents as

$$\vec{T}(t) = -\hat{\mathbf{m}}(t) \times \vec{\nabla} U(t),$$
(17.11)

where $\hat{\mathbf{m}}(t)$ is the unit vector in the direction of the magnetization vector at the instant of time t.

There is one exception to this, however. Spin transfer torque generated by the passage of a spin-polarized current through the magnet has a non-conservative component which cannot be captured within a potential energy term. The torque due to spin-polarized current has been calculated in several references (see, for example, [20]) and is given by

$$\vec{T}_{STT}(t) = \frac{\hbar}{2e}\zeta I\left[b\left(\hat{\mathbf{m}}(t) \times \hat{\mathbf{s}}\right) + c\left(\hat{\mathbf{m}}(t) \times \left(\hat{\mathbf{m}}(t) \times \hat{\mathbf{s}}\right)\right)\right],$$
(17.12)

where I is the current with spin polarization ζ, c is the coefficient of the in-plane (or Slonczewski) component of the torque, and b is the out-of-plane (or field-like) component of the torque. The quantities $\hat{\mathbf{m}}$ and $\hat{\mathbf{s}}$ are the unit vectors in the directions of the magnetization vector and spin polarization of the current, respectively.

Finally, there is one more torque to consider. This is a random torque that will act on the magnetization vector of a magnet at non-zero temperatures owing to thermal noise. It is given by

$$\vec{T}_{th}(t) = \mu_o \Omega \left[\vec{M}(t) \times \vec{H}_{th}(t) \right] = \mu_0 M_s \Omega \hat{m}(t) \times \vec{H}_{th}(t), \qquad (17.13)$$

where $\vec{H}_{th}(t) = h_x(t)\hat{x} + h_y(t)\hat{y} + h_z(t)\hat{z}$ and

$$h_i(t) = \sqrt{\frac{2\alpha kT}{|\gamma| \left(1 + \alpha^2\right) \mu_0 M_s \Omega \Delta t}} G_{(0,1)}(t) \quad (i = x, y, z), \qquad (17.14)$$

where Δt is the inverse of the attempt frequency of the thermal field and $G_{(0,1)}(t)$ is a normalized Gaussian distribution in time with zero mean and unit standard deviation [21].

Once all the torques acting on a magnetization vector have been determined, the magnetization dyanmics, or the time evolution of the magnetization vector, under the various influences can be obtained by solving Equation (17.4).

17.2   Why nanomagnetism?

Many grand challenge computing problems (e.g., decoding the human genome, protein folding, global weather forecasting, etc.) call for massive computational resources that cannot be mustered without significant advances in computing machinery. Todays state-of-the-art CMOS transistor-based computing technology suffers from excessive energy dissipation that prevents packing increasingly more devices on a chip to raise computational prowess. The celebrated Moores law that envisioned doubling the density of devices per unit area on a chip every 18 months cannot be sustained without decreasing the energy dissipation in logic operation. Nanomagnet-based logic processors might dissipate considerably less energy than, and hence have been studied as alternatives to, transistor-based processors. The energy dissipated to switch a nanomagnet with strain at room temperature in ~1 ns can be as low as 100 kT (0.4 aJ) [22, 23], which makes it more energy-efficient than most transistor switches. Low density processors with 10^8 switches per cm^2, operating at 1 GHz, will dissipate a mere 4 mW/cm^2 power, if we assume a 10% activity level, i.e., 10% of the switches are switching at any given time. Such low-power devices can operate by harvesting energy from the surroundings (cable TV, 4G networks) without requiring an independent power source or battery. Therefore, they are very suitable for wearable electronics [24] (Google Glass, Fitbit, Apple Watch) and medically implanted devices such as pacemakers [25]. They are particularly apt for processors, implanted, say, in an epileptic patient's brain, that continuously monitor brain signals to warn of an impending seizure, powered entirely by the patient's head movements. Other potential applications are in structural health monitoring of tall buildings and bridges, where sensor/processors are mounted in relatively inaccessible areas. These sensors, if fashioned out of such low-power magnetic switches, can operate without a battery by harvesting energy from wind or passing traffic.

Perhaps the most significant advantage of magnet-based processors over transistor based processors is that the former can be *non-volatile*. A magnetic Boolean logic device can process binary information and store the output bits in the magnetization states of magnets, thereby doubling as both "logic" and "memory." This affords immense flexibility in architecture design and enables such constructs as computers with no boot delay since the instruction sets and programs are located in the arithmetic logic unit and do not have to be fetched from a remote memory. This improves speed, energy-efficiency and reliability.

Many non-volatile magnetic logic device proposals have appeared in the literature, but they do not often possess the seven essential characteristics of a Boolean logic gate: concatenability, non-linearity, isolation between input and output, gain, universal logic implementation, scalability, and error-resilience [26]. Concatenability means that the output of a preceding gate can be fed directly to the input of the succeeding gate without any intervening hardware. This requires that the input and output bits are encoded in the same physical quantity (both voltages, both currents, both magnetization orientations, etc.). Non-linearity is needed for logic level restoration [27]. If the input signals are corrupted by noise, resulting in level broadening, passage of these signals through the gate must restore the discrete signal levels, i.e., there is automatic error correction in every stage. Isolation between input and output is required so that the input determines the output and not the other way around. Gain is needed so that the energy to switch the bits comes from an external source and not the input signals (the input signals determine which way a bit will switch but will not provide the energy to switch). Universal logic implementation is needed for general purpose universal computation (any Boolean operation can be implemented). Scalability means that there are no fundamental obstacles to reducing gate sizes down to the limit where the gate properties cease to be functional. Gates switched with magnetic fields are not scalable since the magnetic fields cannot be confined to arbitrarily small areas or volumes easily. Finally, error-resilience is needed for fault tolerance.

To the knowledge of the authors, there are two prominent proposals for non-volatile magnetic logic that fulfill all the criteria enumerated above and are therefore promising. These are discussed next.

17.2.1   All-spin logic

"All-spin logic" is a paradigm that uses shape-anisotropic (e.g., elliptical) nanomagnets to encode binary bits 0 and 1 in the two stable magnetization

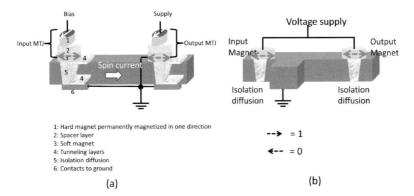

1: Hard magnet permanently magnetized in one direction
2: Spacer layer
3: Soft magnet
4: Tunneling layers
5: Isolation diffusion
6: Contacts to ground

(a)

(b)

FIGURE 17.8

(a) A Bennett clocked all-spin logic element for COPY and NOT operation.
(b) An all-spin logic implementation of COPY and NOT operation that does
not use a Bennett clock and has built-in isolation between input and output.

orientations of the nanomagnets [28]. To make an inverter, one can use the
structure in Fig. 17.8(b) with the left (input) magnet connected to the voltage
supply, the ground terminal opened, and the right (output) magnet connected
to ground (instead of the voltage supply). If the voltage supply is negative,
then electrons from the left magnet (with their spins polarized in the direction
of that magnet's magnetization orientation) will be injected into the right
magnet and exert a spin transfer torque that will turn the right magnet's
magnetization in the direction of the left's. Since the two magnet's will then
have *parallel* magnetizations, they will encode the *same* bit and hence this is
a COPY operation whereby the input bit is copied into the output bit.

Next, consider the situation when the supply voltage is positive. In that
case, electrons will be extracted from the right magnet. Since the left magnet
preferentially transmits those electrons whose spins are parallel to its own
magnetization, the spins that are extracted from the right magnet will mostly
have orientations parallel to the left magnet's magnetization. These spins
will be *depleted* in the right magnet so that ultimately the right magnet's
magnetization will turn in the direction *anti-parallel* to that of the left magnet.
This is the NOT operation since the input and output magnets will encode
complementary bits.

There are many ways to enforce isolation between input and output, i.e.,
make the input bit determine the output bit and not the other way around.
One obvious way is to make the input magnet bigger in size than the output
magnet, but this is not elegant since it makes fabrication more complicated.
Another way is shown in Fig. 17.8(b) where a ground terminal is interposed
asymmetrically between the input and output magnets such that it is closer

to the input magnet. When both magnets are connected to the same voltage suppy as shown, the current through the input magnet driven by the voltage supply to ground is larger than that through the output magnet. This enforces the isolation between the input and output. It has been shown that in this case, if the supply voltage is negative, then once again, the COPY operation will be implemented and if the supply voltage is positive, then the NOT operation will be implemented [29].

In order to have gain, it is preferable to Bennett clock the nanomagnets, i.e., an external energy source will put the output nanomagnet in the maximum energy state (unstable). After that, the influence of the input magnet will drive the output magnet to one its the two stable states – which one is determined by the influence of the input. This makes the output bit conditioned on the input bit and elicits logic operations, while the energy for the operation comes not from the input, but from the Bennett clock that places the output magnet temporarily in its highest energy state. This results in "gain."

Bennett clocking is implemented by exploiting the structure in Fig. 17.8(a) which shows two skewed magneto-tunneling junction (MTJ) stacks with the top layer made of a hard magnet and the bottom layer made of a soft magnet. Both the hard and soft layers are elliptical and their major axes are mutually perpendicular. The hard magnets are permanently magnetized along one of their stable orientations, shown by the white arrows in the figure. If a negative supply voltage is applied to the right (output) MTJ, then the hard layer in that MTJ will inject spins oriented along its magnetization into the soft layer and turn the latter's magnetization by 90°, placing it along the in-plane hard axis (direction of white arrow), which is an unstable state. Next, consider that the bias voltage connected to the input MTJ on the left is negative and small. In that case, not enough spins are injected from the hard to the soft magnet in the left MTJ to exert sufficient spin torque to alter the magnetization orientation of the soft magnet. However, the left (input) soft magnet will now inject a spin current into the right (output) soft magnet because of the spin imbalance between the two, and this exerts a slight spin transfer torque on the right soft magnet that tips its magnetization (left in an unstable state by the Bennett clock) in the direction of the magnetization of the input soft magnet, resulting in the COPY operation. The reader can easily deduce that if the bias voltage attached to the left MTJ is positive, then the NOT operation will result.

In order to make an AND or OR gate, one can use the structure in Fig. 17.9. The input MTJs and the bias MTJ are connected to bias sources and the output MTJ is connected to a supply source. Assume that the supply voltage is negative and that leaves the magnetization of the soft magnet in the output MTJ in an unstable state (pointing in the direction of its hard magnet's magnetization, i.e., the white arrow). The two input soft magnets and the bias element's soft magnet all inject spin currents into the output MTJ's soft magnet. Let us assume that magnetization pointing to the right

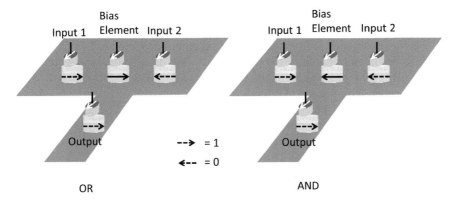

FIGURE 17.9

(a) An all-spin OR gate. (b) An all-spin AND gate.

encodes bit 1 and that pointing to the left encodes bit 0, as shown in Fig. 17.9. If the bias element's soft magnet is magnetized to the right [bit 1], then the sum of the three spin currents injected from the input and bias MTJs into the output MTJ's soft magnet will prod the latter's magnetization to the right [bit 1] as long as at least one of the two input bits is 1, i.e., at least one of the input soft magnet's magnetization is oriented to the right. Therefore, the output bit will be 1 if either or both input bits are 1. If both input bits are 0, then the sum of the spin currents injected into the output soft magnet will prod it to the left. Hence the output bit will be 0 when both input bits are 0. This is the OR operation and is shown in the left panel of Fig. 17.9.

This gate is *reconfigurable* into an AND gate by simply flipping the magnetization of the bias element's soft magnet to the left, as shown in the right panel of Fig. 17.9. In this case, only when both input bits are 1, i.e., the magnetizations of both input MTJs' soft layers are magnetized to the right, the sum total spin current injected into the output soft magnet will prod the latter's magnetization to point to the right. Otherwise, the injected spin current will make the output soft magnet's magnetization point to the left. Thus, the output bit is 1 only when both input bits are 1, and it is 0 otherwise. This is the AND operation.

The combination of NOT, OR, and AND results in a complete Boolean set and universal Boolean logic.

The reason this nanomagnetic logic paradigm is interesting is because it fulfills all the major requirements of logic. It is non-volatile and has a reasonable energy-delay product, but its error resilience has not been examined. Magnetization dynamics is usually very vulnerable to thermal noise and therefore it is important to assess the reliability of magnetic logic gates before adopting them for specific applications.

17.2.2   Magneto-elastic magneto-tunneling junction logic

The first known idea of using magneto-tunneling junctions (MTJ) for Boolean logic gates is most likely due to Ney et al. [30]. Later ideas came from Lee, et al. [31] followed by experimental demonstration of an MTJ-based Boolean logic gate by Wang et al. [32].

The basic idea is the following: An MTJ has a hard magnetic layer with fixed magnetization orientation and the soft layer's magnetization is usually anti-parallel to that of the hard layer's owing to dipole coupling*, so the MTJ resistance is normally high. The output bit is encoded in the resistance state of the MTJ and the high resistance state encodes logic bit 1. Voltage or current inputs that generate magnetic fields, spin transfer torque, or strain rotate the soft layer's magnetization, making the MTJ resistance go low (output bit = 0). The voltage and current levels (or directions of current) act as input bits. If either of two inputs being high is sufficient to rotate the soft layer's magnetization, then the NOR function is realized, whereas if both inputs need to be high to rotate the soft layer's magnetization, then the NAND function is realized.

We will describe the MTJ gate of [33] since it fulfills all the requirements of Boolean logic. It is a magneto-elastic gate since the magnetization of the soft layer of the MTJ is rotated with electrically generated mechanical strain. The gate structure is shown in Fig. 17.10. The MTJ consists of a *skewed* stack where the hard magnet's and soft magnet's major axes are not collinear, but bear an angle. A permanent magnetic field B applied along the minor axis of the soft elliptical magnet causes its magnetization's two stable states to come out of the major axis and lie in the plane of the magnet subtending an angle of $\sim 90°$ between them (shown as Ψ_0 and Ψ_1 in the bottom panel of the figure). The hard magnet's major axis is placed along the direction of Ψ_1 and this magnet is magnetized *anti-parallel* to Ψ_1. Hence, when the soft magnet is in the stable state Ψ_1, the MTJ resistance is high and encodes the bit state 1, whereas when the soft magnet is in the stable state Ψ_0, the MTJ resistance is low and encodes the bit state 0. The MTJ resistance state is converted into the output voltage state by the bias current source I_B. Since the potential drop between the soft magnet and the conducting substrate is negligible (there are only fringing electric fields in this region), $V_0 \approx I_B R_{MTJ}$, where R_{MTJ} is the resistance of the MTJ. Therefore, whenever the MTJ resistance is high, the gate's output voltage (output bit) is high. Two (electrically shorted) electrodes A and A' are placed on the piezoelectric layer to generate biaxial strain in the latter upon application of a voltage between the electrodes and the conducting substrate [34]. The line joining the centers of A and A' lies along ψ_1.

---

*The dipole coupling is weak since the hard magnet is made out of a synthetic anti-ferromagnet.

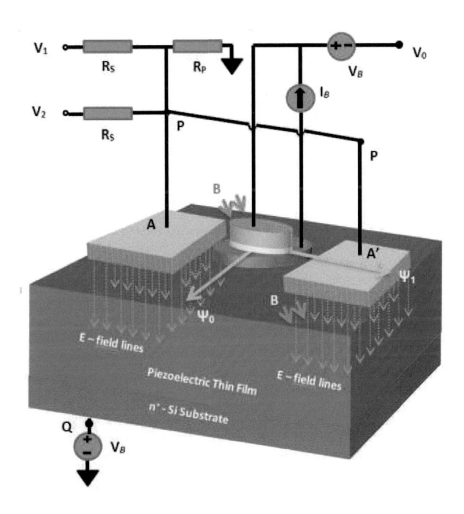

FIGURE 17.10
A magneto-elastic MTJ-based NAND gate.

The gate works as follows. Asssume that the biasing resistors are all equal, i.e., $R_s = R_p$. Input bits are encoded in voltages V_1, V_2 and the output bit is encoded in V_0. Let us assume that the voltage levels for logic 1 and 0 are V and $V/2$, respectively. The bias voltage V_B is set to $5V/12$. Every logic operation is preceded by a RESET operation where the two input voltages V_1 and V_2 are set to $V/4$. This makes the potential drop across the piezoelectric thin film (between nodes P and Q) equal to $-V_0/4$, which generates in-plane tensile strain in the direction of the line joining the two electrodes A and A' (i.e., along Ψ_1) and compressive strain in the direction perpendicular to this line (i.e., along Ψ_0). This biaxial strain is transferred to the soft magnet resting on the piezoelectric and rotates its magnetization via the Villari effect to align the latter along the direction of tensile stress (i.e., Ψ_1) as long as the soft layer has positive magnetostriction (such as the material Terfenol-D). This makes the MTJ resistance "high." When the input voltages are subsequently withdrawn by grounding the inputs and shorting the bias voltage source V_B, the voltage across the piezoelectric layer (and hence the strain in that layer) vanishes, but since Ψ_1 is a stable state for the magnetization of the soft layer, the magnetization remains at Ψ_1 and the MTJ resistance remains high. Therefore, the RESET operation always leaves the MTJ in the high resistance state.

In the logic operation stage, the following scenarios occur: (1) if both inputs are low (i.e., $V_1 = V_2 = V/2$), then the drop across the piezoelectric layer is $-V/12$; (2) if either input is low (i.e., $V_1 = V$ and $V_2 = V/2$, or vice versa), then the drop across the piezoelectric is $+V/12$. The negative drop leaves the soft layer's magnetization in the same state Ψ_1 and the positive drop is so small that it cannot rotate the soft layer's magnetization to Ψ_0. Therefore, the MTJ resistance remains high and the output voltage remains high. In other words, $V_0 = V$ if either V_1 or V_2 is $V/2$, which means that the output bit is 1 if either input bit is 0. However, if both input bits are high (i.e., $V_1 = V_2 = V$), then the drop across the piezoelectric becomes $+V/4$. This is sufficient to generate enough compressive strain in the soft magnet in the direction of the line joining A and A' (i.e., in the direction of Ψ_1) and tensile strain in the direction perpendicular to this line (i.e., in the direction of Ψ_0), to make the magnetization rotate to the orientation Ψ_0, at which point the MTJ resistance goes low. Therefore, the output bit becomes 0 when both input bits are 1. This implements a NAND gate and since the NAND gate is universal, this is sufficient for universal computation. An MTJ-based strain-switched logic gate has been demonstrated experimentally at the time of writing this chapter [35].

Gates such as the above possess the coveted property of *non-volatility* which will make the same device act as both logic and memory. This allows for unprecedented flexibility in architecture design (e.g., non-von-Neumann architectures), elimination of processor-memory communication which is the infamous von-Neumann bottleneck, and "instant-on" computers with no boot delay. However, magnetic gates also have three shortcomings: First, ferromagnets are typically larger than transistors (if they are too small, then they will be super-paramagnetic at room temperature), magnetization switches relatively slow ($\sim$ 1 ns), and magnetization dynamics is very vulnerable to thermal noise. The gate error probability for the above gate at room temperature was found to be $\sim 10^{-6}$. Reliability seems to be the Achilles' heel of magnetic gates and significant research is being devoted to overcome this shortcoming.

17.3 Problems

- **Problem 17.1**

  Consider an elliptical nanomagnet subjected to uniaxial stress along the major axis, as shown in Fig. 17.11.

  We will assume that the nanomagnet has a single domain so that the magnetization is homogeneous in space. The magnitude of the magnetization vector $\vec{M}(t)$ is invariant in time but the direction changes.

  Using Equations (17.7), (17.8), and (17.10) show that the total potential energy can be written as

  $$U(t) \equiv U\left(\theta(t), \phi(t)\right) = A(\phi(t))sin^2\theta(t) + B(t),$$

  where

  $$A(\phi(t)) = A_{sh}(\phi(t)) + A_{st}(t)$$
  $$A_{sh}(\phi(t)) = \frac{\mu_0}{2}M_s^2\Omega\left[N_{d-xx}cos^2\phi(t) + N_{d-yy}sin^2\phi(t) - N_{d-zz}\right]$$
  $$A_{st}(t) = (3/2)\lambda_s\sigma(t)\Omega$$
  $$B(t) = \frac{\mu_0}{2}M_s^2\Omega N_{d-zz} - (3/2)\lambda_s\sigma(t)\Omega.$$

- **Problem 17.2**

  Using Equation (17.6), show that

  $$\vec{H}_{eff}(t) = -\frac{1}{\mu_0 M_s\Omega}\vec{\nabla}_{\theta,\phi}U(t)$$
  $$= -\frac{1}{\mu_0 M_s\Omega}\left[\frac{\partial U(t)}{\partial\theta(t)}\hat{a}_\theta(t) + \frac{1}{sin\theta(t)}\frac{\partial U(t)}{\partial\phi(t)}\hat{a}_\phi(t)\right],$$

  where M_s is the saturation magnetization of the nanomagnet and $\hat{a}_\theta(t)$ and $\hat{a}_\phi(t)$ are the unit vectors along the $\theta(t)$ and $\phi(t)$ directions in spherical coordinates.

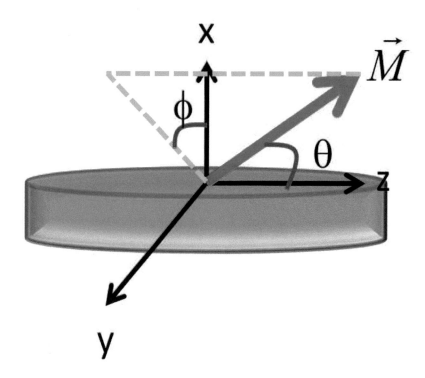

FIGURE 17.11

The magnetization vector of a nanomagnet shaped like an elliptical disk. The major axis of the ellipse is aligned along the z-direction and the minor axis along the y-direction. The polar angle of the magnetization vector is θ and the azimuthal angle is ϕ.

Note that, for the above problem, the magnetization vector is always in the radial direction and hence $\vec{M}(t) = M_s \hat{a}_r(t)$. Using Equation (17.11), show that

$$\vec{T}(t) = -A(\phi(t))sin(2\theta(t))\hat{a}_\phi(t) - A_{sh}(\phi(t))sin\theta(t)\hat{a}_\theta(t).$$

• **Problem 17.3**

Show that the Landau-Lifshitz-Gilbert equation can be written as

$$\frac{d\hat{m}(t)}{dt} + \alpha\left[\hat{m}(t) \times \frac{d\hat{m}(t)}{dt}\right] = \frac{\gamma}{\mu_0 M_s \Omega}\vec{T}(t),$$

where $\hat{m}(t)$ is the unit vector in the direction of magnetization at time t.

Since in spherical coordinates

$$\frac{d\hat{m}(t)}{dt} = \frac{d\theta(t)}{dt}\hat{a}_\theta(t) + sin\theta(t)\frac{d\phi(t)}{dt}\hat{a}_\phi(t),$$

show that the coupled equations describing the temporal evolution of the polar and azimuthal angles (and therefore the direction of the magnetization vector) are

$$\left(1 + \alpha^2\right)\frac{d\theta(t)}{dt} = \frac{\gamma}{\mu_0 M_s \Omega}\left[\frac{\partial A_{sh}(\phi(t))}{\partial t}sin\theta(t) - \alpha A(\phi(t))sin(2\theta(t))\right]$$

$$\left(1 + \alpha^2\right)\frac{d\phi(t)}{dt} = \frac{\gamma}{\mu_0 M_s \Omega}\left[\alpha\frac{\partial A_{sh}(\phi(t))}{\partial t} + 2A(\phi(t))cos(\theta(t))\right].$$

Also show that $\vec{M}(t) \cdot \frac{d\vec{M}(t)}{dt} = 0$.

• **Problem 17.4**

Assume that the thickness of the magnet approaches zero. In that nearly two-dimensional magnet, the magnetization vector will almost always stay in the plane of the magnet and hardly lift out of the plane or dip down below the plane. In other words, $\phi \approx 90°$ always. In that case, show that the time taken for the magnetization to rotate from a location θ_1 to a location θ_2 under a constant stress σ is given by

$$\tau = -\left(1 + \alpha^2\right)\frac{\mu_0 M_s \Omega}{\gamma}\int_{\theta_1}^{\theta_2}\frac{d\theta}{asin\theta + bsin(2\theta)}$$

$$= -\left(1 + \alpha^2\right)\frac{\mu_0 M_s \Omega}{\gamma}\frac{1}{2\left(a^2 - 4b^2\right)} \times$$

$$\left[(a - 2b)ln\left|\frac{cos\theta_2 - 1}{cos\theta_1 - 1}\right| - (a + 2b)ln\left|\frac{cos\theta_2 + 1}{cos\theta_1 + 1}\right| + 4bln\left|\frac{a + 2bcos\theta_2}{a + 2bcos\theta_1}\right|\right],$$

where

$$a = \frac{\mu_0}{2} M_s^2 \Omega \left[N_{d-yy} - N_{d-zz}\right]$$
$$b = \alpha \left[a + (3/2)\lambda_s \sigma \Omega\right].$$

In a strictly two-dimensional nanomagnet, $N_{d-yy} = N_{d-zz} = 0$. Show that in such a nanomagnet

$$\tau = \left(1 + \alpha^2\right) \frac{\mu_0 M_s}{3\gamma\alpha\lambda_s\sigma} ln \left|\frac{tan\theta_1}{tan\theta_2}\right|.$$

Assume that the initial orientation of the magnetization is $\theta_1 = 0°$ or $180°$. The above result shows that starting from this initial orientation, the time taken to reach any orientation θ $(0° < \theta < 180°)$ is infinity. Explain this result. (Hint: What is the torque acting on the magnetization vector at its initial location?).

17.4 References

[1] R. P. Cowburn, D. K. Koltsov, A. O. Adeyeye, M. E. Welland and D. M. Tricker, "Single-domain circular nanomagnets", Phys. Rev. Lett., **83**, 1042 (1999).

[2] S. Salahuddin and S. Datta, "Interacting systems for self-correcting low power switching", Appl. Phys. Lett., **90**, 093503 (2007).

[3] R. P. Cowburn and M. E. Welland, "Room temperature magnetic quantum cellular automata", Science, **287**, 1466 (2000).

[4] S. Bandyopadhyay and M. Cahay, "Electron spin for classical information processing; A brief survey of spin-based logic devices, gates and circuits", Nanotechnology, **20**, 412001 (2009).

[5] C. H. Bennett, "The thermodynamics of computation", Intl. J. Theor. Phys., **21**, 905 (1982).

[6] G. Csaba, A. Imre, G. H. Bernstein, W. Porod and V. Metlushko, IEEE Trans. Nanotech., **1**, 209 (2002).

[7] J. Atulasimha and S. Bandyopadhyay, "Bennett clocking of nanomagnetic logic using multiferroic single-domain nanomagnets", Appl. Phys. Lett., **97**, 173105 (2010).

[8] M. Salehi-Fashami, J. Atulasimha and S. Bandyopadhyay, "Energy-dissipation and error probability in fault-tolerant binary switching", Sci Rep., **3**, 3204 (2013).

[9] M. Salehi-Fashami, K. Munira, S. Bandyopadhyay, A. W. Ghosh and J. Atulasimha, "Switching of dipole-coupled multiferroic nanomagnets in the presence of thermal noise: Reliability of nanomagnetic logic", IEEE Trans. Nanotech., **12**, 1206 (2013).

[10] K. Munira, Y. Xie, S. Nadri, M. B. Forgues, M. Salehi-Fashami, J. Atulasimha, S. Bandyopadhyay and A. W. Ghosh, "Reducing error rates in straintronic multiferroic dipole coupled nanomagnetic logic by pulse shaping", arXiv:1405.4000.

[11] Md. M. Al-Rashid, J. Atulasimha and S. Bandyopdhyay, "Geomtery effects in switching of nanomagnets with strain: Relibility, energy-dissipation and clock speed in dipole-coupled nanomagnetic logic", arXiv:1412.0046.

[12] math.nist.gov/oommf/.

[13] http://www.magpar.net/static/magpar-0.9/doc/html/.

[14] L. D. Landau and E. M. Lifshitz, "Theory of the dispersion of magnetic permeability in ferromagnetic bodies", Phys. Z. Sowietunion, **8**, 153 (1935).

[15] T. L. Gilbert, "A phenomenological theory of damping in ferromagnetic materials", IEEE Trans. Magn., **40**, 3443 (2004).

[16] L. Liu, C-F Pai, Y. Li, H. W. Tseng, D. C Ralph and R. A. Buhrman, "Spin-torque switching with the giant spin Hall effect of Tantalum", Science, **336**, 555 (2012).

[17] D. Bhowmik, L. You and S. Salahuddin, "Spin Hall effect clocking of nanomagnetic logic without a magneic field", Nature Nanotech., **9**, 59 (2014).

[18] I. M. Miron, G. Gaudin, S. Auffret, B. Rodmacq, A. Schuhl, S. Pizzini, J. Vogel and P. Gambardella, "Current driven spin torque induced by the Rashba effect in a ferromagnetic metal layer", Nature Mater., **9**, 230 (2010).

[19] Y. G. Semenov, X. Duan and K. W. Kim, "Electrically controlled magnetization in ferromaghetic-topological insulator heterostructures", Phys. Rev. B., **86**, 161406(R), (2012).

[20] S. Salahudddin, D. Datta and S. Datta, "Spin transfer torque as a non-conservative pseudo-field", arXiv:0811.3472.

[21] G. Brown, M. A. Novotny and P. A. Rikvold, "Langevin simulation of thermally activated magnetization reversal in nanoscale pillars", Phys. Rev. B., **64**, 134422 (2001).

[22] K. Roy, S. Bandyopadhyay and J. Atulasimha, "Hybrid spintronics and straintronics: A magnetic technology for ultralow energy computing and signal processing", Appl. Phys. Lett., **99**, 063108 (2011).

[23] K. Roy, S. Bandyopadhyay and J. Atulasimha, "Energy dissipation and switching delay in strain-induced switching of multiferroic nanomagnets in the presence of thermal fluctuations", J. Appl. Phys., **112**, 023914 (2012).

[24] T. Starner, "Human powered wearable computing", IBM Systems Journal, **35**, 618 (1996).

[25] L. A. Geddes, "Historical highlights in cardiac pacing", IEEE Engineering in Medicine and Biology Magazine, 12 (1990).

[26] R. Waser (ed). *Nanoelectronics and Information Technology*, Ch III (Wiley-VCH, 2003).

[27] D. A. Hodges and H. G. Jackson, *Analysis and Design of Digital Integrated Circuits*, 2nd. ed., (McGraw Hill, New York, 1988).

[28] B. Behin-Aein, D. Datta, S. Salahuddin and S. Dattta, "Proposal for an all-spin logic device with built-in memory", Nature Nanotech., **5**, 266 (2010).

[29] S. Srinivasan, A. Sarkar, B. Behin-Aein and S. Datta, "All-spin logic device with built-in non-reciprocity", IEEE Trans. Nanotech., **47**, 4026 (2011).

[30] A. Ney, C. Pampuch, R. Koch and K. H. Ploog, "Programmable computing with a single magnetoresistive element", Nature, **425**, 485 (2003).

[31] S. Lee, S. Choa, S. Lee and H. Shin, "Magneto-logic device based on single layer magnetic tunnel junction", IEEE Trans. Elec. Dev., **54**, 2040 (2007).

[32] J. Wang, H. Meng and J-P Wang, "Programmable spintronics logic device based on a magneto-tunneling junction element", J. Appl. Phys., **97**, 10D509 (2005)

[33] A. K. Biswas, J. Atulasimha and S. Bandyopadhyay, "An error-resilient non-volatile magneto-elastic universal logic gate with ultralow energy-delay product", Scientific Reports, **4**, 7553 (2014).

[34] J. Cui, J. L. Hockel, P. K. Nordeen, D. M. Pisani, C-Y Liang, G. P. Carman and C. S. Lynch, "A method to control magnetism in individual strain-mediated magnetoelectric islands", Appl. Phys. Lett., **103**, 232905 (2013).

[35] P. Li, A. Chen, D. Li, Y. Zhao, S. Zhang, L. Yang, Y. Liu, M. Zhu, H. Zhang and X. Han, "Electric field manipulation of magnetization

rotation and tunneling magnetoresistance of magnetic tunnel junctions at room temperature", Adv. Mater., **26**, 4320 (2014).

18

A Brief Quantum Mechanics Primer

This chapter should be treated as an appendix. It is included for completeness and covers certain fundamentals of quantum mechanics that will aid the understanding of topics discussed in this textbook. The reader is urged to refer to many of the good introductory textbooks on the subject to learn or revisit the fundamental concepts [1, 2, 3, 4].

First, we briefly discuss three key ingredients of quantum mechanics, i.e., the blackbody radiation spectrum, the concept of the photon and de Broglie wavelength, and the related wave-particle duality concept. Next, we briefly restate the four major postulates of quantum mechanics used in the so-called *orthodox* or Copenhagen interpretation of quantum mechanics that is in vogue. We continue on to describe a generalized form of the Heisenberg Uncertainty Principle which is illustrated with a few familiar bound state problems (particle-in-a-box and one-dimensional harmonic oscillator) and a few spin related problems.

18.1 Blackbody radiation and quantization of electromagnetic energy

18.1.1 Blackbody radiation

A watershed event in quantum physics was Max Planck's derivation of the correct expression for the experimentally measured energy per unit volume per frequency interval (or energy spectral distribution) of the blackbody radiation

$$u = \frac{8\pi h\nu^3}{c^3} \frac{1}{e^{\frac{h\nu}{k_B T}} - 1}, \qquad (18.1)$$

where h is Planck's constant, ν is the frequency of the electromagnetic radiation in the cavity, c is the speed of light in vacuum, k_B is Boltzmann's constant, and T is the temperature inside the cavity in Kelvin. Planck was able to derive this important relation by assuming that the energy exchange between the electromagnetic waves inside a cavity and its walls occurs via emission and absorption of discrete quanta of energy. As a result, the energy of the electromagnetic radiation with frequency ν inside the blackbody cavity exists only in multiples of $h\nu$.

Exercises:

- Show that the maximum of the energy spectral density u in Equation (18.1) occurs when

$$3 - 3exp(-x) = x, \qquad (18.2)$$

where $x = h\nu/k_B T$.

- Solve the equation above numerically and show that the maximum occurs for x around 2.82, i.e., at a frequency ν_{max} given by

$$h\nu_{max}/k_B T \sim 2.82. \qquad (18.3)$$

This last relation is referred to as *Wien's displacement law*.

- According to the previous exercise, the peak in the energy spectral distribution curve shifts upward in frequency as the temperature increases. What is the equilibrium temperature inside a blackbody cavity when the maximum in the energy spectral distribution is to be located at wavelengths 1,000 Å, 5,000 Å, and 10 μm, which correspond to the ultraviolet, visible, and mid-infrared region of the electromagnetic spectrum, respectively?

18.2   Concept of the photon

Following Planck's seminal work, Einstein postulated that electromagnetic radiation consists of particle-like discrete bundles of energy called *photons*. Furthermore, each photon has an energy E that depends only on its frequency ν according to the relation

$$E = h\nu = \frac{hc}{\lambda}, \qquad (18.4)$$

where the last equality is simply obtained from the connection between wavelength and frequency for an electromagnetic wave propagating in vacuum.

The concept of the photon was put on an even more solid ground as Einstein successfully refined the concept in 1905 to give the first correct explanation of the photoelectric effect, one of the major puzzles in physics in the early 1900s. Incredibly, during the same year, Einstein introduced to the world of physics his special theory of relativity in which all photons (in vacuum) travel at the speed of light, irrespective of their frequency and the frame of reference

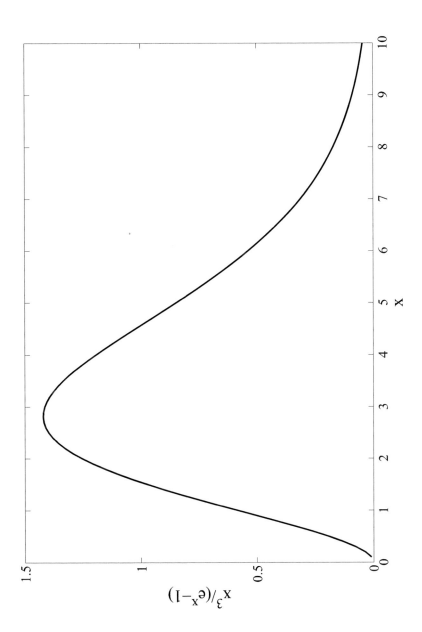

FIGURE 18.1
Plot of u, the blackbody radiation energy per unit frequency interval per unit volume (in units of $\frac{8\pi}{h^2 c^3}(k_B T)^3$); the quantity x is $h\nu/k_B T$.

used (for frames related by the so-called Lorentz transformation). According to the special theory of relativity, photons have a zero rest mass and their energy is entirely kinetic. The relativistic energy momentum relationship for a photon is given by

$$E = pc, \tag{18.5}$$

i.e., a photon in vacuum has a momentum simply related to its wavelength according to the formula

$$p = \frac{E}{c} = \frac{h\nu}{c} = \frac{h}{\lambda}. \tag{18.6}$$

Thus, according to the quantum point of view, a beam of electromagnetic radiation in free space can be thought of as a beam of photons traveling at the speed of light c. The intensity of the beam is proportional to the number of photons crossing a unit area per unit time. For a monochromatic beam, the intensity I is given by

$$I = N h\nu/(At), \tag{18.7}$$

where N is the number of photons falling on an area A during the time interval t.

For convenience in calculations, the following expressions in non-standard units can be used

$$h = 4.136 \times 10^{-15} \text{eV} - \text{s}, \tag{18.8}$$

$$hc = 12.4 \text{keV} - \text{Å}, \tag{18.9}$$

and

$$E = \frac{hc}{\lambda} = \frac{1.24 \text{eV}}{\lambda(\mu m)}, \tag{18.10}$$

where the wavelength must be expressed in μm in the last equality to get the energy in eV.

Exercises:

- What is the wavelength and frequency associated with a photon with 1 GeV energy?

  Answer: $\nu = 2.42 \times 10^{23}$ Hz; $\lambda = 1.24 \times 10^{-5}$ Å= 1.24 fm.

- Determine the photon flux associated with a beam of monochromatic light at 5,000 Å and with an intensity of 100 W/m^2.

  Solution

$$E = h\nu = \frac{hc}{\lambda} = \frac{(6.63 \times 10^{-34} Js)(3 \times 10^{-8} m/s)}{5 \times 10^{-7} m} = \frac{6.63 \times 3}{5} 10^{-19} J. \tag{18.11}$$

  Hence

$$N/(At) = \frac{I}{h\nu} = 25 \times 10^{19} \frac{\text{photons}}{m^2 - s}. \tag{18.12}$$

18.3   Wave-particle duality and the De Broglie wavelength

Planck's and Einstein's work established beyond doubt that electromagnetic radiation must have a particle nature in order to explain certain experimental observations, such as the photoelectric effect [5]. This is reminiscent of Newton's corpuscular theory of light, although the two are not connected. However, interference and diffraction experiments showed that electromagnetic radiation can also behave like a wave. Hence, electromagnetic radiation has a dual character; it exhibits a wave-particle duality. In certain circumstances it behaves like a wave, while in other circumstances, it behaves as a beam of particles.

Bohr postulated the orthodox *complementarity principle* that made the particle character and the wave character mutually exclusive [6]. It asserted that in situations where the particle character is manifested, the wave character is completely suppressed and vice versa. This orthodox principle has since been disproved by a remarkable set of experiments. Ghose, Home and Agarwal [7] had proposed a biprism experiment schematically depicted in Fig. 18.2. A single photon source emits a single photon which is split into orthogonal states ψ_r and ψ_t by a 50:50 beam splitter. They are detected by two photon detectors D_r and D_t. If the photon behaves truly as a "particle," then it should be detected at either D_r or D_t (but never at both) since a particle cannot traverse two paths simultaneously. That is, there should be perfect anti-coincidence between D_r and D_t, or, in other words, *either D_r or D_r* will click but *both* will never click simultaneously in between the arrival of two successive photons. The clever twist in this experiment, motivated by an experiment performed in the 19th century by Jagadish Chandra Bose [8], is the placement of the biprism with a small tunneling gap in the path of the transmitted photon. If D_t clicks *and D_r does not*, then we have made a "which path" (*welcher weg*) determination (the particle took the path of transmission as opposed to reflection) and a sharp *particle nature* is demonstrated [9]. Yet, to arrive at D_t, the particle must have tunneled through the biprism and tunneling is a sharp *wave* attribute. Therefore, this gedanken experiment presents a conundrum: the photon is behaving as both a particle and a wave in the *same experiment*, disproving the strict Bohr Complementarity Principle.

This experiment was actually conducted by Mizobuchi and Ohtake [10], using a single photon source. Perfect anti-coincidence was found between D_t and D_r, demonstrating the particle nature. Yet, the very fact that D_t ever clicked required tunneling and hence the existence of a wave nature. It was claimed that, in this experiment, a photon was behaving both sharply as a particle and as a wave in violation of orthodox Bohr's Complementarity Principle. A slight modification of this experiment has been proposed by Rangwala and Roy [11] where interference is used instead of tunneling to showcase the wave-like behavior. They claimed that quantum mechanics does not prohibit the

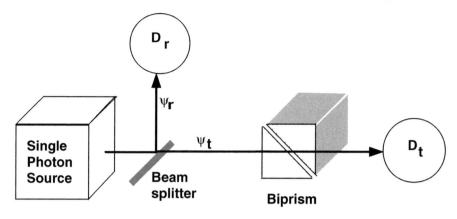

FIGURE 18.2
An experiment to disprove the orthodox Bohr's complementarity principle.

demonstration of simultaneous wave and particle behavior; rather, it prohibits their simultaneity only when the wave and particle attributes are "comple-mentary" [12] in the sense that projection operators associated with them do not *commute*. Their view is that in the experiments of [7, 10, 11], the wave and particle behavior supposedly are not truly complementary and hence not subject to the restrictions of the orthodox Complementarity Principle.

A somewhat tempered version of the Complementarity Principle is the *du-ality principle* due to Englert [13] which states that a quantum system can simultaneously exhibit wave and particle behavior, but sharpening of the wave character blurs the particle character and vice versa. This is an extension of earlier work by Wootters and Zurek [14], Greenberger and Yasin [15], and Mandel [16]. In fact, Englert derives an inequality

$$P^2 + V^2 \leq 1, \tag{18.13}$$

where P is a measure of the particle attribute and V is a measure of the wave attribute. Equation (18.13) immediately shows that stronger wave or stronger particle behavior can be manifested only at the expense of each other.

Returning to the wave-particle duality, Louis de Broglie made another breakthrough in 1924 by postulating that if electromagnetic radiation could have a dual wave-particle character, then material objects will also behave like waves. In other words, de Broglie posited that, if a particle passes through a slit whose width is comparable to the wavelength associated with it, then it should undergo diffraction just as photons do in a single-slit experiment.

Louis de Broglie conjectured that the wavelength associated with a material body would satisfy the same relation as photons, a conjecture which Einstein considered brilliant. Therefore, a material body with a momentum p will have a wavelength given by

$$\lambda = h/p. \tag{18.14}$$

However, there is one important difference between photons and massive objects in the way their wave and particle properties are related. Because $\lambda\nu = c$ for a photon, only one relation is required to get both wavelength and frequency from a photon's particle properties of energy and momentum. A massive object, on the other hand, requires separate equations to calculate its wavelength and frequency. For a relativistic particle, the total energy E (kinetic + rest mass) is related to momentum p according to the relation

$$E = \sqrt{(pc)^2 + E_0{}^2}, \tag{18.15}$$

where $E_0 = m_0c^2$ and m_0 is the rest mass of the particle. We use this relation in Chapter 2 to ultimately derive the Dirac equation.

The total energy E is also related to the kinetic energy K via the relation

$$E = K + m_0c^2. \tag{18.16}$$

Therefore, according to de Broglie, the wavelength associated with a relativistic particle must be determined as follows:

$$\lambda_r = \frac{hc}{pc} = \frac{hc}{\sqrt{E^2 - E_o{}^2}} = \frac{hc}{\sqrt{(K + E_0)^2 - E_0{}^2}}. \tag{18.17}$$

Exercises:

- What is the accelerating potential needed to give an electron a de Broglie wavelength of 5 Å, the size of the typical unit cell in many of the commonly studied semiconductors?

Solution

For an electron (with a initial zero kinetic energy) accelerated by a potential difference ΔV, its kinetic energy K becomes

$$K = e\Delta V = \frac{p^2}{2m_0} = \frac{1}{2m_0}(\frac{h}{\lambda})^2. \tag{18.18}$$

For $\lambda = 5$ Åand m_0 equal to free electron mass, the accelerating potential is $\Delta V \sim 6$ V.

- Calculate the de Broglie wavelength of an electron with a kinetic energy equal to the longitudinal optical phonon energy ($\hbar\omega_o = 36$ meV) in GaAs. The effective mass of the electron is assumed to be $m^* = 0.067m_o$, m_o being the free electron mass.

Solution

For an electron with kinetic energy $\frac{p^2}{2m^*} = \hbar\omega_o = 36$ meV and $m^* = 0.067m_o$, the de Broglie wavelength is

$$\lambda = \frac{h}{p} = \frac{h}{\sqrt{2m^*\hbar\omega_o}} \simeq 250\text{Å}. \tag{18.19}$$

- The non-relativistic limit of expression (18.17) is given by

$$\lambda_{nr} = \frac{h}{p} = \frac{h}{\sqrt{2m_0 K}}. \tag{18.20}$$

For what value of the ratio K/E_0 will the ratio of the non-relativistic and relativistic expressions for the de Broglie wavelength differ by 10%? What kinetic energy does this amount to for an electron? (The rest mass of an electron is 511 keV.)

Solution

We get

$$\frac{\lambda_{nr}}{\lambda_r} = \sqrt{1 + \frac{1}{2}(\frac{K}{E_0})}. \tag{18.21}$$

This ratio is equal to 1.1 when $\frac{K}{E_0} = 0.42$. This corresponds to a kinetic energy of 215 keV for an electron.

18.4 Postulates of quantum mechanics

This section closely follows the treatment of Nielsen and Chuang in their book titled *Quantum Computation and Quantum Information* [17]. A recap of fundamental concepts such as vector space, inner product, outer product, Hilbert space, linear operators, matrices, eigenvectors, adjoint, unitary and Hermitian operators, tensor products, operator functions, commutators and anti-commutators is given in Chapter 2, Section 2.1 of [17].

Postulate 1: Associated with any isolated physical system is a complex vector space with inner product (Hilbert space) known as the *state space* of the system. The system is completely described by its *state vector*, which is a unit vector in the state space.

In many chapters of this textbook, we are mostly concerned with a single qubit and will work in the Hilbert space $\mathcal{C}^2$. Any state vector in that Hilbert space is of the general form $|\psi>= a|0> +b|1>$, where a and b are complex numbers and the kets $|0>$ and $|1>$ are the 2×1 column vectors

$$|0>= \begin{bmatrix} 1 \\ 0 \end{bmatrix}, \tag{18.22}$$

and

$$|1> = \begin{bmatrix} 0 \\ 1 \end{bmatrix}. \tag{18.23}$$

Since $|\psi>$ is a unit vector in that Hilbert space, we must have $<\psi|\psi>=1$, i.e., $|a|^2 + |b|^2 = 1$, which is referred to as the normalization condition.

The extension of Postulate 1 to a multiparticle system (referred to as **Postulate 4** of quantum-mechanics) is based on the concept of tensor product. A rigorous treatment of the exchange interaction between two spins (recall Chapter 10) representing two qubits (recall the square-root-of-swap operation in Chapter 16) would have required the use of tensor products of the two qubits.

Postulate 2: The evolution of a closed quantum system is described by a unitary transformation, i.e., the state $|\psi>$ of the system at time t_1 is related to the state $|\psi>$ at time t_2 by a unitary operator U which depends only on the initial (t_1) and end (t_2) times:

$$|\psi(t_2)> = U(t_2, t_1)|\psi(t_1)>. \tag{18.24}$$

Note that $<\psi(t_2)|\psi(t_2)>=<\psi(t_1)|U^+U|\psi(t_1)>=<\psi(t_1)|\psi(t_1)>=1$, i.e., the state vector stays normalized at all times.

In non-relativistic quantum-mechanics, postulate 2 is sometimes formulated as a prelude to the Schrödinger equation which describes the time evolution of the state of a closed system characterized by a Hamiltonian H

$$i\hbar \frac{d}{dt}|\psi> = H|\psi>. \tag{18.25}$$

If the Hamiltonian does not depend explicitly on time, a formal solution of the Schrödinger equation can be written as

$$|\psi(t_2)> = e^{\frac{-i}{\hbar}H(t_2-t_1)}|\psi(t_1)>, \tag{18.26}$$

and since H is Hermitian, the operator U defined as

$$U(t_2, t_1) = e^{\frac{-i}{\hbar}H(t_2-t_1)} \tag{18.27}$$

is unitary.

Exercises:

- Prove that $U(t_2, t_1) = e^{\frac{-i}{\hbar}H(t_2 - t_1)}$ is unitary if H is Hermitian.

- Calculate the explicit form of the unitary operator for the case of a spin-1/2 particle (electron) in a spatially uniform and time-independent magnetic field along the z-axis.

Solution

The Hamiltonian for this problem is given by

$$H = -\vec{\mu}.\vec{B}, \tag{18.28}$$

where $\vec{\mu}$ is the magnetic moment of the electron, which, in a bulk semi-conductor with gyromagnetic factor g, is given by

$$\vec{\mu} = -\frac{g}{2}\left(\frac{e}{m_0}\right)\vec{S}, \tag{18.29}$$

where

$$\vec{S} = \frac{\hbar}{2}\vec{\sigma}. \tag{18.30}$$

With $\vec{B}$ along the z-axis, the Hamiltonian becomes

$$H = -\vec{\mu}.\vec{B} = \frac{g\mu_B B}{2}\sigma_z = \frac{g}{2}\mu_B B \begin{pmatrix} 1 & 0 \\ 0 & -1 \end{pmatrix}, \tag{18.31}$$

where $\mu_B = \frac{e\hbar}{2m_0}$ is the Bohr magneton. So

$$U(t_2 - t_1) = e^{-\frac{i}{\hbar}H(t_2 - t_1)} = e^{-\frac{i}{\hbar}\frac{g\mu_B B}{2}\sigma_z(t_2 - t_1)}. \tag{18.32}$$

Using the operator identity (3.52) proven in Chapter 3, we get

$$\exp(\theta\sigma_z) = \begin{pmatrix} e^\theta & 0 \\ 0 & e^{-\theta} \end{pmatrix}, \tag{18.33}$$

where

$$\theta = -\frac{i}{\hbar}\frac{g\mu_B B}{2}\Delta t, \tag{18.34}$$

with $\Delta t = t_2 - t_1$.

Therefore,

$$U(t_2, t_1) = U(t_2 - t_1) = U(\Delta t) = \begin{pmatrix} e^{-\frac{i}{\hbar}\frac{g\mu_B B}{2}\Delta t} & 0 \\ 0 & e^{\frac{i}{\hbar}\frac{g\mu_B B}{2}\Delta t} \end{pmatrix}. \tag{18.35}$$

U is unitary, as easily checked.

Postulate 3: Quantum Measurements

The quantum projective approach of Von Neumann describes the wavefunction collapse associated with measurements performed on quantum-mechanical systems. According to this approach, a projective measurement is described by an observable M, which is a Hermitian operator on the state space of the system. Measurements obey the following set of rules:

- Since the operator M is Hermitian, it has a spectral decomposition

$$M = \sum_m m P_m, \tag{18.36}$$

  where P_m is the projector onto the eigenspace of M with eigenvalue m and corresponding eigenvector $|m>$. P_m is explicitly given by the outer product

$$P_m = |m><m|. \tag{18.37}$$

- The possible outcomes of the measurement of the observable M correspond to the eigenvalues m of the observable.

- Upon measuring M on a quantum-mechanical system characterized by the ket $|\psi>$, the probability of getting result m is given by

$$p(m) = <\psi|P_m|\psi> = <\psi|m><m|\psi> = |<\psi|m>|^2. \tag{18.38}$$

- After the measurement is made, if outcome m has occurred, the state of the quantum system immediately after the measurement is given by (or has collapsed to)

$$P_m|\psi> / \sqrt{p(m)} = \frac{<m|\psi>|m>}{\sqrt{|<\psi|m>|^2}}, \tag{18.39}$$

  where $\sqrt{p(m)}$ is introduced to renormalize the state after collapse occurs.

A few problems on quantum projective measurements involving the spin components of a single qubit are discussed in Chapters 3 and 4.

Exercise: A useful property of projection measurements

In order to calculate the average value of repeated measurements of the observable M on a quantum-mechanical system in the state $|\psi>$, we must perform the summation

$$E(M) = \sum_m m p(m), \tag{18.40}$$

which is what we expect from the definition of average value associated with a random variable using probability theory. Plugging in the definition of $p(m) =< \psi|P_m|\psi >$, we get

$$E(M) = \sum_m m < \psi|P_m|\psi >=< \psi| \sum_m mP_m|\psi >=< \psi|M|\psi > . \qquad (18.41)$$

The latter expression can usually be calculated easily for a given operator M and state $|\psi >$.

Similarly, the standard deviation of a large number of measurements of the observable M is defined by

$$\Delta(M) = [< \psi|(M- < M >)^2|\psi >]^{\frac{1}{2}}, \qquad (18.42)$$

where

$$< M >\doteq< \psi|M|\psi > . \qquad (18.43)$$

Therefore,

$$\Delta(M) = \sqrt{< \psi|(M^2 - 2 < M > M+ < M >^2)|\psi >} = \sqrt{< M^2 > - < M >^2}, \qquad (18.44)$$

where $< M^2 >=< \psi|(M^2|\psi >$.

The standard deviation $\Delta(M)$ can therefore also be calculated easily once the operator M and the ket $|\psi >$ are known.

Heisenberg Uncertainty Principle

For any two Hermitian operators A, B associated with physical observables, the average value of the product AB in a quantum state $|\psi >$ will be some complex number $< \psi|AB|\psi >= x + iy$ where x, y are real. Since AB is Hermitian, $< \psi|BA|\psi >= x - iy$ and the following equalities hold:

$$< \psi|[A, B]|\psi >=< \psi|AB|\psi > - < \psi|BA|\psi >= (x + iy) - (x - iy) = 2iy, \qquad (18.45)$$

and

$$< \psi|\{A, B\}|\psi >=< \psi|AB|\psi > + < \psi|BA|\psi >= (x + iy) + (x - iy) = 2x. \qquad (18.46)$$

Hence,

$$| < \psi|[A, B]|\psi > |^2 + | < \psi|\{A, B\}|\psi > |^2 = 4| < \psi|AB|\psi > |^2 = 4(x^2 + y^2). \qquad (18.47)$$

Using the Cauchy-Schwartz inequality

$$| < v|w > |^2 \leq< v|v >< w|w >, \qquad (18.48)$$

with $|v >= A|\psi >$ and $|w >= B|\psi >$, we get

$$| < \psi|AB|\psi > |^2 \leq< \psi|A^2|\psi >< \psi|B^2|\psi >, \qquad (18.49)$$

which leads to

$$| < \psi |[A, B]|\psi > |^2 \le 4 < \psi |A^2|\psi >< \psi |B^2|\psi > . \qquad (18.50)$$

Using the definition of the standard deviation associated with a measurement of an observable M and defining two new operators C and D such that

$$C = A- < A >, D = B- < B >, \qquad (18.51)$$

we get

$$< C >= 0, \Delta(C) = \sqrt{< C^2 >} = \Delta(A), \qquad (18.52)$$

and

$$< D >= 0, \Delta(D) = \sqrt{< B^2 >} = \Delta(B). \qquad (18.53)$$

Furthermore, $[A, B] = [C, D]$ and therefore the inequality (18.50) can be rewritten

$$\Delta(C)\Delta(D) \ge \frac{| < \psi |[C, D]|\psi > |}{2}. \qquad (18.54)$$

This is the generalized form of the Heisenberg uncertainty principle.

Exercises:

In the derivation above, it was assumed that, for any two Hermitian operators A and B, the average value of the product AB in a quantum state $|\psi >$ is equal to some complex number $< \psi |AB|\psi >= x + iy$ where x, y are real. Find the values of x and y.

- For the case where $A = \sigma_x$ and $B = \sigma_y$, and the ket $|\psi >$ equal to $|0 >$, $|1 >$ and $\frac{1}{\sqrt{2}}(|0 > +|1 >)$.

- Repeat the previous exercise if $A = \mathbf{H}$, the Hadamard matrix (see Equation (3.43)) and $B = \sigma_y$.

- Find the value of $x + iy$ for an electron in the ground state of the one-dimensional harmonic oscillator in a box taking $A = x$ (position) and $B = p_x$ (momentum). The wavefunction in the ground state of a one-dimensional harmonic oscillator described by a potential energy $V(x) = \frac{m}{2}\omega^2 x^2$ is given by

$$\phi_0(x) = (\frac{m\omega}{\pi\hbar})^{1/4}e^{-\frac{1}{2}\frac{m\omega}{\hbar}x^2}. \qquad (18.55)$$

- Repeat the previous problem if the electron is in the first excited state of the one-dimensional harmonic oscillator. The wavefunction in the first excited state of the one-dimensional harmonic oscillator is given by

$$\phi_1(x) = [\frac{4}{\pi}(\frac{m\omega}{\hbar})^3]^{1/4}xe^{-\frac{1}{2}\frac{m\omega}{\hbar}x^2}. \qquad (18.56)$$

18.4.1 Interpretation of the Heisenberg Uncertainty Principle

From a practical standpoint, the meaning of the inequality (18.54) is as follows:

- We must first prepare a physical system a large number of times (or ensemble) in state $|\psi>$,

- Then, perform measurements of the observable C on a fraction of the ensemble and of the observable D on the rest of the ensemble.

- When a meaningful number of measurements have been performed so that a standard deviation of these measurements can be calculated, the experimental results for the standard deviation $\Delta(C)$ of the measurements of the observable C and the standard deviation $\Delta(D)$ of the measurements of the observable D will satisfy the inequality

$$\Delta(C)\Delta(D) \geq \frac{|<\psi|[C,D]|\psi>|}{2}. \qquad (18.57)$$

Example 1: Applying the inequality in the case of the position and momentum operators x and p_x, and using the fact that $[x, p_x] = i\hbar$, we get

$$\Delta(x)\Delta(p_x) \geq \frac{\hbar}{2}, \qquad (18.58)$$

which is one of the early forms of the Uncertainty Principle published by Heisenberg.

Example 2:
Consider the Pauli matrices σ_x and σ_y and show that the inequality (18.54) leads to

$$\Delta(S_x)\Delta(S_y) \geq \frac{\hbar^2}{4}. \qquad (18.59)$$

This can be easily shown starting with the inequality (18.54) and the following property of the Pauli matrices proved in Chapter 2:

$$[\sigma_x, \sigma_y] = 2i\sigma_z. \qquad (18.60)$$

So, $\Delta(S_x), \Delta(S_y)$ must both be strictly greater than 0. This can be verified by direct calculation, as shown in the next concrete example.

Calculate both sides of the inequality (18.54) for an ensemble of qubits prepared in the quantum state $|0>$, eigenvector of the Pauli matrix σ_z with eigenvalue $+1$, i.e.,

$$|0> = \begin{bmatrix} 1 \\ 0 \end{bmatrix}, \qquad (18.61)$$

for which

$$< \sigma_x >= (1\ 0) \begin{pmatrix} 0 & 1 \\ 1 & 0 \end{pmatrix} \begin{pmatrix} 1 \\ 0 \end{pmatrix} = 0 \tag{18.62}$$

and

$$< \sigma_y >= (1\ 0) \begin{pmatrix} 0 & -i \\ i & 0 \end{pmatrix} \begin{pmatrix} 1 \\ 0 \end{pmatrix} = 0. \tag{18.63}$$

Furthermore, since $< \sigma_x{}^2 > = < \sigma_y{}^2 > = \frac{\hbar^2}{4}$, we get

$$\Delta S_x \Delta S_y = \frac{\hbar^2}{4}, \tag{18.64}$$

and the inequality (18.54) is an equality in this case.

Example 3: Heisenberg Uncertainty Principle for the 1D particle in a box

Starting with the eigenstates and corresponding eigenvalues for the simple problem of a particle in a one-dimensional quantum box of size W,

- (a) Calculate the average values $< x >$ and $< p_x >$ for an electron prepared in each of the eigenstates of the particle in a box.

- (b) Calculate the standard deviations Δx, Δp_x for an electron prepared in each of the eigenstates of the particle in a box.

- (c) Show that, for large values of the quantum number n characterizing the eigenstates, the standard deviation Δx reduces to its classical value.

- (d) Calculate $\sqrt{\Delta x \Delta p_x}$ for each of the eigenstates found in part (a) and show that the Heisenberg Uncertainty relation is satisfied.

Solution

The solution to the Schrödinger equation for a particle in a one-dimensional box is one of the simplest problems in quantum mechanics and is given by

$$\phi_n(x) = \sqrt{\frac{2}{W}} sin(\frac{n\pi x}{W}), \tag{18.65}$$

where n is an integer ($n = 1,2,3,...$) and W is the width of the box. The corresponding eigenvalues are given by

$$E_n = n^2 \frac{\hbar^2 \pi^2}{2mW^2}. \tag{18.66}$$

(a) Using these results, the following average values are found.

- $$< x > = \int_0^W x|\phi_n(x)|^2 dx = \frac{W}{2}. \tag{18.67}$$

This is to be expected since all eigenstates produce symmetrical probability distributions $|\phi_n|^2$ with respect to the center of the box.

- $$< p_x > = \int_0^W dx\phi_n(x)\frac{\hbar}{i}\frac{d}{dx}\phi_n(x) = 0. \tag{18.68}$$

This is expected since $\phi_n(x)$ and $\frac{d}{dx}\phi_n(x)$ are sine and cosine functions, respectively, and are therefore orthogonal to each other. Classically, this corresponds to the fact that a particle traveling back and forth between the two walls of the box will have the same magnitude of its momentum but opposite signs when traveling from left to right and right to left. Therefore, the average momentum will be zero when trapped inside a box.

(b) Standard deviations

- $$< \Delta x^2 > = \int_0^W dx x^2 \phi_n^2(x) - \left[\frac{W}{2}\right]^2. \tag{18.69}$$

So

$$(\Delta x)^2 = \frac{2}{W}\int_0^W dx x^2 \sin^2\left(\frac{n\pi x}{W}\right) - \frac{W^2}{4}. \tag{18.70}$$

Therefore

$$(\Delta x)^2 = \frac{W^2}{12}\left[1 - \frac{6}{(n\pi)^2}\right]. \tag{18.71}$$

- $$(\Delta p_x)^2 = < p_x^2 > - < p_x >^2 = < p_x^2 >. \tag{18.72}$$

Hence

$$(\Delta p_x)^2 = \int_0^W dx\phi_n(x)p_x^2\phi_n(x)dx. \tag{18.73}$$

But $p_x^2\phi_n = 2mE_n\phi_n(x)$, so

$$(\Delta p_x)^2 = 2mE_n\int_0^W dx|\phi_n|^2 = 2m\left(\frac{\hbar^2 n^2\pi^2}{2mW^2}\right) = \frac{\hbar^2 n^2\pi^2}{W^2}. \tag{18.74}$$

(c) As n increases, we have from part (b)

$$(\Delta x)^2 = \frac{W^2}{12}. \tag{18.75}$$

We compare this to its classical limit which is equal to

$$\Delta x^2 = \frac{1}{W} \int_0^W (x - \frac{W}{2})^2 dx = \frac{W^2}{12}. \qquad (18.76)$$

So indeed the quantum-mechanical result converges to its classical value for eigenstates with large quantum numbers.

(d) Value of $\sqrt{\Delta x \Delta p_x}$:

Using the previous results, we get

$$\sqrt{\Delta x \Delta p_x} = \frac{\hbar}{2} \frac{n\pi}{\sqrt{3}} [1 - \frac{6}{n^2 \pi^2}]^{\frac{1}{2}}, \qquad (18.77)$$

which can be shown to be greater than $\frac{\hbar}{2}$ for all integers ≥ 1. The smallest value of the right hand side in the above equation is for $n = 1$, in which case

$$\sqrt{\Delta x \Delta p_x} = 0.568\hbar, \qquad (18.78)$$

which is slightly larger than $\frac{\hbar}{2}$.

18.4.2 Time evolution of expectation values: Ehrenfest theorem

Ehrenfest was the first to derive an equation to describe the time evolution of the expectation value of a time-dependent observable for a quantum-mechanical system described by a ket $|\phi(t)>$. Using Dirac's notations,

$$\frac{d}{dt} < \phi(t)|A(t)|\phi(t) >= [\frac{d}{dt} < \phi(t)|]A(t)|\phi(t) > + < \phi(t)|A(t)[\frac{d}{dt}|\phi(t) >]$$
$$+ < \phi(t)|\frac{dA(t)}{dt}|\phi(t) > . \qquad (18.79)$$

Taking into account that $|\phi(t)>$ satisfies the Schrödinger equation, we get

$$\frac{d}{dt}|\phi(t) >= \frac{1}{i\hbar} H(t)|\phi(t) > . \qquad (18.80)$$

The conjugate transpose of this last equation is given by

$$\frac{d}{dt} < \phi(t)| = -\frac{1}{i\hbar} < H(t)\phi(t)|. \qquad (18.81)$$

Using the last two equations, Equation (18.79) can be rewritten as follows

$$\frac{d}{dt} < \phi(t)|A(t)|\phi(t) >= \frac{1}{i\hbar} \frac{d}{dt} < \phi(t)|[A(t)H(t) - H(t)A(t)]|\phi(t) >$$
$$+ < \phi(t)|\frac{dA(t)}{dt}|\phi(t) > \qquad (18.82)$$

or

$$\frac{d}{dt} < A >= \frac{1}{i\hbar} < [A, H(t)] > + < \frac{dA}{dt} > . \tag{18.83}$$

This last equation is referred to as *Ehrenfest's theorem*.

Example 1:

We first illustrate the application of Ehrenfest's theorem to a spinless particle whose Hamiltonian is given by

$$H = \frac{p^2}{2m} + V(r), \tag{18.84}$$

where $V(r)$ is the potential energy of the particle.

Applications of Ehrenfest's theorem with operator A equal to the three components of the position operator lead to

$$\frac{d}{dt} < r >= \frac{1}{i\hbar} < [r, H] >= \frac{1}{i\hbar} < [r, \frac{p^2}{2m}] > . \tag{18.85}$$

Similarly, if A is equal to the three components of the momentum operator, we get

$$\frac{d}{dt} < p >= \frac{1}{i\hbar} < [p, H] >= \frac{1}{i\hbar} < [p, V(r)] > . \tag{18.86}$$

Using appropriate identities to calculate commutators of function of operators [17, 18]

$$[r, \frac{p^2}{2m}] = \frac{i\hbar}{m} p, \tag{18.87}$$

and

$$[p, V(r)] = -i\hbar \nabla V(r) \tag{18.88}$$

and combining that with the Ehrenfest theorem ultimately lead to

$$\frac{d}{dt} < \vec{r} >= \frac{1}{m} < p >, \tag{18.89}$$

and

$$\frac{d}{dt} < \vec{p} >= - < \nabla V(r) > . \tag{18.90}$$

The last two equations are similar to the classical equations of motion (Newton's equations) describing the time evolution of the momentum of a particle except that the latter is replaced by the average value of the corresponding quantum-mechanical operator over the state $|\phi(t) >$ of the system. The force acting on the particle is calculated as the average over the state of the particle of the gradient of the potential energy.

Example 2:

In Chapter 4, we invoked Ehrenfest's theorem to describe the time evolution of a spin-1/2 particle in a spatially uniform time-dependent magnetic field whose Hamiltonian is given by $-(g/2)\mu_B\vec{\sigma} \cdot \vec{B}$. Here, we provide the proof.

According to Ehrenfest's theorem, the time evolutions of the averages of the spin components are given by

$$\frac{d}{dt} <\sigma_x> = \frac{1}{i\hbar}\frac{g}{2}\mu_B(B_y < [\sigma_x, \sigma_y] > +B_z < [\sigma_x, \sigma_z] >), \qquad (18.91)$$

$$\frac{d}{dt} <\sigma_y> = \frac{1}{i\hbar}\frac{g}{2}\mu_B(B_x < [\sigma_y, \sigma_x] > +B_z < [\sigma_y, \sigma_z] >), \qquad (18.92)$$

and

$$\frac{d}{dt} <\sigma_z> = -\frac{1}{i\hbar}\frac{g}{2}\mu_B(B_x < [\sigma_z, \sigma_x] > +B_y < [\sigma_z, \sigma_y] >). \qquad (18.93)$$

Using the relations in Problem 2.4, we get

$$\frac{d}{dt} <\sigma_x> = \frac{g\mu_B}{\hbar}(B_y < \sigma_z > -B_z < \sigma_y >), \qquad (18.94)$$

$$\frac{d}{dt} <\sigma_y> = \frac{g\mu_B}{\hbar}(B_z < \sigma_x > -B_x < \sigma_z >), \qquad (18.95)$$

and

$$\frac{d}{dt} <\sigma_z> = \frac{g\mu_B}{\hbar}(B_x < \sigma_y > -B_y < \sigma_x >). \qquad (18.96)$$

The reader can easily show that the last three equations can be written as

$$\frac{d <\vec{\sigma}>}{dt} = \frac{g\mu_B\vec{B}}{\hbar} \times <\vec{\sigma}>, \qquad (18.97)$$

which leads to the familiar Larmor precession equation

$$\frac{d\vec{S}}{dt} = \frac{g\mu_B\vec{B}}{\hbar} \times \vec{S}, \qquad (18.98)$$

since $\vec{S}_{op} = (\hbar/2)\vec{\sigma}$.

18.5 Some elements of semiconductor physics: Particular applications in nanostructures

18.5.1 Density of states: Bulk (3-D) to quantum dot (0-D)

Consider the quantum confined geometries shown in Fig. 18.3 (2D: two-dimensional electron gas, 1D: one-dimensional electron gas, 0D: three-dimensional quantum box). Calculate the energy dependence of the *density of states* in these structures and compare them to that of the 3D bulk sample shown in the upper left corner in Fig. 18.3.

Solution

3D: Consider a uniform homogeneous bulk piece of semiconductor whose conduction band has a parabolic $E - \vec{k}$ relationship with the bottom at E_{co}, as shown in Figure 18.4:

$$E(\vec{k}) = E_{co} + \frac{\hbar^2 k^2}{2m^*}. \tag{18.99}$$

The solutions of the 3D effective mass Schrödinger equation are of the form of plane waves

$$\phi_k(\vec{r}) = \frac{1}{\sqrt{\Omega}} e^{i\vec{k}\vec{r}}, \tag{18.100}$$

normalized over a volume $\Omega = L^3$, where L is the length of the side of a cube large compared to the lattice unit cell. Assuming periodic boundary conditions for $\phi_k(\vec{r})$, i.e.,

$$\phi_k(x + L, y + L, z + L) = \phi_k(x, y, z). \tag{18.101}$$

The allowed values of $\vec{k} = (k_x, k_y, k_z)$ are given by

$$k_x = n_x \frac{2\pi}{L}, \tag{18.102}$$

$$k_y = n_y \frac{2\pi}{L}, \tag{18.103}$$

and

$$k_z = n_z \frac{2\pi}{L}, \tag{18.104}$$

where n_x, n_y, n_z are integers.

The ewquilibrium density of electrons at a location $\vec{r}$ is given by

$$\rho(\vec{r}) = \frac{N}{\Omega} = \sum_{\vec{k}} f(E_k) |\phi_k(\vec{r})|^2, \tag{18.105}$$

where $f(E_k)$ is the Fermi-Dirac distribution function. We can assume that the carrier statistics is governed by the Fermi-Dirac distribution as long as the system is in *equilibrium* (e.g., no current flows and no light is shining on it generating electron-hole pairs).

Each electron eigenstate occupies a volume $\frac{(2\pi)^3}{\Omega}$ in $\vec{k}$-space. Therefore, in a volume of size $d^3\vec{k}$, we have a number of electron eigenstates equal to

$$2\frac{\Omega}{(2\pi)^3} d^3\vec{k}, \tag{18.106}$$

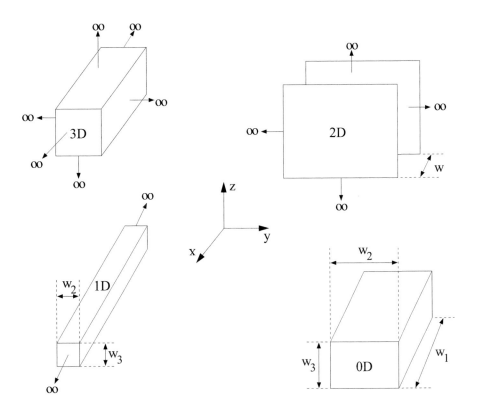

FIGURE 18.3

Illustration of the formation of a quantum dot (bottom right figure) through the gradual squeezing of a bulk piece of semiconductor (upper left). When the dimension of the bulk structure is reduced in one direction to a size comparable to the de Broglie wavelength, the resulting electron gas is referred to as a two-dimensional electron gas (2-DEG) because the carriers are free to move in the y and z directions only. If quantum confinement occurs in two directions, as illustrated in the bottom left figure, the resulting electron gas is referred to as a one-dimensional electron gas (1 DEG) since an electron in this structure is free to move in the x direction only. If confinement is imposed in all three directions, we get a quantum dot (0 DEG).

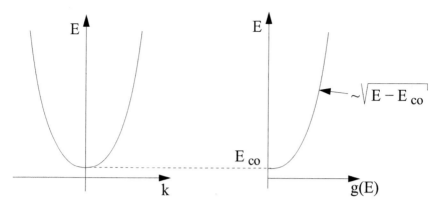

FIGURE 18.4
(Left) Parabolic energy dispersion relation close to the bottom of the conduction band (E_{c0}) of a typical semiconductor. (Right) Corresponding energy dependence of the three-dimensional density of states in a bulk semiconductor.

where the extra factor 2 has been added to take into account the spin degeneracy of each eigenstate in $\vec{k}$-space, as required by the Pauli Exclusion principle. For a large value of Ω, the $\sum_{\vec{k}}$ in Equation (18.105) can be replaced by an integral and we obtain

$$\rho(\vec{r}) = \frac{1}{4\pi^3} \int d^3\vec{k} f(E_k), \qquad (18.107)$$

which is spatially invariant.

Since $f(E_k)$ is spherically symmetric in $\vec{k}$-space, the last integration can be easily performed using spherical coordinates leading to

$$\rho = \int_{E_{co}}^{+\infty} dE g_{3D}(E) f(E), \qquad (18.108)$$

where $g_{3D}(E)$ is by definition the three-dimensional density of states and is given by

$$g_{3D}(E) = \frac{k^2}{\pi^2 \left(\frac{dE}{dk}\right)} = \frac{m^* k}{\pi^2 \hbar^2}, \qquad (18.109)$$

where we have used the dispersion relation in Equation (18.99) to arrive at the last equality.

Using the $E - k$ relationship (18.99) once again to express k in terms of E, we get the well-known result for the 3-D density of states:

$$g_{3D}(E) = \frac{m^*}{\pi^2 \hbar^3} \sqrt{2m^*(E - E_{co})}, \qquad (18.110)$$

whose energy dependence is illustrated in Figure 18.4.

Using the above equation in Equation (18.107), the electron density in a bulk sample is given by

$$\rho = \frac{2}{\pi} N_c F_{\frac{1}{2}}(\xi), \tag{18.111}$$

where

$$N_c = \frac{1}{4\hbar^3} \left(\frac{2m^* k_B T}{\pi} \right)^{\frac{3}{2}}, \tag{18.112}$$

and

$$\xi = \frac{(E_F - E_{co})}{k_B T}. \tag{18.113}$$

In equation (18.111), $F_{\frac{1}{2}}$ is the Fermi-Dirac integral of index $\frac{1}{2}$:

$$F_{\frac{1}{2}}(\xi) = \int_{E_{co}}^{+\infty} \frac{dE \sqrt{E - E_{co}}}{[1 + e^{\frac{E - E_F}{kT}}]}. \tag{18.114}$$

2D: Next, we generalize the derivation above to determine the two-dimensional density of states in a 2-DEG. In this case, the electron density is given by

$$\rho(\vec{r}) = \sum_m \sum_{k_y, k_z} f_0(E_m) |\phi_{m, k_y, k_z}(\vec{r})|^2. \tag{18.115}$$

The eigenfunctions and corresponding eigenvalues of the Schrödinger equation are given by:

$$\phi_{m, k_y, k_z}(\vec{r}) = \frac{1}{\sqrt{A}} e^{ik_y y} e^{ik_z z} \xi_m(x), \tag{18.116}$$

where

$$E_{m, k_y, k_z} = E_m + \frac{\hbar}{2m^*} (k_y^2 + k_z^2), \tag{18.117}$$

and

$$A = L_y L_z \tag{18.118}$$

is a normalization area to describe the in-plane free motion of carriers in the (y, z) directions, x being the direction of quantum confinement of the well. The wavefunctions $\xi_m(x)$ are solutions of the one-dimensional Schrödinger equation and depend on the potential confinement $E_c(x)$ in the x-direction

$$-\frac{\hbar^2}{2m^*} \frac{d^2 \xi_m(x)}{dx^2} + E_c(x) \xi_m(x) = E_m \xi_m(x). \tag{18.119}$$

Each $\xi_m(x)$ is assumed to be normalized and has a corresponding eigenvalue E_m.

Therefore,

$$\rho(\vec{r}) = \sum_m \sigma_m |\xi_m(x)|^2, \tag{18.120}$$

where

$$\sigma_m = \sum_{k_y, k_z} \frac{1}{A} f_o(E_{m,k_y,k_z}). \tag{18.121}$$

Converting the $\sum_{k_y,k_z}$ to an integral following the 3-D case, we get

$$\sum_{k_y,k_z} = 2 \left(\frac{A}{4\pi^2} \right) \int d^2\vec{k}. \tag{18.122}$$

Using polar coordinates in the (k_y, k_z) plane

$$\sigma_m = \int_0^{2\pi} \frac{d\phi}{2\pi^2} \int_0^{+\infty} dk k f_o(E_{m,k_y,k_z}), \tag{18.123}$$

and since $f_o(E_{m,k_y,k_z})$ is independent of ϕ,

$$\sigma_m = \int_0^{+\infty} \frac{kdk}{\pi} f_o(E_{m,k_y,k_z}). \tag{18.124}$$

Using the dispersion relationship of the subbands in the well, we get

$$dE_{m,k_y,k_z} = \frac{\hbar}{m^*} kdk, \tag{18.125}$$

and σ_m becomes

$$\sigma_m = \int_{E_m}^{+\infty} dE g_{2D}(E) f_o(E), \tag{18.126}$$

where

$$g_{2D}(E) = \frac{m^*}{\pi\hbar^2} \tag{18.127}$$

is *independent of energy* and is the density of states in each subband in the well.

Substituting the expression for the Fermi-Dirac factor $f_o(E)$, σ_m can be calculated exactly,

$$\sigma_m = \frac{m^*}{\pi\hbar^2} k_B T ln(1 + e^{-(\frac{E_m - E_F}{k_B T})}). \tag{18.128}$$

This analytical expression for σ_m is valid for any shape of the confining potential in the x direction. This quantity determines the sheet electron concentration in a 2-DEG.

1D:
If we have confinement in the y-z plane and free motion of carriers is allowed in the x-direction, then

$$\rho(\vec{r}) = \sum_{k_x} \sum_n \sum_m f_o(E_{n,m}) |\phi_{k_x,n,m}(\vec{r})|^2 \tag{18.129}$$

where

$$\phi_{k_x,n,m}(\vec{r}) = \frac{1}{\sqrt{L}}e^{ik_x x}\xi_{n,m}(y,z),$$ (18.130)

where L is a normalization factor of the plane wave moving along the x-direction and $\xi_{n,m}(y,z)$ are the solutions of the two-dimensional Schrödinger equation

$$-\frac{\hbar^2}{2m^*}(\frac{d^2}{dy^2} + \frac{d^2}{dz^2})\xi_{n,m}(y,z) + E_c(y,z)\xi_{n,m}(y,z) = E_{n,m}\xi_{n,m}(x,y).$$ (18.131)

Here, n, m are quantum numbers characterizing the quantization in the y- and z-directions. They are also called *transverse subband indices*.

The energy dispersion relationship in each subband characterized by the two quantum numbers (n, m) is given by

$$E_{k_x,n,m} = E_{n,m} + \frac{\hbar^2}{2m^*}k_x^2.$$ (18.132)

In this 1-DEG, the electron density is invariant in the x-direction and is given by

$$\rho(y,z) = \sum_{n,m} \sigma_{n,m}|\xi_{n,m}(y,z)|^2,$$ (18.133)

where

$$\sigma_{n,m} = \sum_{k_x} \frac{1}{L}f_o(E_{k_x,n,m}).$$ (18.134)

Converting the sum over k_x into an integral, i.e.,

$$\sum_{k_x} = 2\left(\frac{L}{2\pi}\right)\int dk_x,$$ (18.135)

we get

$$\sigma_{n,m} = \frac{1}{\pi}\int_{-\infty}^{+\infty} dk_x f_o(E_{k_x,n,m}) = \frac{2}{\pi}\int_0^{+\infty} dk_x f_o(E_{k_x,n,m}).$$ (18.136)

(18.137)

Using the dispersion relation ($E - k_x$ relation), we get

$$dE_{k_x,n,m} = \frac{\hbar}{m^*}k_x dk_x;$$ (18.138)

hence

$$\sigma_{n,m} = \int_{E_{n,m}}^{+\infty} dE g_{1D}(E)f_o(E),$$ (18.139)

where

$$g_{1D}(E) = \frac{1}{\pi}\sqrt{\frac{2m^*}{\hbar^2}}\frac{1}{\sqrt{E - E_{n,m}}},\tag{18.140}$$

which is the expression for the one-dimensional density of states in each subband in the quantum wire. It diverges at $E = E_{n,m}$, the threshold energy for free propagation in that subband.

0D: In this case, we are dealing with a quantum box with quantum confinement in all three directions.

$$\rho(\vec{r}) = \sum_{n,m,l} f_o(E_{n,m,l})|\phi_{n,m,l}(\vec{r})|^2,\tag{18.141}$$

where $\phi_{n,m,l}$ are the solutions of the three-dimensional Schrödinger equation for the $E_c(x, y, z)$ representing the quantum confinement in all three directions. The indices (n, m, l) are three quantum numbers characterizing the eigenstates of the Schrödinger equation.
We can write

$$\rho(\vec{r}) = \sum_{n,m,l} \sigma_{n,m,l}|\phi_{n,m,l}(\vec{r})|^2,\tag{18.142}$$

with

$$\sigma_{n,m,l} = \int_0^{+\infty} dE g_{0D}(E)f_o(E).\tag{18.143}$$

Therefore the 0-dimensional density of states is simply

$$g_{0D}(E) = 2 \sum_{n,m,l} \delta(E - E_{n,m,l})\tag{18.144}$$

where δ is the Dirac delta function and the factor 2 has been included since each $E_{n,m,l}$ state can be occupied by two electrons with opposite spin.

Example 1: Electron sheet concentration in a quantum well

(a) Show that the sheet carrier concentration in a high-electron-mobility-transistor (HEMT) device (which has a 2-DEG in the channel) is given by

$$n_s = \frac{m^*}{\pi\hbar^2}k_B T ln\left[(1 + e^{\frac{E_F - E_1}{k_B T}})(1 + e^{\frac{E_F - E_2}{k_B T}})\right],\tag{18.145}$$

when only two subbands are occupied. Here, E_1 and E_2 are the bottom energies of the first two subbands.

(b) Starting with the result of the part (a), show that at low temperature

$$n_s = \frac{m^*}{\pi\hbar^2}(E_F - E_1)\tag{18.146}$$

when the second subband is unoccupied, and

$$n_s = \frac{m^*}{\pi\hbar^2}(E_2 - E_1) + 2\frac{m^*}{\pi\hbar^2}(E_F - E_2) \tag{18.147}$$

when both subbands are occupied.

Solution

The electron concentration in the 2-DEG formed at the heterointerface between the high and low bandgap materials in a HEMT structure (see Fig. 13.7) is given by

$$\rho(x) = \sum_m \sigma_m |\xi_m(x)|^2, \tag{18.148}$$

where

$$\sigma_m = \frac{m^*(k_B T)}{\pi\hbar^2} ln(1 + e^{-\frac{(E_m - E_F)}{k_B T}}). \tag{18.149}$$

The sheet carrier concentration in the 2-DEG is given by

$$n_s = \int_{-\infty}^{+\infty} \rho(x)dx. \tag{18.150}$$

If the wavefunctions $\xi_m(x)$ are normalized, i.e.,

$$\int_{-\infty}^{+\infty} |\xi_m(x)|^2 dx = 1, \tag{18.151}$$

the sheet carrier concentration is given by the simple formula

$$n_s = \sum_m \sigma_m. \tag{18.152}$$

If only one subband in the 2-DEG is occupied,

$$n_s = k_B T ln(1 + e^{\frac{E_F - E_1}{k_B T}}). \tag{18.153}$$

If $k_B T << E_F - E_1$,

$$n_s = \frac{m^*}{\pi\hbar^2}(E_F - E_1). \tag{18.154}$$

When the second subband is occupied (but the third one unoccupied),

$$n_s = \frac{m^*}{\pi\hbar^2} k_B T \left[ln(1 + e^{\frac{E_F - E_1}{k_B T}}) + ln(1 + e^{\frac{E_F - E_2}{k_B T}}) \right], \tag{18.155}$$

which can also be written as

$$n_s = \frac{m^*}{\pi\hbar^2} k_B T ln \left[(1 + e^{\frac{E_F - E_1}{k_B T}})(1 + e^{\frac{E_F - E_2}{k_B T}}) \right]. \tag{18.156}$$

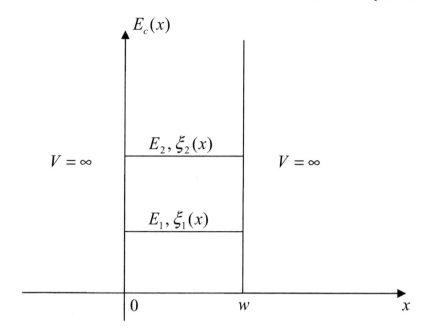

FIGURE 18.5
Confined states in a quantum well (2-DEG) of width w.

If $k_B T << E_F - E_1$, $E_F - E_2$, (i.e., at low temperature), then

$$ln\left(1 + e^{\frac{E_F - E_1}{k_B T}}\right) = \frac{E_F - E_1}{k_B T}, \tag{18.157}$$

and

$$ln\left(1 + e^{\frac{E_F - E_2}{k_B T}}\right) = \frac{E_F - E_2}{k_B T}. \tag{18.158}$$

Hence

$$n_s = \frac{m^*}{\pi\hbar^2}(E_F - E_1) + \frac{m^*}{\pi\hbar^2}(E_F - E_2) = \frac{m^*}{\pi\hbar^2}(2E_F - E_1 - E_2). \tag{18.159}$$

Example 2: Fermi level location in a quantum well

Consider a 100 Å wide potential well (quantum well or 2-DEG) with infinite walls at $T = 0$ K. Assume all impurities are ionized (i.e., neglect carrier freeze out). Assume $m^* = 0.067m_0$ and calculate the location of the Fermi level for

- $N_D = 10^{17}$ cm^{-3},
- $N_D = 10^{19}$ cm^{-3}.

Solution

Assuming that all impurities are ionized, the sheet carrier concentration in the well is given by

$$n_s = N_D W. \tag{18.160}$$

Therefore, for $N_D = 10^{17}$ and 10^{19} cm^{-3}, n_s is equal to 10^{11} and $10^{13} cm^{-2}$, respectively.

At zero temperature, if the Fermi level E_F is between the Nth and $(N+1)^{th}$ subbands in the well, then

$$n_s = \frac{m^*}{\pi \hbar^2} \sum_{i=1}^{N} (E_F - E_i). \tag{18.161}$$

Hence

$$n_s = \frac{m^*}{\pi \hbar^2} \left(N E_F - \sum_{i=1}^{N} E_i \right), \tag{18.162}$$

which is a generalization of the results found in the previous example. Solving for E_F, we get

$$E_F = \frac{1}{N} \left[\frac{N_D W}{\left(\frac{m^*}{\pi \hbar^2} \right)} + \sum_{i=1}^{N} E_i \right]. \tag{18.163}$$

For a well surrounded by an infinite wall (particle in a box problem), the different eigenstate energies are given by

$$E_i = \frac{\hbar^2}{2m^*} (\frac{i\pi}{W})^2, \tag{18.164}$$

where i is an integer.

For $m^* = 0.067 m_o$ and $W = 100$ Å, we find

$$E_i \simeq i^2 56 \text{meV}. \tag{18.165}$$

Therefore, the subband energy bottoms due to the particle-in-a-box confinement in two dimensions are given by the above equation.

For $N_D = 10^{17}$ cm^{-3}, if we assume E_F is between E_1 and E_2, and $N = 1$ in Equation (18.163) above, we get

$$E_F = 59.78 meV, \tag{18.166}$$

which tells us that only one subband is occupied, as assumed.

For $N_D = 10^{19}$ cm^{-3}, assuming E_F is between E_2 and E_3, the Fermi level is found to be

$$E_F = 329.4 \text{meV}, \tag{18.167}$$

which is below E_3. In this case, only two subbands are occupied. The number of subbands that are occupied is usually found by trial and error.

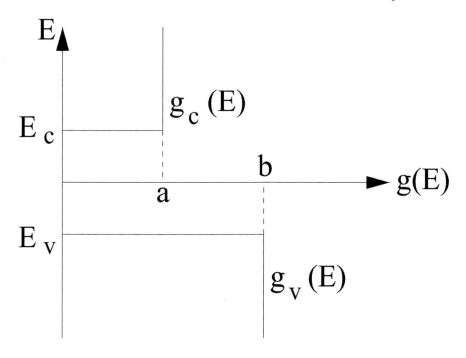

FIGURE 18.6
Density of states of electrons or holes in a two-dimensional electron or hole gas (2-DEG or 2-DHG).

Example 3: Intrinsic carrier concentration in a 2-DEG

Consider the density of states as shown in Fig. 18.6 representing the two-dimensional density of states of electrons and holes in a quantum well.

(a) Assuming the well is undoped, obtain an expression for the Fermi level E_F at room temperature in terms of a, b and temperature T. Assume Boltzmann statistics to be valid. When is E_F exactly equal to the midgap energy, $\frac{(E_c + E_v)}{2}$?

(b) Obtain the expression for n_i, the intrinsic carrier concentration.

Hint: Start with the approximate expressions for the electron (n) and hole (p) concentrations in terms of $g_c(E)$ and $g_v(E)$ [subscripts c and v denote conduction and valence bands], and assume Boltzmann statistics of carriers:

$$f(E) = e^{\frac{E_F - E}{k_B T}}, \qquad (18.168)$$

where k_B is Boltzmann's constant.

Solution

(a) The electron and hole concentrations are given by

$$n = \int_{E_c}^{\infty} g_c(E)f(E)dE, \tag{18.169}$$

$$p = \int_{-\infty}^{E_v} g_v(E)[1 - f(E)]dE. \tag{18.170}$$

Use the Boltzmann approximation, $f(E) = e^{\frac{(E_F - E)}{k_B T}}$ and $g_c(E) = a\theta(E - E_c)$ and $g_v(E) = b\theta(E_v - E)$ where $\theta(x) = 1$ for $x > 0$ and $\theta(x) = 0$ for $x \leq 0$ (Heaviside function). We get

$$n = \int_{E_c}^{\infty} ae^{\frac{(E_F - E)}{k_B T}} dE = ak_B T e^{\frac{E_F - E_c}{k_B T}}. \tag{18.171}$$

Similarly,

$$p = \int_{-\infty}^{E_v} be^{\frac{(E - E_F)}{k_B T}} dE = bk_B T e^{\frac{E_v - E_F}{k_B T}}. \tag{18.172}$$

If the sample is intrinsic, then $n = p = n_i$. Therefore,

$$ak_B T e^{(\frac{E_F - E_c}{k_B T})} = bk_B T e^{(\frac{E_v - E_F}{k_B T})}, \tag{18.173}$$

from which we derive

$$E_F = \frac{E_v - E_c}{2} + \frac{k_B T}{2} ln(\frac{b}{a}). \tag{18.174}$$

Hence, $E_F = \frac{E_v + E_c}{2}$ whenever $a = b$.

(b) The intrinsic carrier concentration is given by $n_i = \sqrt{np}$. Hence, using Equations (18.171) and (18.172) above, we obtain

$$n_i = k_B T \sqrt{ab} e^{\frac{(E_c - E_v)}{2k_B T}} = k_B T \sqrt{ab} e^{-\frac{E_g}{2k_B T}}. \tag{18.175}$$

Example 4: Connection between 2D and 3D density of states

The density of states in the conduction band of a bulk sample is given by Equation (18.110).

If a 2D quantum well (of width W) is formed with infinite barriers on both sides, show that

$$g_{3D}(E_{c0} + E_n) = \frac{n}{W} g_{2D}(E_{c0} + E_n), \tag{18.176}$$

where $g_{2D}(E)$ is the two-dimensional density of states in each subband of the 2-DEG, given by Equation (18.127).

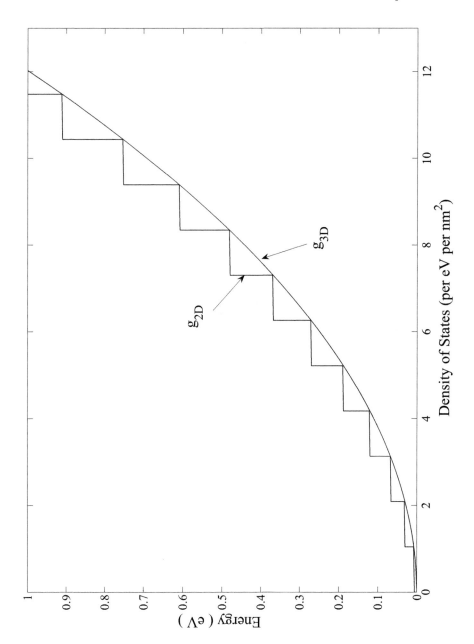

FIGURE 18.7

Illustration of the 2D density of states staircase touching the 3D density of states curve. For this illustration, the effective mass of electrons was assumed to be $m^* = 0.5m_o$ and the quantum well width was assumed equal to 100 Å.

Solution

For a particle in a box with a constant potential energy E_{co}, the eigenenergies are given by

$$E_n = E_{co} + \frac{n^2 \hbar^2 \pi^2}{2m^* W^2}. \tag{18.177}$$

Hence, using Equation (18.110), we get

$$g_{3D}(E_n) = \frac{m^*}{\pi^2 \hbar^3} \sqrt{2m^* (\frac{n^2 \hbar^2 \pi^2}{2m^* w^2})}, \tag{18.178}$$

i.e.,

$$g_{3D}(E_n) = \frac{m^*}{\pi^2 \hbar^2} \frac{n}{w} = \frac{n}{w} g_{2D}(E_{co} + E_n). \tag{18.179}$$

A plot of $W g_{3D}(E_n)$ and $g_{2D}(E_{co} + E_n)$ is shown in Fig. 18.7. This figure shows that the corners of the staircase representing the $\frac{m^*}{\pi \hbar^2}$ jump for each appearance of a new subband in the 2-DEG touch the curve $W g_{3D}(E_n)$. As the well width is increased, the energy levels for the particle-in-a-box are more closely spaced and the staircase becomes closer and closer to the $W g_{3D}$ curve, i.e.,

$$g_{3D}(E) = \lim_{W \to \infty} \frac{1}{W} g_{2D}(E). \tag{18.180}$$

─────────

18.6   Rayleigh–Ritz variational procedure

The Rayleigh-Ritz variational principle is often used when the wavefunctions of the ground state and the first few excited states of a Hamiltonian cannot be found analytically, and when perturbation theory is too poor an approximation to calculate the energies of the lower eigenstates of the Hamiltonian. In that case, physical arguments are invoked to *guess* an analytical form for the eigenstates of the Hamiltonian in terms of some variational parameters. The latter are then varied until the expectation value of the Hamiltonian (or the energy eigenvalue) is minimized. More precisely, the Rayleigh-Ritz variational procedure is based on the premise that, if $|\phi>$ is a guess for the ground state wavefunction, then the actual ground state energy will satisfy the inequality:

$$E_o \leq E_{min} = min(\frac{< \phi|H|\phi >}{< \phi|\phi >}), \tag{18.181}$$

where $|\phi>$ is the trial wavefunction which contains variational parameters. The denominator appearing on the right hand side makes sure that the trial wavefunction is normalized. The variational parameters are varied until the expression in the right hand side is minimized. This yields the best guess for $|\phi>$.

The variational method can also be applied to obtain the eigenfunctions of the higher excited states but, in that case, the trial function for the excited state must be selected such that it is orthogonal to the trial eigenfunctions selected to describe the wavefunctions of the lower energy eigenstates.

A more thorough discussion of the Rayleigh-Ritz variational principle can be found in some quantum mechanics textbooks. In this chapter, we illustrate this principle with a few sample problems, including an estimation of the ground state energy in a triangular well, which approximates the potential energy profile in a HEMT close to the heterointerface [see Fig. 13.7]. Since all spin field effect transistors that we discussed in this book (and a few others) employ HEMT structures, this result is important for the detailed analysis of these transistors.

Examples

Example 1: The harmonic oscillator

Suppose we start with the following trial wavefunction for the harmonic oscillator ground state:

$$\phi(x) = \frac{N}{x^2 + a^2}, \tag{18.182}$$

where a is a variational parameter and N is an extra factor needed to normalize the wavefunction.

This choice seems reasonable since the wavefunction is peaked at $x = 0$ and decays for large values of x. Moreover, this form of $\phi(x)$ allows an exact calculation of the functional E_{min} appearing in the inequality (18.181).

Now, apply the variational principle and find E_{min}. How does it compare with the true eigenvalue, which is $\frac{\hbar\omega}{2}$, where ω is the angular frequency appearing in the expression for the potential energy $V(x) = \frac{m\omega^2 x^2}{2}$?

Solution

The normalization coefficient N is found to be

$$N = \sqrt{\frac{2a^3}{\pi}}, \tag{18.183}$$

and the energy E_{min} on the right hand side of inequality (18.181) is given by

$$E_{min} = \frac{\hbar^2 + 2a^4 m^{*2}\omega^2}{4m^* a^2}. \tag{18.184}$$

The latter reaches a minimum for

$$a_{min} = \frac{1}{2^{1/4}}\sqrt{\frac{\hbar}{m^*\omega}}. \tag{18.185}$$

The corresponding minimum energy E_{min} is

$$E_{min}(a_{min}) = \sqrt{2}\frac{\hbar\omega}{2},\qquad(18.186)$$

which is $\sqrt{2}$ times larger than the exact result $\hbar\omega/2$.

Example 2: The triangular well problem

The following two variational functions are used to describe the two lowest states in the two-dimensional electron gas formed at the interface of a HEMT. Close to the heterointerface, the conduction band in the semiconductor substrate is modeled as a triangular potential well [see Fig. 13.7]. The trial wavefunctions for the ground and first excited states are:

$$\xi_o(x) = (\frac{b_o^3}{2})^{\frac{1}{2}}xe^{\frac{-b_o x}{2}},\qquad(18.187)$$

$$\xi_1(x) = (\frac{3b_1^5}{2[b_o^2 + b_1^2 - b_o b_1]})^{\frac{1}{2}}x(1 - \frac{b_o + b_1}{6}x)e^{-\frac{b_1 x}{2}},\qquad(18.188)$$

where b_o and b_1 are two variational parameters, and $x = 0$ at the heterointerface.

Notice that both $\xi_o(x)$ and $\xi_1(x)$ are zero at $x = 0$, as they should be since neither wavefunction can penetrate the barrier at the heterointerface if it is very high. Since the potential energy increases linearly with distance into the substrate, the choice of exponentially decaying trial wavefunctions seems appropriate.

- Show that $[\xi_o(x), \xi_1(x)]$ is an appropriate set of trial wavefunctions for the ground state and first excited states, i.e., $\xi_o(x)$ and $\xi_1(x)$ are both normalized, so that $\int_0^\infty |\xi_i(x)|^2 dx = 1$, and ξ_o and ξ_1 are orthogonal, meaning $\int_0^\infty \xi_o(x)\xi_1(x)dx = 0$.

Solution
Performing the integration over the probability in the ground state (magnitude of the square of the wavefunction), we get

$$\int_0^{+\infty} |\xi_o(x)|^2 dx = \int_0^{+\infty} (\frac{b_o^3}{2})x^2 e^{-b_o x}dx = \frac{b_o^3}{2}\frac{2}{b_o^3} = 1.\qquad(18.189)$$

So ξ_o is indeed normalized.
Similarly,

$$\int_0^{+\infty} |\xi_1(x)|^2 dx = \frac{3b_1^5}{2(b_o^2 + b_1^2 - b_o b_1)}[I_2 - I_3 + I_4],\qquad(18.190)$$

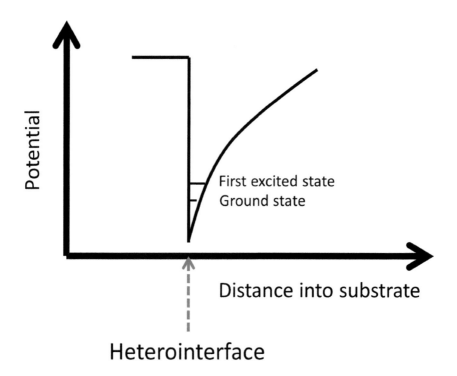

FIGURE 18.8
A triangular shape approximates the conduction band profile in the channel of a HEMT. The bound states and the wavefunctions of these states determine the channel carrier concentration and the threshold voltage.

where

$$I_2 = \int_0^{+\infty} x^2 e^{-b_1 x} dx, \tag{18.191}$$

$$I_3 = \frac{(b_o + b_1)}{3} \int_0^{+\infty} x^3 e^{-b_1 x} dx, \tag{18.192}$$

and

$$I_4 = (\frac{b_o + b_1}{6})^2 \int_0^{+\infty} x^4 e^{-b_1 x} dx. \tag{18.193}$$

We find

$$I_2 = \frac{2}{b_1^3}, \tag{18.194}$$

$$I_3 = \frac{2(b_o + b_1)}{b_1^4}, \tag{18.195}$$

and

$$I_4 = \frac{2}{3} \frac{(b_o + b_1)^2}{b_1^5}. \tag{18.196}$$

Hence

$$\int_0^{+\infty} |\xi_1(x)|^2 dx = \frac{3}{2} \frac{b_1^5}{[b_o^2 + b_1^2 - b_o b_1]} [\frac{2}{b_1^3} - \frac{2(b_o + b_1)}{b_1^4} + \frac{2(b_o + b_1)^2}{3b_1^5}]. \tag{18.197}$$

Simplifying

$$\int_0^{+\infty} |\xi_1(x)|^2 dx = \frac{3}{2} \frac{1}{[b_o^2 + b_1^2 - b_o b_1]} \frac{2}{3} [b_o^2 + b_1^2 - b_o b_1] = 1. \tag{18.198}$$

So $\xi_1(x)$ is also normalized.

To prove the orthogonality of ξ_o and ξ_1, we must show that the following integral is equal to zero:

$$\int_0^{+\infty} \xi_o^*(x)\xi_1(x)dx = (\frac{b_o^3}{2})^{\frac{1}{2}} (\frac{3b_1^5}{2[b_o^2 + b_1^2 - b_o b_1]})^{\frac{1}{2}} (J_2 - J_3), \tag{18.199}$$

where

$$J_2 = \int_0^{+\infty} x^2 e^{-(\frac{b_o + b_1}{2})x} dx, \tag{18.200}$$

and

$$J_3 = (\frac{b_o + b_1}{6}) \int_0^{+\infty} x^3 e^{-(\frac{b_o + b_1}{2})x} dx. \tag{18.201}$$

We find

$$J_2 = J_3 = \frac{16}{(b_o + b_1)^3}. \tag{18.202}$$

Hence

$$\int_0^{+\infty} \xi_o(x)\xi_1(x)dx = 0, \tag{18.203}$$

and ξ_o, ξ_1 are indeed orthogonal.

- Using the results above, calculate an upper bound for the energy of the ground state in a triangular well in which the potential is assumed to be equal to infinity for $x < 0$ and for which the potential energy is given by

$$E_c(x) = \beta x, \qquad (18.204)$$

for $x > 0$.

This problem can actually be solved exactly but requires the use of Airy functions, which are rather cumbersome to deal with. The exact result for the ground state energy is given by [19]

$$E_0 = 1.857 \left(\frac{\beta^2 \hbar^2}{m^*}\right)^{1/3}. \qquad (18.205)$$

Compare your answer obtained using the trial wavefunction with the exact value above.

Solution
The expectation value E_{min} appearing in the inequality (18.181) is easily calculated using MATHEMATICA [20] or other software tools such as MAT-LAB that allow symbolic manipulation. The minimum of E_{min} occurs for

$$b_{o,min} = \left(\frac{12\beta m^*}{\hbar^2}\right)^{1/3}, \qquad (18.206)$$

leading to an upper bound on the ground state energy

$$E_{min}(b_{0,min}) = 1.966 \left(\frac{\beta^2 \hbar^2}{m^*}\right)^{1/3}, \qquad (18.207)$$

which is only about 6% larger than the exact result.

18.7   The transfer matrix formalism

The transfer matrix method is a powerful technique used to solve tunneling problems in quantum mechanics. It can calculate the transmission and reflection coefficients of an electron wave impinging on an arbitrary conduction band energy profile, as shown in Fig 18.9, provided there are no dissipative interactions such as phonon or electron-electron scattering that destroy the electron's phase memory. Non-phase-breaking scattering potentials (i.e., time-independent scattering potentials due to impurities, etc.) can be handled within this formalism since they simply add (or subtract) a spatially varying potential to the conduction band profile. The transfer matrix method is used

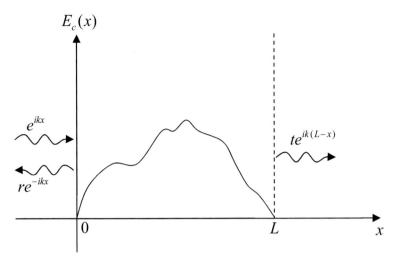

FIGURE 18.9
Arbitrary conduction band energy profile $E_c(x)$ in a region of length L (the device) sandwiched between two contacts; r and t are the reflection and transmission amplitudes, respectively, for an electron incident from the left contact.

to study quantum transport (including spin dependent transport through such devices as Spin Field Effect Transistors) and in modeling passive devices such as spin valves (see Chapter 11).

We wish to calculate the transmission and reflection coefficients of an electron wave incident on an arbitrary conduction band energy profile in a region of length L ("the device") sandwiched between a left ($x < 0$) and a right ($x > L$) contact, as illustrated in Fig. 18.9. The structure is shown under zero bias with similar contacts. Electrons are assumed to travel across the device without losing energy, i.e., the transport of carriers is assumed to be coherent throughout the structure. We proceed as follows.

18.7.1 Linearly independent solutions of the Schrödinger equation

We consider the one-dimensional time-independent Schrödinger equation assuming that only the potential energy $E_c(x)$ is varying in the x-direction, as shown in Fig. 18.9. If the effective mass of the electron is assumed to be independent of x, the Schrödinger equation becomes

$$-\frac{\hbar^2}{2m^*}\ddot{\phi}(x) + E_c(x)\phi(x) = E_p\phi(x), \tag{18.208}$$

where $\ddot{\phi}(x)$ stands for $\frac{d^2}{dx^2}\phi(x)$, the second derivative with respect to x, and E_p is the longitudinal kinetic energy, i.e., the kinetic energy in the x-direction, assumed to be the direction of current flow.

The general solution of this second order differential equation for $\phi(x)$ can be written as a linear combination of two linearly independent solutions [21]. Two solutions ϕ_1 and ϕ_2 of a differential equation are linearly independent if the equality $c_1\phi_1 + c_2\phi_2 = 0$ cannot be satisfied $\forall(x)$ for any choice of (c_1, c_2) except for $c_1 = c_2 = 0$. If non-zero solutions (c_1, c_2) exist, then ϕ_1 and ϕ_2 are said to be linearly dependent.

18.7.2   Concept of Wronskian

If ϕ_1 and ϕ_2 are linearly dependent, $c_1\phi_1 + c_2\phi_2 = 0$ and $c_1\dot{\phi}_1 + c_2\dot{\phi}_2 = 0$, where the dot stands for the first derivative with respect to x. Hence, in a matrix form, we have

$$\begin{bmatrix} \dot{\phi}_1 & \dot{\phi}_2 \\ \phi_1 & \phi_2 \end{bmatrix} \begin{bmatrix} c_1 \\ c_2 \end{bmatrix} = \begin{bmatrix} 0 \\ 0 \end{bmatrix}. \qquad (18.209)$$

This means that the Wronskian $(W(x) = \dot{\phi}_1\phi_2 - \phi_1\dot{\phi}_2)$, which is the determinant of the 2×2 matrix in the last equation, must be identically zero or otherwise only the trivial solution $c_1 = c_2 = 0$ in the last equation would be admissible. Stated otherwise, for two solutions to be linearly independent, their Wronskian must be non-zero for all x.

Even if $E_c(x)$ has finite discontinuities, $\ddot{\phi}(x)$ exists throughout and hence $\phi(x)$ and $\dot{\phi}(x)$ are continuous. Since (ϕ_1, ϕ_2) satisfy the Schrödinger equation, we have

$$\ddot{\phi}_1\phi_2 - \ddot{\phi}_2\phi_1 = 0, \qquad (18.210)$$

or

$$\frac{d}{dx}(\dot{\phi}_1\phi_2 - \dot{\phi}_2\phi_1) = \frac{dW}{dx} = 0. \qquad (18.211)$$

Thus, $W(x)$ is a constant independent of x.

In summary, if the Wronskian $W(x) = \dot{\phi}_1\phi_2 - \phi_1\dot{\phi}_2 = 0$, (ϕ_1, ϕ_2) are linearly dependent and, if $W(x) \neq 0$, it is a constant and the two solutions are linearly independent.

If we can find these two linearly independent solutions, their linear combination is the most general solution to the Schrödinger equation,

$$\phi(x) = c_1\phi_1(x) + c_2\phi_2(x). \qquad (18.212)$$

If we select $\phi_1(x)$ and $\phi_2(x)$ such that

$$\phi_1(0) = 0, \dot{\phi}_1(0) = 1, \qquad (18.213)$$

and

$$\phi_2(0) = 1, \dot{\phi}_2(0) = 0, \qquad (18.214)$$

they will be linearly independent since their Wronskian, which is independent of x, is equal to

$$W(x) = W(0) = \dot{\phi}_1(0)\phi_2(0) - \phi_1(0)\dot{\phi}_2(0) = 1. \qquad (18.215)$$

18.7.3 Concept of transfer matrix

Since ϕ_1 and ϕ_2 satisfy the conditions (18.213) and (18.214), we have

$$c_1 = \dot{\phi}(0), \qquad (18.216)$$

and

$$c_2 = \phi(0). \qquad (18.217)$$

Hence, from Equation (18.212) and its first derivative with respect to x, we get

$$\phi(L) = \dot{\phi}(0)\phi_1(L) + \phi(0)\phi_2(L), \qquad (18.218)$$

and

$$\dot{\phi}(L) = \dot{\phi}(0)\dot{\phi}_1(L) + \phi(0)\dot{\phi}_2(L), \qquad (18.219)$$

which can be rewritten in a matrix form:

$$\begin{bmatrix} \dot{\phi}(L) \\ \phi(L) \end{bmatrix} = \begin{bmatrix} \dot{\phi}_1(L) & \dot{\phi}_2(L) \\ \phi_1(L) & \phi_2(L) \end{bmatrix} \begin{bmatrix} \dot{\phi}(0) \\ \phi(0) \end{bmatrix}. \qquad (18.220)$$

The matrix appearing on the right hand side is called the transfer matrix because it relates (or connects) the column vector $(\dot{\phi}(x), \phi(x))^{\dagger}$ (where the $\dagger$ stands for the transpose operation) at location $x = L$ to its value at location $x = 0$.

18.7.4 Cascading rule for transfer matrices

An arbitrary spatially varying potential energy profile $E_c(x)$ can always be decomposed into a series of steps where the potential energy in each step is approximated by a spatially invariant constant which is its average value over that interval. The accuracy of this approximation increases with decreasing interval size. Since the potential within each section is constant, the transfer matrix of each small section can be derived exactly. Once the individual transfer matrix for each small segment is known, the overall transfer matrix W_{TOT} needed to relate the wavefunction on the right to the wavefunction on the left (i.e., describe the tunneling process of a particle incident from the left contact) is the product of the individual transfer matrices associated with each small segment, i.e.,

$$W_{TOT} = W_N W_{N-1}...W_2 W_1, \qquad (18.221)$$

where W_i is the transfer matrix associated with the ith segment counted from the left contact.

Since matrices do not commute in general, it makes a difference whether we start cascading from the left or the right. A little introspection should convince the reader that the cascading should proceed from right to left as shown in the equation above. The tunneling problem for a particle incident from the right contact can be treated using the same formalism by flipping the structure around, which is equivalent to taking the product of the different segments in the reverse order.

From now on, we use the notation W for the overall transfer matrix W_{TOT} and label its elements W_{ij} with $(i, j) = 1$ or 2. As shown in Fig. 18.9, the boundary conditions for a particle incident from the left contact are given by

$$\phi(x) = e^{ikx} + re^{-ikx}, \tag{18.222}$$

and

$$\phi(x) = te^{ik(L-x)}, \tag{18.223}$$

where we have assumed that the contacts are identical and that there is no bias across the device (the case of non-zero bias is an easy extension).

From Equation (18.220), we get the following two equations for the reflection (r) and transmission (t) *amplitudes*:

$$\begin{pmatrix} ikt \\ t \end{pmatrix} = \begin{pmatrix} W_{11} & W_{12} \\ W_{21} & W_{22} \end{pmatrix} \begin{pmatrix} ik(1-r) \\ 1+r \end{pmatrix}. \tag{18.224}$$

These two equations can be solved explicitly, leading to

$$t = \frac{2ik}{(k^2 W_{21} - W_{12}) + ik(W_{11} + W_{22})}, \tag{18.225}$$

and

$$r = \frac{(k^2 W_{21} + W_{12}) + ik(W_{11} - W_{22})}{(k^2 W_{21} - W_{12}) + ik(W_{11} + W_{22})}. \tag{18.226}$$

Exercises:

- Each of the transfer matrices appearing in the product is unimodular, i.e., their Wronskian is unity. Prove that it is also true for the overall transfer matrix W_{TOT}.

- Derive the expressions of the transmission and reflection amplitudes given above by solving the system of two equations with two unknowns (r, t) in equation (18.224).

- Prove that the sum of the reflection coefficient, $|r|^2$, and transmission coefficient, $|t|^2$, is unity since the incident particle must be either reflected or transmitted.

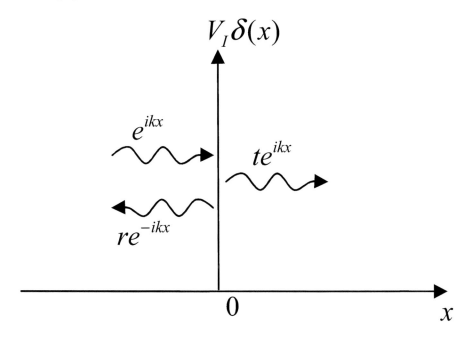

FIGURE 18.10

Scattering problem for an electron incident from the left on a one-dimensional delta-scatterer.

q

Next, we calculate explicitly the transfer matrix and transmission coefficient for a specific problem.

Example

Tunneling through a one-dimensional delta scatterer

Consider the one-dimensional scattering problem shown in Fig. (18.10) where the scattering potential is modeled as a one-dimensional delta scatter, i.e., $V(x) = V_I \delta(x)$, located at $x = 0$. The parameter V_I is the strength of the delta-scatterer and has S.I. units of J-m. However, it is often given in the literature in non-S.I. units of eV-Å.

Starting with the Schrödinger equation

$$-\frac{\hbar^2}{2m^*}\ddot{\phi}(x) + V_I \delta(x)\phi(x) = E_p \phi(x), \tag{18.227}$$

and integrating it on both sides from 0_- to 0_+, we get

$$-\frac{\hbar^2}{2m^*}\left[\dot{\phi}(0_+) - \dot{\phi}(0_-)\right] + V_I\phi(0) = 0, \tag{18.228}$$

which leads to

$$\dot{\phi}(0_+) = \dot{\phi}(0_-) + \frac{2m^*V_I}{\hbar^2}\phi(0_+). \tag{18.229}$$

In addition, the wavefunction is assumed to be continuous across the delta scatterer, i.e.,

$$\phi(0_+) = \phi(0_-). \tag{18.230}$$

Regrouping the last two equations in a matrix form, we get

$$\begin{bmatrix} \dot{\phi}(0_+) \\ \phi(0_+) \end{bmatrix} = \begin{bmatrix} 1 & \frac{2m^*V_I}{\hbar^2} \\ 0 & 1 \end{bmatrix} \begin{bmatrix} \dot{\phi}(0_-) \\ \phi(0_-) \end{bmatrix}. \tag{18.231}$$

So, the transfer matrix across a delta scatterer is given by

$$W = \begin{bmatrix} 1 & 2k_\delta \\ 0 & 1 \end{bmatrix}, \tag{18.232}$$

where $k_\delta = \frac{m^*V_I}{\hbar^2}$.

Exercise: Find the reflection and transmission amplitudes for an electron incident from the left on the delta scatterer shown in Fig. 18.10 and prove that $|r|^2 + |t|^2 = 1$.

Solution

Equation (18.230) leads to

$$1 + r = t, \tag{18.233}$$

whereas Eq.(18.229) leads to

$$ikt - ik(1 - r) = 2k_\delta t. \tag{18.234}$$

Solving these last two equations for the two unknowns t and r, we get

$$t = \frac{ik}{ik - k_\delta}, \tag{18.235}$$

and

$$r = \frac{k_\delta}{ik - k_\delta}. \tag{18.236}$$

Using the last two results, it is readily shown that

$$|r|^2 + |t|^2 = 1. \tag{18.237}$$

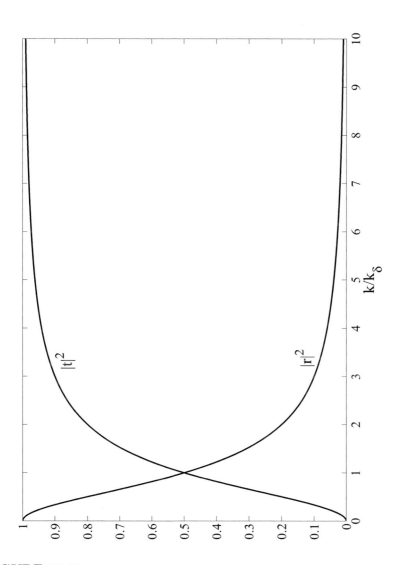

FIGURE 18.11

Plot of the transmission probability $|t|^2$ and reflection probability $|r|^2$ across a delta-scatterer as a function of the reduced eigenvector k/k_δ. $|t|^2 = |r|^2 = 0.5$ when $k/k_\delta = 1$.

18.8   Peierls' transformation

We show that, to account for an external magnetic field of flux density $\vec{B}$ and an electrostatic potential ϕ in the Schrödinger equation of a free particle

$$H\psi = \frac{|\vec{p}|^2}{2m}\psi = E\psi, \tag{18.238}$$

we must make the following two substitutions

$$\vec{p} \to \vec{p} - q\vec{A}(\vec{r}, t), \tag{18.239}$$

referred to as Peierls' substitution or transformation and

$$E \to E - q\phi(\vec{r}, t), \tag{18.240}$$

where $\vec{A}$ is a vector potential whose curl is the magnetic flux density, i.e., $\vec{B} = \vec{\nabla} \times \vec{A}$ and ϕ is the electrostatic potential. The latter is defined as the work done per unit charge to bring the charge to the location $\vec{r}$ from some reference point, typically selected at infinity.

The new Hamiltonian of the Schrödinger equation then becomes

$$H = \frac{\left|\vec{p} - q\vec{A}\right|^2}{2m} + q\phi. \tag{18.241}$$

The easiest way to derive the two substitutions described above is to start with a Lagrangian description of charge dynamics. In the presence of an electrostatic potential alone, the Lagrangian of a charged particle is given by

$$L = T - V = (1/2)m\vec{v} \cdot \vec{v} - q\phi. \tag{18.242}$$

The charge dynamics are then described by Euler's equation

$$\frac{\partial L}{\partial x_i} - \frac{d}{dt}\left(\frac{\partial L}{\partial \dot{x}_i}\right) = 0, \tag{18.243}$$

where x_i-s are spatial coordinates of the particle, a single dot superscript denotes the first derivative with respect to time and a double dot superscript denotes the second derivative with respect to time.

With the Lagrangian above, Euler's equation becomes

$$-q\frac{\partial \phi}{\partial x_i} - \frac{d}{dt}(m\dot{x}_i) = 0. \tag{18.244}$$

which is equivalent to Newton's equation of motion

$$m\ddot{x}_i = -q\frac{\partial \phi}{\partial x_i}. \tag{18.245}$$

In the presence of an external magnetic field, the force acting on the charge q should contain the Lorentz contribution, and the total force acting on the particle is given by

$$\vec{F} = q\vec{E} + q\vec{v} \times \vec{B}, \tag{18.246}$$

where the electric field $\vec{E} = -\vec{\nabla}\phi - \partial\vec{A}/\partial t$.

Next we show that a modification of the Lagrangian in Equation (18.242) to the new form

$$L = (1/2)m\vec{v} \cdot \vec{v} - q\phi + q\vec{v} \cdot \vec{A} \tag{18.247}$$

leads to Newton's equation where the force term includes both the Lorentz force due to the magnetic field and the force due to the electric field.

Starting with this new expression for the Lagrangian, we get the canonical momentum component

$$p_i = \frac{\partial L}{\partial \dot{x}_i} = m\dot{x}_i + qA_i, \tag{18.248}$$

or in a vector form

$$\vec{p} = m\vec{v} + q\vec{A}. \tag{18.249}$$

Now,

$$\frac{d}{dt}\left(\frac{\partial L}{\partial \dot{x}_i}\right) = m\ddot{x}_i + q\frac{dA_i}{dt}. \tag{18.250}$$

Since in the most general case, $\vec{A}$ is a function of position and time, we have

$$\frac{dA_i}{dt} = \frac{\partial A_i}{\partial t} + \vec{v} \cdot \vec{\nabla}A_i. \tag{18.251}$$

Moreover, because

$$\frac{\partial L}{\partial x_i} = -q\frac{\partial \phi}{\partial x_i} + q\vec{v} \cdot \frac{\partial \vec{A}}{\partial x_i}, \tag{18.252}$$

Euler's equation becomes

$$= -q\frac{\partial \phi}{\partial x_i} + q\vec{v} \cdot \frac{\partial \vec{A}}{\partial x_i} - m\ddot{x}_i - q\left(\frac{dA_i}{dt} + \vec{v} \cdot \vec{\nabla}A_i\right) = 0. \tag{18.253}$$

In vector notation, the right hand side of the last equation can be expressed as a general force $\vec{F}$ equal to

$$\vec{F} = q\left(-\vec{\nabla}\phi - \frac{\partial \vec{A}}{\partial t}\right) + q\left[\vec{\nabla}\left(\vec{v} \cdot \vec{A}\right) - \left(\vec{v} \cdot \vec{\nabla}\right)\vec{A}\right], \tag{18.254}$$

which reduces to

$$\vec{F} = q\vec{E} + q\vec{v} \times (\vec{\nabla} \times \vec{A}) = q\vec{E} + q\vec{v} \times \vec{B}, \tag{18.255}$$

where we have used the relation between the magnetic field and the vector potential

$$\vec{B} = \vec{\nabla} \times \vec{A}. \tag{18.256}$$

Starting with the modified Lagrangian, the Hamiltonian of the charge particle can then be derived as

$$H = \sum_{i=1}^{3} p_i \dot{x}_i - L = \sum_{i=1}^{3} (m\dot{x}_i + qA_i)\dot{x}_i - \frac{1}{2}m(\dot{x}_i)^2 + q\phi - qA_i\dot{x}_i), \tag{18.257}$$

or

$$H = \sum_{i=1}^{3} \frac{1}{2}m(\dot{x}_i)^2 + q\phi, \tag{18.258}$$

which is equivalent to

$$H = \frac{\left|\vec{p} - q\vec{A}\right|^2}{2m} + q\phi, \tag{18.259}$$

where we have used Equation (18.249).

18.9   Problem

- **Problem 18.1**

  Calculate the right hand side of the generalized Heisenberg inequality (18.54) for the case where the two operators C and D are given by

  $$C = \sigma_+, D = \sigma_-, \tag{18.260}$$

  where $\sigma_+ = \sigma_x + i\sigma_y$ and $\sigma_- = \sigma_x - i\sigma_y$.

- Prove that, for the case where the operators C and D are equal to x and p_x, the inequality becomes an equality sign for a one-dimensional harmonic oscillator in its ground state.

- Repeat the previous problem for an harmonic oscillator in its first excited state. What is the value of $\Delta x \Delta p_x$ in this case?

Solution

For the ground state and first excited states,

$$< x >= 0, \tag{18.261}$$

and

$$< p_x >= 0. \tag{18.262}$$

In the ground state,

$$< x^2 > = \frac{\hbar}{2m\omega}, \tag{18.263}$$

$$< p_x^{\,2} > = \frac{\hbar m\omega}{2}, \tag{18.264}$$

and

$$\Delta x \Delta p_x = \frac{\hbar}{2}. \tag{18.265}$$

In the first excited state,

$$< x^2 > = \frac{3\hbar}{2m\omega}, \tag{18.266}$$

$$< p_x^{\,2} > = \frac{3\hbar m\omega}{2}, \tag{18.267}$$

and

$$\Delta x \Delta p_x = \frac{3}{2}\hbar. \tag{18.268}$$

18.10   References

[1] L. I. Schiff, *Quantum Mechanics*, (McGraw-Hill, New York, 1955).

[2] J. J. Sakurai, *Modern Quantum Mechanics* (Addison Wesley, Redwood City, CA, 1985).

[3] D. J. Griffiths, *Introduction to Quantum Mechanics*, (Prentice Hall, Englewood Cliffs, New Jersey, 1995).

[4] H. Kroemer, *Quantum Mechanics for Engineering, Materials Science and Applied Physics*, (Prentice Hall, Englewood Cliffs, New Jersey, 1994).

[5] G. Gamow and J. M. Cleveland, *Physics*, (Prentice Hall, Englewood Cliffs, NJ, 1969).

[6] N. Bohr, "The quantum postulate and the recent development of atomic theory", *Naturwissenschaften*, **16**, 245 (1928).

[7] P. Ghose, D. Home and G. S. Agarwal, "An experiment to throw more light on light", Phys. Lett. A, **153**, 403 (1991).

[8] J. C. Bose, in: *Collected Physical Papers*, (Longman and Green, London, 1927) pp. 44-49.

[9] P. Grangier, G. Roger and A. Aspect, "Experimental evidence for a photon anti-correlation effect on a beam splitter: A new light on single photon interferences", Europhys. Lett., **1**, 173 (1986).

[10] Y. Mizobuchi and Y. Ohtaké, "An 'experiment to throw more light on light' ", Phys. Lett. A, **168**, 1 (1992).

[11] S. Rangwala and S. M. Roy, "Wave behaviour and non-complementary particle behaviour in the same experiment", Phys. Lett. A, **190**, 1 (1994).

[12] P. A. M. Dirac, *Principles of Quantum Mechanics*, 4th edition, (Oxford University Press, Oxford, 1958), p. 8.

[13] B-G Englert, "Fringe visibility and which way information: An inequality", Phys. Rev. Lett., **77**, 2154 (1996).

[14] W. K. Wootters and W. K. Zurek, "Complementarity in the double slit experiment: Quantum non-separability and a quantitative statement of Bohr's principle", Phys. Rev. D, **19**, 473 (1979).

[15] D. M. Greenberger and A. Yasin, "Simultaneous wave and particle knowledge in a neutron interferometer", Phys. Lett. A, **128**, 391 (1988).

[16] L. Mandel, "Coherence and indistinguishability", Opt. Lett., **16**, 1882 (1991).

[17] M. A. Nielsen and I. L. Chuang, *Quantum Computation and Quantum Information*, (Cambridge University Press, 2000).

[18] C. Cohen-Tannoudji, B. Diu and F. Laloe, *Quantum Mechanics*, (Hermann, Paris, France, 1977).

[19] D. ter Haar, *Problems in Quantum Mechanics*, (Pion, 1975).

[20] http://www.wolfram.com

[21] E. Kreyszig, *Advanced Engineering Mathematics*, (John Wiley & Sons, New York, 2006).

Index